Lecture Notes in Computer Science 14491

The series Lecture Notes in Computer Science (LNCS), including its subseries Lecture Notes in Artificial Intelligence (LNAI) and Lecture Notes in Bioinformatics (LNBI), has established itself as a medium for the publication of new developments in computer science and information technology research, teaching, and education.

LNCS enjoys close cooperation with the computer science R & D community, the series counts many renowned academics among its volume editors and paper authors, and collaborates with prestigious societies. Its mission is to serve this international community by providing an invaluable service, mainly focused on the publication of conference and workshop proceedings and postproceedings. LNCS commenced publication in 1973.

Zahir Tari · Keqiu Li · Hongyi Wu
Editors

Algorithms and Architectures for Parallel Processing

23rd International Conference, ICA3PP 2023 Tianjin, China, October 20–22, 2023 Proceedings, Part V

Editors
Zahir Tari
Royal Melbourne Institute of Technology
Melbourne, VIC, Australia

Keqiu Li
Tianjin University
Tianjin, China

Hongyi Wu
University of Arizona
Tucson, AZ, USA

ISSN 0302-9743 ISSN 1611-3349 (electronic)
Lecture Notes in Computer Science
ISBN 978-981-97-0807-9 ISBN 978-981-97-0808-6 (eBook)
https://doi.org/10.1007/978-981-97-0808-6

This Springer imprint is published by the registered company Springer Nature Singapore Pte Ltd.
The registered company address is: 152 Beach Road, #21-01/04 Gateway East, Singapore 189721, Singapore

Paper in this product is recyclable.

Preface

On behalf of the Conference Committee, we welcome you to the proceedings of the 2023 International Conference on Algorithms and Architectures for Parallel Processing (ICA3PP 2023), which was held in Tianjin, China from October 20–22, 2023. ICA3PP2023 was the 23rd in this series of conferences (started in 1995) that are devoted to algorithms and architectures for parallel processing. ICA3PP is now recognized as the main regular international event that covers the many dimensions of parallel algorithms and architectures, encompassing fundamental theoretical approaches, practical experimental projects, and commercial components and systems. This conference provides a forum for academics and practitioners from countries around the world to exchange ideas for improving the efficiency, performance, reliability, security, and interoperability of computing systems and applications.

A successful conference would not be possible without the high-quality contributions made by the authors. This year, ICA3PP received a total of 503 submissions from authors in 21 countries and regions. Based on rigorous peer reviews by the Program Committee members and reviewers, 193 high-quality papers were accepted to be included in the conference proceedings and submitted for EI indexing. In addition to the contributed papers, six distinguished scholars, Lixin Gao, Baochun Li, Laurence T. Yang, Kun Tan, Ahmed Louri, and Hai Jin, were invited to give keynote lectures, providing us with the recent developments in diversified areas in algorithms and architectures for parallel processing and applications.

We would like to take this opportunity to express our sincere gratitude to the Program Committee members and 165 reviewers for their dedicated and professional service. We highly appreciate the twelve track chairs, Dezun Dong, Patrick P. C. Lee, Meng Shen, Ruidong Li, Li Chen, Wei Bao, Jun Li, Hang Qiu, Ang Li, Wei Yang, Yu Yang, and Zhibin Yu, for their hard work in promoting this conference and organizing the reviews for the papers submitted to their tracks. We are so grateful to the publication chairs, Heng Qi, Yulei Wu, Deze Zeng, and the publication assistants for their tedious work in editing the conference proceedings. We must also say "thank you" to all the volunteers who helped us at various stages of this conference. Moreover, we were so honored to have many renowned scholars be part of this conference. Finally, we would like to thank all speakers, authors, and participants for their great contribution to and support for the success of ICA3PP 2023!

October 2023

Jean-Luc Gaudiot
Hong Shen
Gudula Rünger
Zahir Tari
Keqiu Li
Hongyi Wu
Tian Wang

Organization

General Chairs

Jean-Luc Gaudiot	University of California, Irvine, USA
Hong Shen	University of Adelaide, Australia
Gudula Rünger	Chemnitz University of Technology, Germany

Program Chairs

Zahir Tari	Royal Melbourne Institute of Technology, Australia
Keqiu Li	Tianjin University, China
Hongyi Wu	University of Arizona, USA

Program Vice-chair

Wenxin Li	Tianjin University, China

Publicity Chairs

Hai Wang	Northwest University, China
Milos Stojmenovic	Singidunum University, Serbia
Chaofeng Zhang	Advanced Institute of Industrial Technology, Japan
Hao Wang	Louisiana State University, USA

Publication Chairs

Heng Qi	Dalian University of Technology, China
Yulei Wu	University of Exeter, UK
Deze Zeng	China University of Geosciences (Wuhan), China

Workshop Chairs

Laiping Zhao	Tianjin University, China
Pengfei Wang	Dalian University of Technology, China

Local Organization Chairs

Xiulong Liu	Tianjin University, China
Yitao Hu	Tianjin University, China

Web Chair

Chen Chen	Shanghai Jiao Tong University, China

Registration Chairs

Xinyu Tong	Tianjin University, China
Chaokun Zhang	Tianjin University, China

Steering Committee Chairs

Yang Xiang (Chair)	Swinburne University of Technology, Australia
Weijia Jia	Beijing Normal University and UIC, China
Yi Pan	Georgia State University, USA
Laurence T. Yang	St. Francis Xavier University, Canada
Wanlei Zhou	City University of Macau, China

Program Committee

Track 1: Parallel and Distributed Architectures

Dezun Dong (Chair)	National University of Defense Technology, China
Chao Wang	University of Science and Technology of China, China
Chentao Wu	Shanghai Jiao Tong University, China

Chi Lin	Dalian University of Technology, China
Deze Zeng	China University of Geosciences, China
En Shao	Institute of Computing Technology, Chinese Academy of Sciences, China
Fei Lei	National University of Defense Technology, China
Haikun Liu	Huazhong University of Science and Technology, China
Hailong Yang	Beihang University, China
Junlong Zhou	Nanjing University of Science and Technology, China
Kejiang Ye	Shenzhen Institute of Advanced Technology, Chinese Academy of Sciences, China
Lei Wang	National University of Defense Technology, China
Massimo Cafaro	University of Salento, Italy
Massimo Torquati	University of Pisa, Italy
Mengying Zhao	Shandong University, China
Roman Wyrzykowski	Czestochowa University of Technology, Poland
Rui Wang	Beihang University, China
Sheng Ma	National University of Defense Technology, China
Songwen Pei	University of Shanghai for Science and Technology, China
Susumu Matsumae	Saga University, Japan
Weihua Zhang	Fudan University, China
Weixing Ji	Beijing Institute of Technology, China
Xiaoli Gong	Nankai University, China
Youyou Lu	Tsinghua University, China
Yu Zhang	Huazhong University of Science and Technology, China
Zichen Xu	Nanchang University, China

Track 2: Software Systems and Programming Models

Patrick P. C. Lee (Chair)	Chinese University of Hong Kong, China
Erci Xu	Ohio State University, USA
Xiaolu Li	Huazhong University of Science and Technology, China
Shujie Han	Peking University, China
Mi Zhang	Institute of Computing Technology, Chinese Academy of Sciences, China

Track 3: Distributed and Network-Based Computing

Kaiping Xue	University of Science and Technology of China, China
Laurent Lefevre	National Institute for Research in Digital Science and Technology, France
Lanju Kong	Shandong University, China
Lei Zhang	Henan University, China
Li Duan	Beijing Jiaotong University, China
Lin He	Tsinghua University, China
Lingling Wang	Qingdao University of Science and Technology, China
Lingjun Pu	Nankai University, China
Liu Yuling	Institute of Information Engineering, China
Meng Li	Hefei University of Technology, China
Minghui Xu	Shandong University, China
Minyu Feng	Southwest University, China
Ning Hu	Guangzhou University, China
Pengfei Liu	University of Electronic Science and Technology of China, China
Qi Li	Beijing University of Posts and Telecommunications, China
Qian Wang	Beijing University of Technology, China
Raymond Yep	University of Macau, China
Shaojing Fu	National University of Defense Technology, China
Shenglin Zhang	Nankai University, China
Shu Yang	Shenzhen University, China
Shuai Gao	Beijing Jiaotong University, China
Su Yao	Tsinghua University, China
Tao Yin	Beijing Zhongguancun Laboratory, China
Tingwen Liu	Institute of Information Engineering, China
Tong Wu	Beijing Institute of Technology, China
Wei Quan	Beijing Jiaotong University, China
Weihao Cui	Shanghai Jiao Tong University, China
Xiang Zhang	Nanjing University of Information Science and Technology, China
Xiangyu Kong	Dalian University of Technology, China
Xiangyun Tang	Minzu University of China, China
Xiaobo Ma	Xi'an Jiaotong University, China
Xiaofeng Hou	Shanghai Jiao Tong University, China
Xiaoyong Tang	Changsha University of Science and Technology, China
Xuezhou Ye	Dalian University of Technology, China
Yaoling Ding	Beijing Institute of Technology, China

Track 4: Big Data and Its Applications

Track 5: Parallel and Distributed Algorithms

Track 6: Applications of Parallel and Distributed Computing

Track 7: Service Dependability and Security in Distributed and Parallel Systems

Ze Zhang	University of Michigan/Cruise, USA
Ravishka Rathnasuriya	University of Texas at Dallas, USA

Track 8: Internet of Things and Cyber-Physical-Social Computing

Yu Yang (Chair)	Lehigh University, USA
Qun Song	Delft University of Technology, The Netherlands
Chenhan Xu	University at Buffalo, USA
Mahbubur Rahman	City University of New York, USA
Guang Wang	Florida State University, USA
Houcine Hassan	Universitat Politècnica de València, Spain
Hua Huang	UC Merced, USA
Junlong Zhou	Nanjing University of Science and Technology, China
Letian Zhang	Middle Tennessee State University, USA
Pengfei Wang	Dalian University of Technology, China
Philip Brown	University of Colorado Colorado Springs, USA
Roshan Ayyalasomayajula	University of California San Diego, USA
Shigeng Zhang	Central South University, China
Shuo Yu	Dalian University of Technology, China
Shuxin Zhong	Rutgers University, USA
Xiaoyang Xie	Meta, USA
Yi Ding	Massachusetts Institute of Technology, USA
Yin Zhang	University of Electronic Science and Technology of China, China
Yukun Yuan	University of Tennessee at Chattanooga, USA
Zhengxiong Li	University of Colorado Denver, USA
Zhihan Fang	Meta, USA
Zhou Qin	Rutgers University, USA
Zonghua Gu	Umeå University, Sweden
Geng Sun	Jilin University, China

Track 9: Performance Modeling and Evaluation

Zhibin Yu (Chair)	Shenzhen Institute of Advanced Technology, Chinese Academy of Sciences, China
Chao Li	Shanghai Jiao Tong University, China
Chuntao Jiang	Foshan University, China
Haozhe Wang	University of Exeter, UK
Laurence Muller	University of Greenwich, UK

Lei Liu	Beihang University, China
Lei Liu	Institute of Computing Technology, Chinese Academy of Sciences, China
Jingwen Leng	Shanghai Jiao Tong University, China
Jordan Samhi	University of Luxembourg, Luxembourg
Sa Wang	Institute of Computing Technology, Chinese Academy of Sciences, China
Shoaib Akram	Australian National University, Australia
Shuang Chen	Huawei, China
Tianyi Liu	Huawei, China
Vladimir Voevodin	Lomonosov Moscow State University, Russia
Xueqin Liang	Xidian University, China

Reviewers

Dezun Dong
Chao Wang
Chentao Wu
Chi Lin
Deze Zeng
En Shao
Fei Lei
Haikun Liu
Hailong Yang
Junlong Zhou
Kejiang Ye
Lei Wang
Massimo Cafaro
Massimo Torquati
Mengying Zhao
Roman Wyrzykowski
Rui Wang
Sheng Ma
Songwen Pei
Susumu Matsumae
Weihua Zhang
Weixing Ji
Xiaoli Gong
Youyou Lu
Yu Zhang
Zichen Xu
Patrick P. C. Lee
Erci Xu
Xiaolu Li
Shujie Han
Mi Zhang
Jing Gong
Radu Prodan
Wei Wang
Himansu Das
Rong Gu
Yongkun Li
Ladjel Bellatreche
Meng Shen
Ruidong Li
Bin Wu
Chao Li
Chaokun Zhang
Chuan Zhang
Chunpeng Ge
Fuliang Li
Fuyuan Song
Gaopeng Gou
Guangwu Hu
Guo Chen
Guozhu Meng
Han Zhao
Hai Xue
Haiping Huang
Hongwei Zhang
Ioanna Kantzavelou

Jiawen Kang
Jie Li
Jingwei Li
Jinwen Xi
Jun Liu
Kaiping Xue
Laurent Lefevre
Lanju Kong
Lei Zhang
Li Duan
Lin He
Lingling Wang
Lingjun Pu
Liu Yuling
Meng Li
Minghui Xu
Minyu Feng
Ning Hu
Pengfei Liu
Qi Li
Qian Wang
Raymond Yep
Shaojing Fu
Shenglin Zhang
Shu Yang
Shuai Gao
Su Yao
Tao Yin
Tingwen Liu
Tong Wu
Wei Quan
Weihao Cui
Xiang Zhang
Xiangyu Kong
Xiangyun Tang
Xiaobo Ma
Xiaofeng Hou
Xiaoyong Tang
Xuezhou Ye
Yaoling Ding
Yi Zhao
Yifei Zhu
Yilei Xiao
Yiran Zhang
Yizhi Zhou
Yongqian Sun
Yuchao Zhang
Zhaoteng Yan
Zhaoyan Shen
Zhen Ling
Zhiquan Liu
Zijun Li
Li Chen
Alfredo Cuzzocrea
Heng Qi
Marc Frincu
Mingwu Zhang
Qianhong Wu
Qiong Huang
Rongxing Lu
Shuo Yu
Weizhi Meng
Wenbin Pei
Xiaoyi Tao
Xin Xie
Yong Yu
Yuan Cao
Zhiyang Li
Wei Bao
Jun Li
Dong Yuan
Francesco Palmieri
George Bosilca
Humayun Kabir
Jaya Prakash Champati
Peter Kropf
Pedro Soto
Wenjuan Li
Xiaojie Zhang
Chuang Hu
Hang Qiu
Ang Li
Daniel Andresen
Di Wu
Fawad Ahmad
Haonan Lu
Silvio Barra
Weitian Tong
Xu Zhang
Yitao Hu

Zhixin Zhao
Wei Yang
Dezhi Ran
Hanlin Chen
Jun Shao
Jinguang Han
Mirazul Haque
Simin Chen
Wenyu Wang
Yitao Hu
Yueming Wu
Zhengkai Wu
Zhiqiang Li
Zhixin Zhao
Ze Zhang
Ravishka Rathnasuriya
Yu Yang
Qun Song
Chenhan Xu
Mahbubur Rahman
Guang Wang
Houcine Hassan
Hua Huang
Junlong Zhou
Letian Zhang
Pengfei Wang
Philip Brown
Roshan Ayyalasomayajula
Shigeng Zhang
Shuo Yu
Shuxin Zhong
Xiaoyang Xie
Yi Ding
Yin Zhang
Yukun Yuan
Zhengxiong Li
Zhihan Fang
Zhou Qin
Zonghua Gu
Geng Sun
Zhibin Yu
Chao Li
Chuntao Jiang
Haozhe Wang
Laurence Muller
Lei Liu
Lei Liu
Jingwen Leng
Jordan Samhi
Sa Wang
Shoaib Akram
Shuang Chen
Tianyi Liu
Vladimir Voevodin
Xueqin Liang

Contents – Part V

EQFF: An Efficient Query Method Using Feature Fingerprints

Xiaolei Zhou[1,2], Yuelin Hua[1,2,3(✉)], Shan Huang[1,2], Qiang Fan[1,2], Hao Yan[1,2], and Shuai Wang[1,2,4]

[1] The 63rd Research Institute, National University of Defense Technology, Nanjing 210007, China
huayl0618@126.com

[2] Laboratory for Big Data and Decision, National University of Defense Technology, Changsha 410073, China

[3] School of Computer Science, Nanjing University of Information Science and Technology, Nanjing 210044, China

[4] School of Computer Science and Technology, Southeast University, Nanjing 211189, China

Abstract. The amount of data features is growing rapidly in the era of big data, posing challenges to both the security and efficiency of feature query. Most existing encryption-based retrieval approaches are limited by the significant computational overhead and merely support precise query, which might fail to handle the incomplete keywords and misspellings in the query. To achieve both query efficiency and privacy-preserving for large-scale data, this paper presents EQFF, an Efficient Query Method Using Feature Fingerprints. It converts varying-length features into fingerprints in the form of fixed-length vectors, and hence turns semantic information invisible to ensure query security. Based on the feature fingerprints, we further present the corresponding precise and fuzzy query approaches, design the inverted index library and propose a compression storage mechanism to improve query efficiency. Extensive experiments are conducted based on real-world datasets. Experimental results show that our EQFF takes only 6.4% memory compared with raw data, reduces the time cost from minutes to tens of milliseconds, and achieves an accuracy of 98% above.

Keywords: Data features · Data fingerprint · Privacy-preserving · Data query · Bloom filter

1 Introduction

In the past decade, the application of big data has made great progress, generating massive data and corresponding data features. According to the 2025 White Paper [1], the total amount of global data is expected to reach 175 zettabytes by 2025. Owing to recent developments in machine learning, data feature has become a critical technique for intelligent data applications such as data analysis,

Z. Tari et al. (Eds.): ICA3PP 2023, LNCS 14491, pp. 1–20, 2024.
https://doi.org/10.1007/978-981-97-0808-6_1

prediction, and decision-making. However, due to the rich semantic information contained in data characteristics, there are increasing concerns about personal privacy and the security of sensitive information [2]. Since traditional encryption algorithms can complicate queries against ciphertexts [3], the rapid growth in data volume further challenges efficient encrypted query and storage solutions under data security.

In order to ensure the privacy of data query, traditional methods [4] usually upload data to the cloud after local encryption, and then return the corresponding query result through ciphertext calculation. However, when the amount of data becomes great, the computational overhead of these methods makes them impossible to apply in real systems. In addition, most of these methods only support the precise query mode and may not return the expected results if there are problems with incomplete keywords or misspellings. One way to enable fuzzy query is hashing, which greatly compresses data features while maintaining data similarity. However, existing hashing based methods [5] can only handle unencrypted data features. In addition, the widely used tree-based index structure is designed for small-scale data. Massive feature data not only brings pressure to storage, but also poses a severe challenge to the efficiency of feature query.

This paper asks a natural question that can we achieve query efficiency and privacy-preserving simultaneously to facilitate the rapid feature query in both precise and fuzzy modes. Our insight is inspired by the Bloom filter [6], a widely-used probabilistic data index structure for set membership checking. It uses one-way hash functions to store elements in the set. As a result, a set of elements are converted into a fixed-length bit vector. Such an operation is irreversible, i.e., the bit vector cannot be restored by those hash functions. This advantage prompts us to design a feature query system based on Bloom filter to satisfy the concerns on privacy-preserving.

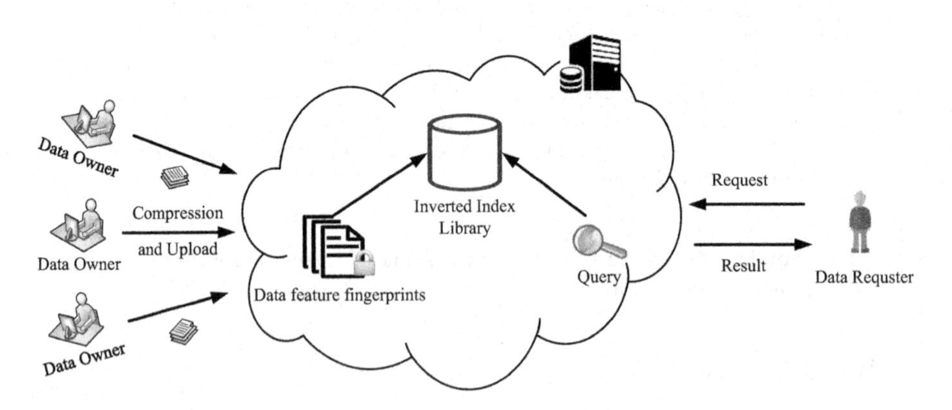

Fig. 1. An illustrative example of EQFF.

We present our key idea in Fig. 1. Data owners compress the original feature sets into feature fingerprints with Bloom filter locally and then upload those fingerprints to the cloud side. An inverted index library is constructed with compressed fingerprints collected from multiple data owners. When a user's query arrives, the query is also compressed in the same way to access the inverted index library. As a result, the corresponding features are returned to the requester. During the entire process, the semantics information in the feature set is converted into a set of bit vectors, and hence is invisible from the cloud's view. Based on the above considerations, this paper proposes the Efficient Query Method using Feature Fingerprints (EQFF), which is lightweight, rapid, and secure. EQFF fully takes advantages of cryptographic hashing to generate the unique fingerprints that represents the data features, and hence can support both the precise and fuzzy query modes. Besides, we design the corresponding data index library to further improve the query efficiency.

The contributions of this paper can be summarized as follows:

- We present EQFF, an efficient query method using feature fingerprints. It converts the long and sparse feature table into a series of bit vectors with constant length, resulting in higher space utilization, lower computational complexity, better scalability, and data security.
- We design the corresponding precise/fuzzy query and compression storage approaches for EQFF. Compared to pairwise comparisons, EQFF reduces the time cost from several minutes to several tens of milliseconds and seldom make mistakes, greatly improving the efficiency of feature query.
- We conduct extensive experiments based on a real-world dataset to validate the correctness and performance of EQFF. The results show that the heap memory usage is reduced to 6.4% of the original when EQFF is executed. Features can be queried with accuracy greater than 98% in milliseconds.

The rest of this paper is organized as follows. Section 2 summarizes the most relevant work in this field. Section 3 provides an overview of EQFF. Section 4 describes the construction of encrypted feature fingerprints. Section 5 constructs the inverted index library for feature fingerprint storage and proposes the precise query algorithm. Section 6 implements fuzzy query and measures the similarity of the results with a parallel approach for speedup. Section 7 verifies the effectiveness of EQFF through extensive experiments. Finally, we conclude this paper in Sect. 8.

2 Related Work

This section summarizes the related works of Bloom filter and fuzzy query.

2.1 Privacy of Bloom Filters

How to protect data privacy is an interesting and challenging problem. In recent years, a large number of solutions have emerged to protect the privacy of private data query [7–9]. Most of the above schemes use direct encryption algorithm, which has high security, but it is difficult to provide efficient privacy data query service for users.

Relatively speaking, the above problems can be solved by using indirect encryption methods. Bloom filter is a common file query structure. It achieves efficient query of elements by generating bit vectors through a series of random hash functions. Bloom filters offer a high degree of data privacy since they do not contain raw data. It has been proposed for improving the privacy of searches, recording links or authentication [10], and many other uses. Many et al. [11] apply Bloom filters to build a multi-party secure sharing model that significantly reduced computational overhead. Goh [12] proposes a Bloom filter-based z-index method, which adds a corresponding BF data structure for each document. Since only encrypted fingerprints are submitted for query, the user's query keywords are not exposed to the cloud server. Furthermore, the server does not need to decrypt the encrypted documents during the query process, which better protects data privacy and security. However, this method requires traversing all filters in the document to search for matching documents, which greatly affects the performance of the system.

There are some concerns about the security of Bloom filters [12]. However, existing cryptanalysis attacks are not practical. They assume that the attacker knows certain parameter settings used in the Bloom filter encoding process and are computationally expensive.

2.2 Fuzzy Query

In actual query scenarios, it is very common for users to submit queries with wrongly formatted or spelled keywords. However, the precise query does not have the corresponding fault tolerance, so the functional requirement of fuzzy query is proposed. Fuzzy query is an important technology in the field of information retrieval. How to perform fast and accurate fuzzy query on massive feature fingerprint data is a problem that must be faced when dealing with large-scale data. Fuzzy query is mainly applied to scenarios such as spelling errors and similar queries. It calculates the similarity between fingerprints and encrypted documents based on some similarity measures, such as edit distance, Jaccard distance, Euclidean distance, etc. Eventually, it returns results containing keywords with high similarity to the query. KD-Tree [13] is a binary search tree mainly used for nearest neighbor query in high-dimensional space. It reduces the amount of search spaces and finds the query target faster by dividing the high-dimensional space into many small regions. Although KD-Tree can be used for precise and fuzzy query on high-dimensional data structures, it can only handle numeric or continuous data. For discrete binary feature fingerprints, KD-Tree will not be able to achieve fuzzy query.

Wang et al. [14] first propose a fuzzy query scheme based on Bloom filter. This approach utilizes the adjacency of character positions in a string to map the hash results of two adjacent characters into a Bloom filter. As one of the key methods for fuzzy query, the hash function can significantly compress high-dimensional data while maintaining data similarity. However, traditional hashing methods can only cope with non-encrypted data, and it is difficult to process encrypted feature fingerprints. He Hen et al. [5] use the minhash algorithm to reduce the dimension of keywords in a multi-keyword query scheme, which reduces the computational and spatial complexity. Then the keywords are subdivided and vectorized into binary vectors, and the vectors are mapped into the counting Bloom filter of each document according to Jaccard similarity distance. The counting Bloom filter can reflect the number of keywords corresponding to the block, which can be used to calculate the relevance of the document. According to the characteristics of counting Bloom filter, dynamic updating of the index can be achieved by changing the value of the Bloom filter.

However, as mentioned above, mapping fuzzy sets into Bloom filters brings great computational and storage overheads. Besides, due to the random hash characteristics of Bloom filters, the edit distance does not reflect the similarity characteristics among fingerprints well.

3 System Overview

In this section, we propose the system architecture of EQFF in Fig. 2. We first construct a compressed data feature fingerprint for rapidly and secure feature query. Such fingerprints not only ensure the security of data features, but also support fast feature queries. EQFF starts with the extracted data features as input. Note that, data files are usually processed with data cleaning [15,16] as a prior operation to remove the erroneous, incomplete or duplicate data. Then, the data features are extracted from the documents using natural language processing approaches such as term frequency-inverse document frequency (TF-IDF) [17,18], Chinese concept dictionary (CCD) [19], and other techniques [20, 21]. It should be noted that both data cleaning and feature engineering are essential steps for data analysis, which are regarded as primary operations and are beyond the scope of this paper.

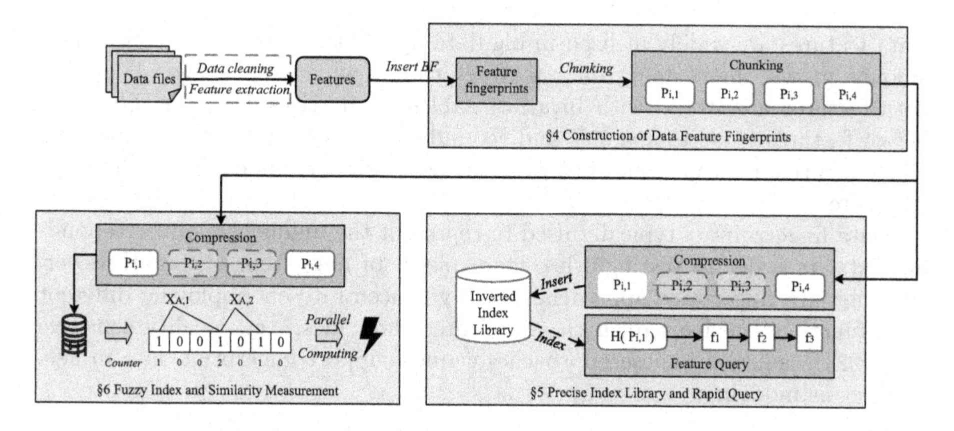

Fig. 2. The system architecture of the EQFF.

EQFF contains three major modules, namely construction of data fingerprints, precise index library and rapid query, and fuzzy index and similarity measurement.

Firstly, the extracted data features are inserted into Bloom filters to generate fingerprints through the feature fingerprint construction module. Each data file or input is transformed into a fixed-length bit vector through encryption hash functions. These bit vectors are considered as the feature fingerprints of the corresponding data. Furthermore, we divide them into blocks with fixed length.

Since the original data feature fingerprints contain a significant amount of redundant information, we compress the feature fingerprints. Meanwhile, the semantic information of those features is no longer accessible. Furthermore, we establish an inverted index library based on compressed fingerprints to enhance the query efficiency of big data. For a feature query request, we first convert the feature into a Bloom filter fingerprint and compress it. Then we search for matches in the inverted index library and return the results.

In practical applications, fuzzy query is a common and important requirement in distributed scenarios. However, precise query can only return specific results and cannot meet the above requirement. Therefore, we propose a similarity measurement for compressed fingerprints. For compressed feature fingerprints, EQFF uses a counting Bloom filter to determine the number of similar features and optimizes query efficiency by parallel frameworks.

With the above three modules, the features of data are represented as compressed fingerprints in format of fixed-length bit vector. The corresponding similarity measurement, inverted index library and feature query are designed to achieve both space efficiency and time efficiency simultaneously. Besides, since the fingerprint is constructed through encryption hash operations, the fingerprint no longer contains the semantics of the raw data. Hence, the concern about data privacy protection is satisfied.

4 Construction of Data Feature Fingerprints

Data features are widely utilized in big data applications. However, as the data amount grows, maintaining data features can become a challenge. Typically, data features are stored with bivariate tables or matrices. The variable number of features leads to sparsity and redundancy in storage. Besides, updating newly identified features into bivariate tables requires modification to the table structure.

Data fingerprint is typically used to represent the uniqueness of corresponding data, in order to distinguish certain pieces of data from others. However, existing data fingerprint approaches mostly concentrate on employing different hash functions to distinguish the duplicated data files, e.g., locality sensitive hash [22]. Meanwhile, these approaches cannot support fingerprint data in heterogeneous modality.

4.1 Construction and Chunking of Data Feature Fingerprints

Based on the above considerations, this paper proposes the EQFF, which constructs fingerprints from the perspective of data features. Bloom filter [6] is employed as a fundamental structure of feature fingerprint. It is a compact probabilistic data structure used to test whether an element belongs to a set, and is widely used in network routing and distributed systems. Following this thought, we consider the extracted features as elements of a feature set.

Given a feature set $DS = \{ds_1, ds_2, ..., ds_n\}$, where ds_i represents a specific feature. The feature fingerprint can be represented in the form of a Bloom filter which is denoted as $BF(DS)$. The Bloom filter is initialized as a fixed-length bit vector where each bit in $BF(DS)$ is a binary bit with a value of 0 or 1. When all the features are mapped to bit vectors, a feature fingerprint of the data is obtained.

Algorithm 1 describes the data feature fingerprint construction algorithm. Steps 1 to 3 initialize the bit vector with length m, steps 5 to 14 set value in the bit vector with hash function mapping, and BF is the constructed feature fingerprint. It is easy to prove that the time complexity of data feature fingerprint construction algorithm is $O(n \cdot k)$, where n is the number of data features. We omit the derivation process in this paper.

Algorithm 1. Constructing data feature fingerprints.

Require: Bit vector length m, Hash function k, Data feature set DS, Data feature fingerprint f_i
Ensure: Data Feature Fingerprint
1: BF is a bit vector of length m;
2: **for** $(i = 0; i \leq m - 1; i++)$ **do**
3: $BF[i] = 0$;
4: **end for**
5: **for** each $ds \in DS$ **do**
6: **for** $(i = 0; i \leq k - 1; i++)$ **do**
7: $j = f_i$;
8: **if** $(j \leq m - 1)$ **then**
9: $BF[i] = BF[j] \,||\, 1$;
10: **else**
11: break;
12: **end if**
13: **end for**
14: **end for**
15: **Output** BF as feature fingerprint.

It should be noted here that the MD5 hash algorithm is one of the widely used classic cryptographic hash functions. The MD5 hash algorithm compresses messages of arbitrary length into fixed-length bit vectors, which greatly improves the efficiency of message storage. As hash algorithm can be considered as a type of compression algorithm, the final digest is typically shorter than the length of the original message. After using MD5 hashing, it is difficult to restore the original data even if the digest information is known. The same result can only be obtained by using the same data for encryption, so MD5 hashing has a unique advantage in terms of privacy of data storage.

The trade-off of utilizing Bloom filter as fingerprint is false positive rate generated by hash collisions. Bloom filters may feedback erroneous answers to queries, as the same bit may be mapped by different elements. The false positive rate p can be calculated by Eq. (1).

$$p = \left(1 - \left(1 - \frac{1}{m}\right)^{n \cdot k_{BF}}\right)^{k_{BF}}, \tag{1}$$

where n denotes the size of the feature set, k_{BF} denotes the number of hash functions. To reduce the probability of false positives, the length of the bit vector m can be expanded to reduce hash conflicts. We will further discuss the parameter selection of the Bloom filter through theoretical analysis in Sect. 7. After generating encrypted fingerprints, we use a fixed-length chunking to divide the entire fingerprint into several equal-length bit blocks.

4.2 Fingerprint Compression Storage Algorithm

Bloom filter use k independent hash functions to map features into a bit vector, creating a unique feature fingerprint for efficient storage and retrieval. However, according to the Bloom filter's false positive rate formula, the smaller the value of $\frac{k}{m}$, the lower the false positive rate. To ensure the uniqueness of the feature fingerprint, k is usually set to a value much smaller than m. Consequently, feature fingerprint generated by Bloom filter tends to contain a high proportion of 0 elements and a low proportion of 1. However, 0 has no meaning in the feature fingerprint. A large number of 0 will greatly reduce the space efficiency of the feature fingerprint. In order to solve this problem, EQFF designs a chunked storage mechanism for feature fingerprints based on compressed bitmaps.

The chunked storage mechanism based on compressed bitmap fingerprints first slices the feature fingerprint into chunks of length c. The chunked feature fingerprint is shown in Fig. 3. Due to the feature bits in the chunks of bits 2, 3, 5 and 8 are all 0, these chunks are meaningless in the feature fingerprint. To improve the space efficiency of feature fingerprinting, EQFF uses a placeholder fingerprint to represent those blocks of feature fingerprints that are meaningless.

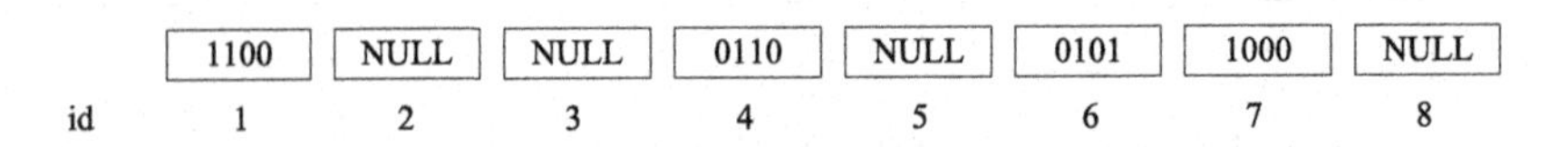

Fig. 3. A chunked fingerprint after compression.

5 Precise Query Library and Rapid Query

This section focuses on how to perform a precise query for encrypted feature fingerprints. Since the semantics of data are filtered out by Bloom filter, we cannot directly conclude whether the fingerprint contains the same features. However, we can use Bloom filter to approximate the presence or absence of the features with a certain degree of accuracy. Note that for identical features in two different feature sets, the hash values remain the same when using a same set of hash functions. As a result, the corresponding positions in the fingerprints are set to be 1. Then, we can perform a precise query on the encrypted feature fingerprints to confirm whether the data contains the exact features we are searching for.

5.1 Acceleration Algorithm Based on Inverted Index Library

EQFF uses Bloom filter to construct the secure query, which improves the efficiency of ciphertext search. However, it still has the following drawbacks. Firstly, each document needs to store an additional secure query structure. When the document content is small, the query structure may be larger than the document. Secondly, the query process still needs to judge the documents one by one. The search time overhead is proportional to the number of documents with the time complexity of $O(n)$, which is inefficient for large datasets. So we design an inverted index library to speed up the query.

Inverted index is an effective way to speed up data query. Generally speaking, it turns the index structure of $(document, words)$ to $(word, documents)$ to improve the efficiency of retrieving a particular word. As for the scenario of feature query in Bloom filter based fingerprints, the positions mapped by hash functions are usually separated. It is difficult to tell which bit is corresponding to the specific feature. In this paper, we build an inverted index library using a chunking based approach. Assume that we have obtained a data feature fingerprint of length m for each data file. EQFF uses a fixed-length chunking mechanism to divide the entire fingerprint into several equal-length bit chunks. Algorithm 2 shows the process of constructing an inverted index library for the chunked feature fingerprints.

Assuming that the fingerprint consists of m bits and is divided into chunks of c bits each. Then the entire fingerprint is divided into $\lceil m/c \rceil$ chunks. The inverted index library consists of two parts. The former is the value of fingerprint block (denoted as v) as the key, while the latter is the identification of data (denoted as ID). As a result, a key-value pair of (v, ID) is obtained.

We show an illustrative example of an inverted index library in Fig. 4. The length of the fingerprint is $m = 32$ and we take the chunk size as $c = 4$. Then the fingerprint is divided into 8 equal-length data chunks. By recording each data block at the same position, the feature fingerprints representing 0–3 bits are "1100", "1101" and "0100", and bits 4–7 are "0101" and "1101". These fingerprints are stored in the inverted index library as index keys. In the first paragraph, "1100" corresponds to the first fingerprint of data A and data B, so A and B are stored in the inverted index database as index values.

Algorithm 2. Inverted index construction.

Require: Feature fingerprint set F_n, Number of fingerprint bits n, Number of bits per block c
Ensure: Feature fingerprint index table $\langle key, value\rangle$
1: **for** $(i = 0; i < n; i++)$ **do**
2: divide F_n into $\lceil m/c \rceil$;
3: **end for**
4: **for** $(j = 0; j < \lceil m/c \rceil; j++)$ **do**
5: key=$Fingerprint\,[F_n, j]$;
6: value=$Data\,[F_n, j]$;
7: **end for**
8: **Output** $\langle key, value\rangle$

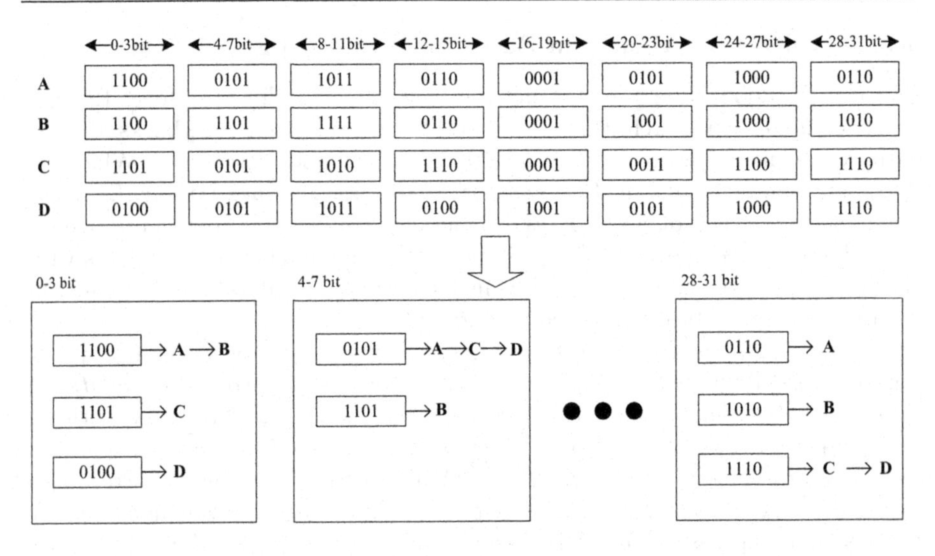

Fig. 4. Construction of data feature fingerprint inverted index library.

5.2 Precise Query Algorithm Based on Bloom Filter

When performing a precise query operation, the data user generates a compressed chunked feature fingerprint using the same hash function. Then, the newly generated fingerprint is compared bit by bit with the feature fingerprint in the library. If the corresponding bits are the same, it is considered as the same document.

Here we show an illustrative example of a feature query. Assuming that the file feature set queried by the user is $DS = \{ds_1, ds_2, ..., ds_n\}$, and the inverted index database is constructed in the form of Fig. 4. Then, EQFF utilizes the Bloom filter to create encrypted fingerprints of the data features and divides them into chunks. Assuming that the block feature fingerprint that we obtained is 1100 1101 1111 0110 0001 1001 1000 0101. Next, we index each fingerprint chunk in the inverted index library and obtain the corresponding candidate set. In this example, the candidate sets can be denoted as $Set_1 = \{A, B\}$, $Set_2 = \{B\}$,..., $Set_8 = \{B\}$. Finally, we take the intersection of these candidate sets and obtain the final index result as B.

6 Fuzzy Query and Similarity Measurement

Most of the existing searchable encryption schemes can only provide precise match query. However, there are some application scenarios where users may require more advanced similarity data management schemes, such as medical information systems, e-commerce, etc. [23]. At the same time, users may misspell the requested entries during the query. In most cases, minor spelling errors should not affect the final output of the retrieval system. User satisfaction can be enhanced by delivering query results that are similar to the data being searched for.

6.1 Similarity Measurement Based on Counting Bloom Filters

First, we construct data feature fingerprints using the chunking mechanism in precise query for data features. Since hash functions filter out the semantics in the original features, the constructed fingerprint can effectively prevent the leakage of sensitive and private data. A common limitation of Bloom filters is the inability to perform direct similarity comparisons. This is due to the presence of hash conflicts, where feature bits of different features are coupled to each other. However, similarity comparison is a very common requirement in the scenario of feature fingerprint queries. Such a requirement can be supported by some variants of Bloom filter, e.g. the counting Bloom filter [24]. When a new feature is inserted, we can update the counters of the feature bits to record the latent information of the feature.

This property of the counting Bloom filter makes it possible to compare similarities between fingerprints. Therefore, EQFF uses the counting Bloom filter as Fig. 5. It adds one more row of counters than the traditional Bloom filter in order to count the feature mappings.

Feature fingerprints constructed by Bloom filters for similar features are not similar in m-dimensional space. Therefore, the traditional similarity measurement scheme based on geometric distance is not applicable to our scheme. Based on this consideration, we design a novel similarity measurement for feature fingerprints, which we call the bit-based similarity measurement.

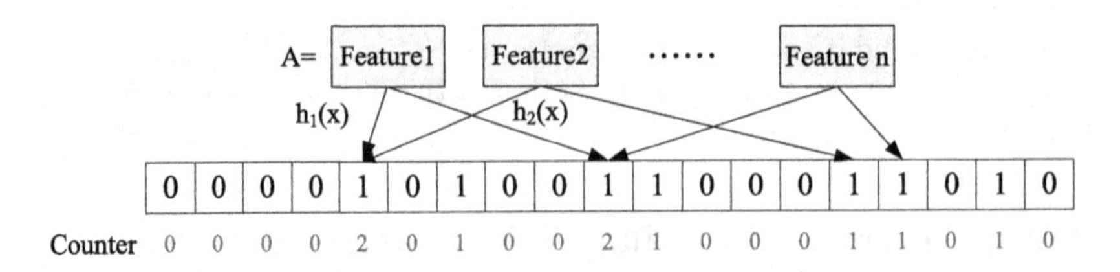

Fig. 5. Counting Bloom filter guarantees the maintainability of the feature fingerprints.

Due to the property of Bloom filter, the mapping of same features in the bit vector is the same. For the feature set, we define a completely new similarity measurement $S(A,B) = \frac{num}{max(l_A,l_B)}$. In this formula, $S(A,B)$ denotes the similarity between A and B, $num(A,B)$ denotes the number of identical features between A and B, and l_i denotes the number of features of A and B. For Bloom filter fingerprints, we cannot get the specific features contained in the data. Therefore, we modify the similarity measurement to a bit-based form $S_{bit}(A,B) = \frac{same_{bit}(A,B)}{max(q_A,q_B)}$. In this equation, $same_{bit}(A,B)$ denotes the same number of bits between A and B, q_i denotes the total number of bits contained in A or B that have been set to 1.

6.2 Search Algorithm Acceleration Using Parallel Computing

When a user makes a fuzzy query, the query system generates an encrypted feature fingerprint based on the data features. The user also needs to pre-define a threshold value in the system, and the data items with similarity higher than this threshold are considered to be similar. For the generated feature fingerprint, we first calculate the counters to obtain the maximum number of feature bits $max(q_A, q_A)$. Then, we take the intersection of the counting Bloom filters to get the number of bits $same_{bit}(A,B)$ with the same mapping.

When dealing with a large amount of data in the database, traditional traversal query can be very time-consuming. To improve query efficiency, we propose a parallel query scheme based on document partitioning. Firstly, we split the fingerprint database into a number of sub-database of the same size. Then, EQFF uses the Fork/Join framework to perform parallel query on split sub-databases. The Fork/Join framework splits the task and uses the system's scheduling algorithm to speed up the process. Compared to traditional traversal query algorithms, parallel query based on splitting can improve the processing efficiency of the program to a great extent.

7 Evaluation

In this section, we empirically evaluate the performance of EQFF in this paper using a real-world dataset. We describe our experimental setup and then present extensive experimental results. The results show that the precise query algorithm of EQFF has higher accuracy and efficiency, and occupies less memory than the comparison method. Besides, the similarity measurement proposed in this paper has validity and can handle fuzzy indexing scenarios.

7.1 Experimental Settings and Dataset

In our experiments, we use an Alienware Laptop, equipped with 11th Gen Intel(R) Core(T.M.) i7-11800H processor with eight 2.80GHz cores and 16GB RAM. The machine runs Windows system.

Datasets. We use a real-world dataset MLRSNet [25] to evaluate the performance of EQFF. MLRSNet is a dataset designed for semantic scene understanding in remote sensing applications. The dataset has high spatial resolution and multiple labels. It provides different perspectives of the world captured from satellites. MLRSNet contains 109,161 remote sensing images that are annotated into 46 categories, and the number of sample images in a category varies from 1,500 to 3,000. The distribution of the number of images for each scene in the dataset is shown in Fig. 6.

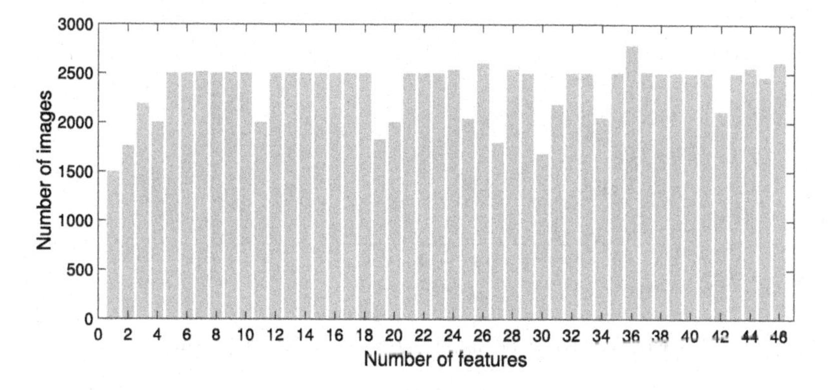

Fig. 6. Distribution of images with different number of features.

Comparison Method and Metrics. We consider comparison methods for precise and fuzzy query respectively. For precise query, we use the KD-Tree as the comparison method. For fuzzy query we use traditional traversal query and perform the similarity calculation according to the formula in Sect. 6.

To evaluate the performance of EQFF, we use accuracy and recall as evaluation metrics. The accuracy evaluates whether the query will return or drop out the data incorrectly. The higher the accuracy rate, the higher the relevance of the results the user gets to the query. Recall evaluates the ability of the algorithm to return the correct answer. The higher the recall, the more complete answers the user will receive. In addition, we evaluate the time efficiency of query by query time. The shorter the query time, the higher the user satisfaction.

7.2 Storage Overhead and Parameter Selection for Feature Fingerprints

Storage Overhead. When constructing data feature fingerprints, we choose a Bloom filter-based approach. It has the advantage that the space efficiency and query time are much higher than the general algorithms. And the storage space and insertion time of Bloom filter are constant $O(k)$, where k is the number of hash functions.

However, when using text files for data feature fingerprint, the storage space is usually required $O(xn)$, where x is the amount of data and n is the number of data features. In this dataset, a storage space of 15201 Kb ($\approx$ 14.8 Mb) is required despite the use of *csv* files for storage. In contrast, when using Bloom filters to construct feature fingerprints, the storage space is determined by the length of the bit vectors corresponding to the feature fingerprints. The storage overhead increases with the increase of bit size. For practical reasons, we set the feature fingerprint bitsize $m = 128$, which occupies 1706 Kb of memory.

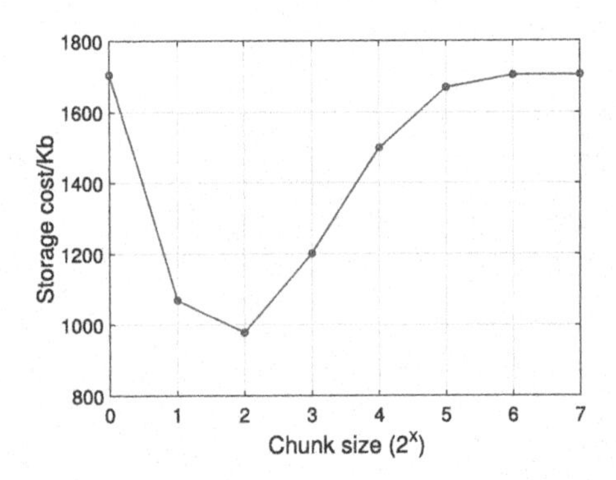

Fig. 7. Overhead for different bit vector length.

In fact, there are a large number of invalid bits in the traditional Bloom filter fingerprint. To improve the space efficiency of feature fingerprints, we use the compressed bitmap-based fingerprint storage mechanism as described in Sect. 4. Figure 7 exhibits the feature fingerprint storage overhead under different compression sizes. When the compression size is 4, the storage occupation decreases from $1706Kb$ to about $978Kb$, which is only 6.4% of the original *csv* file and 57.3% of the traditional Bloom filter fingerprint. The result indicates that the compressed fingerprint mechanism in EQFF greatly improves the space utilization of the feature fingerprints.

Parameter Selection. The false positive of Bloom filter can affect the accuracy of query. The probability of false positive is calculated by Eq. 1. It can be concluded that the false positive rate is lowest when $k = \frac{m \cdot ln2}{n}$.

Figure 8 depicts the false positive rate for different bit vector length when $k = 4$. The false positive rate decreases as the length of the bit vector increases. However, this will lead to an increase in the storage overhead of the feature fingerprint. Figure 9 exhibits the false positive rate for different number of hash functions when $m = 128$. The false positive rate decreases with the increase of the number of hash functions. However, when the number of hash functions continues to increase, the false positive rate will increase instead of decrease. This is due to the fact that inserting a feature will map to multiple bits, increasing the probability of hash conflicts.

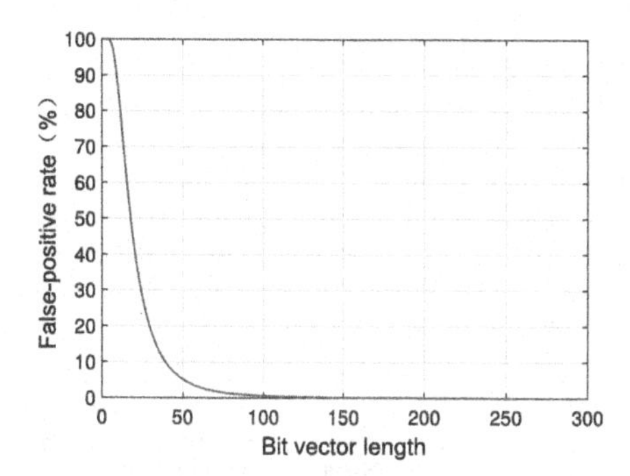

Fig. 8. False positive rate with different bit vector length.

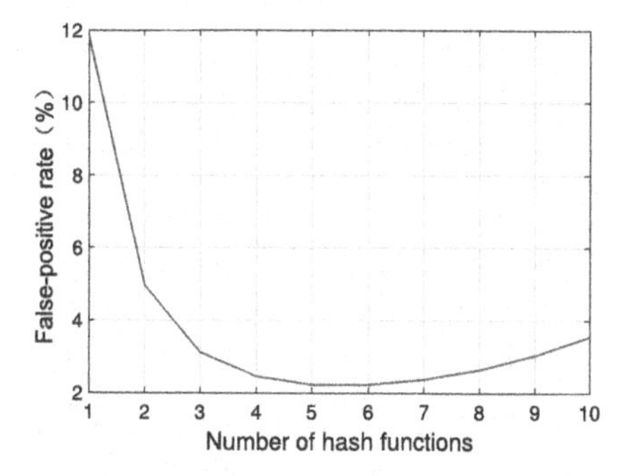

Fig. 9. False positive rate with different number of hash functions.

To trade off the query accuracy and storage efficiency, we set the bit vector length $m = 128$ and the number of hash functions $k = 4$. Assuming the maximum number of features of dataset is 12, the false positive rate will be lower than 0.96% according to Eq. 1. For compressed storage of feature fingerprints, a suitable compression size will result in low storage overhead. Based on the experimental results in Fig. 7, we set the compression size to $c = 4$.

7.3 Performance for Precise Query

Accuracy and Recall. We demonstrate the feasibility and effectiveness of using Bloom filter to generate data feature fingerprints and perform precise query. In this paper, we first insert data features into the Bloom filter to obtain data feature fingerprints. However, since the Bloom filter itself has a probability of false positives, it is important to consider whether the generated data feature fingerprints can perform accurate query.

In this paper, we randomly select η images in the dataset for query, η is set to 10, 20, 50 and 100 respectively. This can verify the accuracy and stability of the query algorithm under different loads. To eliminate randomness, for each η we performed 20 rounds of query experiments. We use the KD-Tree as the baseline and compare with it to verify the precise, recall and query time of EQFF.

As mentioned before, we set the bit vector length $m = 128$ and the number of hash functions $k = 4$ in our experiments. Figure 10 exhibits the query accuracy of 20 rounds under different loads. It can be seen that the accuracy of Bloom filter fingerprints under different loads is above 99%. Figure 11 depicts the query accuracy comparison between our algorithm and KD-Tree. The accuracy of the conventional KD-Tree algorithm fluctuates between 87% and 96%, while the accuracy of our algorithm is consistently close to 100%. This suggests that EQFF rarely returns incorrect index results compared to the traditional KD-Tree.

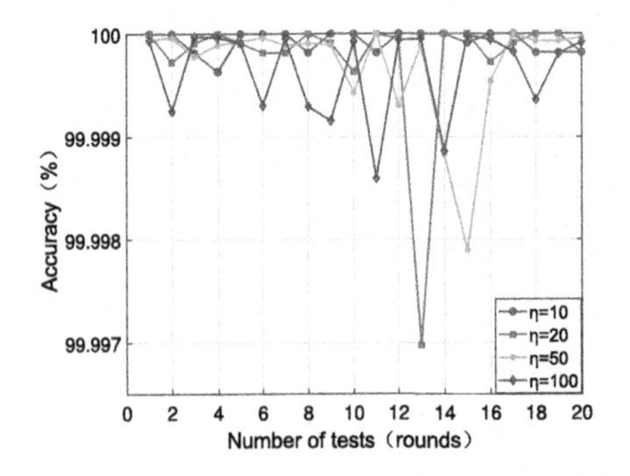

Fig. 10. Accuracy for different numbers of query features.

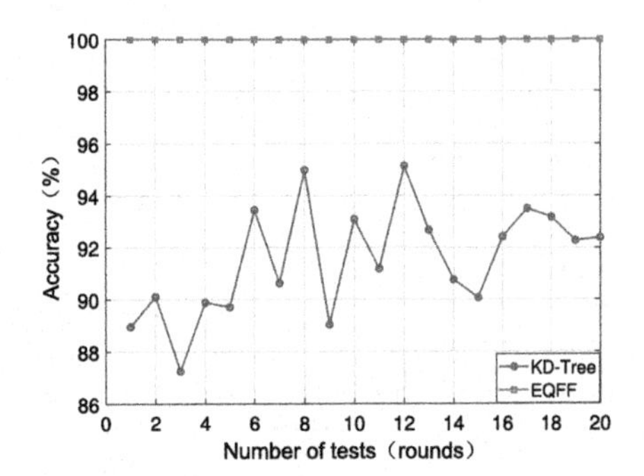

Fig. 11. Accuracy of query using chunk compressed query and KD-Tree query.

Then we test the query recall of Bloom filter fingerprints and compared it with the KD-Tree. Figure 12(a) exhibits the query recall of 20 rounds under different loads. The recall of query is higher than 98% under different loads. Figure 12(b) exhibits the query recall compared with KD-Tree. In most scenarios, the query recall of our algorithm is higher than KD-Tree and can reach 100%. It is slightly lower than the KD-Tree in a few cases but also exceeds 99%. The experimental results show that EQFF is able to return all query results as completely as possible for precise query.

Time Efficiency. Figure 13 evaluates the time efficiency of EQFF. Figure 13(a) compares the query time of traversal query, chunked inverted query, and compressed bitmap query. Due to our inverted index library for the chunked data fingerprint, the query time is 66.7% faster than the traditional traversal query. Additionally, the query time is further reduced to approximately 3 ms as a result of the data feature fingerprint compression process. This is due to the fact that invalid blocks are skipped during the query and the number of comparisons is reduced.

Figure 13(b) compares the query time between compressed bitmap query and KD-Tree query. The query time of EQFF is less than half of that of the KD-Tree. The result shows that the query based on compressed bitmaps has a significant efficiency improvement.

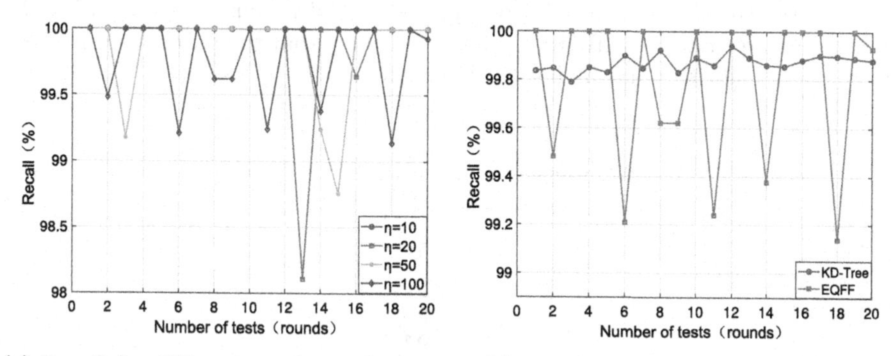

(a) Recall for different numbers of query features.

(b) Recall compared with KD-Tree.

Fig. 12. Recall of precise query.

7.4 Performance for Fuzzy Query

To verify the validity of EQFF, we conduct an experiment to calculate the similarity of the pairwise data feature sets. We use the TF-IDF method [17] to extract features from the dataset. Then we use the similarity formula defined in Sect. 6 with the traversal algorithm as the baseline. We test the performance of EQFF in terms of accuracy, recall and query time respectively. We randomly generate a set of image features and use Bloom filters to generate fingerprints. Then we query the fingerprint with fuzzy approach. Finally, the obtained results are compared with the baseline. To eliminate the randomness of the query, we perform 20 rounds of queries and set different similarity thresholds.

Accuracy and Recall. The accuracy and recall are exhibited in Fig. 14. Figure 14(a) and Fig. 14(b) evaluate the accuracy and recall of fuzzy query under different similarity thresholds respectively. It can be seen that, the fuzzy query has high accuracy and recall when the similarity threshold is high. The accuracy is higher than 98% while the recall is higher than 94%. The accuracy performance of fuzzy indexing decreases slightly as the similarity threshold decreases. However, in most cases, the accuracy is higher than 96% and recall is higher than 95%. The decrease in performance is due to the fact that different similarity definitions may produce different results at lower similarity thresholds. Therefore, in practical applications EQFF is able to return results that are as satisfactory as possible to the user.

Time Efficiency. To verify the efficiency of the fuzzy query approach, we record the query time for each round of experiments and compare it with the traversal query. Figure 15 evaluates the query time of the traditional traversal approach and the partitioned paralleled computing approach. It can be seen that the traditional traversal approach consumes 80 ms to process each query. By using the partitioned parallel computation approach, EQFF consumes only 10 ms to process each query. The experimental results reflect that the partitioned parallel fuzzy query algorithm has higher time efficiency.

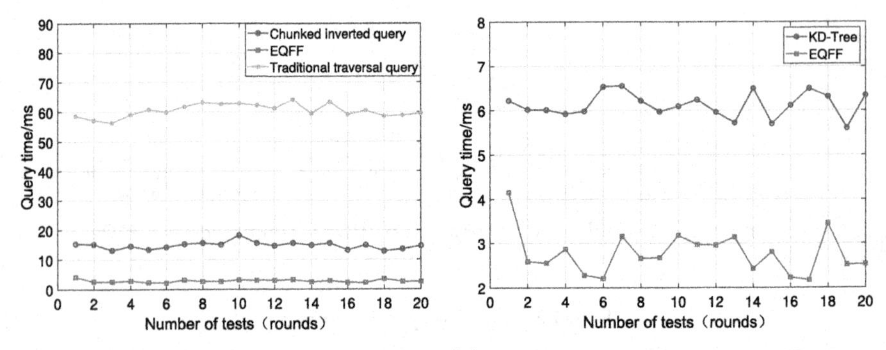

(a) Query time with different query approaches. (b) Query time compared with KD-Tree.

Fig. 13. Query time of precise query.

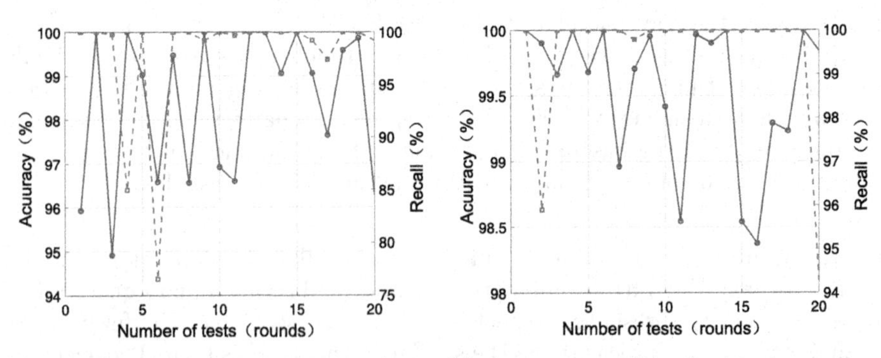

(a) Accuracy and Recall with the similarity threshold 0.6. (b) Accuracy and Recall with the similarity threshold 0.7.

Fig. 14. Fuzzy query accuracy with different similarity threshold settings.

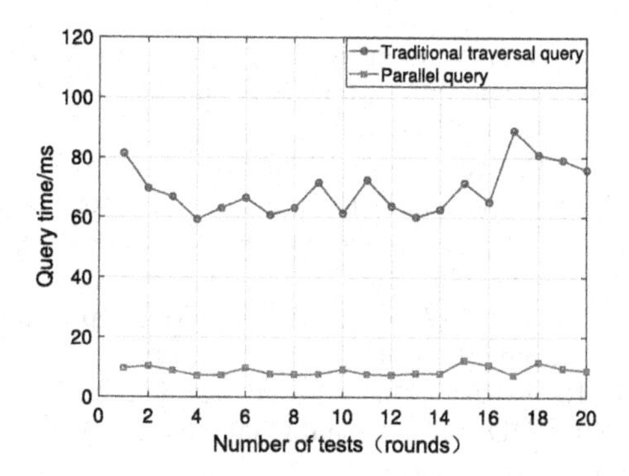

Fig. 15. Query time using parallel fuzzy query algorithm.

8 Conclusion

In the era of big data, how to query and store data features securely and efficiently is an urgent problem. This paper presents EQFF, an Efficient Query method using Feature Fingerprints. We design EQFF mainly to achieve the following two goals: 1) to construct a feature fingerprint mechanism to protect feature privacy and 2) to achieve fast and accurate query of data feature fingerprints. To this end, we develop the corresponding precise and fuzzy query with an inverted index and compression storage mechanism for EQFF. Extensive experimental results show that EQFF can effectively reduce the storage cost of index structure in massive data while using privacy-protected feature fingerprints. Besides, it has higher accuracy, recall and query efficiency compared to traditional query methods.

References

1. Reinsel, D., Gantz, J., Rydning, J.: Data age 2025: the evolution of data to life-critical. In: Don't Focus on Big Data, vol. 2 (2017)
2. Aba, B., Ad, C., Yca, D.: A review of privacy-preserving techniques for deep learning. Neurocomputing **384**, 21–45 (2020)
3. Wu, J.: Dynamic similarity queries on encrypted data. Ph.D. thesis, Xidian University (2019)
4. Cheng, K., et al.: Secure k k-nn query on encrypted cloud data with multiple keys. IEEE Trans. Big Data **7**(4), 689–702 (2017)
5. He, H., Xia, W., Zhang, J., Jin, Y., Li, P.: A fuzzy multi-keyword retrieval scheme for ciphertext data in cloud environment. Comput. Sci. **44**(5), 7 (2017)
6. Bloom, Burton, H.: Space/time trade-offs in hash coding with allowable errors. Commun. ACM **13**(7), 422–426 (1970)
7. Reviriego, P., Sanchez-Macian, A., Walzer, S., Merino-Gomez, E., Liu, S., Lombardi, F.: On the privacy of counting bloom filters. IEEE Trans. Depend. Secure Comput. **20**(2), 1488–1499 (2023). https://doi.org/10.1109/TDSC.2022.3158469
8. Calderoni, L., Palmieri, P., Maio, D.: Probabilistic properties of the spatial bloom filters and their relevance to cryptographic protocols. IEEE Trans. Inf. Forensics Secur. **13**(7), 1710–1721 (2018). https://doi.org/10.1109/TIFS.2018.2799486

9. Miao, Y., et al.: Efficient privacy-preserving spatial range query over outsourced encrypted data. IEEE Trans. Inf. Forensics Secur. **18**, 3921–3933 (2023). https://doi.org/10.1109/TIFS.2023.3288453
10. Malhi, A., Batra, S.: Privacy-preserving authentication framework using bloom filter for secure vehicular communications. Int. J. Inf. Secur. **15**(4), 433–453 (2016)
11. Many, D., Burkhart, M., Dimitropoulos, X.: Fast private set operations with sepia. ETZ G93 (2012)
12. Bösch, C., Hartel, P., Jonker, W., Peter, A.: A survey of provably secure searchable encryption. ACM Comput. Surv. **47**(2), 1–51 (2014)
13. Ram, P., Sinha, K.: Revisiting KD-tree for nearest neighbor search. In: Proceedings of the 25th ACM SIGKDD International Conference on Knowledge Discovery and Data Mining, pp. 1378–1388 (2019)
14. Wang, Z.F., Jing, D., Wei, W., Shi, B.L.: Fast query over encrypted character data in database. Lect. Notes Comput. Sci. **04**(4), 289–300 (2004)
15. Wang, Q., Guo, Y., Yu, L., Chen, X., Li, P.: Deep q-network-based feature selection for multisourced data cleaning. IEEE Internet Things J. **8**(21), 16153–16164 (2020)
16. Rahul, K., Banyal, R.K.: Detection and correction of abnormal data with optimized dirty data: a new data cleaning model. Int. J. Inf. Technol. Decis. Mak. **20**(02), 809–841 (2021)
17. Ni, J., Cai, Y., Tang, G., Xie, Y.: Collaborative filtering recommendation algorithm based on tf-idf and user characteristics. Appl. Sci. **11**(20), 9554 (2021)
18. Xiang, L.: Application of an improved TF-IDF method in literary text classification. Adv. Multim. (2022)
19. Yu, J.S., Yu, S.W.: The structure of Chinese concept dictionary. J. Chin. Inf. Process. **16**(4), 12–20 (2002)
20. Zhang, L., Lu, X.: Feature extraction based on support vector data description. Neural Process. Lett. **49**(2), 643–659 (2019)
21. Li, B., Wang, J., Liu, X.: Parallel cleaning algorithm for similar duplicate Chinese data based on Bert. Sci. Program. **2021**, 1–11 (2021)
22. Yang, C., Deng, D., Shang, S., Shao, L.: Efficient locality-sensitive hashing over high-dimensional data streams. In: 2020 IEEE 36th International Conference on Data Engineering (ICDE), pp. 1986–1989 (2020). https://doi.org/10.1109/ICDE48307.2020.00220
23. Zezula, P.: Scalable similarity search for big data: challenges and research objectives. In: Jung, J.J., Badica, C., Kiss, A. (eds.) INFOSCALE 2014, LNICST, vol. 139, pp. 3–12. Springer, Cham (2015). https://doi.org/10.1007/978-3-319-16868-5_1
24. Reviriego, P., Sánchez-Macian, A., Walzer, S., Merino-Gómez, E., Liu, S., Lombardi, F.: On the privacy of counting bloom filters. IEEE Trans. Depend. Secure Comput. **20**(2), 1488–1499 (2023). https://doi.org/10.1109/TDSC.2022.3158469
25. Qi, X., et al.: Mlrsnet: a multi-label high spatial resolution remote sensing dataset for semantic scene understanding. ISPRS J. Photogramm. Remote Sens. **169**, 337–350 (2020)

Joint Controller Placement and Flow Assignment in Software-Defined Edge Networks

Shunpeng Hua[1], Baoliu Ye[1](✉), Yue Zeng[1], Zhihao Qu[2], and Bin Tang[2]

[1] Department of Computer Science and Technology, Nanjing University, Nanjing 210023, China
yebl@nju.edu.cn

[2] School of Computer and Information, Hohai University, Nanjing 211100, China

Abstract. Software-Defined Networking (SDN) has been introduced into edge networks as a popular paradigm, leveraging its high programmability, where SDN controllers are enabled to centralize network configuration and management. However, frequent flow fluctuations at the edge can result in a large backlog of local flow requests in the processing queue of the controllers, leading to high response delays. Although optimized controller placement and assignment can reduce flow-setup delay, existing approaches are limited in their ability to jointly optimize controller placement and fine-grained flow assignment and address high queuing delay of flow requests. In this paper, we investigate how to jointly optimize controller placement and flow assignment under limited controller capacity, to reduce the propagation delay of data nodes and controllers and the queuing delay of flow requests, and therefore, reducing flow-setup delay. We systematically model the problem and propose a traffic segmentation-based controller placement and flow assignment algorithm. Simulation experimental results demonstrate that our scheme can reduce the flow-setup delay by up to 21.6% compared to existing solutions.

Keywords: Edge computing · software-defined networks · controller placement and flow assignment · latency

1 Introduction

In recent years, the rise of the Internet of Things has led to the emergence of edge computing as a new computing framework. As an significant technology, edge computing brings computing resources closer to users, resulting in lower latency and reduced energy consumption [1]. As another key technology, Software-Defined Networking (SDN) enables the separation of control and data planes, allowing for centralized control logic on the controller [2,3]. Thanks to these two technologies, network and computing resources can be flexibly managed and configured [4,5].

In an SDN-enabled edge network, the control plane and the data plane are decoupled, rather than tightly coupled as in traditional networks. The controllers

Z. Tari et al. (Eds.): ICA3PP 2023, LNCS 14491, pp. 21–38, 2024.
https://doi.org/10.1007/978-981-97-0808-6_2

in the control plane manage the traffic in the network by setting flow rules for the switches in the data plane [6,7]. The specific process of flow setup is as follows. When a new flow arrives at a switch and there is no corresponding forwarding rule, the switch notifies the controller to plan the routing path by setting flow rules. The controller then computes the forwarding path for that flow and pushes the corresponding flow rules to the switches in the data plane. The aforementioned flow rule setup process introduces flow-setup delay. As networks become larger and more complex, a single controller may not be able to handle the massive flow requests and ensure reasonable flow-setup delay, and a single point of failure can lead to network paralysis. As a result, the control plane is typically implemented with multiple controllers that are logically centralized but physically distributed. Thus, a critical issue is how to coordinate multiple controllers to manage the entire network.

Many works have studied controller placement and switch assignment [4,8–18], where controller placement refers to which nodes the controllers are placed at, and switch assignment refers to which controllers the switches are assigned to. First, several works [8,9] have studied switch assignment, which determines which controllers to assign switches to, with a goal of minimizing delay or balancing load. However, edge network is dynamic while static controller placement is lack of flexibility. Therefore, some works [4,10–18] further study controller placement and switch assignment. Switch assignment is a coarse-grained node-based assignment method compared to flow-based assignment, which bind data forwarding nodes to a certain controller for management over a period of time. The problem with this approach is that if the data forwarding nodes assigned to a certain controller generate a large number of flow requests during a slot, these flow requests will be assigned to the bound controller. At the edge, many nodes have limited computing and storage resources, and the controller's request processing ability is relatively small. This results in a large backlog of requests in the controller request queue, and the controller response delay for these flow requests will be high. Thus, several studies [19,20] have focused on fine-grained flow-based assignment that determines which controllers to assign the flows to. However, they only consider flow assignment and ignore controller placement, which may lead to inefficient controller plane layout.

Therefore, a critical issue is controller placement and flow assignment. Next, we give an example to show its advantages in detail. As shown in Fig. 1, there are four nodes (A, B, C, and D) that can be used to place controllers. As shown in Fig. 1(a), the controllers are randomly placed in A and B, which leads to an inefficient control plane layout. In this case, the propagation delay between the controllers and the data plane forwarding devices is as high as 10 ms. Furthermore, the node-based assignment scheme assigns switch s_1 to controller c_1 and switch s_2 to controller c_2. In this case, the queuing delay of flow requests handled by controller c_1 is as high as 11 ms, while controller c_2 remained idle and underutilized. Figure 1(b) shows the optimized controller placement and flow-based assignment scheme, which places controllers on nodes C and D. In this case, the propagation delay between the controllers and the forwarding devices is no

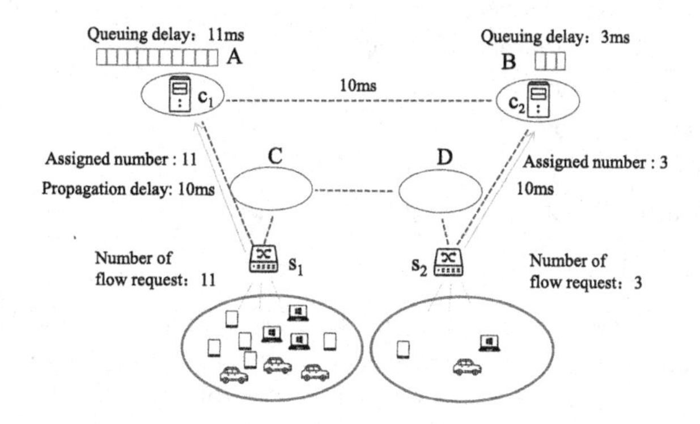

(a) Random controller placement and switch assignment.

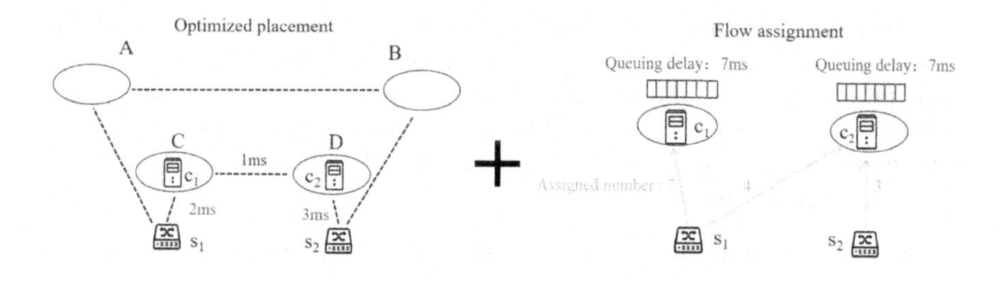

(b) Optimized controller placement and flow assignment.

Fig. 1. A motivational example to show the importance of controller placement and flow assignment.

higher than 3 ms, which is lower than random controller placement. In addition, flow-based assignment scheme dynamically allocates flow requests of s_1 to the two controllers, which can effectively balance the workload between them. This optimization reduces the queuing delay of the controller processing requests to 7 ms, which can improve the efficiency and service performance of the controllers.

Consequently, in this paper, we explore how to dynamically place controllers and assign flows in edge networks to optimize the average flow-setup delay. We present a traffic segmentation-based controller placement and flow assignment scheme to address the high flow-setup delay challenge in edge networks. To the best of our knowledge, this is the first work to combine optimized controller placement and fine-grained flow assignment to tackle the problem of dynamic controller placement and assignment in edge networks. Our proposed scheme is a two-stage algorithm: a) Controller placement, in which we employ traffic segmentation to divide the network into multiple traffic slices with approximately equal traffic loads, and then determine a suitable location for controller

Related Work	Decision		Assignment granularity
	Placement	Assignment	Flow-based
[8,9]	✗	✓	✗
[19,20]	✗	✓	✓
[4,10–18]	✓	✓	✗
Our scheme	✓	✓	✓

placement within each traffic slice, and b) Flow assignment, involving hierarchical flow assignment based on controller positions in the network. The main contributions of this paper can be summarized as follows.

- We formulate the controller placement and flow assignment problem in SDN-enabled edge networks, with the objective of minimizing the average flow-setup delay, which we prove to be NP-hard.
- We introduce a traffic segmentation-based controller placement and flow assignment algorithm to optimize the average flow-setup delay.
- Our proposed scheme outperforms existing approaches in several key indicators, such as average flow-setup delay, percentage of high queuing delay flows, and percentage of flow-setup delay constraint violated flows, as demonstrated by experimental results.

The remainder of this paper is organized as follows. Section 2 discusses related work and highlights the limitations of these works in SDN-enabled edge networks. We introduce the system model and problem formulation in Sect. 3. In Sect. 4, we propose a traffic segmentation-based controller placement and hierarchical flow assignment algorithm. Section 5 presents the evaluation of our proposed scheme.

2 Related Work

Heller et al. [10] first proposed the Controller Placement Problem (CPP) in SDN. Afterwards, a large number of studies on controller placement and assignment emerged. We divide the existing work into three categories: assignment schemes, controller placement and assignment schemes, and controller placement and assignment schemes in software-defined edge networks.

2.1 Assignment Schemes

Sun et al. [8] proposed a dynamic workload balancing scheme based on multi-agent reinforcement learning, which outperformed optimization algorithms to manipulate the mapping relationship between controllers and switches. Huang et al. [9] devised a predictive online switch-controller association and control devolution scheme, which reduced request latency significantly. Xie et al. [19] proposed a light-weight and load-ware switch-to controller selection scheme and designed a general delay-aware switch-to-controller selection scheme to cut the long-tail response latency and provide higher system throughput. Bera et al. [20] proposed a dynamic scheme to assign the flow to the controller to minimize the flow-setup delay and control overhead, which was more fine-grained than the scheme of allocating the switch. This kind of work does not consider the influence of controller position on the system.

2.2 Placement and Assignment Schemes

Guo et al. [11] jointly considered the deployment of SDN switches and controllers to find the locations of updated switches, the locations of deployed controllers, and the mappings between the controllers and the upgraded switches for hybrid SDNs. The authors proposed MapFirst, which returned a high-quality solution with a significant less CPU time. Wu et al. [12] used a deep Q-network to optimize the network delay and load in the dynamic network, optimizing the location of the controller, and dynamically adjusting the mapping between switches and controllers. Basu et al. [13] took the controller placement problem and hypervisor placement problem into consideration at the same time and propose an approach of dynamically deploying controller-hypervisor pairs to provide a variety of network functions with low latency. Bouzidi et al. [14] studied which controllers are selected and enabled in a separate control plane how to partition the set of data plane switches into clusters and assign them to these controllers. The network span studied in the above work is relatively small, and the propagation delay of communication between data nodes and controllers is not fully considered in the decision of controller layout.

2.3 Placement and Assignment Schemes in SDN-Enabled Edge Network

Qin et al. [4] placed the controllers in the static edge nodes in SDN-enabled edge networks. The authors took the inter-controller and controller-node overheads as two important factors to formulate the controller placement and assignment problem at the edge and proposed exact and approximate algorithms to optimize delay and overheads. Chen et al. [15] studied the adaptive controller placement and assignment problem to adapt to the dynamic topology and time-varying workload in SDN-based low-earth-orbit satellite networks. The authors proposed control relation graph (CRG) and a CRG-based algorithm which outperforms existing schemes in terms of response time and load balancing. Li et al. [16] proposed a Louvain algorithm-based controller placement policy to solve the CPP in SDN-IoV and a controller replacement policy to adapt to the dynamic topology of IoV. The objectives are to optimize the control load, control delay, intra-cluster delay and throughput. Li et al. [17] proposed an edge service mesh architecture with distributively deployed controllers and investigate the controller placement problem toward communication cost minimization. Soleymanifar et al. [18] introduce two novel maximum entropy based clustering algorithms to address controller placement problem in wireless edge networks. However, the

Table 1. Summary of Notations

Symbol	Descriptions
G	Undirected Network Graph
S	Set of switches
C	Set of controllers
F	Set of flows
P	Placement policy
A	Assignment policy
$\delta_{s,j}$	Propagation delay of the shortest path between the switch s and the controller placed at s_j
$\alpha_j(t)$	Flow-request arrival rate at the controller placed at s_j
μ_j	Execution rate of the controller placed at s_j
w_i	Number of CPU cycles required by the controller to calculate the forwarding path of the flow f_i
D_i	Flow-setup delay of the flow f_i
D	Average flow-setup delay in time slot t
k	Number of controllers
γ	flow-setup delay bound
Decision variables	Descriptions
x_i^t	Binary variable indicating whether a controller is placed at switch
y_{ij}^t	Binary variable indicating whether the flow f_i is assigned to the controller c_j placed at s_j

above work adopts a node-based assignment method, which can easily cause long waits for flow requests.

3 System Model and Problem Formulation

In order to realize the optimal placement and assignment of controllers, in this section, we model the problem as a multi-constraint optimization problem, and then show that it is a NP hard problem.

3.1 System Model

We consider an SDN-enabled edge network as depicted in Fig. 2, which has multiple controllers to be deployed in the system. The SDN-enabled edge network is modeled as an undirected graph $G(V, E)$ with n nodes, where V represents the set of edge nodes, and E represents the set of network links between edge nodes. Let $S = \{s_1, s_2, ..., s_m\} \in V$ denote the switch set and $C = \{c_1, c_2, ..., c_k\}$ denote the controller set, where $m, k \in \mathbb{N}^+$. Additionally, the set of flows in the network

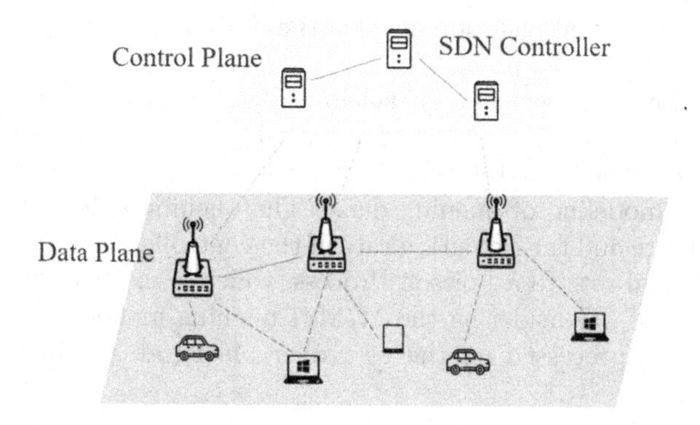

Fig. 2. An example of an SDN-enabled edge network.

is represented as $F = \{f_1, f_2, ..., f_n\}$ where n is a positive integer denoting the number of flows within a time slot. Assuming that the set of flows assigned to the controller c_j is denoted with C_j, we have $C_1 \cup C_2 \cup ... \cup C_k = F$.

To study the placement and assignment policy, We introduce two binary optimization variable x_i^t and y_{ij}^t. The variable x_i^t indicates whether a controller is placed at switch s_i ($x_i^t = 1$) or not ($x_i^t = 0$). These variables constitute the controller placement policy as

$$P = (x_i \in \{0, 1\} : i = 0, 1, ..., n). \tag{1}$$

The other variable y_{ij}^t represents whether the flow f_i is assigned to the controller c_j placed at s_j ($y_{ij}^t = 1$) or not ($y_{ij}^t = 0$). These variables constitute the flow assignment policy as

$$A = (y_{ij} \in \{0, 1\} : i = 0, 1, ..., n, j = 0, 1, ..., m). \tag{2}$$

3.2 Flow-Setup Delay Model

When a switch receives a flow, it sends a Packet-in message to the controller, and the controller processes and replies to it. The time taken for this entire process is referred to as the flow-setup delay. The flow-setup delay we consider consists of three delays: propagation delay, queuing delay and processing delay.

The communication between the controller and the switch needs to propagate through the channel between each other. Therefore, the propagation delay depends on the path delay which includes the process to send and to receive the request. According to the shortest path algorithm [21], we can get the shortest path between the switch and the controller. We assume that the switch associated with the flow request f_i is s. Consequently, the propagation delay can be defined as follows

$$D_{ij}^{\text{prop}}(t) = 2 * \delta_{s,j}. \tag{3}$$

where $\delta_{s,j}$ denotes the propagation delay of the shortest path between the switch s and the controller placed at s_j.

When the controller receives a Packet-in message, it puts it in its own message processing queue. We need to account for the delay that the message experiences while waiting to be processed in the queue, known as queuing delay. We refer to the [20] for the modeling of queuing delay. The queuing delay depends on the request arrival rate and the execution rate of the controller, and approximate the flow-request arrival rate to a Poisson Process, which is mathematically denoted as $\alpha_j(t) = \sum_{i=0}^{k} y_{ij}^t$. Considering the M/M/1 queuing model,the queuing delay of flow request f_i processed on the controller placed at s_j can be defined as follows

$$D_{ij}^{que}(t) = \frac{1}{\mu_j - \alpha_j(t)}. \tag{4}$$

where μ_j denotes the execution rate of the controller placed at s_j.

The processing delay is determined by the time it takes for the controller to calculate the forwarding path for the flow. Let's assume that the number of CPU cycles required by the controller to calculate the forwarding path of the flow f_i is w_i. The processing delay can be defined as follows

$$D_{ij}^{proc}(t) = \frac{w_i}{\mu_j}. \tag{5}$$

When a packet arrives at a switch and the switch does not find a corresponding entry in its flow table, the switch will send a flow request to the controller to establish the forwarding path for that flow. The controller then accepts and processes the request and sends the calculated forwarding path back to the switch. Therefore, the control response delay of the controller placed at s_j for a flow f_i is calculated as

$$D_{ij} = D_{ij}^{prop}(t) + D_{ij}^{que}(t) + D_{ij}^{proc}(t) \tag{6}$$

So the flow-setup delay of the flow f_i is presented as follows

$$D_i = \sum_{j=0}^{m} y_{ij}^t x_j^t D_{ij}. \tag{7}$$

Consequently, the average flow-setup delay in time slot t can be quantified as follows

$$D = \sum_{i=1}^{n} D_i. \tag{8}$$

3.3 Problem Formulation

Our objective is to minimize the average flow-setup delay in the network. Based on the flow-setup delay model above, We formulate the optimization problem as flows

$$\min_{\boldsymbol{P},\boldsymbol{A}} \quad D$$

$$\text{s.t. } \forall i \in [0,m], x_i^t \in \{0,1\} \tag{9}$$

$$\forall i \in [0,n], j \in [0,m], y_{ij}^t \in \{0,1\} \tag{10}$$

$$\forall i \in [0,n], j \in [0,m], y_{ij}^t \leq x_j^t \tag{11}$$

$$\forall i \in [0,n], \sum_{j=0}^{m} y_{ij}^t = 1 \tag{12}$$

$$\sum_{i=0}^{m} x_i^t = k \tag{13}$$

$$\forall j \in [0,m], \alpha_j^t < \mu_j \tag{14}$$

$$\forall i \in [0,n], D_i < \gamma \tag{15}$$

where constraint (9)–(11) ensure that each switch can have at most one controller placed on it, and each flow is assigned to exactly one node where a controller is placed. Constraint (12) guarantees a flow is assigned to only one controller. Constraint (13) specifies the total number of controllers in the system. Constraint (14) ensures that the arrival rate of flow requests should be less than the processing rate of the controller. Lastly, constraint (15) takes into account the delay bound for a flow request.

It can be demonstrated that the above problem is a generalization of the well-studied p-median problem [22] which is NP-hard by setting the flow-setup delay bound and controller processing capacity to infinite and setting the queuing delay and the processing delay to zero. As a result, direct utilization of mathematical programming solvers such as CPLEX [23] can be computationally expensive. To address this, we propose an approximation algorithm as an alternative approach.

4 Controller Placement and Flow Assignment

The decisions regarding controller placement and flow allocation are interdependent in this problem. To address this interdependency, we propose a two-stage algorithm. In the first stage, we optimize controller placement by traffic segmentation. This helps in effectively distributing the controllers across the network. In the second stage, we combine hop-count and controller request queue length considerations to assign flows to controllers. This approach ensures that flows are assigned to controllers in a way that reduces both communication delay and queuing delay. By dividing the optimization process into these two stages, we can achieve an efficient and effective solution for both controller placement and flow assignment.

4.1 Controller Placement

The placement of controllers should consider the traffic distribution across the network. In our algorithm design 1, we utilize the traffic distribution to divide

Algorithm 1. Algorithm for Controller Placement

Input: Network topology G, Flow set F, Number of controllers k
Output: Controller placement scheme P
1: $P \leftarrow \emptyset$
2: **for** $i = 1$ to k **do**
3: $a \leftarrow$ a random node of $G.nodes$
4: Start breadth first search from a to get a largest subgraph g, the sum of flows in the subgraph $\leq \frac{|F|}{k}$
5: $G \leftarrow G - g$
6: $a \leftarrow$node with the highest degree in g
7: $P.add(a)$
8: **end for**

the network, employing a breadth-first traversal approach to partition the graph into traffic slices. Each traffic slice aims to have a total traffic load of nodes that is close to the average load of the controllers. The controller is then placed at the node with the highest degree in the traffic slice.

Although not all flow requests from switches in the traffic slice may be assigned to the controller placed within that specific slice during the final flow assignment, this approach leads to a more evenly distributed traffic slicing in the network. As a result, the traffic load around each controller's placement location becomes nearly balanced, reducing the likelihood of high queuing delays for flow requests.

Furthermore, this approach helps prevent excessive controller placement in one particular area. If too many controllers are placed in a concentrated area, some controllers may remain idle while others become heavily overloaded. This resource waste can be avoided by distributing controllers evenly across different areas, ensuring that flow request queues are not backlogged and controllers can respond promptly to incoming requests.

4.2 Flow Assignment

When considering the assignment of a flow request, the optimal choice is to select the controller with the lowest response delay. Controllers that are closer to the switch that sends the flow request have shorter propagation delays between them. Additionally, assigning flow requests to more idle controllers helps minimize the delays in the flow request queues.

Our assignment algorithm 2 is based on a greedy strategy. It operates by assigning flow requests in a layer-by-layer manner, starting from the nodes where the controllers are placed. Each switch's flow requests are assigned based on the number of hops between the switch and the controllers and request queue length of the controllers. Specifically, for each flow request originating from a switch, we choose the controller with the smallest number of hops from that switch. If there are multiple controllers with the same number of hops, we prioritize the most idle controller for assignment.

Algorithm 2. Algorithm for Flow Assignment

Input: Network topology G, Controller placement scheme P
Output: Flow assignment Scheme

```
1: q ← P
2: l = 0
3: for all n ∈ G do
4:     n.S ← ∅
5: end for
6: while q ≠ ∅ do
7:     for i = 1 to q.size do
8:         v ← q.pop()
9:         if |v.S| = 1 then
10:            assign v's all flows to the c ∈ v.S
11:        else
12:            for each flow f of v do
13:                c ← the most idle controller in v.S
14:                assign f to c
15:            end for
16:        end if
17:        for each neighbour e of v do
18:            if e has not been visited then
19:                q.put(e)
20:                e.S ← v.S
21:                e.l ← l + 1
22:            else
23:                if e.l = l + 1 then
24:                    e.S ← e.S ∪ v.S
25:                end if
26:            end if
27:        end for
28:    end for
29:    l ← l + 1
30: end while
```

By following this approach, we aim to minimize the response delay by selecting controllers that are physically close to the requesting switches and have lower levels of workload. This strategy ensures efficient flow assignment and reduces the overall delay experienced by flow requests in the network.

5 Experiment and Performance Evaluation

We conduct Python-based simulations to evaluate the performance of the our scheme. The AttMpls network topology we used in the simulation is from the Internet topology Zoo [24]. Different parameters and their values are presented in Table 2. We compare our proposed scheme (CPFA) with two existing schemes: the Random Greedy algorithm (RG) [4] and Dynamic Controller Assignment

Table 2. Simulation Parameters

Parameter	Value
Number of controllers	[2,3,4,5]
Number of flows	10000–30000
Network topology	AttMpls [24]
Controller exec.rate	50–150
Delay bound	50–200 ms

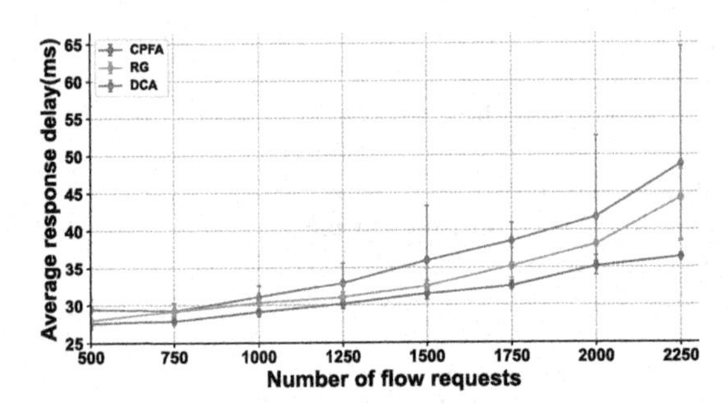

Fig. 3. Average controller response time with different number of flows.

(DCA) [20]. In the rest of the paper, we refer to CPFA, RG, and DCA to represent the proposed scheme, random greedy algorithm, and dynamic controller assignment scheme, respectively. The RG algorithm calculates the probability of placing controllers on nodes based on the optimization objective function, and obtains the corresponding controller-node assignment scheme according to the calculated set of controllers. On the other hand, DCA assigns flows to controllers according to preference lists. In contrast, our proposed scheme(CPFA) takes into account the influence of controller placement, effectively utilizing traffic distribution to partition traffic in the network. It also adopts a more fine-grained flow assignment method based on the greedy strategy. To evaluate the performance, we use the following metrics: average control response delay, maximum control response delay, the percentage of flows with high queuing delay, and the percentage of flows that violate the delay constraint. These metrics demonstrate the effectiveness and superiority of our proposed scheme. The experimental results are discussed in the following subsections.

5.1 Controller Response Time

We measure the average controller response delay and the maximum controller response delay for different numbers of flows and controllers. As shown in the Figs. 3, 4, and 5, our proposed scheme outperforms the existing schemes (RG and DCA) in terms of controller response delay. When the number of flows in

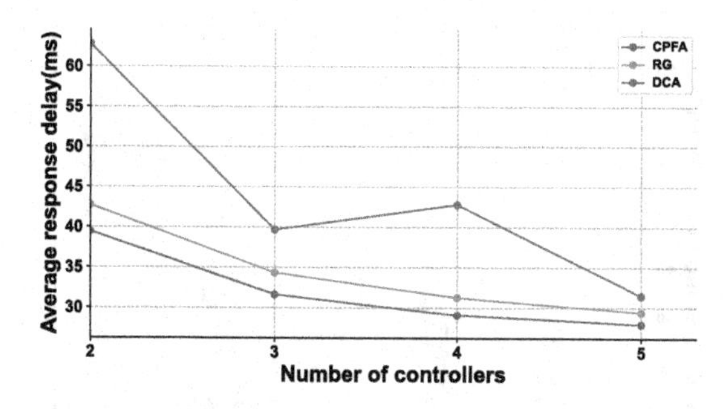

Fig. 4. Average controller response time with different number of controllers.

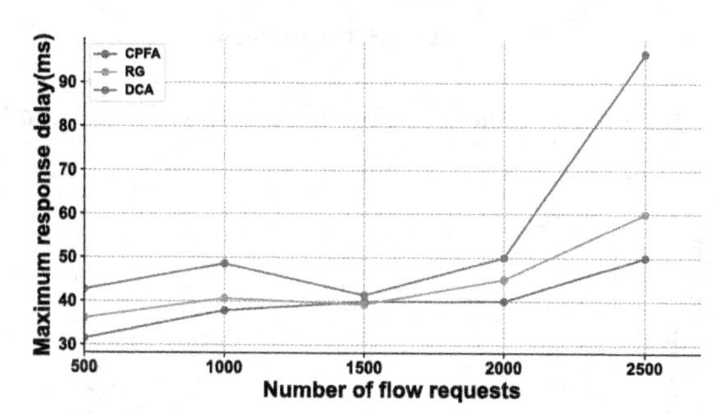

Fig. 5. Maximum controller response time with different number of flows.

the network is 2250, our scheme achieves a 34.0% improvement over DCA and a 21.6% improvement over RG in terms of average flow-setup delay.

The superiority of our scheme can be attributed to two key factors. Firstly, our scheme leverages the traffic distribution in the network to determine the optimized controller locations. This enables more efficient distribution of flow requests across controllers. In contrast, the DCA scheme does not consider the impact of controller location on system performance. As a result, it may lead to a scenario where only a few controllers handle a high traffic load, while other controllers remain idle. This imbalance reduces the overall working performance of the control plane, with overloaded controllers experiencing increased response delay and idle controllers wasting resources. Secondly, our scheme optimizes flow request assignment based on delay requirements and controller workload. This approach reduces the likelihood of response time timeouts caused by controller overload and ensures that flow requests are assigned in a way that helps minimize queuing delay. In contrast, the existing RG scheme assigns edge nodes to controllers, and each controller handles all the flow requests from the assigned nodes within a given period. If a controller receives a large number of flow requests from managed nodes in a short period of time, the queuing delay of flow requests can significantly increase, resulting in high controller response delay.

Furthermore, it is evident that the controller response delay increases with an increase in the number of flows and decreases with an increase in the number

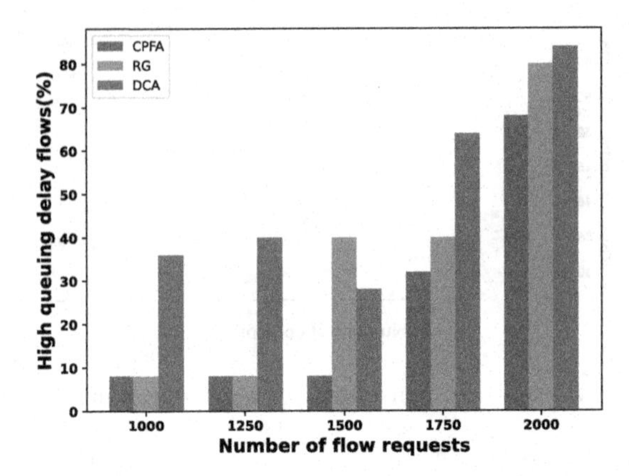

Fig. 6. Percentage of high queueing delay flows with different number of flows.

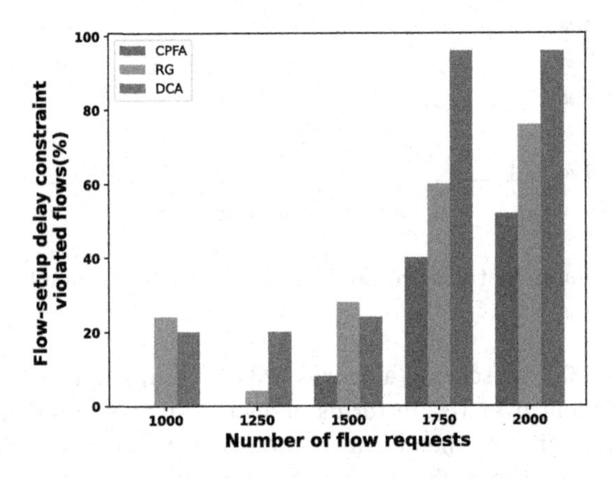

Fig. 7. Percentage of delay constraint violated flows with different number of flows.

of controllers from Figs. 3, 4, and 5. This relationship can be attributed to the fact that a higher number of flows results in a single controller handling a larger volume of flow requests, leading to increased queuing delay and, consequently, higher controller response delay. On the other hand, deploying a larger number of controllers in the network reduces the average number of flow requests that each controller needs to handle. This allows for more efficient assignment of flow requests to controllers with lower propagation delay and higher levels of idleness within the network. As a result, the controller response delay is reduced. Surprisingly, when the number of controllers changes from 3 to 4, the average response delay of DCA increased as shown in Fig. 4. We investigate the reason and find that it is due to the DCA algorithm randomly generating a terrible controller plane layout when the number of controllers is 4.

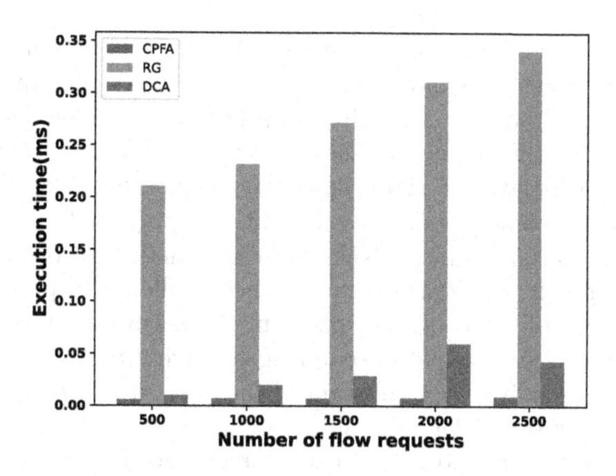

Fig. 8. Algorithm execution time with different number of flows.

5.2 High Queuing Delay Flows

We also conduct an experiment to measure the percentage of flows experiencing high queuing delay, and the results demonstrated that our scheme outperforms RG and DCA. RG is a static assignment scheme that assigns all flow requests from an edge node to its designated controller. However, when the number of flow requests from an edge node increases significantly within a time slot, it leads to a backlog of flow requests in the controller's request processing queue, resulting in high queuing delays. On the other hand, DCA does not take into account the placement of controllers in the network. It is influenced by the convergence of transmission delay and queuing delay while maintaining the preference list of flows. This may result in higher priority in the priority queue of controllers with relatively high queuing delay due to low propagation delay, leading to increased delays in request queuing (Fig. 6).

In contrast, our scheme is based on dynamic flow assignment. After optimizing controller placement, for controllers with close transmission delays, we prioritize selecting relatively idle controllers to process requests. This approach reduces the queuing time of flow requests and enhances the efficiency of the controller plane.

5.3 Folw-Setup Delay Constraint Violated Flows

We also measure the percentage of flows in the network that violated the flow-setup delay constraint. The results clearly demonstrate that our proposed scheme outperforms the existing schemes RG and DCA in this aspect. Our proposed scheme takes the constraints of both the controller response delay and the flow delay bound into consideration. By incorporating these constraints, we are able to ensure that a larger number of flows receive responses with low delays, within the specified constraints. Consequently, only a small number of flows exceed the setup delay bound (Fig. 7).

In contrast, the RG scheme does not consider the delay bound, and its node-based assignment method is more prone to overlooking the controller response delay of individual flows. As a result, some flow requests end up queuing for extended periods in the request queues of certain controllers, leading to high controller response delays for those flows. Consequently, these flows are more likely to violate the flow-setup delay bound. Although DCA takes the flow-setup delay bound into account, it lacks a decision-making process for controller placement. As a result, random placement of controllers can potentially lead to an inefficient layout of the controller plane in the network. This can result in a significant number of flow requests requiring relatively high propagation delays during processing, leading to a substantial increase in the percentage of flows that violate the flow-setup delay bound.

Furthermore, it is worth noting that the percentage of flows violating the flow-setup delay threshold increases with the number of flows, while the RG scheme decreases at 1250 due to the randomness of the algorithm. This is because as the number of flows increases, the controller response delay also tends to increase. Consequently, more flows exceed the specified delay bound.

5.4 Algorithm Execution Time

We also compare the execution time of the algorithms, and the results indicate that our algorithm achieves a more efficient execution process by optimizing the algorithm design and considering the specific requirements of the problem. This observation is also supported by the theoretical analysis of the algorithm. Our algorithm is based on traversal and greedy strategies, which has a linear complexity. In contrast, DCA involves calculating preference lists for controllers and performing bidirectional matching for flow requests, resulting in higher computational complexity. The number of iterations of the RG algorithm depends on the number of nodes that can play the role host for a controller, and in each iteration, the optimal assignment strategy under the current placement strategy needs to be calculated, resulting in relatively high algorithm complexity (Fig. 8).

6 Conclusion

To enhance the programmability and scalability of edge networks, multiple controllers need to be placed in the network to effectively manage the network traffic in the data plane. Well-placed controllers in the network and fine-grained assignment of flows to controllers can effectively shorten flow-setup delay in the network, thereby improving network performance. In this paper, we consider the problem of controller placement and flow assignment in edge networks. To address the issue of high controller response delay caused by excessively long request queues of some controllers, we propose a controller placement and fine-grained flow assignment scheme based on traffic segmentation, with the objective of minimizing the average flow setup delay in the network. The evaluation results show that compared with existing solutions, our proposed scheme has better performance in controller response time, percentage of high queuing delay flows, percentage of flow-setup delay constraint violated flows, and algorithm execution time.

References

1. Shi, W., Cao, J., Zhang, Q., Li, Y., Xu, L.: Edge computing: vision and challenges. IEEE Internet Things J. **3**(5), 637–646 (2016). https://doi.org/10.1109/JIOT.2016.2579198
2. Zeng, Y., Guo, S., Liu, G.: Comprehensive link sharing avoidance and switch aggregation for software-defined data center networks. Futur. Gener. Comput. Syst. **91**, 25–36 (2019)
3. Li, P., Liu, G., Guo, S., Zeng, Y.: Traffic-aware efficient consistency update in NFV-enabled software defined networking. Comput. Netw. **228**, 109755 (2023)
4. Qin, Q., Poularakis, K., Iosifidis, G., Tassiulas, L.: SDN controller placement at the edge: optimizing delay and overheads. In: IEEE Conference on Computer Communications (IEEE INFOCOM 2018), pp. 684–692. IEEE (2018)
5. Zeng, Y., Guo, S., Liu, G., Li, P., Yang, Y.: Energy-efficient device activation, rule installation and data transmission in software defined DCNs. IEEE Trans. Cloud Comput. **10**(1), 396–410 (2019)
6. Zeng, Y., et al.: Mobility-aware proactive flow setup in software-defined mobile edge networks. IEEE Trans. Commun. **71**(3), 1549–1563 (2023)
7. Li, P., Guo, S., Pan, C., Yang, L., Liu, G., Zeng, Y.: Fast congestion-free consistent flow forwarding rules update in software defined networking. Future Gen. Comput. Syst. **97**, 743–754 (2019)
8. Sun, P., Guo, Z., Wang, G., Lan, J., Hu, Y.: Marvel: enabling controller load balancing in software-defined networks with multi-agent reinforcement learning. Comput. Netw. **177**, 107230 (2020)
9. Huang, X., Bian, S., Shao, Z., Hong, X.: Predictive switch-controller association and control devolution for SDN systems. IEEE/ACM Trans. Netw. **28**(6), 2783–2796 (2020)
10. Heller, B., Sherwood, R., McKeown, N.: The controller placement problem. ACM SIGCOMM Comput. Commun. Rev. **42**(4), 473–478 (2012)
11. Guo, Z., Chen, W., Liu, Y.-F.., Yang, X., Zhang, Z.-L.: Joint switch upgrade and controller deployment in hybrid software-defined networks. IEEE J. Select. Areas Commun. **37**(5), 1012–1028 (2019)
12. Wu, Y., Zhou, S., Wei, Y., Leng, S.: Deep reinforcement learning for controller placement in software defined network. In: IEEE Conference on Computer Communications Workshops (INFOCOM WKSHPS 2020), pp. 1254–1259. IEEE (2020)
13. Basu, D., Jain, A., Ghosh, U., Datta, R.: A reverse path-flow mechanism for latency aware controller placement in VSDN enabled 5G network. IEEE Trans. Indust. Inf. **17**(10), 6885–6893 (2020)
14. El Hocine, B., Outtagarts, A., Langar, R., Boutaba, R.: Dynamic clustering of software defined network switches and controller placement using deep reinforcement learning. Comput. Netw. **207**, 108852 (2022)
15. Chen, L., Tang, F., Li, X.: Mobility-and load-adaptive controller placement and assignment in LEO satellite networks. In: IEEE Conference on Computer Communications (IEEE INFOCOM 2021), pp. 1–10. IEEE (2021)

16. Li, B., Deng, X., Deng, Y.: Mobile-edge computing-based delay minimization controller placement in SDN-IOV. Comput. Netw. **193**, 108049 (2021)
17. Li, Y., Zeng, D., Chen, L., Gu, L., Ma, W., Gao, F.: Cost efficient service mesh controller placement for edge native computing. In: 2022 IEEE Global Communications Conference (GLOBECOM 2022), pp. 1368–1372. IEEE (2022)
18. Soleymanifar, R., Beck, A.S.C., Salapaka, S.: A clustering approach to edge controller placement in software-defined networks with cost balancing. IFAC-PapersOnLine **53**(2), 2642–2647 (2020)
19. Xie, J., Guo, D., Li, X., Shen, Y., Jiang, X.: Cutting long-tail latency of routing response in software defined networks. IEEE J. Sel. Areas Commun. **36**(3), 384–396 (2018)
20. Bera, S., Misra, S., Saha, N.: Traffic-aware dynamic controller assignment in SDN. IEEE Trans. Commun. **68**(7), 4375–4382 (2020)
21. Dijkstra, E.W., et al.: A note on two problems in connexion with graphs. Numer. Math. **1**(1), 269–271 (1959)
22. Daskin, M.S., Maass, K.L.: The p-median problem. In: Laporte, G., Nickel, S., Saldanha da Gama, F. (eds.) Location Science, pp. 21–45. Springer, Cham (2015). https://doi.org/10.1007/978-3-319-13111-5_2
23. IBM ILOG. Cplex optimizer. En ligne (2012). http://www-01ibm.com/software/commerce/optimization/cplex-optimizer
24. Knight, S., Nguyen, H.X., Falkner, N., Bowden, R., Roughan, M.: The internet topology zoo. IEEE J. Select. Areas Commun. **29**(9), 1765–1775 (2011)

Distributed Latency-Efficient Beaconing for Multi-channel Asynchronous Duty-Cycled IoT Networks

Peng Long[1], Yuhang Wu[1], Quan Chen[1(✉)], Lianglun Cheng[1], and Yongchao Tao[2]

[1] School of Computer Science and Technology, Guangdong University of Technology, Guangzhou, China
{2112005252,2112105283}@mail2.gdut.edu.cn, {quan.c,llcheng}@gdut.edu.cn
[2] Shenzhen Academy of Aerospace Technology, Shenzhen, China
taoyongchao@chinasaat.com

Abstract. Beaconing is a fundamental task in IoT networks where each node tries to locally broadcast a packet to all its neighbors. Unfortunately, the problem of Minimum Latency Beaconing Schedule (MLBS), which tries to obtain a fastest and collision-free schedule is not well studied when the IoT devices employ the duty-cycled working mode. The existing works have rigid assumptions that there exists a single channel in networks, and can only work in a centralized fashion. Aiming at making the work more practical and general, in this paper, we investigate the first distributed method for the MLBS problem in Multi-channel asynchronous duty-cycled IoT networks (MLBSMD problem). The MLBSMD problem is firstly formulated and proved to be NP-hard. To avoid the collisions locally, several special structures are designed, which works in $O(\Delta)$ time, where Δ denotes the maximum node degree in the network. Then, a distributed beaconing scheduling method which can compute a low-latency and collision-free schedule is proposed with a theoretical bounded, taking the active time slots of each node into account. Finally, the extensive simulation results demonstrate the effectiveness of the proposed algorithm in terms of latency.

Keywords: Beaconing schedule · low latency · distributed algorithm · multi-channel · duty-cycled wireless networks

1 Introduction

In IoT networks, beaconing is essential for communication between nodes, which requires that each node to broadcast a message locally to all its neighboring nodes

This research was supported by China University Industry University Research Innovation Fund under Grant No. 2021FNA02010, the NSFC under Grant No. U2001201, U20A6003, 62372118, the Guangdong Basic and Applied Basic Research Foundation under Grant No. 2022A1515011032, the Guangzhou Science and Technology Plan under Grant 2023A04J1701, and the Guangdong Provincial Key Laboratory of Cyber-Physical System under Grant 2020B1212060069.

Z. Tari et al. (Eds.): ICA3PP 2023, LNCS 14491, pp. 39–55, 2024.
https://doi.org/10.1007/978-981-97-0808-6_3

[2,10,13,26]. Assume that all nodes within the network operate synchronously, and the network time is split into time slots which is enough for one transmission. Then, the latency of beaconing refers to the amount of required time slots such that all the nodes receive the beaconing messages from their neighboring nodes. In node-always-awake networks, the MLBS problem which seeks to find the fastest and collision-free beaconing schedule while the beaconing latency is minimized, has been well studied by [9,11,15,17–22]. Even under the simplest assumption of uniform beaconing and interference radius ρ=1, the MLBS problem has been proved to be NP-hard in [19]. To reduce the beaconing latency, various solutions have been proposed by minimizing the interference between wireless transmissions.

To assist in conserving energy for the IoT devices, the duty-cycled working scheme has been implemented as an effective approach [8,14,25]. This scheme involves two states for each node: the active state and the dormant state. During the dormant state, all functional modules of the node are powered down to save energy, and the node alternates cyclically between the active and dormant states. This mechanism allows nodes to receive data packets only when they are in active state in the duty-cycled network. In such a network, the definition of beaconing latency is referred to the latest time slot at which the last node receives a packet from one of its neighbors [6]. Recently, researchers have studied the MLBS problem in duty-cycled networks [6,24] with the objective of minimizing the beaconing latency. Several constant-approximation algorithms based on graph coloring and label coloring techniques have been proposed, which can offer high latency efficiency.

However, all these existing works assumed the beaconing schedule is computed in a centralized fashion, which is not suitable for the dynamic networking scenario. Compared to the centralized algorithms, where the computed node has the complete view of the whole network, in the distributed algorithm all the nodes need to compute the conflict-free beaconing schedule locally with only its neighboring nodes' information. This makes the MLBS problem more challenging. In addition, the existing works assume there exists only one single channel in the whole network, which is not a general model [12].

This paper focuses on addressing the above challenges by introducing the first distributed algorithm for the MLBS problem in Multi-channel Duty-cycled IoT networks (MLBSMD problem). In the proposed distributed algorithm, several special structures and mechanisms are designed to avoid the collisions locally, which works in $O(\Delta)$ time. Then, a distributed beaconing scheduling algorithm has been developed that considers the active time slots of each node to compute a collision-free and low-latency schedule with a theoretical bounded.

The remainder of this paper is structured as follows. Section 2 provides an overview of related works on the MLBS problem, while Sect. 3 outlines the problem definition and network model. Sections 4 and 5 explain the proposed distributed algorithm and the theoretical analysis of the algorithm. Section 6 shows the simulations results, and Sect. 7 concludes this work.

2 Related Works

In traditional networks, where the nodes are presumed to be continuously active, the prior works [9,11,15,17,19,20] study the MLBS problem and its variants by considering a simplest assumption that all nodes have the same transmission and interference radius. Aiming to decrease the beaconing latency, the authors in [18] introduced a centralized algorithm using the first-fit coloring technique. Wan et al. [23] proved that the approximation of the proposed beaconing algorithm is at most 7. After that, the authors in [22] developed a strip-coloring algorithm for scenarios where all nodes have a matching interference radius of $\rho \geq 1$. They demonstrated that when $\rho = 1$, the proposed algorithm has an approximation bound of at most 5, while the approximation bound varies between 3 and 6 when $\rho \geq 1$. The work in [21] explores a more comprehensive scenario where each node u possesses a distinct interference radius, i.e., $\rho(u) \geq 1$ and introduces an approximation algorithm that uses the first-fit coloring strategy. The algorithm's approximation bound is proven to be no greater than 61 under the protocol interference model. Additionally, Wan et al. [21] introduced an approximate algorithm to solve the MLBS problem under physical interference model. However, it's important to note that all of these algorithms operate in a centralized way and assume that all nodes are continuously active.

Recently, the works in [6,24] have examined the MLBS problem in duty-cycled networks. In [24], Wang et al. explored the MLBS problem in the duty-cycled network and demonstrated that it's an NP-hard problem. They first devised a 74-approximation algorithm that uses centralized first-fit coloring under the protocol interference model. Subsequently, they introduced an algorithm that utilizes strip-coloring technology, which has an approximation bound of no more than 10 when the interference radius is $\rho = 1$. However, they assumed quite strictly that every node can wake up only once throughout its working period. The MLBS problem in duty-cycled network in which every node owns multiple active time slots during each working period is investigated in [6]. In such a network, Chen et al. [6] firstly proposed an edge-based scheduling framework. The edge-based approach involves scheduling communication links between nodes instead of individual nodes themselves. Then proposed a new label coloring technique to solve the MLBS problem. With the aim of minimizing beaconing latency, their proposed algorithm achieves an approximation ratio of $(\rho + 1)^2 \times \mathcal{W}$. In addition, the authors propose two approximate algorithms designed to solve the MLBS problem under the physical interference model. However, all the above algorithms are all centralized, which are not suitable for the dynamic network. In addition, these existing algorithms assume there exist only one single channel in the whole network. Thus, we investigate the distributed algorithm for MLBS problem in Multi-channel Duty-cycled IoT networks in this paper.

Besides the beacon scheduling problem in duty-cycled wireless networks, the fast data aggregation and data collection scheduling problem was studied by [3,5,12,27]. The minimum latency broadcast scheduling and multicast scheduling problems in duty-cycled networks were investigated in [4,7] respectively.

3 Problem Definition

Consider a duty-cycled wireless network $G = (V, E)$ with multiple channels, where V represents the set of all nodes, and E represents the set of all network edges. Note that if two nodes are within each other's transmission radius, there is an edge between them, which allows for direct communication. For any node u, we denote its set of one-hop neighboring nodes as $NB(u)$.

As in [12], it is assumed that the network has m orthogonal channels to prevent collisions during transmission. Similarly to other beacon scheduling algorithms [6,24], this paper adopts the protocol interference model. A protocol interference model shows that the node $u's$ interference range is a circle with radius $\rho(u)$, where $\rho(u)$ is greater than or equal to 1. Therefore, all nodes v that are interfered by node u satisfy the condition that $d(v, u) \leq \rho(u)$, where the $d(v, u)$ denotes the distance between node v and u.

Each sensor node in duty-cycled networks operates in two distinct states, *i.e.*, active state and dormant state, and can alternate between them on a regular basis. Assuming a working period that has been partitioned into time slots of equal size, denoted by $T = \{0, 1, 2, \ldots, |T| - 1\}$. There is a working schedule known as $W(u)$ for each node u, which can be described as the collection of time slots that are active throughout a working period. Node u can only accept data packets from its neighbors if and only if it is awake, hence it can only do so within time slot t ($t\%|T| \in W(u)$). It's worth noting that in order to send a data packet, the sensor node u has the flexibility to move into the active state at any time slot. The duty cycle of node u is calculated by $|W(u)|/|T|$. For example, Fig. 1 illustrates an example of a duty-cycled network, where the numbers on the nodes indicate their corresponding active time slots, and $|T| = 3$ here.

Given that each node must send its data packet to all neighboring nodes, it means we need to compute a fast and conflict-free node transmission schedule for each node. Prior to formally introducing the MLBSMD problem, we first outline the definition of a link transmission schedule as follows.

Definition 1. (Link Transmission Schedule) Given any node $u \in V$ and a neighbor $v \in NB(u)$, let $lts(u, v) = [u, v, t_{uv}, c_{uv}]$ denote the link transmission schedule of node u to its neighbor v, where t_{uv} ($t_{uv}\%|T| \in W(v)$) denotes the transmitting time and c_{uv} ($1 \leq c_{uv} \leq m$) denotes the transmitting channel.

As for any node u, its node transmission schedule means the link transmission schedule to all of its neighbors, which can be defined as:

Definition 2. (Node Transmission Schedule) Given any node $u \in V$, let $nts(u) = \{\ lts(u, v) \mid \forall v \in NB(u)\}$ denote the node transmission schedule of node u. To avoid the interference between transmissions, same as other distributed algorithm [1,5], we take into account the protocol interference model and make simple assumption that the interference range and transmission range are equivalent. Under protocol interference model, the interference occurs when a node receives two and more data packets simultaneously. Given two link transmission schedules $lts(u, v) = [u, v, t_{uv}, c_{uv}]$ and $lts(u\prime, v\prime) = [u\prime, v\prime, t_{u\prime v\prime}, c_{u\prime v\prime}]$ ($u \neq u'$), they are conflicted if one of the following conditions is satisfied:

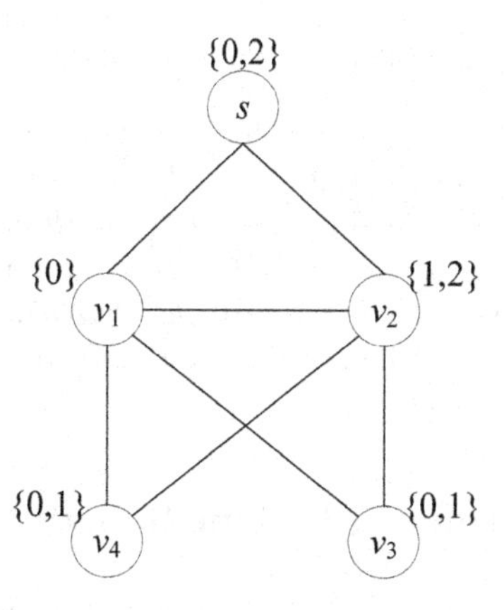

Fig. 1. An example of duty-cycled network.

1) ($v = v'$ or $u = v'$ or $v = u'$) & ($t_{uv} = t_{u'v'}$);
2) ($u \neq u' \neq v \neq v'$) & ($v' \in NB(u)$ or $v \in NB(u')$) & ($t_{uv} = t_{u'v'}$ & $c_{uv} = c_{u'v'}$).

The problem of the distributed beaconing schedule involves each node locally determining its own node transmission schedule, with the goal of minimizing the beaconing latency $\mathcal{L}$. The MLBSMD problem has been proven to be NP-hard, as demonstrated in Theorem 1.

Theorem 1. *The MLBSMD problem under protocol interference model is NP-hard.*

Proof. According to [6], the MLBS problem in single-channel duty-cycled networks is NP-hard. Since the aforementioned problem is a specific instance of the MLBSMD problem, it can be concluded that the MLBSMD problem is also NP-hard. The theorem is proved.

Next, we will present a distributed latency-efficient scheduling algorithm for the MLBSMD problem. This algorithm is capable of calculating a collision-free transmission schedule for each node locally.

In the distributed scheduling algorithm, each node can only compute the node transmission schedule with the local information. Aiming to compute the node transmission schedule locally, we assume the following information are maintained for each node u:

- $u's$ unique ID, and $W(u)$;
- $u's$ level information, l;
- $u's$ special structures for collision avoidance, i.e., $RC(u)$, $TC(u)$, $HT(u,t,c)$ and $FT(u,t,c)$ which will be introduced later;
- $u's$ neighboring nodes' information;
- $u's$ channel used at time slot t, i.e., $CS(u,t)$, which is initialized to 0;

– $u's$ node state and its node transmission schedule $nts(u)$.

The information level l represents the shortest distance from the sink node and can be acquired through a breadth-first search conducted from the sink node. It is assumed that the working schedule, level information, and neighboring nodes' information are obtained in advance. Note that, the node state and the special structures employed to avoid collisions ($RC(u)$, $TC(u)$, $FT(u,t,c)$, $HT(u,t,c)$), and $CS(u,t)$), are configured when executing the proposed algorithm, which will be introduced later.

4 Distributed Latency-Efficient Scheduling Algorithm

In this section, we firstly introduce the special structures used to avoid collisions and how to calculate the candidate node transmission schedule for each node in network. Subsequently, we provide a comprehensive explanation of the distributed algorithm.

4.1 Special Structures for Collision Avoidance

To prevent collisions between multiple transmissions, we take a deep look into the occurrence of collisions. Assume there is a link transmission schedule $lts\,(x_1, y_1) = [x_1, y_1, 0, 1]$ which has been established as in Fig. 2 and Fig. 3. Under protocol interference model, the collisions may happen if: 1) node x_1 in Fig. 2(a) tries to receive data packets from another neighbor z_1 at time slot 0 on channel 2; 2) node y_1 in Fig. 2(b) tries to receive data packets from another neighbor z_1 at time slot 0 on channel 2; 3) node y_1 int Fig. 2(c) tries to transmit data packets to another neighbor z_1 at time slot 0 on channel 2; 4) node x_2 in Fig. 3(a) tries to transmit data packets at time slot 0 on channel 1; 5) node y_3 in Fig. 3(b) tries to receive data packet at time slot 0 on channel 1 when two neighbors are transmitting data packets. As one can see that the above mentioned conflict scenarios 1), 2) and 3) occur when two link transmission schedules share one common node are assigned the same time slot. In this paper, we call them structural collision. In addition, when two link transmission schedules share no common node, the above mentioned conflict scenarios 4) and 5) occur when one of the receiver node lies in the transmission range of another transmitting node. It is worth noting that in this scenario, the two link transmission schedules will only conflict if they are allocated to the same channel and time slot, which is referred to as interference collision. In the following, we will discuss methods to avoid these two types of collisions.

To avoid structural collision, each node u maintains two special sets, i.e., $RC(u)$ and $TC(u)$, which respectively indicate the set of time slots when node u is receiving and transmitting data packets. They are updated when a link transmission schedule $lts(v,u)$ or $lts(u,z)$ is established. Actually, they are computed as follows:

$$RC\,(u) = \{\, t \mid t = lts\,(v,u)\,.t_{vu}, \forall lts(v,u) \in S\}$$

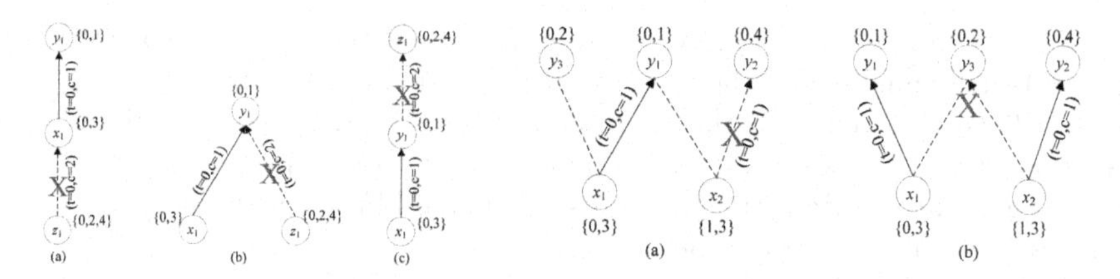

Fig. 2. Structural Collision.

Fig. 3. Interference Collision.

$TC(u) = \{\ t \mid t = lts(u, z).t_{uz}, \forall lts(u, z) \in S\}$

Note that, for any time slot t ($t \in RC(u)$), node u can neither receive nor send data packets at time slot t. And for any time slot t ($t \in TC(u)$), node u cannot receive any data packets. To be noticed that node u can still transmit data packets to other neighbors at time slot t by exploiting the broadcasting nature of wireless transmission.

With the above $RC(u)$ and $TC(u)$ structures, then we can forbid node u to transmit or receive data packets at time slots in $RC(u)$ and $TC(u)$ for avoiding the above structural collisions. As shown in the Fig. 2(a), when x_1 is scheduled to transmit the data packet to y_1 at time slot 0 on channel 1, the set $TC(x_1)$ is updated as $TC(x_1) = TC(x_1) \cup \{0\}$. In this case, node x_1 cannot receive the data packet from node z_1 at the same time slot 0. By this way, we can avoid the conflict scenario in Fig. 2(a) that one node receives and send data packets simultaneously. As for the case in Fig. 2(b), when x_1 is scheduled to send the data packet to y_1 at time slot 0 on channel 1, the "Link Transmission Schedule" message is also broadcasted by x_1. After y_1 receives the message, it adds time slot 0 into its $RC(y_1)$, i.e., $RC(y_1) = RC(y_1) \cup \{0\}$. In this case, node y_1 cannot receive the data packet from node z_1 at the same time slot 0. By this way, we can avoid the conflict scenario that one node receives multiple data packets simultaneously. As for the case in Fig. 2(c), after y_1 update $RC(y_1)$, i.e., $RC(y_1) = RC(y_1) \cup \{0\}$, y_1 also cannot send data packets at the time slot 0.

Second, in order to avoid interference collision, such as [4], we design two special sets, i.e., $FT(u,t,c)$ and $HT(u,t,c)$. The set $FT(u,t,c)$ stores the set of neighbors which is scheduled to receive data packets at time slot t on channel c. If $|FT(u,t,c)| \geq 1$, then node u cannot transmit data packets at time slot t on channel c. Otherwise, it would result in collisions to the node in $FT(u,t,c)$. The set $FT(u,t,c)$ is updated when node u receives one of its neighbor v broadcasting the "Link Transmission Schedule" message, i.e., $lts(z,v)$, which can be computed as:

$FT(u,t,c) = \{\ v \mid t = lts(z,v).t_{zv}\ \&\ c = lts(z,v).c_{zv}\ \&\ u \neq z\ \&\ v \in NB(u), \forall lts(z,v) \in S\}$

As illustrated in Fig. 3(a), when node x_1 is assigned to transmit a data packet to node y_1 at time slot 0 on channel 1, the "Link Transmission Schedule" message including $[x_1, y_1, 0, 1]$ is also broadcasted by the receiving node y_1. After x_2 receives the message, it adds node y_1 to its $FT(x_2, 0, 1)$, i.e.,

$FT(x_2, 0, 1) = FT(x_2, 0, 1) \cup \{y_1\}$. By this way, x_2 cannot transmit data packets at time slot 0 on channel 1, and then interference collision in Fig. 3(a) is avoided.

The set $HT(u,t,c)$ refers to the neighbors of node u who are transmitting data packets at time slot t on channel c. The set $HT(u,t,c)$ is updated when $u's$ neighbor has generated a link transmission schedule and broadcasted it to all of its neighbors. Actually, it can be updated as:

$HT(u,t,c) = \{ z \mid lts(z,v).t_{zv} = t \;\&\; lts(z,v).c_{zv} = c \;\&\; z \in NB(u), \forall lts(z,v) \in S\}$

To avoid interference collisions, we must carefully select the time slot and channel for node u to receive data packets based on the hearing set $HT(u,t,c)$, which can be categorized into the following three cases:

- If $|HT(u,t,c)| = 0$, it indicates that none of node $u's$ neighbors are transmitting data packets at time slot t on channel c, which allows node u to receive data packets on channel c at time slot t without generating collisions;
- If $|HT(u,t,c)| = 1$, it shows that there is only one neighbor sending data packets at time slot t on channel c, then node u can receive the data packet from the node in $HT(u,t,c)$ at time slot t on channel c;
- If $|HT(u,t,c)| \geq 2$, it implies that there are at least two distinct neighbors transmitting data packets at time slot t on channel c, then node u cannot receive data packets at time slot t on channel c. Otherwise, it causes collisions.

4.2 Computing Candidate Node Transmission Schedule

In the proposed algorithm, we attempt to determine the node transmission schedule for each node level by level in a bottom-up fashion. Let R stand for the highest level coming from the sink node. Initially, each node u at level R begin to calculate its candidate node transmission schedule which consists of multiple candidate link transmission schedules. After that, each candidate link transmission schedule of u will be broadcasted to a subset of its neighboring nodes. If all neighboring nodes support a candidate link transmission schedule $CLTS(u,v)$, then the link transmission schedule $lts(u,v)$ will be established. If all the candidate link transmission schedules of node u are established, which shows that node u finishes beaconing transmission task, and then node u turns to the scheduled state. Once all the nodes at level R have been scheduled, the algorithm begin to schedule the nodes at level $R-1$.

In order to calculate the node transmission schedule for node u locally, we divide $u's$ neighbors into two non-overlapped sets: 1) the neighbors to which the link transmission schedule of node u have been scheduled, denoted as $NB_s(u)$; 2) the neighbors to which the link transmission schedule of node u have not been scheduled, denoted as $NB_u(u)$. Initially, the $NB_u(u) = NB(u)$ and the $NB_s(u) = \emptyset$. According to the definition of beaconing transmission task, node u only needs to send data packages to nodes in $NB_u(u)$. Then the node $u's$ Candidate Node Transmission Schedule (CNTS) is denoted by $CNTS(u) = \{ CLTS(u,v) \mid \forall v \in NB_u(u)\}$ where $CLTS(u,v)$ will be introduced later.

Actually, the $CNTS(u)$ means the candidate link transmission schedules to all of its neighbors in $NB_u(u)$.

Let $CLTS\,(u,v) = (u, v, t_{uv}, c_{uv}, w_{uv}, n_{uv})$ denote $u's$ Candidate Link Transmission Schedule (CLTS) to neighbor v, where t_{uv} and c_{uv} denotes the candidate transmitting time slot and channel. The weight w_{uv} serves as a means of competition against the candidate link transmission schedules of other nodes and n_{uv} denotes the size of a special subset of its neighboring nodes, which will be introduced later.

First, since each node can only utilize a single channel for data packet transmission at any given time slot, the channel used at time slot t must be recorded for each node u. Thus, we design a data structure $CS(u,t)$ which is initialized to 0.

Second, to reduce the latency, we need to compute the minimum available collision-free time slot and channel for candidate link transmission schedule $CLTS\,(u,v)$. By utilizing the previously mentioned special structures and $CS\,(u,t)$, we can calculate the smallest and collision-free time slot and channel for candidate link transmission schedule $CLTS\,(u,v)$. This is represented by (t_{min}, c_{min}), which can be calculated as follows:

$(t_{min}, c_{min}) = min\{\ (t,c) \mid \psi = 1\ \&\ \chi = 1\ \&\ \xi = 1\ \&\ |FT\,(u,t,c)| = 0\ \&\ t\%\,|T| \in W\,(v)\}$

Where $\psi = (CS\,(u,t) = c\ or\ CS\,(u,t) = 0)$, $\chi = (t \notin RC\,(u)\ \&\ t \notin RC\,(v)\ \&\ t \notin TC\,(v))$ and $\xi = (|HT\,(v,t,c)| = 0)\ or\ (|HT\,(v,t,c)| = 1\ \&\ u \in HT\,(v,t,c))$. The condition $\chi = 1$ and $\xi = 1\ \&\ |FT\,(u,t,c)| = 0$ is utilized to avoid the aforementioned structural collision and interference collision respectively. The condition $t\%|T| \in W(v)$ is to ensure that the node v is active during the time slot t, and the condition ψ is used for avoiding that node u uses multiple different channels at the same time slot.

After (t_{min}, c_{min}) is obtained, then $u's$ CLTS to neighbor v, i.e., $CLTS\,(u,v)$, can be calculated as $CLTS\,(u,v) = [u, v, t_{min}, c_{min}, 0, 0]$.

Next, we will describe how to calculate the weight w_{uv} and the size n_{uv} of $CLTS(u,v)$.

Since the node may broadcast the data packet to multiple neighbors at the same time slot on the same channel, we call the number of candidate link transmission schedule at time slot t on channel c as "Link Transmission Amount" of node u at time slot t on channel c. Then we set "Link Transmission Amount" of node u at time slot t_{uv} on channel c_{uv} as the weight of $CLTS\,(u,v)$.

After obtaining the weight, each $CLTS\,(u,v)$ in $CNTS\,(u)$ is sent to the corresponding set of neighboring nodes called as the Inquiry Set [5]. The Inquiry Set of $CLTS\,(u,v)$ is presented by $INQ(u,v)$. Actually, the size of $INQ(u,v)$ is equal to the $CLTS\,(u,v)\,.n_{uv}$. The set $INQ(u,v)$ can be calculated as:

$INQ\,(u,v) = \{\ z \mid z \in NB\,(u)\ \&\ t_{uv}\%\,|T| \in W\,(z)\ \&\ t_{uv} \notin RC\,(z)\ \&\ t_{uv} \notin TC(z)\ \&\ |HT(z, t_{uv}, c_{uv})| < 2\}$

Actually, $INQ(u,v)$ denotes the set of $u's$ neighboring nodes which can receive data packets at the time slot $CLTS\,(u,v)\,.t_{uv}$ on channel $CLTS\,(u,v)\,.c_{uv}$. Since nodes in $INQ(u,v)$ have the potential to receive CLTSs from other nodes, the following theorem ensures that if all nodes in $INQ(u, v)$ support only $u's$ $CLTS(u, v)$, then establishing $CLTS\,(u, v)$ can be achieved without any collisions. The proof is omitted here.

Theorem 2. *If all the nodes in $INQ(u, v)$ only support one node's $CLTS$, then link transmission schedule can be established without any collisions.*

4.3 The Distributed Beaconing Schedule Algorithm

In the following, we will provide a detailed introduction to the DBSDA algorithm utilized for the MLBSMD problem. During DBSDA execution, each node owns nine different states: Not Ready, Synchronized, Ready, Wait, Update, Received, Wait2, Update2 and Scheduled. Initially, all the nodes at level R are programmed to the Synchronized state and other nodes located at level l ($l < R$) are set to the Not Ready state. The detailed description of DBSDA algorithm is as follows.

1) At the start, each node at level R is configured to be in the Synchronized state, while the nodes located at level l ($l < R$) are programmed to Not Ready state. Additionally, it is necessary for each node u to maintain two specific variables α and β. The α presents the number of neighboring nodes located at level $u.l + 1$ that have not yet completed their beaconing transmission task. It is initially set to the number of neighboring nodes located at level $u.l + 1$. The DBSDA algorithm is executed in a bottom-up manner level by level, which means that each node u keeps Not Ready state until all of its next level neighbors are scheduled, and then node u transfers to the Synchronized state.
2) Upon entering the Synchronized state, node u disseminates its "Synchronization" messages to notify all neighboring nodes at the same level. Once node u successfully obtains "Synchronization" messages from all neighboring nodes at the same level, it transfers to the Ready state.
3) Upon entering the Ready state, node u calculates its $CNTS(u)$ which consists of multiple candidate link transmission schedules and the candidate link transmission schedule can be calculated by the Algorithm 1 mentioned at the above subsection. After that, node u broadcasts each $CLTS(u, v)$ in $CNTS(u)$ to all its neighbors and then it goes to Wait state to wait for receiving the replying messages from its neighbors.
4) Upon receiving a Replying message from a neighboring node z, for each node u, the following actions will be taken.
   (a) if u is in the Wait state, it stores the replying message for further use and stays in the Wait state until all replying messages are received. Once all replying messages are received, u switches to Received state.
   (b) Otherwise, ignore the message.
5) When node u is in Received state, for each received replying message $RM = [s, r, c, CLTS(u, v)]$, it does as follows:
   (a) if the replying message is "Support" message, then do
      i. $CLTS\,(u, v)\,.n_{uv} \leftarrow CLTS\,(u, v)\,.n_{uv} - 1$.

ii. if $CLTS(u,v).n_{uv} = 0$, then it sets its link transmission schedule to neighbor v as $lts(u,v) = [u, v, CLTS(u,v).t_{uv}, CLTS(u,v).c_{uv}]$ and the set $NB_u(u)$ is updated as $NB_u(u) \leftarrow NB_u(u) - \{v\}$. After that, it incorporates the link transmission schedule into a data package called as "Schedule", and then broadcasts this "Schedule" package to all neighboring nodes.

iii. if the set $NB_u(u) = \emptyset$, then node u finishes its beaconing transmission task. After that, the variable β is set as 1.

(b) if the replying message received is "Refuse" message, which means that the candidate link transmission schedule $CLTS(u,v)$ fail to win the competition and the link transmission schedule between node u and v need to be recalculated, the the variable β is set as 0.

6) If the variable β is 1 which means that node u has transmitted the data packet to all of its neighbors, then node u switches to the Scheduled state and broadcasts the "Finished" message to all its neighboring nodes. Otherwise, the node u switches to Wait2 state to wait for receiving "Schedule" messages. If no scheduling message is received, node u switches to Ready state for calculating a new candidate node transmission schedule.

7) Upon receiving a "Schedule" message which includes the link transmission schedule $lts(x,y)$ from a neighboring node z, for each node u, the following actions will be taken.

(a) If u is Wait or Wait2 state, then do

i. u switches to Update or Update2 state.

ii. If $z = x$, which means $lts(x,y)$ is broadcasted by the node x itself. It firstly saves $lts(x,y)$ for further use, then

A. If $u = y$, $RC(u) \leftarrow RC(u) \cup \{lts(x,y).t_{xy}\}$, then the "Schedule" message is retransmitted through broadcasting by node u.

B. If $u \neq y$, $HT(u,t,c) \leftarrow HT(u,t,c) \cup \{x\}$ where $t = lts(x,y).t_{xy}$ and $c = lts(x,y).c_{xy}$.

iii. Otherwise, if $z = y$, which means it is broadcasted by the receiver y,

A. If $u = x$, $TC(u) \leftarrow TC(u) \cup \{lts(x,y).t_{xy}\}$ and $CS(u,t) \leftarrow lts(x,y).c_{xy}$ where $t = lts(x,y).t_{xy}$ and $NB_s(u) \leftarrow NB_s(u) \cup \{y\}$.

B. If $u \neq x$, $FT(u,t,c) \leftarrow FT(u,t,c) \cup \{y\}$ where $t = lts(x,y).t_{xy}$ and $c = lts(x,y).c_{xy}$.

iv. u switches back to Wait or Wait2 state.

(b) Otherwise, ignore the message.

8) Upon receiving a "Finished" message from a neighboring node z, for each node u, the following actions will be taken.

(a) if $u.l + 1 = z.l$, do

i. $\alpha \leftarrow \alpha - 1$.

ii. if $\alpha = 0$, which indicates the all of neighboring nodes that located at the next level have been scheduled and then node u switches to Synchronized state from Not Ready state.

(b) Otherwise, ignore the message.

Now, the complete distributed algorithm is introduced. The resulted schedule after executing DBSDA algorithm is shown in Table 1.

Table 1. The result of beaconing schedule for Fig. 1.

Node	Node transmission schedule
s	$nts(s) = \{[s, v_1, 9, 1], [s, v_2, 5, 1]\}$
v_1	$nts(v_1) = \{[v_1, s, 2, 2], [v_1, v_2, 4, 1], [v_1, v_3, 4, 1], [v_1, v_4, 1, 2]\}$
v_2	$nts(v_2) = \{[v_2, s, 0, 2], [v_2, v_1, 6, 1], [v_2, v_3, 3, 2], [v_2, v_4, 0, 2]\}$
v_3	$nts(v_3) = \{[v_3, v_1, 0, 1], [v_3, v_2, 1, 1]\}$
v_4	$nts(v_4) = \{[v_4, v_1, 3, 1], [v_4, v_2, 2, 1]\}$

5 Performance Analysis

Theorem 3 gives the lower bound for MLBSMD problem under protocol interference model.

Theorem 3. *The upper bound of the DBSDA algorithm is at most* $\lceil \frac{3\Delta-1}{\mathcal{W}} \rceil \times |T| + \lceil \frac{2(\Delta-1)\times\Delta+1}{m\times\mathcal{W}} \rceil \times |T|$.

Proof. Let link $lts(u, v)$ be the one with largest latency in the computed beaconing schedule of DBSDA algorithm. In the following, we will prove that the latency of the link transmission schedule $lts(u, v)$ is at most $\lceil \frac{3\Delta-1}{\mathcal{W}} \rceil \times |T| + \lceil \frac{2(\Delta-1)\times\Delta+1}{m\times\mathcal{W}} \rceil \times |T|$, i.e., $lts(u, v).t_{uv} \le \lceil \frac{3\Delta-1}{\mathcal{W}} \rceil \times |T| + \lceil \frac{2(\Delta-1)\times\Delta+1}{m\times\mathcal{W}} \rceil \times |T|$.

First, consider the Structural Collision problem for link $lts(u, v)$, we prove that it can be conflicted by at most $\lceil \frac{3\Delta-1}{\mathcal{W}} \rceil \times |T|$ time. As the example shown in Fig. 2, the structural collision occurs when two link transmission schedules share a common node. In this case, these two link transmission schedules can not transmit at the same time slot, even using different channels. For the receiver v, since it has at most Δ neighbors, then there exist at most $2\Delta-1$ link transmission schedules which are conflicted with $lts(u, v)$, including the link transmission schedules for link $lts(w, v)$ ($w \in NB(v) \& w \neq u$) and $lts(v, z)$ ($z \in NB(v)$). Similarly, for the sender u, there exist at most $\Delta - 1$ link transmission schedules which are conflicted with $lts(u, v)$. Note that, we do not need to consider the link $lts(v, u)$ twice here. Thence, to avoid the structural collision, it needs at most $\lceil \frac{2\Delta-1+\Delta-1+1}{\mathcal{W}} \rceil \times |T| = \lceil \frac{3\Delta-1}{\mathcal{W}} \rceil \times |T|$ time.

Second, consider the Interference Collision problem for link $lts(u, v)$, we prove that it can be conflicted by at most $\lceil \frac{2(\Delta-1)\times\Delta+1}{m\times\mathcal{W}} \rceil \times |T|$ time. As the example shown in Fig. 3, the interference collision arises when a receiver node hears two and more data packets at the same time slot on the same channel. In this case, one needs to find either a different time slot or a different channel to avoid the collisions. For the receiver v, there exist at most $(\Delta - 1) \times \Delta$ such links which are conflicted with $lts(u, v)$, i.e., for each neighbor node w ($w \in NB(v) \& w \neq u$), it cannot transmit data packets at the same time slot on same channel. For the sender u, there also exist at most $(\Delta - 1) \times \Delta$ such links which are conflicted with $lts(u, v)$, i.e., for each neighbor node w ($w \in NB(u) \& w \neq v$), it cannot receive data packets at the same time slot on same channel. Thence, to avoid the interference collision problem, it needs at most $\left\lceil \frac{2(\Delta-1)\times\Delta+1}{m\times W} \right\rceil \times |T|$ time.

Therefore, according to the above analysis, for any link $lts(u,v)$, its latency is at most $\lceil \frac{3\Delta-1}{W} \rceil \times |T| + \left\lceil \frac{2(\Delta-1)\times\Delta+1}{m\times W} \right\rceil \times |T|$.

Theorem 4. *The time complexity of the algorithm of computing candidate node transmission schedule is at most* $O(\Delta^2 \times |T|)$.

Proof. In the algorithm of computing candidate node transmission schedule, to compute $u's$ candidate link transmission schedule to v, one needs to compute the smallest and collision-free time slot and channel. First, it takes at most $O(\lceil \frac{3\Delta-1}{W} \rceil \times |T|)$ time to avoid structural collision. Second, to avoid interference collision, it takes at most $O(\left\lceil \frac{2(\Delta-1)\times\Delta+1}{m\times W} \right\rceil \times |T|)$ time. Therefore, the time complexity of Algorithm 1 is at most $O(\lceil \frac{3\Delta-1}{W} \rceil \times |T|) + O(\left\lceil \frac{2(\Delta-1)\times\Delta+1}{m\times W} \right\rceil \times |T|) = O((\frac{3\Delta-1}{W} + \frac{2(\Delta-1)\times\Delta+1}{m\times W} + 2) \times |T|) = O((2\Delta^2 + \Delta + 2) \times |T|) = O(\Delta^2 \times |T|)$.

6 Simulation Results

In this section, we extensively experiment with a protocol interference model to evaluate the effectiveness of the proposed algorithm.

When utilizing the protocol interference model, two algorithms have been used in previous work to solve the MLBS problem in duty-cycle networks, i.e., FFBSD [24] and SCBSD [24]. These algorithms operate under the rigid assumption that each node only has one active time slot during each working period. In addition, there are two other algorithms (GFFC [6] and DLFBS [6]) currently available to solve the MLBS problem in duty-cycled networks. These algorithms assume that nodes can have any time slot during each working period. It is worth noting that the algorithms mentioned above are centralized and make an assumption that each node can only communicate using a single unique channel. In the experiments, our proposed algorithm is verified by comparing the above four centralized algorithms.

Our primary focus is on evaluating the performance of the proposed algorithms in terms of beaconing latency and beaconing throughput through simulations conducted on various network topologies. First, 100 nodes are randomly distributed throughout a space, and their performance was evaluated under the assumptions that each node has the same beaconing transmission radius which is equal to interference radius equal. To generate a diverse range of network topologies containing 50 to 300 sensor nodes, we make full use of the Networkx tool [16]. We assess the performance of different configurations in all simulations by varying the duty cycle between 10% and 35%, and by randomly generating active time slots for each sensor node. Each point depicted in the simulation represents the average value obtained from 100 executions.

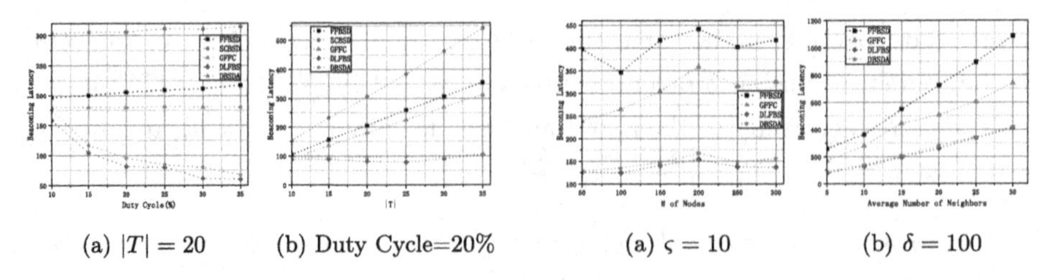

(a) $|T| = 20$ (b) Duty Cycle=20%

Fig. 4. The Beaconing Latency Under Different Duty Cycle and $|T|$.

(a) $\varsigma = 10$ (b) $\delta = 100$

Fig. 5. The Beaconing Latency Under Different Network Topologies.

6.1 Performance Comparison in Single-channel Duty-Cycled Scenarios

Beaconing Latency Comparison: First, we test the beaconing latency of proposed algorithms when 100 sensor nodes are randomly placed in an $200m * 200m$ area. In this simulation, we test different Duty Cycle and $|T|$.

Figure 4 illustrates how beaconing latency performs. Figures 4(a) and 4(b) show that in all cases, the DBSDA algorithm's beaconing latency is comparable to that of DLFBS and notably lower than the other three algorithms. The full utilization of each node's available time slots and broadcast function are primarily responsible for the DBSDA algorithm's exceptional performance, which maximizes the beaconing throughput throughout each scheduling period.

Second, we use Networkx tool to examine beaconing latency for various network topologies, including different network size (i.e., δ) and average neighbor density (i.e., ς). The SCBSD algorithm cannot be used in these group experiments because it operates on the assumption that all nodes in the network have knowledge of their own location.

Figure 5 shows the beaconing latency of four algorithms under different network size δ and ς. First of all, as you can see from Fig. 5(a), the beaconing latency of all algorithms hardly change with the network size when we fixed ς. Secondly, the beaconing latency increases with the increase of ς when we fixed δ. What has been mentioned above shows that the beaconing latency is mostly influenced by the average degree and is not dependent on network size. In all the network topologies examined, DBSDA's beaconing latency is nearly equivalent to DLFBS's latency and significantly lower than that of other algorithms, which suggests that our algorithm performs very well in reducing the beaconing latency.

6.2 Performance Analysis in Multi-channel Duty-Cycled Scenarios

Figure 6 illustrates the number of messages produced by DBSDA. As depicted in Fig. 6(a), the number of messages expands as the network size increases. This is mainly because more link transmissions need to be scheduled when the network scale becomes larger. Figure 6(b) shows that the number of messages increase significantly with the increase of Duty Cycle when we fix the length of $|T|$, this is due to "Replying" message is also increased when the number of avaiable time

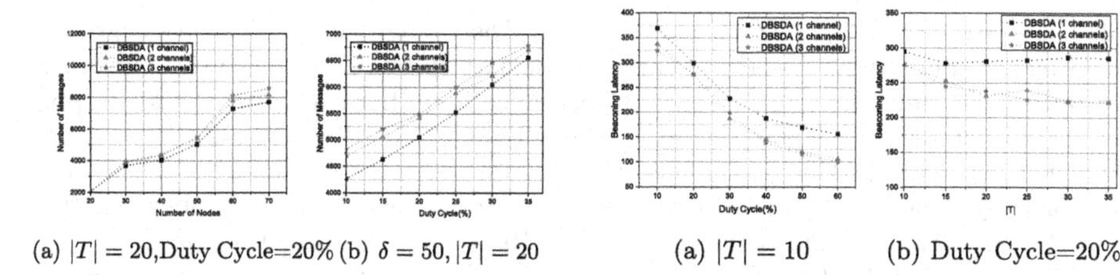

(a) $|T| = 20$,Duty Cycle=20% (b) $\delta = 50, |T| = 20$

Fig. 6. Number of Messages.

(a) $|T| = 10$ (b) Duty Cycle=20%

Fig. 7. The Beaconing Latency Under Different Number of Channels and Duty Cycle.

slots increase. Lastly, as you can see from Fig. 6, the number of messages of DBSDA hardly change with the number of channel when we fix the network topology.

Figure 7 shows the beaconing latency under different $|T|$ and duty cycle where we set $\delta = 100$ and $\varsigma = 20$. The Fig. 7(a) shows that the beaconing latency decrease when we increase Duty Cycle and fix $|T|$. This is mainly attributed to the number of active time slots increase and our algorithm can make good use active time slot to reduce the latency. Figure 7(b) shows that the beaconing latency keep stable when we increase $|T|$ and fix Duty Cycle. In addition, as shown in Fig. 7, when there is only one channel in the network, the DBSDA algorithm exhibits relatively high latency. The two pictures in Fig. 7 all demonstrate that the number of channels has a little impact on beaconing latency. This is because DBSDA algorithm has scheduled interferencing links at different time slot. When interferencing links have the same transmission time slot, they will also be assigned to two different channels.

7 Conclusions

In this paper, we investigate the MLBS problem in Multi-channel Duty-cycled IoT networks (MLBSMD). We prove the MLBSMD problem is NP-hard. To avoid the collisions locally, several special structures are designed. Then, we propose the distributed beaconing scheduling method which can compute a low-latency and collision-free schedule with a theoretical bounded. Finally, the extensive simulation results demonstrate the effectiveness of the proposed algorithm in terms of latency.

References

1. Bagaa, M., Younis, M., Djenouri, D., Derhab, A., Badache, N.: Distributed low-latency data aggregation scheduling in wireless sensor networks. ACM Trans. Sensor Netw. **11**(3), 1–36 (2015)
2. Cai, Z., Chen, Q., Shi, T., Zhu, T., Chen, K., Li, Y.: Battery-free wireless sensor networks: a comprehensive survey. IEEE Internet Things J. **1** (2022). https://doi.org/10.1109/JIOT.2022.3222386

3. Chen, K., Gao, H., Cai, Z., Chen, Q., Li, J.: Distributed energy-adaptive aggregation scheduling with coverage guarantee for battery-free wireless sensor networks. In: IEEE Conference on Computer Communications (IEEE INFOCOM 2019), pp. 1018–1026 (2019). https://doi.org/10.1109/INFOCOM.2019.8737492
4. Chen, Q., Cai, Z., Cheng, L., Gao, H., Li, J.: Structure-free broadcast scheduling for duty-cycled multihop wireless sensor networks. IEEE Trans. Mob. Comput. **21**(12), 4624–4641 (2022). https://doi.org/10.1109/TMC.2021.3084145
5. Chen, Q., Gao, H., Cai, Z., Cheng, L., Li, J.: Distributed low-latency data aggregation for duty-cycle wireless sensor networks. IEEE/ACM Trans. Netw. **26**(5), 2347–2360 (2018)
6. Chen, Q., Gao, H., Cheng, L., Li, Y.: Label coloring based beaconing schedule in duty-cycled multihop wireless networks. IEEE Trans. Mobile Comput. **19**(5), 1123–1137 (2020)
7. Chen, Q., Gao, H., Cheng, S., Fang, X., Cai, Z., Li, J.: Centralized and distributed delay-bounded scheduling algorithms for multicast in duty-cycled wireless sensor networks. IEEE/ACM Trans. Netw. **25**(6), 3573–3586 (2017)
8. Cheng, L., et al.: Collision-free dynamic convergecast in low-duty-cycle wireless sensor networks. IEEE Trans. Wirel. Commun. **21**(3), 1665–1680 (2022). https://doi.org/10.1109/TWC.2021.3105983
9. Chlamtac, I.: A spatial-reuse tdma/fdma for mobile mlulti-hop radio networks. In: Proceedings of the IEEE INFOCOM (1985)
10. Choudhury, N., Matam, R., Mukherjee, M., Lloret, J.: A beacon and GTS scheduling scheme for IEEE 802.15. 4 DSME networks. IEEE Internet Things J. **9**(7), 5162–5172 (2021)
11. Ephremides, A., Truong, T.V.: Scheduling broadcasts in multihop radio networks. IEEE Trans. Commun. **38**(4), 456–460 (1990)
12. Jiao, X., et al.: Delay efficient scheduling algorithms for data aggregation in multichannel asynchronous duty-cycled WSNs. IEEE Trans. Commun. **67**(9), 6179–6192 (2019)
13. Li, J., et al.: Digital twin-assisted, SFC-enabled service provisioning in mobile edge computing. IEEE Trans. Mobile Comput. **23**(1), 393–408 (2022). https://doi.org/10.1109/TMC.2022.3227248
14. Morillo, R., Qin, Y., Russell, A., Wang, B.: More the merrier: neighbor discovery on duty-cycled mobile devices in group settings. IEEE Trans. Wirel. Commun. **21**(7), 4754–4768 (2022)
15. Nelson, R., Kleinrock, L.: Spatial tdma: A collision-free multihop channel access protocol. IEEE Trans. Commun. **33**(9), 934–944 (1985)
16. Networkx. http://networkx.lanl.gov/
17. Ramanathan, S., Lloyd, E.L.: Scheduling algorithms for multihop radio networks. IEEE/ACM Trans. Netw. **1**(2), 166–177 (1993)
18. Sen, A.: Approximation algorithms for radio network scheduling. In: 35th Allerton Conference on Communication, Control and Computing, September 1997, pp. 573–582 (1997)
19. Sen, A., Huson, M.L.: A new model for scheduling packet radio networks. Wirel. Netw. **3**(1), 71–82 (1997)
20. Stevens, D.S., Ammar, M.H.: Evaluation of slot allocation strategies for TDMA protocols in packet radio networks. In: IEEE Conference on Military Communications, pp. 835–839. IEEE (1990)
21. Wan, P.J., Wang, Z., Du, H., Huang, S.C.H., Wan, Z.: First-fit scheduling for beaconing in multihop wireless networks. In: 2010 Proceedings IEEE INFOCOM, pp. 1–8. IEEE (2010)

22. Wan, P.J., Xu, X., Wang, L., Jia, X., Park, E.K.: Minimum-latency beaconing schedule in multihop wireless networks. In: IEEE INFOCOM 2009, pp. 2340–2346. IEEE (2009)
23. Wan, P.J., Yi, C.W., Jia, X., Kim, D.: Approximation algorithms for conflict-free channel assignment in wireless ad hoc networks. Wirel. Commun. Mob. Comput. **6**(2), 201–211 (2006)
24. Wang, L., Wan, P.J., Young, K.: Minimum-latency beaconing schedule in duty-cycled multihop wireless networks. In: 2015 IEEE Conference on Computer Communications (INFOCOM), pp. 1311–1319. IEEE (2015)
25. Wu, W., Wang, X., Hawbani, A., Liu, P., Zhao, L., Al-Dubai, A.Y.: Flora: fuzzy based load-balanced opportunistic routing for asynchronous duty-cycled WSNs. IEEE Trans. Mobile Comput. (2021)
26. Yao, B., Gao, H., Chen, Q., Li, J.: Energy-adaptive and bottleneck-aware many-to-many communication scheduling for battery-free wsns. IEEE Internet Things J. **8**(10), 8514–8529 (2021). https://doi.org/10.1109/JIOT.2020.3045979
27. Zhang, J., Gao, H., Zhang, K., Chen, Q., Li, J.: A distributed framework for low-latency data collection in battery-free wireless sensor networks. IEEE Internet Things J. **9**(11), 8438–8453 (2022). https://doi.org/10.1109/JIOT.2021.3115824

Improved Task Allocation in Mobile Crowd Sensing Based on Mobility Prediction and Multi-objective Optimization

Zhidong Xie, Tao Peng(✉), Wei You, and Guojun Wang

School of Computer Science and Cyber Engineering, Guangzhou University, Guangzhou, China
{xiezhidong0413,iuvyouwei}@e.gzhu.edu.cn, {pengtao,csgjwang}@gzhu.edu.cn

Abstract. Mobile Crowd Sensing(MCS), a novel data sensing paradigm, its success largely depends on the design of a reasonable and feasible task allocation strategy. Recent research works have increasingly focused on exploring task allocation scenarios that are more realistic and specific, involving heterogeneous tasks and participants, and often incorporating multi-objective optimization techniques. In this paper, we consider the spatial-temporal sensing properties of the tasks and the participants, and design a novel multi-objective multi-task allocation scheme with mobility prediction(M3P). Experiments on the real-world dataset validate the effectiveness of our proposed methods compared against baselines.

Keywords: Mobile Crowd Sensing · Multi-objective optimization · Task allocation · Mobility Prediction

1 Introduction

With the advancement of smartphones and other smart wearable devices, a novel sensing paradigm is emerging. It utilizes these mobile and sensor-equipped nodes for performing sensing tasks known as Mobile Crowd Sensing (MCS) [1]. Gaining popularity, this paradigm is in contrast to the conventional Internet of Things (IoT) sensing model, which requires the fixed deployment of sensing devices. In real-world scenarios, MCS utilizes the sensing capabilities of numerous individuals and their sensor-equipped intelligent devices, including cameras, microphones, and gyroscopes. To sense the actual requirements of the physical world, such as intelligent transportation, public safety, and environmental monitoring, it can offer unparalleled coverage of spatial-temporal sensing and address the challenges associated with high system cost, complex maintenance, and limited flexibility in traditional data collection models.

Mobile Crowd Sensing(MCS), a novel data sensing paradigm, emphasizes the collective power of a group, and its success depends on the ability to select

Z. Tari et al. (Eds.): ICA3PP 2023, LNCS 14491, pp. 56–72, 2024.
https://doi.org/10.1007/978-981-97-0808-6_4

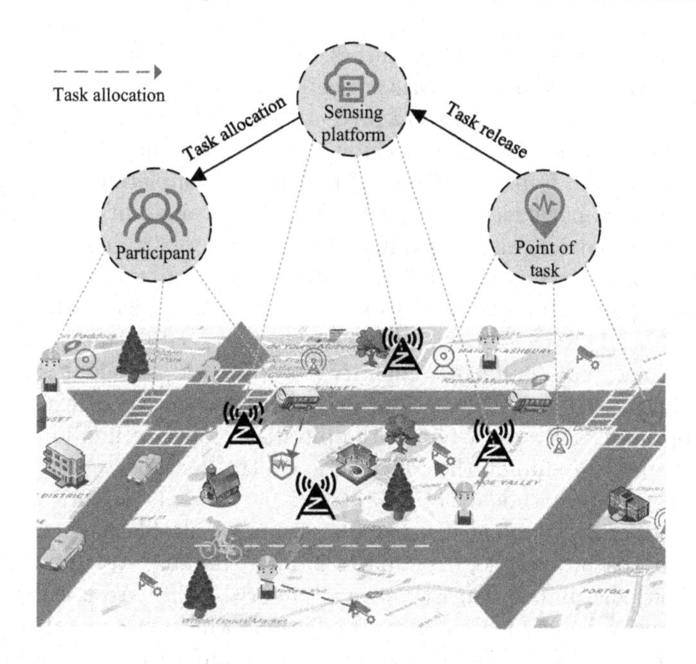

Fig. 1. Scenario of MCS task allocation.

appropriate participants from a large pool of users to perform sensing tasks while ensuring the satisfaction of spatial-temporal constraints for both participants and tasks. A scenario of MCS task allocation is illustrated in Fig. 1. Designing a reasonable and practical task allocation algorithm poses a critical research challenge. On the one hand, recruiting many participants to perform sensing tasks can enhance task coverage within the spatial-temporal domain. On the other hand, the recruited participants need to receive appropriate rewards from the platform to compensate for the costs incurred during task performance, such as energy consumption and participant time consumption, as well as potential privacy risks [2]. Regarding the sensing platform, enhancing the completion rate of sensing tasks and reducing costs are two crucial objectives; however, these objectives often constrain each other. A designed multi-objective optimization algorithm is required to balance these objectives. In recent years, significant progress has been made in the research on task allocation for Mobile Crowd Sensing. Two types of task allocation modes exist based on participants' movement and participation mechanisms: opportunistic mode and participatory mode. In the opportunistic mode, tasks are assigned to participants who can complete them without disrupting their daily routines [3]. Conversely, in the participatory mode, participants must alter their original routes and specifically travel to designated locations to perform sensing tasks [4]. This article primarily focuses on task allocation based on the opportunistic mode. Therefore, it is crucial to develop an attractive and sustainable task allocation scheme that assigns tasks along participants' daily routes by predicting their mobility and providing appropriate rewards upon successful task execution [5]. Recent studies have explored the use of probabilistic trajectory models or statistical analysis of participants' historical trajectories to predict their mobility. Tasks in the Mobile Crowd Sensing platform typically exhibit diverse requirements in the spatial-temporal domain and impose different objectives on completion rates and costs. Moreover, participants have varying time and area availability for task performance, and each participant has a different capacity for carrying out tasks [6,7]. Consequently, finding a practical and feasible allocation solution for all tasks on the sensing platform often involves solving a complex optimization problem, typically NP-hard. Recent studies have employed greedy strategies to design task allocation schemes, aiming to identify suitable task participants iteratively. Alternatively, intelligent optimization algorithms have been utilized to search for global optimal solutions, avoiding the pitfalls of local optima [5,8].

In this paper, we employ the fuzzy logic control method to predict participants' mobility. This method effectively utilizes participants' historical trajectory information, offering valuable insights for selecting suitable participants. As a rule-based approach, fuzzy logic emulates human thinking and transforms professional knowledge into a heuristic control algorithm [9]. We use fuzzy values instead of precise values for mobility prediction, which incorporates practical considerations for Mobile Crowd Sensing (MCS) applications. Consider a scenario where a task on the sensing platform requires a continuous sensing time of 15 min, and a participant's daily route records meet the spatial domain requirements and other task constraints. However, based on past habits, the participant has only stayed at the task location for 14 min and 30 s. For those accurate value prediction methods, this trajectory record would not contribute to mobility prediction. However, in more realistic scenarios, when a participant is assigned a task along their daily route, they tend to complete the task if it does not significantly disrupt their daily life to receive corresponding rewards. The fuzzy logic method incorporates the participant's residence time into the membership function, providing probability values for different degrees of residence time. These probability values help predict the participant's likelihood of task completion. Next, we model the multi-task allocation problem to balance sensing quality (e.g., task completion rate) and cost (e.g., incentive cost) while adhering to the task's budget and participants' availability constraints. Specifically, we design a greedy task allocation algorithm based on the fuzzy logic results, prioritizing allocating tasks to participants with the highest possibility of completion. However, since the proposed multi-task allocation problem in this paper is a multi-objective optimization challenge, the greedy strategy is prone to local optima and requires balancing various optimization objectives. To address this, we employ the NSGA-II multi-objective optimization algorithm based on the initial allocation results from the greedy strategy. It allows us to obtain a Pareto solution that balances the different optimization objectives.

The main contributions of this paper can be summarized as follows:

- We present a mobility prediction method based on fuzzy logic that effectively utilizes participants' historical trajectories and expedites the resolution of the task allocation problem. Unlike previous mobility prediction methods that utilize fuzzy logic, we incorporate both distance and duration as inputs, enabling simultaneous consideration of the spatial-temporal aspects of the task and the participant.

- We put forward two task allocation algorithms for the NP-hard M3P problem. We devise the MPF algorithm as a greedy strategy and utilize its outcome as the initialization for NSGA-II. Given that the M3P problem involves multi-objective optimization, our NSGA-II algorithm achieves a pareto solution to balance the bi-objective.
- We perform comprehensive experiments on a real-world ch1dataset, demonstrating that the proposed mobility prediction and task allocation methods outperform the comparison methods.

The rest of this paper is structured as follows: Sect. 2 reviews related works. In Sect. 3, we introduce the system framework, formulate the M3P problem, and describe the design of the fuzzy control system in Sect. 4. Section 5 presents a detailed demonstration of the two proposed task allocation algorithms. Section 6 presents the experiments conducted on a real-world dataset. Finally, Sect. 7 presents the conclusion of this paper.

2 Related Work

Task allocation is a crucial aspect of MCS, as it enhances the task completion rate on the sensing platform and reduces the incentive cost for task fulfillment. In many existing methods, task allocation is commonly formulated as an optimization problem with various objectives, such as minimize travel distance, expenditure cost, and maximize task completion rate. For instance, Wang et al. [10] proposed a three-layer architecture for participant selection, incorporating data attributes that consider spatial-temporal constraints and calculate the probability of task completion across different scenarios. Shi et al. [11] precisely evaluated the credibility of fine-grained participants in large-scale tasks, proposing an online task allocation mechanism that combines the Gaussian mechanism with the Bayesian probability model. Hu et al. [12] introduced a reinforcement learning framework to select an appropriate group of participants in a continuous perceptual time series. The framework comprehensively considers the candidate's previous spatial-temporal coverage and predicts their movement patterns to enhance selection performance. Dai et al. [13] devised a many-to-many stable matching algorithm. They demonstrated that, in cases where each participant offers a comparable quality of service, it could achieve a minimum of half the optimal system efficiency. Li et al. [2] introduced a caching mechanism and devised offline and online task allocation methods to address the participant selection problem in MCS with heterogeneous tasks. Similarly, Tao et al. [14] investigated offline and online MCS task allocation scenarios. In offline task allocation, they utilized the ant colony algorithm to acquire an approximate optimal solution, whereas, for online task allocation, they leveraged participant history for predictive task allocation. Zhang et al. [15] investigated the problem of multi-task allocation under time constraints and devised two heuristic genetic algorithms to resolve it.

In recent years, several studies have incorporated participant mobility prediction into task allocation research to enhance the accuracy of allocation results and expedite problem-solving. Hu et al. [12] employed reinforcement learning to predict candidates' mobility in continuous sensing time series and select an appropriate set of participants for the task. Wang et al. [16] utilized a semi-Markov process to determine the probability distribution of participants' arrival time at the task location. Yang et al. [17] employed Markov models to predict the likelihood of participants completing tasks, ensuring that the allocation results enable the timely completion of spatial-temporal sensitive tasks while minimizing the number of participants. Wang et al. [18] employed statistical models to predict participants' mobility. They subsequently utilized particle swarm optimization algorithms to address multi-objective optimization scenarios for task allocation, aiming to maximize task completion rate and minimize cost. Other approaches, such as [19,20], used the Poisson distribution [19] or exponential distribution [20] to predict participant mobility. Zhang et al. [5] introduced fuzzy logic to predict participants' mobility, which yielded promising results for time-sensitive tasks. Building upon this prediction, they proposed a heuristic algorithm to resolve the multi-task allocation problem.

Our approach distinguishes itself from existing works by considering the spatial-temporal heterogeneity of tasks and participants when utilizing fuzzy logic for predicting participants' mobility. By leveraging fuzzy logic to process fuzzy values, we effectively utilize participants' historical trajectory information to enhance mobility prediction. Building upon these advancements, we employ the NSGA-II algorithm to address the multi-objective optimization problem associated with task allocation in MCS.

3 System Overview and Problem Formulation

3.1 System Overview

A Mobile Crowd Sensing (MCS) system typically involves three key roles: task requester, sensing platform, and participant. Task requesters submit their task requirements to the sensing platform, and interested users provide their information in response. The sensing platform utilizes task and participant information to execute task allocation algorithms, selecting suitable participants for each task. In this article, we focus specifically on task allocation. Therefore, we assume the task requesters have already submitted their tasks to the sensing platform. As depicted in Fig. 2, our system framework diagram primarily emphasizes mobility prediction and task allocation. This framework encompasses two main modules:

- Mobility Prediction. This module employs fuzzy logic reasoning to evaluate the possibility of participants completing tasks based on their uploaded historical trajectory information.

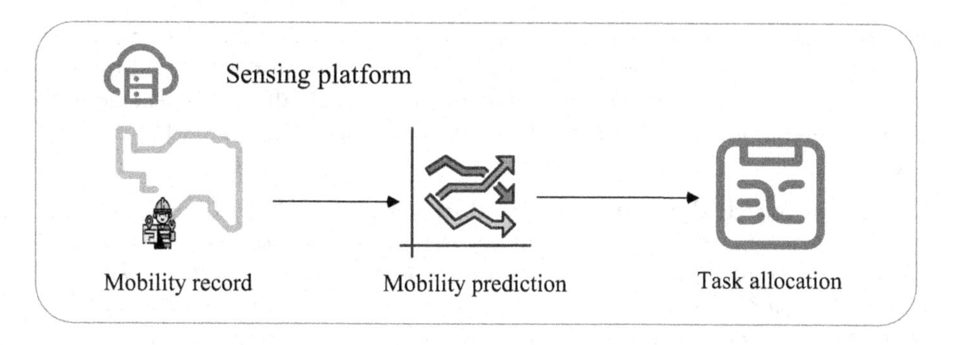

Fig. 2. System framework.

- Task Allocation. The task allocation module initially generates allocation results based on mobility prediction. Subsequently, it designs a multi-objective optimization algorithm to select appropriate participants for each task while considering various constraints such as spatial-temporal factors, cost considerations, and capacity limitations of both tasks and participants. We employ the NSGA-II algorithm for multi-objective optimization in this context.

3.2 Problem Formulation

Definition 1 (Sensing Task). *Let $\mathbb{T} = \{t_1, t_2, \ldots, t_n\}$ represents a set of n tasks uploaded to sensing platform by task requesters. The heterogeneity of task is reflected in the fact that each task $t_i (1 \leq i \leq n)$ is defined as a tuple with three elements, $t_i = (loc_i, dur_i, bud_i)$. Here, loc_i represents the location of the task, dur_i represents the required duration time for completing the task, and bud_i represents the budget constraint associated with the task.*

Definition 2 (Participant). *Let $\mathbb{W} = \{w_1, w_2, \ldots, w_m\}$ represents a set of m participants who are interested in performing tasks. The heterogeneity of participant is reflected in the fact that each participant $w_j (1 \leq j \leq m)$ is defined as a tuple with two elements, $w_j = (\mathbb{R}_j, cap_j)$. Here, $\mathbb{R}_j$ represents the historical trajectory information of participant, including sampling time, location, and duration time, while cap_j indicates the participant's task carrying capacity. We assume that the participant w_j is assigned to the task t_i by the sensing platform. Upon seccessful execution of task t_i, the task requester is responsible for providing a specific reward, denoted as c_{ij}, to compensate for the participant's energy loss, time cost, and potential privacy risks associated with the task execution process.*

Similar to [18], the problem addressed in this study focuses on selecting appropriate participants for task execution, considering two key metrics: crowdsensing quality, measured by the task completion ratio, and incentive budget, represented by the task cost. The task completion ratio refers to the proportion of successfully executed tasks out of the total number of tasks. The task cost represents the overall budget allocated by the task requester. To address this problem, we aim to design a multi-objective multi-task allocation scheme with mobility prediction (M3P) that adheres to the constraints of both tasks and participants.

Definition 3 (M3P Problem). *The M3P problem addresses the assignment of participants from set $\mathbb{W}$ to tasks from set $\mathbb{T}$, where $\mathbb{T}$ contains n tasks and $\mathbb{W}$ contains m participants. The objective is to minimize the total budget cost allocated to participants $w_j \in \mathbb{W}$ while maximizing the overall task completion ratio for all assigned tasks. The problem is subject to two constraints:*

1. the cost of each task $t_i \in \mathbb{T}$ must not exceed its budget bud_i,

2. the number of tasks assigned to participant w_j must not exceed the maximum workload constraint cap_j.

Formally, the M3P problem aims to concurrently optimize both objectives as follow:

$$\min \sum_{i=1}^{n} \sum_{j=1}^{m} x_{ij} * r_{ij} * c_{ij}, \tag{1}$$

$$\max \frac{\sum_{i=1}^{n} \mathrm{com}_i}{n}, \tag{2}$$

$$\text{s.t.} \sum_{j=1}^{m} x_{ij} * c_{ij} \leq \mathrm{bud}_i, \forall t_i \in \mathbb{T}, \tag{3}$$

$$\sum_{i=1}^{n} x_{ij} \leq \mathrm{cap}_j, \forall w_j \in \mathbb{W}, \tag{4}$$

where com_i, x_{ij} and r_{ij} are all binary variables: com_i is set to 1 if task t_i is successfully performed by any of the assigned participants and 0 otherwise. Similarly, x_{ij} is set to 1 if task t_i is assigned to participant w_j and 0 otherwise. Likewise, r_{ij} is set to 1 if participant w_j successfully performs task t_i and 0 otherwise.

3.3 Hardness Analysis of M3P Problem

M3P is an optimization problem with two objectives. Given n tasks and m available participants, the potential combinations of task-participant allocations can become exceedingly numerous, thus the searching all possible allocation solutions can be computationally infeasible($O\left((n+1)^m\right)$) [18]. This is an NP-hard problem, because it can be reduced from the well-known Knapsack problem [21]. Therefore, our M3P problem is also NP-hard.

Given the bi-objective's conflicting nature and multiple constraints in the M3P problem, it is not feasible to directly utilize existing approximation methods for its solution. Therefore, considering the NP-hardness of our M3P problem, we will employ a combination of the mobility prediction method and the NSGA-II multi-objective optimization algorithm to explore Pareto-optimal solutions effectively.

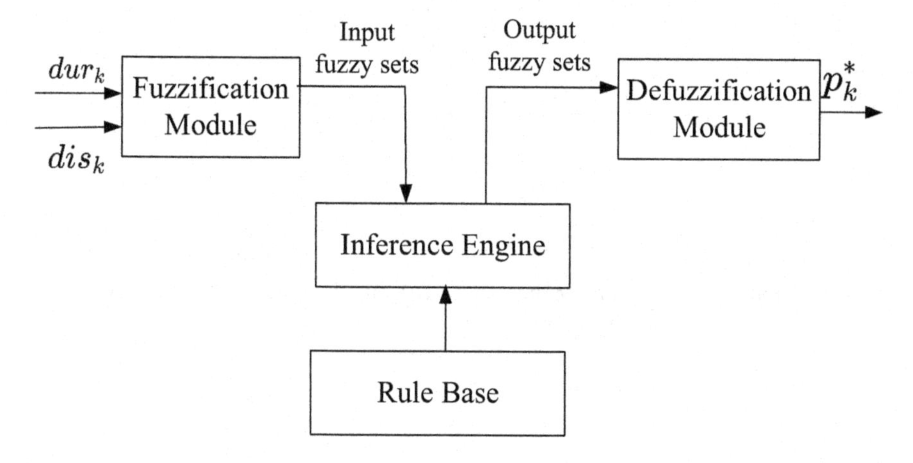

Fig. 3. Fuzzy control system for mobility prediction.

4 Fuzzy Control System

Decision-making processes are often influenced by uncertainties arising from various factors. Real-world information can be incomplete, inaccurate, and lack complete credibility. Fuzzy logic can replicate human decision strategies and make informed choices in uncertain information, eliminating the need for complex models. Given the non-linear and intricate nature of participants' historical trajectory data, fuzzy logic naturally excels in predicting mobility patterns. As illustrated in Fig. 3, a conventional fuzzy logic control system typically comprises three components: Fuzzification, Inference Engine, and Defuzzification. The fuzzy control system takes the trajectory information of participants as input, transforming crisp inputs into fuzzy sets of input variables through the fuzzification module. The fuzzy inference engine then deduces the output fuzzy sets by employing predefined fuzzy rules and membership functions of the output variable. Finally, the defuzzification module employs the center of gravity (COG) strategy to convert the fuzzy output sets into crisp values.

In this work, we present a comprehensive design of the fuzzy logic control system, as outlined below:

4.1 Fuzzy Set and Membership Function

The crisp values of distance dis_k and duration time dur_k serve as inputs to the fuzzy system. Prior to fuzzification, we establish predetermined linguistic values for the two inputs: distance and duration time. Specifically, for distance, we utilize exactly(E), nearly(N), and far(F) as the linguistic values. Regarding duration time, short(S), middle(M), and long(L) are employed as the linguistic values. Similarly, for the output variable p_k^*, we define very unlikely(VU), unlikely(U), likely(L), and very likely(VL) as the linguistic values. The fuzzy sets representing the input and output variables can be described by establishing the membership

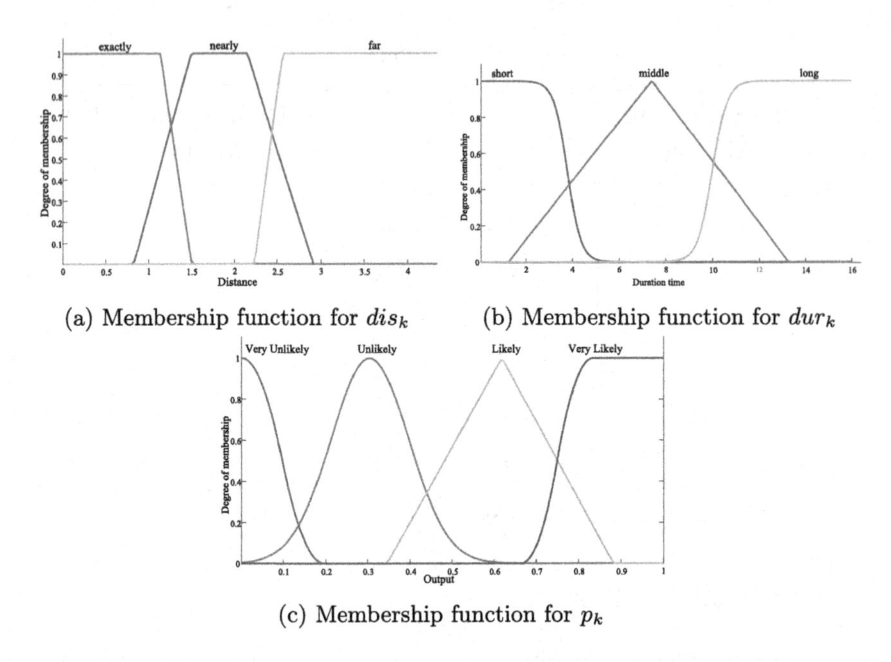

(a) Membership function for dis_k

(b) Membership function for dur_k

(c) Membership function for p_k

Fig. 4. Membership function of fuzzy logic.

functions. Taking into account real-world scenarios, the membership functions of the fuzzy sets for the input and output variables are illustrated in Fig. 4.

4.2 Fuzzy Inference Engine

The fuzzy logic inference employs a rule-based approach, where fuzzy logic rules are formulated as IF-THEN statements with conditions and conclusions. After defining the input and output variables, the next step involves matching them with the IF-THEN rules and aggregating the matching levels [22]. In this paper, Mamdani's fuzzy inference method is utilized. The matrix shown in Fig. 5 depicts the complete set of rules.

4.3 COG Defuzzification

This paper utilizes a defuzzification module to convert the fuzzy output into a precise and non-fuzzy value. Specifically, the chosen strategy is the center of gravity (COG) strategy, which calculates the center of gravity of the membership function curve and cross-coordinate area. The fuzzy control system's output, denoted as p_k^*, effectively expresses the result by employing the COG strategy. This approach enhances the interpretability of the crisp value derived from the fuzzy inference engine, making it suitable for subsequent analysis and decision-making.

$$p_k^* = \frac{\int p_k \mu\left(p_k\right) dp_k}{\int \mu\left(p_k\right) dp_k} \tag{5}$$

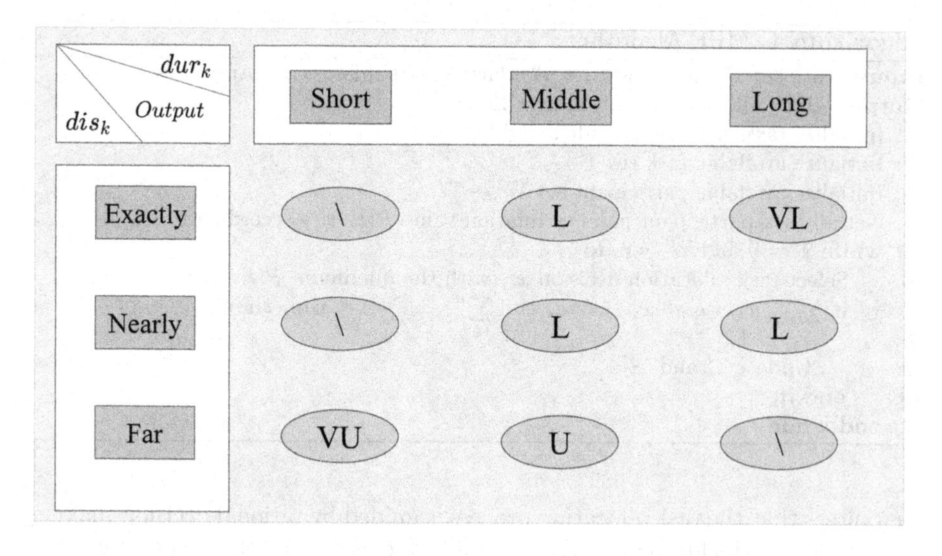

Fig. 5. Fuzzy logic rule matrix.

Then, we can obtain the probability of participant w_j completing task t_i by the following equation.

$$P(t_i, w_j) = \frac{\sum_{k=1}^{|\mathbb{R}_j|} p_{ijk}^*}{|\mathbb{R}_j|} \tag{6}$$

For the historical trajectory $\mathbb{R}_j$ of participant w_j, we use fuzzy logic inference to compute the p_{ijk}^* for each record $\mathbb{R}_j^k (1 \leq k \leq |\mathbb{R}_j|)$ in the historical trajectory, which reflects the likelihood of participant w_j successfully completing task t_i during that specific record.

5 Task Allocation Algorithm

The multi-task allocation problem in MCS is known to be NP-hard, making it impractical to solve using an exhaustive search algorithm due to its vast solution space. To address this challenge, our approach focuses on maximizing the overall task completion rate by initially assigning tasks to participants most likely to complete them, while minimizing the total task cost. First, we employ a greedy strategy to select task participants, to optimize the overall probability of task completion. This strategy adheres to a systematic procedure where the participant with the highest probability of completion is selected for each task. The selection process depends on the mobility prediction results of the fuzzy logic module. By leveraging the principles of fuzzy logic, the approach incorporates various factors, including historical mobility patterns and real-time environmental conditions, to accurately predict participant's mobility. The participant with the highest anticipated mobility and, consequently, the highest completion probability is selected for each task based on these predictions. This greedy strategy

Algorithm 1. MPF Algorithm

Input: Task set $\mathbb{T}$; Participant set $\mathbb{W}$; Participants' historical trajectories set $\mathbb{R}$
Output: Task allocation result matrix $\mathbb{X}$
1: Initialize task allocation result matrix$\mathbb{X} \leftarrow 0$;
2: Initialize available task set $\mathbb{T}' \leftarrow \mathbb{T}$;
3: Initialize available participant set $\mathbb{W}' \leftarrow \mathbb{W}$;
4: Get all task-participant pairs estimation value $P(t_i, w_j)$ according to Eq.(6) ;
5: **while** $\mathbb{T}' \neq \emptyset$ and $\mathbb{W}' \neq \emptyset$ **do**
6: Select task allocation decision x_{ij} with the maximum $P(t_i, w_j)$;
7: **if** $\sum_{j=1}^{m} x_{ij} * c_{ij} + c_{ij} \leq bud_i$ and $\sum_{i=1}^{n} x_{ij} + 1 \leq cap_j$ **then**
8: $x_{ij} = 1$;
9: Update $\mathbb{T}'$ and $\mathbb{W}'$;
10: **end if**
11: **end while**

guarantees that the task allocation process is guided by rational decision-making and maximizes the likelihood of successful task completion. By integrating fuzzy logic into the participant selection process, the approach significantly enhances the efficiency and effectiveness of task allocation, leading to improved overall system performance. However, to avoid getting trapped in an optimal local solution, we further optimize the task allocation using the NSGA-II algorithm, which can handle multiple objectives effectively.

5.1 Most Possible First Algorithm

The MPF (Most Possible First) algorithm efficiently assigns tasks to participants based on the likelihood of a participant successfully performing a given task. Algorithm 1 provides the pseudo-code for the MPF algorithm, outlining the task allocation steps. In each iteration, the algorithm aims to identify the task allocation decision that maximizes the estimated probability of a participant successfully performing a given task. Once the decision is made, the task is assigned to the selected participant, provided all necessary constraints are satisfied. The available task set and participant set are updated throughout the iteration process. If a task's budget is depleted, it is removed from the available task set to prevent further assignment. Similarly, if participants reach their maximum task capacity, they are removed from the participant set to ensure fair task allocation. The iteration process continues until all tasks are assigned or until no available participants remain on the Mobile Crowd Sensing (MCS) platform. Utilizing the MPF algorithm optimizes task assignments, resulting in efficient task completion within the MCS system.

5.2 Improved NSGA-II Algorithm

Considering that MPF has the potential to converge to locally optimal solutions, we propose an enhanced version of the NSGA-II algorithm to obtain globally improved Pareto-optimal solutions.

Algorithm 2. Improved NSGA-II Algorithm

Input: Task set $\mathbb{T}$; Participant set $\mathbb{W}$; MPF's task allocation result matrix $\mathbb{X}$
Output: Task allocation result matrix $\mathbb{X}^{\#}$
1: Initialize population based on $\mathbb{X}$, and calculate objective function ;
2: Fast non-dominated sorting and calculate the crowding distance ;
3: Generate the first generation population by selection, crossover, and mutation operations ;
4: **for** i = 1:gen **do**
5: Produce new offspring by selection, crossover, and mutation operations;
6: Calculate objective function value of offspring population;
7: Form a new parent population;
8: Fast non-dominated sorting and crowding distance calculation;
9: Generation of the next population based on elite retention strategies;
10: **end for**
11: return Pareto optimal set and select an optimal chromosome as the task allocation matrix $\mathbb{X}^{\#}$;

NSGA-II incorporates several mechanisms to enhance the performance of the algorithm. These include utilizing a fast non-dominated sorting algorithm to preserve superior individuals in the population, integrating an elite strategy to expand the search space, and implementing a crowding degree to sustain population diversity [23].

During the initialization phase, the parameters for the NSGA-II algorithm are set, including the population size, number of iterations, probabilities of crossover and mutation, and the number of objective functions $\mathbb{M}$ and decision variables $\mathbb{V}$. Subsequently, the population is initialized, with each task allocation decision represented as a chromosome with gene values ranging from 1 to $\mathbb{V}$. Moving forward, the evaluation step involves assessing the fitness of each individual in the population using fitness functions. Furthermore, a constraint violation value (CV) is introduced to quantify the degree of constraint violation. The Fast Non-dominated Sorting Algorithm is then implemented to evaluate the population, assigning parameters n_p and S_p to each individual. Here, n_p represents the number of individuals that dominate individual p, while S_p denotes the set of individuals dominated by p. Through this algorithm, the population is categorized into non-dominant layers. To assess the diversity among individuals within the same layer based on their objective functions, the calculation of the crowding degree is performed. Subsequently, the selection operation is enhanced by incorporating a logistic distribution-based probability selection operator. Crossover and mutation operations are employed, involving the random selection of two individuals, generation of a crossover position, exchange of values at the crossover position to create a new individual, and subsequent feasibility checks. To generate a new population, the individuals are sorted and classified based on Pareto rank and crowding degree. The most promising individuals are subsequently chosen for the next generation. This entire process is reiterated until the termination condition is satisfied.

6 Evaluation

In this section, we assess the performance of the proposed algorithms using the San Francisco dataset. Initially, we provide an introduction to the dataset and baseline methods. Subsequently, a thorough evaluation was conducted to assess the proposed algorithms and their respective baselines, focusing on two critical aspects: mobility prediction and task allocation. The evaluation results yielded vital insights regarding the performance and effectiveness of the algorithms under scrutiny. A comprehensive understanding of the relative strengths and weaknesses of the proposed algorithms was obtained through a comparative analysis with the baseline approaches. The evaluation process rigorously analyzed accuracy to determine the predictive capabilities of the algorithms concerning mobility patterns. Furthermore, the performance of task allocation was assessed using metrics like task completion ratio and cost.

6.1 Dataset

This study's primary source of information is the San Francisco dataset [24], which encompasses the trajectories of 536 taxis recorded over one month. Each entry within the dataset provides crucial details, including the unique identification number of the taxi, the specific sampling time, and the corresponding geographical coordinates (latitude and longitude) of the sampled position. The research methodology employed in this study closely aligns with a previous investigation referenced as [5], thereby ensuring methodological consistency and comparability. From the available pool of participants, a subset of 100 active taxis was purposefully selected to represent a diverse and representative sample for inclusion in the experiments. Furthermore, based on an analysis of a heat map, ten highly populated areas were identified as target locations for the tasks at hand, facilitating a focused investigation within specific regions. Notably, the original dataset needs more information regarding the duration of each taxi's stay at any given point. We adopted a random generation approach to address this limitation by introducing varying sensing time durations ranging from 1 to 15 min. This approach ensures a realistic portrayal of taxi behavior and captures the temporal dynamics of the studied phenomena.

6.2 Baselines

For mobility prediction, we present a baseline method for comparative studies, as outlined below:

- **Greedy**. This approach adopts an opportunistic mode to minimize the overall cost. It iteratively selects participants whose cost is the lowest.

Regarding task allocation, we provide the following methods to demonstrate the efficiency of our scheme:

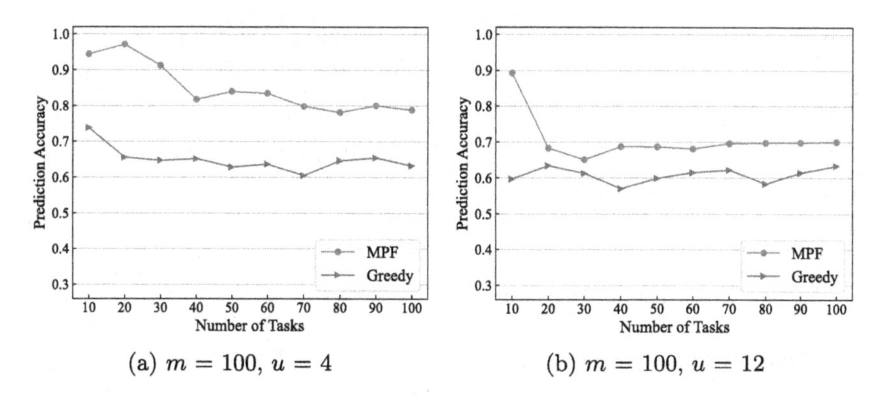

Fig. 6. Prediction accuracy with different mobility prediction methods.

- **Greedy**. This method follows the same approach as the aforementioned greedy algorithm.
- **MPF(Most Possible First)**. As a baseline, we employ the MPF algorithm, which iteratively selects participants based on the mobility prediction results obtained through fuzzy logic. The participant most possible to perform the task is chosen.
- **GGPSO(Greedy-and-Genetic Enhanced Particle Swarm Optimization)**. For comparison purposes, we utilize the GGPSO algorithm proposed in [5]. This algorithm combines greedy and genetic search techniques within a particle swarm optimization framework.

6.3 Performance of Mobility Prediction

Figure 6 illustrates the average prediction accuracy as the number of tasks varies. It is evident that the MPF algorithm consistently outperforms the compared method under all conditions. Notably, when the number of participants is fixed at 100, and the task budget follows a uniform distribution ranging from 2 to 6, the MPF algorithm achieves an average accuracy of 19.9% higher than that of the Greedy method. Subsequently, we increment the task budget. As depicted in Fig. 6a, both methods experience a decrease in prediction accuracy. This decline can be attributed to the larger task budgets, which increase the likelihood of recruiting participants who cannot perform the tasks, ultimately reducing the overall prediction accuracy. Nonetheless, our method consistently outperforms the compared method across all scenarios.

6.4 Performance of Task Allocation

The M3P problem addresses the challenge of multi-task allocation, aiming to maximize the average task completion ratio while minimizing the total budget cost. To assess the effectiveness of our task allocation scheme, we compare it with the Greedy, MPF, and GGPSO approaches.

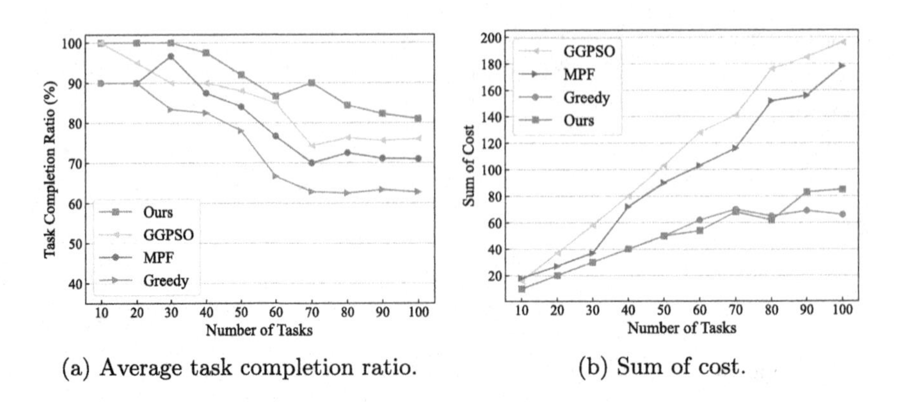

(a) Average task completion ratio.

(b) Sum of cost.

Fig. 7. Performance of Task Allocation with different methods.

As illustrated in Fig. 7a, our method surpasses the MPF algorithm, exhibiting a task completion ratio that is 10.4% higher. Moreover, compared to the baseline methods Greedy and GGPSO, our method achieves a task completion ratio of 17.2% and 6.37% higher, respectively.

Figure 7b further demonstrates the superiority of our method. It outperforms the proposed MPF algorithm by 44.7 and surpasses the baseline GGPSO by 61.2 in terms of lower average cost. Notably, the average cost of our method only slightly exceeds that of the greedy algorithm by two units. This discrepancy arises because the greedy algorithm prioritizes selecting participants with the lowest cost.

In summary, our method consistently outperforms the other approaches in task completion ratio and cost optimization objectives.

7 Conclusion

The examined research paper centers around a multi-task allocation problem aimed at optimizing task assignment to participants by maximizing the task completion ratio and minimizing associated costs. To address this problem, the paper introduces an approach that utilizes fuzzy logic for mobility prediction. This method analyzes the historical trajectories of participants to forecast the possibility of task completion by specific participants. Building upon the insights gained from mobility prediction, the paper presents two allocation algorithms designed to address the multi-task allocation problem effectively. Extensive experimentation confirms that the proposed method excels in prediction accuracy, effectively overcoming the inherent challenges of the problem. Moreover, the method consistently produces superior solutions in the context of task allocation, thereby highlighting its potential for practical implementation and real-world applicability. In future work, we will consider introducing a reinforcement learning approach to solve the dynamic task allocation problem.

Acknowledgment. This work was supported by the National Natural Science Foundation of China (Grant No.62372121) and the Natural Science Foundation of Guangdong Province (Grant No.2023A1515012358, No.2022A1515011386).

References

1. Ganti, R.K., Ye, F., Lei, H.: Mobile crowdsensing: current state and future challenges. IEEE Commun. Mag. **49**(11), 32–39 (2011)
2. Li, H., Li, T., Wang, W., Wang, Y.: Dynamic participant selection for large-scale mobile crowd sensing. IEEE Trans. Mob. Comput. **18**(12), 2842–2855 (2018)
3. Zhang, D., Xiong, H., Wang, L., Chen, G.: Crowdrecruiter: selecting participants for piggyback crowdsensing under probabilistic coverage constraint. In: Proceedings of the 2014 ACM International Joint Conference on Pervasive and Ubiquitous Computing, pp. 703–714 (2014)
4. To, H.: Task assignment in spatial crowdsourcing: challenges and approaches. In: Proceedings of the 3rd ACM SIGSPATIAL PhD Symposium, pp. 1–4 (2016)
5. Zhang, J., Zhang, X.: Multi-task allocation in mobile crowd sensing with mobility prediction. IEEE Trans. Mobile Comput. (2021)
6. Wang, L., Yu, Z., Zhang, D., Guo, B., Liu, C.H.: Heterogeneous multi-task assignment in mobile crowdsensing using spatiotemporal correlation. IEEE Trans. Mob. Comput. **18**(1), 84–97 (2018)
7. Gao, G., Huang, H., Xiao, M., Wu, J., Sun, Y.E., Du, Y.: Budgeted unknown worker recruitment for heterogeneous crowdsensing using cmab. IEEE Trans. Mob. Comput. **21**(11), 3895–3911 (2021)
8. Liu, J., Tan, W., Liang, Z., Ding, K.: Multi-task allocation under multiple constraints in mobile crowdsensing. In: Human Centered Computing: 7th International Conference, HCC 2021, Virtual Event, 9–11 December 2021, Revised Selected Papers, pp. 183–195. Springer (2023). https://doi.org/10.1007/978-3-031-23741-6_17
9. Pérez-Correa, J., Zaror, C.A.: Recent advances in process control and their potential applications to food processing. Food Control **4**(4), 202–209 (1993)
10. Wang, E., Yang, Y., Lou, K.: User selection utilizing data properties in mobile crowdsensing. Inf. Sci. **490**, 210–226 (2019)
11. Shi, Z., Jiang, S., Zhang, L., Du, Y., Li, X.Y.: Crowdsourcing system for numerical tasks based on latent topic aware worker reliability. In: IEEE INFOCOM 2021-IEEE Conference on Computer Communications, pp. 1–10. IEEE (2021)
12. Hu, Y., Wang, J., Wu, B., Helal, S.: Rl-recruiter+: mobility-predictability-aware participant selection learning for from-scratch mobile crowdsensing. IEEE Trans. Mob. Comput. **21**(12), 4555–4568 (2021)
13. Dai, C., Wang, X., Liu, K., Qi, D., Lin, W., Zhou, P.: Stable task assignment for mobile crowdsensing with budget constraint. IEEE Trans. Mob. Comput. **20**(12), 3439–3452 (2020)
14. Tao, X., Song, W.: Profit-oriented task allocation for mobile crowdsensing with worker dynamics: cooperative offline solution and predictive online solution. IEEE Trans. Mob. Comput. **20**(8), 2637–2653 (2020)
15. Li, M., Gao, Y., Wang, M., Guo, C., Tan, X.: Multi-objective optimization for multi-task allocation in mobile crowd sensing. Proc. Comput. Sci. **155**, 360–368 (2019)

16. Wang, E., Yang, Y., Wu, J., Liu, W., Wang, X.: An efficient prediction-based user recruitment for mobile crowdsensing. IEEE Trans. Mob. Comput. **17**(1), 16–28 (2017)
17. Yang, Y., Liu, W., Wang, E., Wu, J.: A prediction-based user selection framework for heterogeneous mobile crowdsensing. IEEE Trans. Mob. Comput. **18**(11), 2460–2473 (2018)
18. Wang, L., Yu, Z., Han, Q., Guo, B., Xiong, H.: Multi-objective optimization based allocation of heterogeneous spatial crowdsourcing tasks. IEEE Trans. Mob. Comput. **17**(7), 1637–1650 (2017)
19. Xiao, M., Wu, J., Huang, L., Wang, Y., Liu, C.: Multi-task assignment for crowdsensing in mobile social networks. In: 2015 IEEE Conference on Computer Communications (INFOCOM), pp. 2227–2235. IEEE (2015)
20. Lai, C., Zhang, X.: Duration-sensitive task allocation for mobile crowd sensing. IEEE Syst. J. **14**(3), 4430–4441 (2020)
21. Martello, S., Toth, P.: Algorithms for knapsack problems. North-Holland Math. Stud. **132**, 213–257 (1987)
22. Yang, G., Zhang, Y., Wang, B., He, X., Wang, J., Liu, M.: Task allocation through fuzzy logic based participant density analysis in mobile crowd sensing. Peer-to-Peer Netw. Appli. **14**, 763–780 (2021)
23. Li, W., Feng, G., Huang, Y., Liu, Y.: Multi-task allocation based on edge interaction assistance in mobile crowdsensing. In: Algorithms and Architectures for Parallel Processing: 21st International Conference, ICA3PP 2021, Virtual Event, 3–5 December 2021, Proceedings, Part III, pp. 214–230. Springer (2022). https://doi.org/10.1007/978-3-030-95391-1_14
24. Michal, P., Sarafijanovic-Djukic, N., Matthias, G.: Crawdad dataset epfl/mobility (v. 2009-02-24) (2009). https://crawdad.org/epfl/mobility/20090224

SMCoEdge: Simultaneous Multi-server Offloading for Collaborative Mobile Edge Computing

Changfu Xu[1,2], Yupeng Li[1], Xiaowen Chu[3], Haodong Zou[1,2], Weijia Jia[2,4], and Tian Wang[2,4(✉)]

[1] Hong Kong Baptist University, Hong Kong, China
{chanfuxu,haodongzou}@uic.edu.cn
[2] BNU-HKBU United International College, Zhuhai, China
wsnman@gmail.com
[3] The Hong Kong University of Science and Technology (Guangzhou), Guangzhou, China
xwchu@ust.hk
[4] BNU-UIC Institute of Artificial Intelligence and Future Networks, Beijing Normal University, Zhuhai, China
jiawj@bnu.edu.cn

Abstract. Collaborative Mobile Edge Computing (MEC) has emerged as a promising solution for low service delay in computation-intensive Internet of Things (IoT) applications. However, current approaches typically perform offline task partitioning and offload each subtask to an Edge Server (ES) for processing. This leads to varying delays in subtask processing across different ESs, resulting in a high make-span of task offloading. To address this issue, we propose a novel approach called SMCoEdge, which utilizes simultaneous multi-ES offloading to minimize the make-span of task offloading for computation-intensive IoT applications. Specifically, we formulate our problem as a mixed integer non-linear programming problem and prove its NP-hardness. We then decompose our problem into two sub-problems of multi-ES selection and task allocation, and propose a Deep Reinforcement Learning-based Simultaneous Multi-ES Offloading (DRL-SMO) algorithm to effectively solve it. Additionally, we analyze the computation complexity of DRL-SMO. Our extensive simulation results demonstrate that SMCoEdge outperforms state-of-the-art approaches by reducing make-span by 18.93% while maintaining a low offloading failure rate.

Keywords: Mobile edge computing · Simultaneous multi-server offloading · Edge-edge collaboration · Task allocation

Z. Tari et al. (Eds.): ICA3PP 2023, LNCS 14491, pp. 73–91, 2024.
https://doi.org/10.1007/978-981-97-0808-6_5

1 Introduction

The proliferation of fifth-generation (5G) communication technology has found numerous applications in modern Internet of Things (IoT) systems, such as automatic driving and augmented/virtual reality [1]. These applications often require intensive computation with stringent demands for low service response time [2,3]. To meet these demands, Mobile Edge Computing (MEC) has been proposed as a computing service that operates at the network edges, such as base stations and access points [4,5]. However, due to the limited resources of Edge Servers (ES), a smart resource scheduling solution is urgently required to ensure low delay for computation-intensive IoT applications [6–8].

To address the above problems, collaborative MEC was proposed as a promising technique to further reduce service delay [9]. According to different collaboration ways, existing collaborative MEC solutions can be divided into two categories as the following: edge-cloud collaboration solutions and edge-edge collaboration solutions. In edge-cloud collaboration solutions, tasks are usually offloaded to an ES or Cloud Server (CS) for further processing, thus reducing service delay compared with traditional MEC solutions [10]. In edge-edge collaboration solutions, tasks are offloaded to multiple ESs by leveraging the idle resources of ESs [11,12]. However, existing edge-edge collaboration solutions are to first conduct all the tasks partition offline and then sequentially select an ES for each subtask [13]. As a result, the processing delays of subtasks of a task at different ESs are obviously not equal due to the heterogeneous subtask workloads and ES resources, which causes a failure in minimizing make-span of task offloading and thus cannot ensure low delay for computation-intensive IoT applications. Therefore, it is desirable to a method that incorporates online task allocation into collaborative MEC.

In this paper, we propose SMCoEdge, a simultaneous multi-ES offloading approach for collaborative MEC. Specifically, we formulate our problem as a Mixed Integer Non-Linear Programming (MINLP) problem. We then decompose our problem into two sub-problems of simultaneous multi-ES selection and online task allocation, and propose a Deep Reinforcement Learning-based Simultaneous Multi-ES Offloading (DRL-SMO) algorithm to effectively solve it. SMCoEdge can minimize the make-span of task offloading to improve the Quality of Service (QoS) for computation-intensive IoT applications. The contributions of our work are summarized as follows.

- To the best of our knowledge, we are the first to study the online simultaneous multi-ES offloading and task allocation problem in the context of collaborative MEC. We formulate our problem as an MINLP problem with the objective of minimizing the make-span of task offloading and prove its NP-hardness.
- We propose a novel approach called SMCoEdge which performs online simultaneous multi-ES selection and task allocation to minimize the make-span of task offloading for computation-intensive IoT applications.
- We decompose our problem into two sub-problems to reduce computational complexity and propose the DRL-SMO algorithm to effectively solve it. Moreover, we analyze the computation complexity of DRL-SMO.

- We evaluate the effectiveness of the proposed SMCoEdge approach through comprehensive simulations. Simulation results show that the SMCoEdge significantly outperforms the state-of-the-art approaches in reducing make-span by 18.93% while maintaining a low offloading failure rate.

The remainder of this paper is organized as follows. Section 2 introduces the related work. Section 3 gives the system model and problem formulation. Section 4 describes the detail of the proposed approach. Section 5 presents our evaluation. The conclusions are shown in Sect. 6.

2 Related Work

In this section, we briefly review the literature of collaborative MEC, mainly including edge-cloud collaboration and edge-edge collaboration.

2.1 Edge-Cloud Collaboration

To improve the QoS for computation-intensive IoT applications, edge-cloud collaboration solution usually enables scheduler to first select a suitable ES for a task offloading. Then, if the selected ES does not meet the task's computation requirement, the task will be offloaded to CS for further processing [14]. For example, considering the limited computing capacity of ES, Ren et al. [15] investigate the collaborative computation between ESs and CS, where the task can be offloaded to an ES or CS according to the workload size of tasks. Thus, the QoS for task offloading in both ES and CS are improved. Additionally, considering the unbalanced workload locality in edge computing systems, Hao et al. [16] propose an efficient consensus solution to achieve fast event ordering for delay-sensitive IoT applications. Han et al. [17] propose a fast scheduling algorithm for dynamic edge-cloud task collaboration to reduce service response time. These works usually focus on the edge-cloud collaborative processing with diverse workload characteristics (e.g., stochastic arrival rate and different resource demands). However, due to the distance of CS and the limited network bandwidth, how ESs and CS collaboration still remains an open issue to achieve the optimal QoS in MEC systems.

2.2 Edge-Edge Collaboration

Edge-edge collaboration solution was recently proposed by leveraging the idle resources of ESs [18], where a task can be offloaded to multiple ESs for collaborative processing via single-ES selection for each task request. For example, by considering the uncertain computational requirements of offloading tasks, Eshraghi et al. [19] propose an efficient algorithm to select a suitable ES for task offloading by load level awareness. Chu et al. [18] present a fine-grained task offloading solution by dividing tasks of each service into multiple sub-types. Qin et al. [20] design a vehicular collaborative edge computing system to make the best use of vehicles' idle and redundant resources. Besides, Zeng et al. [13] propose CoEdge to collaboratively process task over heterogeneous edge devices. This method first partitions the tasks into several subtasks and then offloads subtasks to multiple ESs for respectively processing to reduce processing delay. These existing works well investigate task offloading with edge-edge collaboration for many kinds of network environments to improve the performance of collaborative MEC.

In summary, existing works only study the single-ES offloading problem of collaborative MEC, i.e., only a suitable ES is sequentially decided for each task (or subtask) offloading. Different from existing methods, we propose the SMCoEdge approach that simultaneously selects multiple suitable ESs for each task request and performs task allocation in an online way.

3 System Model and Problem Formulation

This section first presents the system model of the proposed SMCoEdge approach and then formulates our problem into an optimization problem.

3.1 System Model

We consider the system model of SMCoEdge approach as illustrated in Fig. 1. Each Base Station (BS) is equipped with an ES. Different ESs have different resources. Each ES has a scheduler with a queue to store arrived tasks. BSs provide computing service for Mobile Devices (MDs) in the BSs' signal range via wireless communication. All BSs are connected via a wired core network. Similar to existing works (e.g., [13,21,22]), we focus on the computation-intensive tasks that can be partitioned and have to be offloaded to ESs for processing via edge-edge collaboration. The typical examples are object classification, virus scanning, and file compression, where the images or files can be partitioned into several independent subsets to be processed in parallel.

Let B denote the total number of ESs (or BSs). $\mathcal{B} = \{1, ..., B\}$ denotes the set of ESs. $b \in \mathcal{B}$ denotes the b-th ES. We consider that the system time is divided into consecutive time frames with equal length [24]. Let T denote the total number of time slots. $\mathcal{T} = \{1, ..., T\}$ denotes the set of time slots. $t \in \mathcal{T}$ denotes the t-th time slot. The length of time slot t is denoted as Δ (in seconds). Next, we will present the task model, decision variable, and task scheduling model in details as follows.

Task Model. We denote $\mathcal{N}_{b,t} = \{1, 2, ..., N_{b,t}\}$ by a set of tasks arrived to BS b at time slot t. $n \in \mathcal{N}_{b,t}$ denotes as the n-th task. A tuple (d_n, ρ_n, τ_n) denotes the offloading request of a task n, where d_n (in bits), ρ_n (in CPU cycles/bit), and τ_n (in seconds) are the number of bits, required computation density, and deadline of the task n, respectively. We consider that if the task has not been completely processed through τ_n seconds, it will be immediately dropped.

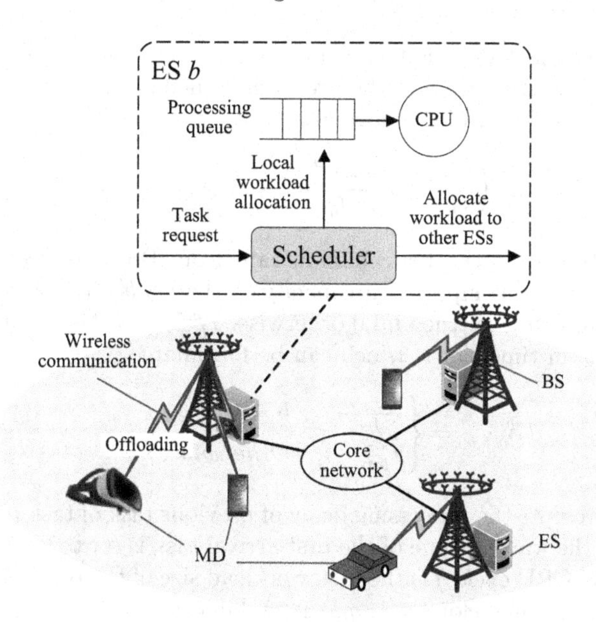

Fig. 1. The system model of the proposed SMCoEdge approach.

Decision Variable. For each time slot, the scheduler decides the ESs to process tasks. Let a vector $\boldsymbol{\alpha}_{b,n,t} = [\alpha_{b,n,t,1}, \alpha_{b,n,t,2}, ..., \alpha_{b,n,t,b'}, ..., \alpha_{b,n,t,B}]$ denote the decision of simultaneous multi-ES selection for the task n at time slot t, where $b' \in \mathcal{B}$ and $\alpha_{b,n,t,b'} \in \{0,1\}$. Moreover, when $\alpha_{b,n,t,b'}$ equals to 1, the ES b' is selected for task processing. Otherwise, $\alpha_{b,n,t,b'} = 0$. Note that SMCoEdge simultaneously selects multiple ESs (rather than one ES) for each task request. Therefore, a feasible constraint is that at least one and at most B ESs are selected for each task request, i.e.,

$$1 \leq \sum_{b' \in \mathcal{B}} \alpha_{b,n,t,b'} \leq B, \ \forall b \in \mathcal{B}, n \in \mathcal{N}_{b,t}, t \in \mathcal{T}. \tag{1}$$

Secondly, after the finishing of the above multi-ES selection, the scheduler also decides the fractions of task workloads that are processed at the selected ESs. Similar to existing works (e.g., [22,25]), let a variable $x_{b,n,t,b'} \in [0,1]$ denote the allocation fraction of task workload that is processed at ES b' in time slot t. Obviously, we have another constraint is that the sum of the allocation fractions of a task workload is equal to 1, i.e.,

$$\sum_{b' \in \mathcal{B}} x_{b,n,t,b'} = 1, \ \forall b \in \mathcal{B}, n \in \mathcal{N}_{b,t}, t \in \mathcal{T}. \tag{2}$$

Task Scheduling Model. Let $T^o_{b,n,t,b'}$ denote the processing delay of task n that arrives to BS b and offloads to ES $b' \in \mathcal{B}$ in time slot t. In this case, the task processing delay includes the task transmission delay over the link from BS b to BS b', the task waiting time in the processing queue at ES b', and the task computing delay on ES b', which is formulated as

$$T^{o}_{b,n,t,b'} = x_{b,n,t,b'} \cdot \left(\frac{d_n}{v_{b,b',t}} + \frac{\rho_n \cdot d_n}{f_{b'}} \right) + \mathbb{I}(x_{b,n,t,b'} > 0) \cdot T^{w}_{b,n,t,b'}, \tag{3}$$

where $v_{b,b',t}$ (in bits/s) is the transmission rate from BS b to BS b' and $f_{b'}$ (in CPU cycles/s) is the computation capacity of the ES b'. $\mathbb{I}(J)$ is a indicator function and equals 1 If J is true and 0 otherwise. $T^{w}_{b,n,t,b'}$ is the waiting time of task n at the ES b' in time slot t, which can be formulated as

$$T^{w}_{b,n,t,b'} = \begin{cases} \frac{q_{t-1,b'}}{f_{b'}}, & b = 1, n = 1, \\ T^{o-1}_{b,n,t,b'}, & Otherwise, \end{cases} \tag{4}$$

where $T^{o-1}_{b,n,t,b'}$ represents the processing delay of previous task of task n. Without loss of generality, the waiting time of the first arrival task is set to 0, i.e., $T^{w}_{1,1,1,b'} = 0$. The $q_{t-1,b'}$ (in CPU cycles) is the task workload size of the processing queue at ES b' at the end of time slot $t-1$. $q_{t,b'}$ is updated as

$$q_{t,b'} = \max\{q_{t-1,b'} + \sum_{b \in \mathcal{B}} \sum_{n \in \mathcal{N}_{b,t}} x_{b,n,t,b'} \cdot d_n \cdot \rho_n - f_{b'} \Delta, 0\}, \tag{5}$$

where without loss of generality, the initial task workload $q_{0,b'}$ in the processing queue is set to 0.

Furthermore, let $T^{s}_{b,n,t}$ denote the make-span of task n processing at time slot t. Then, $T^{s}_{b,n,t}$ equals to the longest processing delay at the selected ESs or the deadline when the task is dropped, i.e.,

$$T^{s}_{b,n,t} = \max\{\max\{T^{o}_{b,n,t,b'}\}_{b' \in \mathcal{B}}, \tau_n\}, \tag{6}$$

where max{} is equal to the biggest value in {}.

3.2 Problem Formulation

The objective of our problem is to minimize the make-span of tasks offloading in collaborative MEC systems by making online simultaneous multi-ES offloading and task allocation, which can be formulated into an optimization problem in the following:

$$\min_{\alpha, X} \quad \lim_{T \to \infty} \frac{1}{T} \sum_{t=1}^{T} \sum_{b=1}^{B} \sum_{n=1}^{N_{b,t}} T^{s}_{b,n,t} \tag{7}$$

$$\text{s.t.} \quad (1) \text{ and } (2), \tag{7(a)}$$

where optimization variables are multi-ES selection $\boldsymbol{\alpha}$ and task allocation $\boldsymbol{X}$. This problem (7) is an MINLP problem and that is proved to be NP-hard by the following **Theorem 1**.

Theorem 1. *The problem (7) is an MINLP problem and that is NP-hard.*

Proof. First, we can see that in the problem (7), each element of the variable $\boldsymbol{\alpha} \in \{0, 1\}$ is an integer, and each element of the $\boldsymbol{X} \in [0, 1]$ is a positive real number. Meanwhile, the equation (3) is non-linear so that the objective function of problem (7) is also non-linear. Thus, the problem (7) is an MINLP problem. Second, this MINLP problem is typical NP-hard as shown in [30]. Therefore, the problem (7) is NP-hard.

4 Proposed Approach

In this section, we present the detail of proposed SMCoEdge approach. In Subsect. 4.1, we first describe the SMCoEdge framework and then decompose our problem into two sub-problems. In Subsects. 4.2 and 4.3, we present the solutions to solve these two sub-problems, respectively. In Subsect. 4.4, the implementation of the proposed DRL-SMO algorithm is presented.

4.1 SMCoEdge Framework

The whole framework of the proposed SMCoEdge approach is shown in Fig. 2. SMCoEdge uses a basic DRL structure with a Deep Q-learning Network (DQN) [26]. DQN is a model-free approach and was widely used to obtain the optimal offloading policy for MEC [27]. We consider that each ES is deployed with a trainer and a scheduler. The trainer mainly includes an Evaluation Q-learning Network (referred to as EQN) and a Target Q-learning Network (referred to as TQN), which are trained using the history data from system environment. The scheduler makes the decisions of multi-ES selection and task allocation.

Based on the above SMCoEdge framework, we can decompose our problem (7) into two sub-problems of multi-ES selection and task allocation. Furthermore, we propose an improved DQN model and a resource-aware closed-form solution to solve these two sub-problems, respectively. The specific details are given in the next two subsections.

4.2 Improved DQN for Multi-ES Selection

Traditional DQN model cannot be directly applied for our SMCoEdge approach since it only can select an ES for each task offloading [27,32]. Therefore, we design an improved DQN model to realize simultaneous multi-ES selection in our approach. The state, action, and reward elements of the DRL model in our problem are given as follows.

State. At each time slot $t \in \mathcal{T}$, each ES observes the system's state information, mainly include the task size and load level of each ES. Then, the state information $s_{b,n,t}$ of task n at ES b is presented as

$$\boldsymbol{s}_{b,n,t} = [d_n, \ \boldsymbol{q}_{t-1}], \tag{8}$$

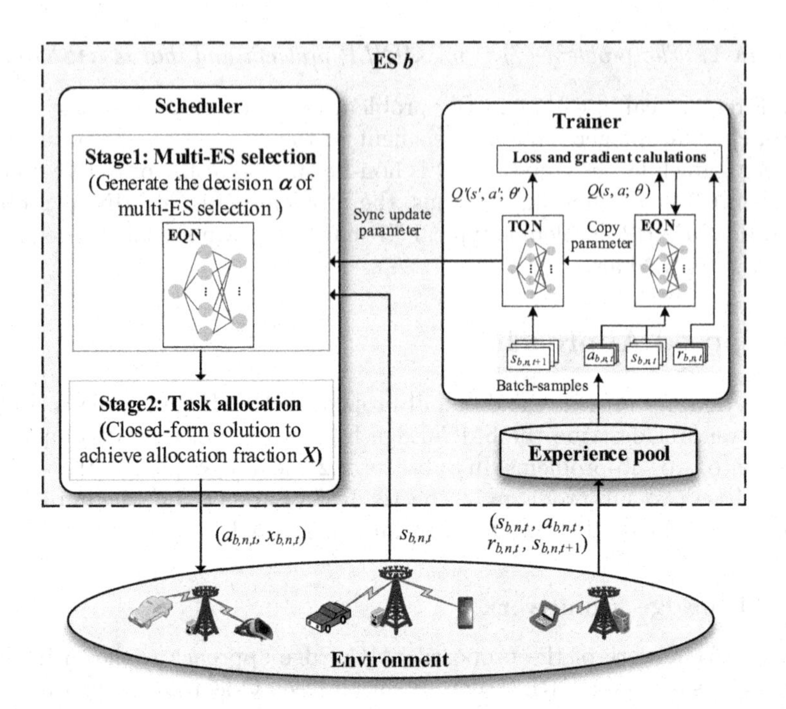

Fig. 2. The framework of the proposed SMCoEdge approach.

where the $\boldsymbol{q}_{t-1} = [q_{t-1,1}, ..., q_{t-1,b'}, ..., q_{t-1,B}]$ represents the ESs' load levels at time slot $t-1$ in the system and can be achieved by the Eq. (5).

Action. For a arrival task n in time slot t, the action includes the indexes of all ESs. A one-hot encoding is introduced to model the action. Then, the action $a_{b,n,t}$ of task n arrived to BS b in time slot t is represented as

$$a_{b,n,t} = [\alpha_{b,n,t,1}, \alpha_{b,n,t,2}, ..., \alpha_{b,n,t,B}]. \tag{9}$$

Here, we use a parameter $1 \leq k \leq B$ as the number of multi-ES selection. That is to say, the EQN at scheduler simultaneously selects top k ESs (i.e., $a_{b,n,t}$ will have k 1 s) for a task offloading. k can be adjusted by user. Note that when $k =$ 1, our approach degrades into the traditional single-ES selection method.

Reward. The objective of our problem is to minimize the make-span of task offloading. Hence, we hope that the higher reward is achieved with the lower make-span. Let the negative make-span of a task offloading denote the reward if the task is successfully processed before the deadline. Otherwise, the reward is set to the negative value $-2 * \tau$ which represents a constant penalty. Therefore, the reward $r_{b,n,t}$ of a task n in time slot t is formulated as

$$r_{b,n,t} = \begin{cases} -T^s_{b,n,t}, & \text{If the task is processed.} \\ -2 * \tau_n, & \text{If the task is dropped.} \end{cases} \tag{10}$$

Afterwards, for a state-action pair s and a, the loss function of DQN model is calculated by

$$Loss = \mathbb{E}[(Q(s,a;\theta) - Q'(s',a';\theta'))^2], \tag{11}$$

where the $Q(s,a;\theta)$ is the Q-value of EQN at trainer. The $Q'(s',a';\theta')$ is the Q-value of TQN at trainer.

4.3 Closed-Form Solution for Task Allocation

The decision variable $\boldsymbol{\alpha}_{b,n,t}$ has been solved in the multi-ES selection stage. Then, the task allocation problem can be formulated as

$$\min_{\boldsymbol{X}_{b,n,t}} \quad \max_{b' \in \mathcal{B}} T^o_{b,n,t,b'} \tag{12}$$

$$\text{s.t.} \quad (2), \tag{12(a)}$$

where $\boldsymbol{X}_{b,n,t} = [x_{b,n,t,1}, x_{b,n,t,2}, ..., x_{b,n,t,B}]$ is the optimization variable. Furthermore, we propose a resource-aware closed-form solution to solve the above problem (12) as follows.

First, according to the generated action $a_{b,n,t}$, we can achieve the selected k ESs of a task offloading. Note that at this time, we just solve the corresponding k fractions of task allocation in the $\boldsymbol{X}_{b,n,t}$. Moreover, the minimal make-span of a task is existed when the processing delays of subtasks of the task are equal at different ESs [28]. Then, we can formulate the $k-1$ equations of equal processing delay of subtasks according to the k ESs' computation resources. Finally, based on the Eq. (12(a)) and the formulated $k-1$ equations, the corresponding k fractions of task allocation can be solved in following cases.

- *Case 1:* When an ES b' is only selected for a task request, the task workload is whole allocated to the ES b' for processing. At this time, the allocation fraction $x_{b,n,t,b'}$ obviously equals to 1, and other elements of $\boldsymbol{X}_{b,n,t}$ are equal to 0.
- *Case 2:* When two different ESs (e.g., b' and $i \in \mathcal{B}$) are simultaneously selected for a task request, the variables $\alpha_{b,n,t,b'}$ and $\alpha_{b,n,t,i}$ are all equal to 1. The task workloads $x_{b,n,t,b'} d_n \rho_n$ and $x_{b,n,t,i} d_n \rho_n$ are allocated to the ESs b' and i for processing, respectively. At this time, the problem (12) is expressed to

$$\min_{\boldsymbol{X}_{b,n,t}} \quad \max\{T^o_{b,n,t,b'}, T^o_{b,n,t,i}\} \tag{13}$$

$$\text{s.t.} \quad x_{b,n,t,b'} + x_{b,n,t,i} = 1. \tag{13(a)}$$

Obviously, the optimal allocation fraction is existed when $T^o_{b,n,t,b'} = T^o_{b,n,t,i}$. Then, according to equation (3), the $T^o_{b,n,t,b'} = T^o_{b,n,t,i}$ is represented as

$$x_{b,n,t,b'} \cdot T_{b,n,t,b'} + T^w_{b,n,t,b'} = x_{b,n,t,i} \cdot T_{b,n,t,i} + T^w_{b,n,t,i}, \tag{14}$$

where $T_{b,n,t,b'} = \frac{d_n}{v_{b,b',t}} + \frac{\rho_n \cdot d_n}{f_{b'}}$ and $T_{b,n,t,i} = \frac{d_n}{v_{b,i,t}} + \frac{\rho_n \cdot d_n}{f_i}$. Afterwards, by combining with Eqs. (13(a)) and (14), the $x_{b,n,t,b'}$ and $x_{b,n,t,i}$ are solved as

$$x_{b,n,t,b'} = \frac{T_{b,n,t,i} + T^w_{b,n,t,i} - T^w_{b,n,t,b'}}{T_{b,n,t,b'} + T_{b,n,t,i}}, \tag{15}$$

and

$$x_{b,n,t,i} = \frac{T_{b,n,t,b'} + T^w_{b,n,t,b'} - T^w_{b,n,t,i}}{T_{b,n,t,b'} + T_{b,n,t,i}},, \tag{16}$$

respectively. Here, all the variables in the right of the Eqs. (15) and (16) are known in the system environment. Meanwhile, other elements (i.e., except $x_{b,n,t,b'}$ and $x_{b,n,t,i}$) of $\boldsymbol{X}_{b,n,t}$ are equal to 0.

- *Case 3:* When three different ESs (e.g., b', i, and $j \in \mathcal{B}$) are selected for a task offloading, the variables $\alpha_{b,n,t,b'}$, $\alpha_{b,n,t,i}$, and $\alpha_{b,n,t,j}$ are all equal to 1. At this time, the problem (12) is expressed to

$$\min_{\boldsymbol{X}_{b,n,t}} \max\{T^o_{b,n,t,b'}, T^o_{b,n,t,i}, T^o_{b,n,t,j}\} \tag{17}$$

$$\text{s.t.} \quad x_{b,n,t,b'} + x_{b,n,t,i} + x_{b,n,t,j} = 1, \tag{17(a)}$$

where the optimal allocation fraction is existed when $T^o_{b,n,t,b'} = T^o_{b,n,t,i} = T^o_{b,n,t,j}$. Then, the problem (17) is equal to solve the variables $\alpha_{b,n,t,b'}$, $\alpha_{b,n,t,i}$, and $\alpha_{b,n,t,j}$ by bellowing equations

$$\begin{cases} T^o_{b,n,t,b'} = T^o_{b,n,t,i} \\ T^o_{b,n,t,i} = T^o_{b,n,t,j} \\ x_{b,n,t,b'} + x_{b,n,t,i} + x_{b,n,t,j} = 1. \end{cases} \tag{18}$$

Accordingly, similar to *Case 2* by combining with the equation (3), the $x_{b,n,t,b'}$, $x_{b,n,t,i}$, and $x_{b,n,t,j}$ also can be solved based on the above three equations.

- *Other cases:* When 4,5, .., or B different ESs are respectively selected for a task offloading, we have corresponding 4,5, ..., or B equations. Then, similar to the *Case 3*, the optimal allocation fraction $\boldsymbol{X}_{b,n,t}$ can be solved based on these corresponding equations.

4.4 DRL-SMO Algorithm

Based on the above two solutions, the DRL-SMO algorithm is implemented as shown in Algorithm 1, which mainly includes four steps as follows:

Algorithm 1: The DRL-SMO Algorithm

Input: The task set $\{\mathcal{N}_{b,t}\}$;
Output: The offloading decisions $\boldsymbol{\alpha}$ and $\boldsymbol{X}$;
1 Initial variables $actStep = 0$, $trainStep = 0$;
2 **for** *each time slot* $t \in \mathcal{T}$ **do**
3 **for** *each* $b \in \mathcal{B}$ *and* $n \in \mathcal{N}_{b,t}$ **do**
4 Observe the system state $s_{b,n,t}$ and make the action $a_{b,n,t}$ by the improved DQN model;
5 Based on the achieved $a_{b,n,t}$, solve the decision $\boldsymbol{X}_{b,n,t}$ of task allocation sub-problem (12) by the resource-aware closed-form solution;
6 Calculate the reward $r_{b,n,t}$ and observe the next system state $s_{b,n,t+1}$;
7 Store the transition tuple $(s_{b,n,t}, a_{b,n,t}, r_{b,n,t}, s_{b,n,t+1})$ to experience pool;
8 $actStep$++;
9 **if** $actStep > 200$ *and* $actStep \ \% \ 10 == 0$ **then**
10 Extract a batch sample from the experience pool to train the EQN of trainer;
11 $trainStep$++;
12 **if** $trainStep \ \% \ 200 == 0$ **then**
13 Update the TQN parameters by copying the EQN parameters of trainer;
14 Update the EQN parameters of scheduler by copying TQN parameters;

Step1: For each time slot $t \in \mathcal{T}$ and task $n \in \mathcal{N}_{b,t}$ arrived to each BS $b \in \mathcal{B}$, by observing system state $s_{m,n,t}$, the decision variable $\boldsymbol{\alpha}_{b,n,t}$ (i.e., $a_{b,n,t}$) of simultaneous multi-ES selection are achieved by the proposed improved DQN model (line 4). Then, based on the achieved $\boldsymbol{\alpha}_{b,n,t}$, the decision variable $\boldsymbol{X}_{b,n,t}$ of task allocation is achieved by the proposed resource-aware closed-form solution (line 5).

Step2: After the finish of task processing, the reward $r_{b,n,t}$ is calculated and the next system state $s_{m,n,t+1}$ is observed from environment (line 6). Accordingly, the transition tuple $(s_{b,n,t}, a_{b,n,t}, r_{b,n,t}, s_{b,n,t+1})$ is sent and stored in the experience pool at the trainer (line 7).

Step3: If the EQN action at the scheduler is executed by more than 200 times, a bath sample from the experience pool is first extracted to train the EQN by the trainer with every 10 times of the action (lines 9 and 10). Here, 200 and 10 are set by user according to real situation [27].

Step4: After every 200 (set by user) times of EQN training (line 12), the TQN parameters are updated by copying the EQN parameters of trainer (line 13). Meanwhile, the EQN parameters of scheduler are synchronously updated by copying the TQN parameters (line 14).

Based on the above steps, the near-optimal offloading policies $\boldsymbol{a}$ and $\boldsymbol{X}$ can be gradually learned.

Theorem 2. *The computation complexity of DRL-SMO is linear with the number of tasks offloading.*

Proof. For each time slot $t \in \mathcal{T}$ and each task $n \in \mathcal{N}_{b,t}$, the proposed DRL-SMO algorithm mainly includes two parts: multi-ES selection and task allocation. In the multi-ES selection part, the time required to multi-ES selection is determined by the EQN inference of scheduler. Since the EQN is previously trained, the computation complexity of multi-ES selection is $O(1)$. In the task allocation part, the allocation decision is directly calculated by Eqs. (e.g., (14 and (15)). Hence, the computation complexity of task allocation part is also $O(1)$. Thus, the overall computation complexity of DRL-SMO algorithm still is $O(1)$ for a task request at time slot t. Furthermore, when $N_{b,t} * B$ tasks are offloaded at time slot t, the computation complexity of DRL-SMO is $O(N_{n,t} * B)$. Therefore, the computation complexity of DRL-SMO is linear with the number of tasks offloading.

5 Performance Evaluation

This section first presents the simulation setting and then shows the performance results of our approach in comprehensive experiments with insightful analysis.

5.1 Setting

We set the parameters in our DRL-SMO algorithm by default as follows, unless specified.

Environment Parameters. Similar to [18,27,31], we consider a collaborative MEC scenario with $B = 5$ BSs. The number $\mathcal{N}_{b,t}$ of tasks arrived to each BS and the task arrival probability P are set in the range [10, 50] and 0.3, respectively [27]. The task size d_n and computation density ρ_n are set in the range [10, 40] Mbits and [100, 300] CPU cycle/bit, respectively [31]. The transmission rate $v_{b,b',t}$ is set in the range [400, 500] Mbits/s [31]. The computation capacity $f_{b'}$ is set in the range [10, 50] GHz [18]. The total number T of time slots is set to 100. The length Δ of time slot t is set to 0.1 s [27]. The processing deadline τ_n is set to 10 time slots (i.e., 1 s) based on our experimental results in Fig. 5, which is consistent with [27].

Training Parameters. For the purpose of easy and convenient implementation that is similar to [27], the DQN model in our SMCoEdge approach is set with a fully connected CNN consisting of one input layer, one hidden layer, and one output layer. The hidden layer is set to 20 neurons. We consider each episode includes T time slots and the number of episode is set to 40. The discount factor of our model is set to 0.9. The ϵ-greedy probability is gradually increasing from 0 to 0.99 with step 0.001. The storage size of experience pool at each ES is set to 1024. The learning rate (denoted as lr), batch size, optimizer, and k parameters are set to 0.001, 32, Adam, and 3, respectively.

To evaluate the performance of our proposed approach, we adopt two metrics: the average make-span [27] and the offloading failure rate [33].

5.2 Baselines

Four competitive baselines are used as a performance comparison with our proposed approach, as listed below.

- *Random Single-ES Offloading*(RSO). RSO method is to randomly select an ES to process for each task offloading. This method is a classic offloading solution that is usually used as comparison in MEC study.
- *Nearest Single-ES Offloading* (NSO). NSO method is to select the nearest ES to process for each task offloading, which is also a classic offloading solution to be used as comparison in MEC study.
- *DRL based Single-ES Offloading* (DRLSO) [27]. DRLSO method is to select an ES to process for each task offloading based on DRL, which achieves the state-of-the-art performance in DRL-based MEC.
- *Fine-grained Multi-ES Offloading* (FMO) [18]. FMO method is first to divide a task into several subtasks offline, and then sequentially select an ES to process for each subtask request. This method has the state-of-the-art performance in collaborative MEC. For fair comparison, the number of partitioned subtasks in the FMO method is set to the same number of selected ESs in our approach.

5.3 Results

The training process of DRL-SMO is presented in Fig. 3. The simulation results of SMCoEdge and baselines are presented in Figs. 4-6 in terms of average makespan and offloading failure rate.

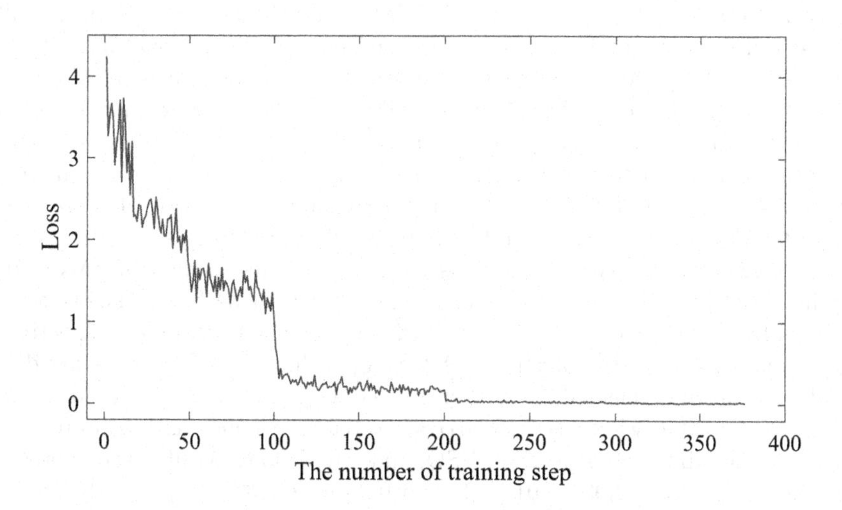

Fig. 3. Training process of the proposed DRL-SMO algorithm.

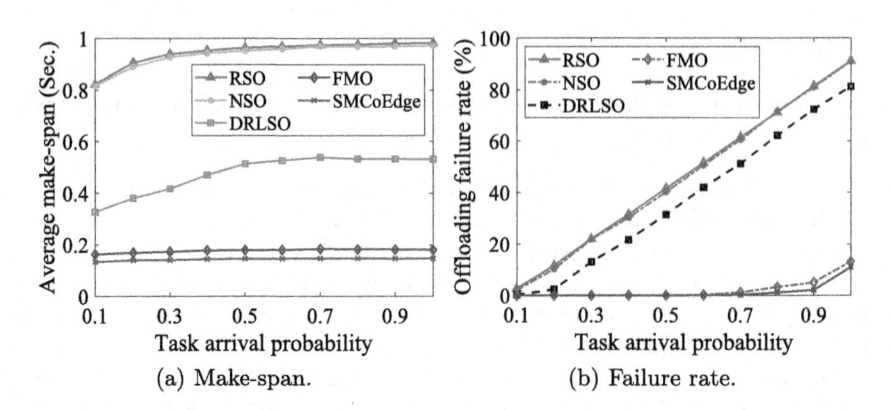

(a) Make-span.
(b) Failure rate.

Fig. 4. Performance comparisons under varying task arrival probability P.

Overall, the loss of DRL-SMO drops gradually and goes to stability with training step goes from Fig. 3. From the Fig. 4–6, the proposed SMCoEdge approach has great improvement on both make-span and failure rate over baselines.

In particular, Fig. 3 illustrates that the loss of the DRL-SMO algorithm initially drops erratically with the increase of training steps from 1 to 90. This behavior can be attributed to the fact that the DRL method is gradually learning to find the near-optimal solution. Subsequently, the loss of DRL-SMO decreases rapidly beyond 90 training steps and stabilizes at 200 training steps. These results demonstrate that the DRL-SMO effectively converges to a stable solution.

The results presented in Fig. 4 are obtained by varying the task arrival probability P from 0 to 1. As shown in Fig. 4(a), when the task arrival probability is increased from 0.1 to 1, the average make-spans of RSO, NSO, and DRLSO methods initially increase and then stabilize. However, the make-spans of FMO and SMCoEdge approaches slightly increase with the increase in task arrival probability from 0.1 to 1. Notably, the make-span of SMCoEdge is consistently the lowest compared to the other four methods. Specifically, with the increase in task arrival probability from 0.1 to 1, the average make-spans of RSO, NSO, DRLSO, and FMO increase from 0.82 to 0.98 s, 0.81 to 0.97 s, 0.32 to 0.53 s, and 0.16 to 0.18 s, respectively. In contrast, the average make-span of SMCoEdge only increases from 0.13 to 0.15 s, which significantly outperforms the other four methods by an average of 84.90%, 84.73%, 70.53%, and **18.93%**, respectively. This is mainly due to the fact that SMCoEdge achieves equal processing delay of subtasks at the selected ESs, thereby reducing the make-span. Moreover, Fig. 4(b) shows that when the task arrival probability is increased from 0.1 to 1, the offloading failure rates of RSO, NSO, and DRLSO are significantly increased from 2.66 to 91.16%, 2.03 to 91.03%, and 0.23 to 81.23%, respectively. In contrast, the offloading failure rates of FMO and SMCoEdge only increase from 0 to 13.23% and 0 to 11.03%, respectively. Furthermore, SMCoEdge has the lowest failure rate compared to the other four methods.

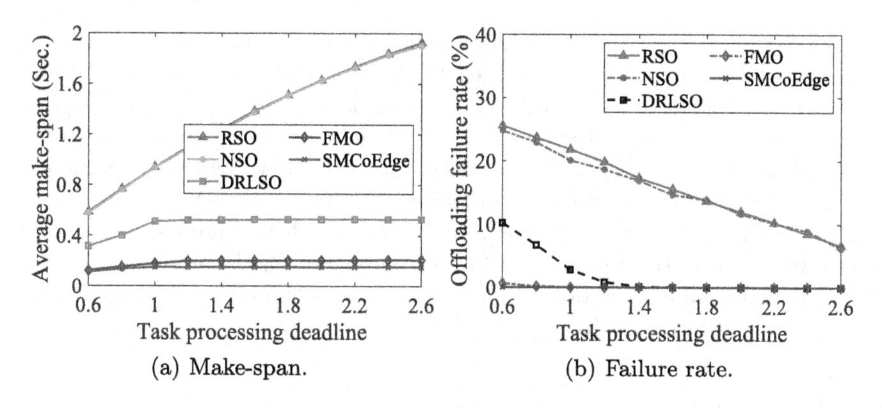

(a) Make-span. (b) Failure rate.

Fig. 5. Performance comparisons under varying task processing deadline τ_n.

The results presented in Fig. 5 were obtained by varying the task processing deadline τ_n from 0.6 to 2.6. Figure 5(a) shows that the average make-span of SMCoEdge is consistently lower than that of the other four methods as the task deadline increases. Furthermore, when the task deadline is set to 1, the make-span of SMCoEdge stabilizes. This is because, when the deadline is large, almost all tasks can be successfully processed. Thus, the make-span is only slightly affected when the deadline is greater than 1. Consequently, τ_n is set to 1 s by default. Secondly, Fig. 5(b) indicates that as the task deadline increases from 0.1 to 1, the offloading failure rate of SMCoEdge remains at 0, which is lower than or equal to that of the other four methods.

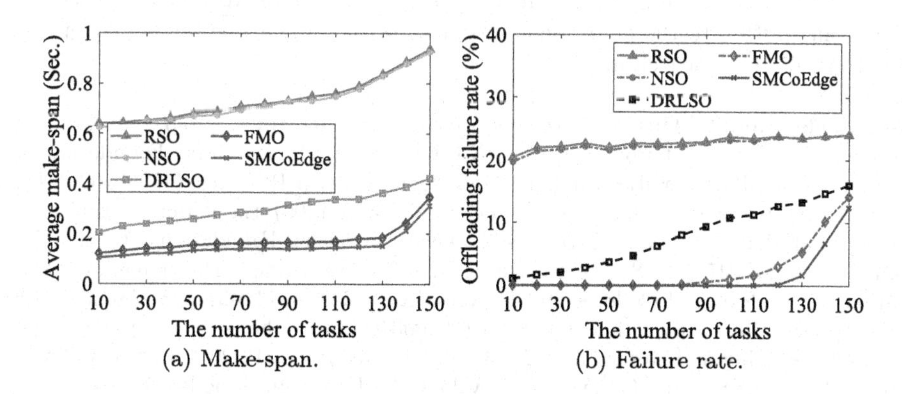

(a) Make-span. (b) Failure rate.

Fig. 6. Performance comparisons under varying the number $\mathcal{N}_{b,t}$ of tasks.

The results presented in Fig. 6 are obtained by varying the number $N_{b,t}$ of tasks arriving at each BS from 10 to 150. As shown in Fig. 6(a), the average make-spans of the five methods first increase slowly with the number of tasks from 10 to 130 and then rapidly increase. However, the make-span of SMCoEdge remains significantly lower than the other four methods as the number of tasks increases from 10 to 150. This can be attributed to the fact that as the number of tasks increases, the workloads at the ESs also potentially increase. Once the workloads exceed a certain threshold, the make-span will increase rapidly due to the limited computation capacity of ESs. Furthermore, Fig. 6(b) shows that as the number of tasks increases from 10 to 150, the offloading failure rate of SMCoEdge is lower than or equal to that of the other four methods. These findings demonstrate that SMCoEdge outperforms the state-of-the-art methods in reducing make-span and maintaining low offloading failure rates, thereby improving the QoS for computation-intensive IoT applications.

6 Conclusions

This paper proposes a novel approach called SMCoEdge that simultaneously selects multiple suitable ESs for each task request and allocates tasks online to the selected ESs to achieve minimal make-span of task offloading, thereby improving QoS for computation-intensive IoT applications. We formulate the problem as an MINLP problem with the objective of minimizing make-span of task offloading and prove its NP-hardness. To reduce computational complexity, we decompose the problem into two sub-problems of multi-ES selection and task allocation and propose a DRL-SMO algorithm to effectively solve it. Additionally, we present a theoretical analysis to show that the computation complexity of DRL-SMO is linear with the number of tasks offloading. Finally, extensive simulation experiments demonstrate that the proposed SMCoEdge approach significantly outperforms state-of-the-art approaches by reducing make-span by 18.93% while maintaining a low offloading failure rate. This work provides a valuable contribution to the field of collaborative MEC and can be further extended to other related research areas.

Acknowledgment. This work is supported in part by grants from the National Natural Science Foundation of China (No. 62272117) and the Joint Foundation of Guangzhou and Universities on Basic and Applied Basic Research (202201020126), the National Key R&D Program of China (2022YFE0201400), the Beijing Natural Science Foundation (No. 4232028), the National Natural Science Foundation of China (No. 62172046, 62372047), the Special Project of Guangdong Provincial Department of Education in Key Fields of Colleges and Universities (2021ZDZX1063), the Zhuhai Basic and Applied Basic Research Foundation (2220004002619), the Joint Project of Production, Teaching and Research of Zhuhai (2220004002686, ZH22017001210133PWC, and 2220004002686), the Guangdong Key Lab of AI and Multi-modal Data Processing, BNU-HKBU United International College (UIC), Zhuhai (No. 2020KSYS007), the UIC General project (No. R0200005-22), the UIC Start-up Research Fund (No. R72021202), the Science and Technology Projects of Social Development in Zhuhai (No. 2320004000213), the Guangdong Basic and Applied Basic Research Foundation (No. 2022A1515011583 and No. 2023A1515011562), the One-off Tier 2 Start-up Grant (2020/2021) of Hong Kong Baptist University (Ref. RC-OFSGT2/20-21/COMM/002), the Startup Grant (Tier 1) for New Academics AY2020/21 of Hong Kong Baptist

University, National Natural Science Foundation of China (No. 62202402), the Germany/Hong Kong Joint Research Scheme sponsored by the Research Grants Council of Hong Kong and the German Academic Exchange Service of Germany (No. G-HKBU203/22), and the Hong Kong RGC Early Career Scheme (No. 22202423)

References

1. Siriwardhana, Y., Porambage, P., Liyanage, M., Ylianttila, M.: A survey on mobile augmented reality with 5G mobile edge computing: architectures, applications, and technical aspects. IEEE Commun. Surv. Tutorials **23**(2), 1160–1192 (2021)
2. Dai, P., Hu, K., Wu, X., Xing, H., Yu, Z.: Asynchronous deep reinforcement learning for data-driven task offloading in MEC-empowered vehicular networks. In: 2021-IEEE Conference on Computer Communications (INFOCOM), pp. 1–10, IEEE, Virtual Conference (2021)
3. Wang, T., Lu, Y., Wang, J., Dai, H.-N., Zheng, X., Jia, W.: EIHDP: edge-intelligent hierarchical dynamic pricing based on cloud-edge-client collaboration for IoT systems. IEEE Trans. Comput. **70**(8), 1285–1298 (2021)
4. Wan, Z., Dong, X., Deng, C.: Deep learning with enhanced convergence and its application in MEC task offloading. In: 21rd International Conference on Algorithms and Architectures for Parallel Processing (ICA3PP), pp. 361–375. Springer, Xiamen, China (2021). https://doi.org/10.1007/978-3-030-95388-1_24
5. Zhang, Y., Liu, T., Zhu, Y., Yang, Y.: A deep reinforcement learning approach for online computation offloading in mobile edge computing. In: 2020 IEEE/ACM 28th International Symposium on Quality of Service (IWQoS), pp. 1–10. IEEE, Hangzhou, China (2020)
6. Li, X., Zhang, X., Huang, T.: Asynchronous online service placement and task offloading for mobile edge computing. In: 18th Annual IEEE International Conference on Sensing, Communication, and Networking (SECON), pp. 1-9. IEEE, Virtual Conference (2021)
7. Wang, X., Ye, J., Lui, J. C.S.: Joint D2D collaboration and task offloading for edge computing: A mean field graph approach. In: 2021 IEEE/ACM 29th International Symposium on Quality of Service (IWQoS), pp. 1-10. IEEE, Tokyo, Japan (2021)
8. Chen, S., Tang, B., Yang, Q., Liu, Y.: Operator placement for IoT data streaming applications in edge computing environment. In: 22rd International Conference on Algorithms and Architectures for Parallel Processing (ICA3PP), pp. 605–619. Springer, Copenhagen, Denmark (2022). https://doi.org/10.1007/978-3-031-22677-9_32
9. Tran, T., Hajisami, A., Pandey, P., Pompili, D.: Collaborative mobile edge computing in 5G networks: new paradigms, scenarios, and challenges. IEEE Commun. Mag. **55**(4), 54–61 (2017)
10. Shi, W., Cao, J., Zhang, Q., Li, Y., Xu, L.: Edge computing: vision and challenges. IEEE Internet Things J. **3**(5), 637–646 (2016)
11. Xu, J., Chen, L., Zhou, P.: Joint service caching and task offloading for mobile edge computing in dense networks. In: 2018-IEEE Conference on Computer Communications (INFOCOM), pp. 207–215. IEEE, Honolulu, HI, USA (2018)
12. Poularakis, K., Llorca, J., Tulino, A. M., Taylor, I., Tassiulas, L.: Joint service placement and request routing in multi-cell mobile edge computing networks. In: 2019-IEEE Conference on Computer Communications (INFOCOM), pp. 10–18. IEEE, Paris, France (2019)

13. Zeng, L., Chen, X., Zhou, Z., Yang, L., Zhang, J.: CoEdge: cooperative DNN inference with adaptive workload partitioning over heterogeneous edge devices. IEEE/ACM Trans. Netw. **29**(2), 595–608 (2021)
14. Han, Y., Shen, S., Wang, X., Wang, S., Leung, V.r C.M.: Tailored learning-based scheduling for kubernetes-oriented edge-cloud system. In: 2021-IEEE Conference on Computer Communications (INFOCOM), pp. 1–10. IEEE, Virtual Conference (2021)
15. Ren, J., Yu, G., He, Y., Li, Geoffrey Y.: Collaborative cloud and edge computing for latency minimization. IEEE Trans. Vehicular Technol. **68**(5), 5031–5044 (2019)
16. Hao, Z., Yi, S., Li, Q.: Nomad: an efficient consensus approach for latency-sensitive edge-cloud applications. In: 2019-IEEE Conference on Computer Communications (INFOCOM), pp. 2539–2547. IEEE, Paris, France (2019)
17. Han, R., Wen, S., Liu, C. H., Yuan, Y., Wang, G., Chen, L. Y.: EdgeTuner: fast scheduling algorithm tuning for dynamic edge-cloud workloads and resources. In: 2022-IEEE Conference on Computer Communications (INFOCOM), pp. 880–889, IEEE, Virtual Conference (2022)
18. Chu, W., Yu, P., Yu, Z., Lui, J.C.S., Lin, Y.: Online optimal service selection, resource allocation and task offloading for multi-access edge computing: a utility-based approach. IEEE Trans. Mobile Compu. (Early access) (2023). https://doi.org/10.1109/TMC.2022.3152493
19. Eshraghi, N. Liang, B.: Joint offloading decision and resource allocation with uncertain task computing requirement. In: 2019-IEEE Conference on Computer Communications (INFOCOM), pp. 1414–1422. IEEE, Paris, France (2019)
20. Qin, P., Fu, Y., Tang, G., Zhao, X., Geng, S.: Learning based energy efficient task offloading for vehicular collaborative edge computing. IEEE Trans. Veh. Technol. **71**(8), 8398–8413 (2022)
21. Gao, M., Shen, R., Shi, L., Qi, W., Li, J., Li, Y.: Task partitioning and offloading in DNN-task enabled mobile edge computing networks. IEEE Trans. Mob. Comput. **22**(4), 2435–2445 (2023)
22. Ma, X., Zhou, A., Zhang, S., Wang, S.: Cooperative service caching and workload scheduling in mobile edge computing. In: 2020-IEEE Conference on Computer Communications (INFOCOM), pp. 2076–2085. IEEE, Toronto, ON, Canada (2020)
23. Wang, Y., Sheng, M., Wang, X., Wang, L., Li, J.: Mobile-edge computing: partial computation offloading using dynamic voltage scaling. IEEE Trans. Commun. **64**(1), 4268–4282 (2016)
24. Yu, S., Chen, X., Zhou, Z., Gong, X., Wu, D.: When deep reinforcement learning meets federated learning: intelligent multitimescale resource management for multiaccess edge computing in 5G ultradense network. IEEE Internet Things J. **8**(4), 2238–2251 (2021)
25. Zhou, R., Wu, X., Tan, H., Zhang, R.: Two time-scale joint service caching and task offloading for UAV-assisted mobile edge computing. In: 2022-IEEE Conference on Computer Communications (INFOCOM), pp. 1189–1198. IEEE, Virtual Conference (2022)
26. Mnih, V., et al.: Others: human-level control through deep reinforcement learning. Nature **518**(7540), 529–533 (2015)
27. Tang, M., Wong, V.W.S.: Deep reinforcement learning for task offloading in mobile edge computing systems. IEEE Trans. Mob. Comput. **21**(6), 1985–1997 (2022)
28. Cao, J.,d Yang, L., Cao, J.: Revisiting computation partitioning in future 5G-based edge computing environments. IEEE Internet of Things J. **6**(2), 2427–2438 (2018)
29. Liu, Y., et al.: Dependency-aware task scheduling in vehicular edge computing. IEEE Internet Things J. **7**(6), 4961–4971 (2020)

30. Sahni, Y., Cao, J., Yang, L., Ji, Y.: Multi-hop multi-task partial computation offloading in collaborative edge computing. IEEE Trans. Parallel Distrib. Syst. **32**(5), 1133–1145 (2020)
31. Fan, W., et al.: Collaborative service placement, task scheduling, and resource allocation for task offloading with edge-cloud cooperation. IEEE Trans. Mobile Comput. (Early access) (2023). https://doi.org/10.1109/TMC.2022.3219261
32. Wang, X., Ye, J., Lui, J.C.S: Decentralized task offloading in edge computing: a multi-user multi-armed bandit approach. In: IEEE INFOCOM 2022-IEEE Conference on Computer Communications, pp. 1199–1208,(2022)
33. Tan, J., Khalili, R., Karl, H., Hecker, A.: Multi-agent distributed reinforcement learning for making decentralized offloading decisions. In: 2022-IEEE Conference on Computer Communications (INFOCOM), pp. 2098–2107. IEEE, Virtual Conference (2022)

Image Inpainting Forensics Algorithm Based on Dual-Domain Encoder-Decoder Network

Dengyong Zhang[1,2], En Tan[1,2], Feng Li[1,2], Shuai Liu[1,2], Jing Wang[1,2], and Jinbin Hu[1,2](✉)

[1] Hunan Provincial Key Laboratory of Intelligent Processing of Big Data on Transportation, School of Computer and Communication Engineering, Changsha University of Science and Technology, Changsha 410004, China
[2] School of Computer and Communication Engineering, Changsha University of Science and Technology, Changsha 410004, China
{zhdy,lif,znwj_cs,jinbinhu}@csust.edu.cn,
{21208051616,csustliushuai}@stu.csust.edu.cn

Abstract. Image inpainting can fill regions of images with plausible content and can also be used to remove specific objects and leave only faint traces in inpainted images, which pose serious security issues. At present, there are relatively few forensic works on image inpainting. Moreover, there is a problem of poor generalization. Therefore, This paper proposes a dual-domain encoder-decoder network (DDEDNet) based on different input types, which is a two-branch network. The first branch is a spatial domain-based encoder network (S-Encoder) used to capture the tampering traces left by image inpainting in the spatial domain; the second branch is an encoder network based on the frequency domain (F-Encoder), which is used to mine the subtle artifacts left in the frequency domain. Then a cross-modal attention fusion module (CMAF) is used to fuse the features of the two encoder networks to obtain rich fused features. Finally, attention-gated (AG) skip connections are utilized to improve localization performance by properly incorporating multi-scale features in the decoder. Experimental results show that in the face of data sets with both deep inpainting and traditional schemes, DDEDNet can locate the inpainting area more accurately, effectively resist JPEG compression and Gaussian noise attacks, and performs better generalization.

Keywords: Image inpainting forensics · Encoder-Decoder network · Frequency domain · Cross-modal attention · Attention gate

This work is supported by the National Natural Science Foundation of China (62172059, 62072055, 62102046, 62072056), the Natural Science Foundation of Hunan Province (2022JJ50318, 2022JJ30621, 2023JJ50331, 2022JJ30618, 2020JJ2029), the Hunan Provincial Key Research and Development Program (2022GK2019), the Scientific Research Fund of Hunan Provincial Education Department (22A0200, 22B0300).

Z. Tari et al. (Eds.): ICA3PP 2023, LNCS 14491, pp. 92–111, 2024.
https://doi.org/10.1007/978-981-97-0808-6_6

1 Introduction

With the rapid development of internet technology and the widespread use of more and more image editing and processing software such as Photoshop, it has become easier to edit and tamper with images. At the same time, the tampered image is not easily recognized by human eyes, and it is difficult for ordinary people to detect the traces of tampering. Therefore, more attention has been paid to the authenticity and reliability of digital images, and image information security has become an urgent problem to be solved. As an important research topic in the field of forensics, image inpainting forensics has attracted extensive attention from researchers[21,25,49].

Image inpainting is the process of recovering damaged or missing regions of a given image based on information from undamaged regions. Traditional image inpainting can be divided into diffusion-based methods [5,19] and exemplar-based methods [2,7]. However, they are only suitable for filling images with similar textures and cannot deeply understand context information. Traditional methods did not work well for complex images. For this reason, researchers have proposed many inpainting methods based on deep learning in recent years [16,27, 34,53]. Deep inpainting methods can not only infer image structure and produce finer details but also create novel objects. Figure 1 (b) shows an inpainted image made from the real image in Fig. 1 (a) by the recent deep inpainting method [53], and Fig. 1 (c) is the inpainting mask.

Existing forensics methods for image inpainting can be broadly classified into two categories: traditional-based methods and deep learning-based methods. Traditional methods rely on hand-crafted features, such as those based on the similarity between image patches [22,48]. Such methods have many disadvantages, such as high computational cost and high false positive rate. Currently, deep learning-based image inpainting forensics methods have been developed [20,28,46]. Li and Huang [21] conducted a pioneering study on deep inpainting detection, designing a high-pass pre-filtering module to obtain image residuals to enhance traces. But its effectiveness is limited to the inpainting methods trained on its network, performing poorly on other unseen inpainting methods. Li et al. [20] found that the noise patterns existing in the original and inpainting regions are different and proposed a method to generate a common training set so that the model can detect other methods. Wu et al. [46] enhanced inpainting traces by combining different special layers and used neural architecture search (NAS) to design an optimal model to extract inpainting features, which performed better on 10 unseen inpainting methods. However, the above methods use a single method of enhancing the traces or combining different special layers to achieve the effect of enhancing the repair traces, all of which use the features left by the inpainting operation in the spatial domain without considering the subtle artifacts in the frequency domain.

To address the above issues, we propose a novel end-to-end image inpainting forensic network to localize inpainted regions with pixel precision. The proposed network has two branches, one branch is the S-Encoder, which is used to capture the tampering traces left by image inpainting in the spatial domain, and in the

(a) Original image (b) Inpainted image (c) Inpainting mask

Fig. 1. (a) The original image; (b) The inpainted image by the deep inpainting method in [53]; (c) The mask that defines the inpainting region.

S-Encoder, an enhancement block is designed, which aims to combine some special layers to enhance the inpainting traces; another branch is the F-Encoder, which is used to mine the subtle artifacts left in the frequency domain for feature supplementation. Then, the features of the two branches are self-attentionally fused, and then the attention-gated decoder is used to gradually upsample to the original image size to achieve pixel-level prediction. The main contributions of this work can be summarized as follows:

1) We propose a dual-domain encoder-decoder network DDEDNet. The spatial domain encoder network S-Encoder is used to capture the tampering traces left by image inpainting in the spatial domain, while F-Encoder is mainly used to mine the subtle artifacts left by inpainting in the frequency domain. With multiple different types of inputs, it is easier to discover more general features of the inpainted image.
2) We design a cross-modal attention fusion module to enhance inpainting traces by exploiting noise features in the spatial domain and inpainting artifacts in the frequency domain, respectively. Using skip connections based on the attention gate, we make full use of contextual information and improve localization performance.
3) DDEDNet performs well on the test sets of 5 deep inpainting schemes and 5 traditional inpainting schemes, and the average AUC and F1 score are better than the state-of-the-art inpainting forensics methods.

2 Related Works

2.1 Image Inpainting Forensics

Many inpainting forensics methods have been proposed to counter the malicious use of inpainting operations. Wu et al. [48] proposed a blind detection method based on zero-connectivity feature and fuzzy set. Liang et al. [25] designed an efficient forgery detection algorithm that integrates center pixel mapping, maximum zero-connectivity component labeling, and fragment stitching detection.

Furthermore, Li et al. [22] proposed a discriminative feature based on the local variance varying within and across channels to detect diffusion-based inpainted regions. However, the above traditional methods are computationally intensive, time-consuming, and have a high false alarm rate. With the rise of deep learning, various network structures and optimization algorithms [12,23,24,29,40,43,50] emerged in endlessly. Then some image inpainting forensics methods based on deep learning have emerged. In order to detect complex forgery combinations, Wu et al. [49] proposed a more general forgery localization network MT-Net, which firstly extracts image tampering trace features, and then recognizes them by evaluating the degree of difference between local features and their reference features abnormal area. Li et al. [20] proposed a novel data generation approach to generate a general training dataset that simulates the noise differences in real and inpainting image content to train a general detector. Wu et al. [46] also proposed a deep learning image inpainting detection network IID-Net. The network consists of an enhancement block, an extraction block, and a decision block. The extraction block is automatically designed by the NAS algorithm, which can accurately detect and locate inpainting regions. Liu et al. [28] presented a progressive spatial channel correlation network PSCC-Net, which exploits the properties of dense cross-connections at different scales to detect and localize image forgery in a coarse-to-fine manner.

Although there have been some research results in image inpainting forensics, however, there are relatively few existing studies on the generalization of inpainting forensics, and there is a lack of methods to mine subtle inpainting artifacts from the frequency domain. In this paper, by exploring different feature information from the spatial domain and then extracting the clues left by different inpainting from the frequency domain, a better performance has been achieved.

2.2 Attention Mechanism

Attention mechanisms aim to focus on salient features in computer vision tasks [39,41,42,44,57]. Hu et al. [14] proposed a squeeze-and-excite SE block to flexibly update channel features. Roy et al. [36] combined channel and spatial attention units in parallel, called the Parallel Spatial and Channel Squeeze and Excitation scSE module. Woo et al. [45] designed a convolutional block attention module, CBAM, which connects channel and spatial attention in a lightweight manner. Vaswani et al. [38] propose a novel self-attention mechanism that focuses on the correlations between different parts in the input features. Although attention mechanisms have been applied to image inpainting forensics, some previous studies have adopted traditional channel and spatial attention. In this paper, a fusion module based on self-attention is used to fuse the features of the S-Encoder and the F-Encoder, and by interacting the features of different modalities, more relevant fusion features are obtained. At the same time, skip connections with attention gate [33] are used in the decoder to suppress irrelevant regions in the feature image of the encoder, making full use of network context information.

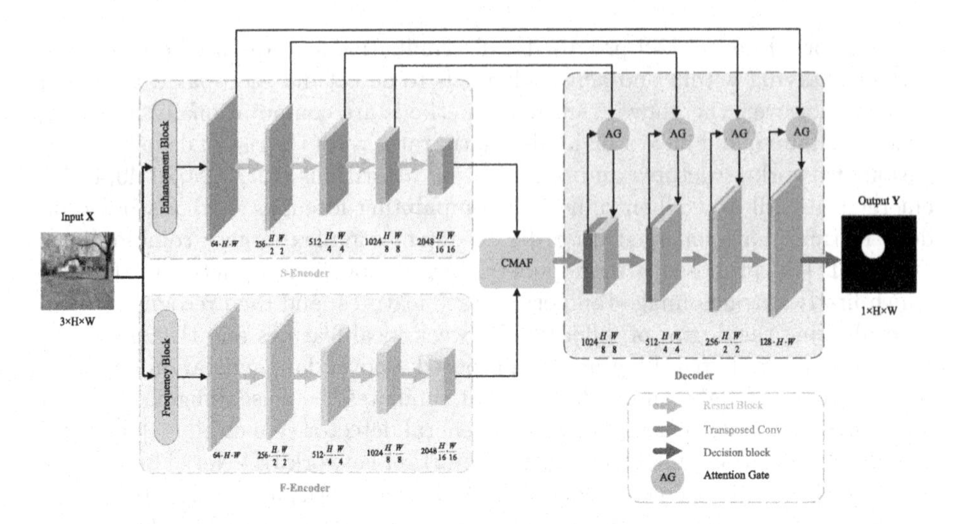

Fig. 2. Overview of DDEDNet.The proposed architecture consists of two encoders and a decoder: S-Encoder for capturing the tampering traces in the spatial domain; F-Encoder for extracting the subtle artifacts left in the frequency domain and CMAF for collaborative feature interaction; the pixel-level prediction is performed through the decoder.

3 Proposed Method

In this section, the dual-domain encoder-decoder network is presented. As shown in Fig. 2, DDEDNet has two encoders, one of which is an S-Encoder, the input image is first enhanced by a combination of different special layers and then sent to the backbone to extract features; the other is F-Encoder, which extracts the inpainting trace features in the frequency domain through an adaptive combination filter. The backbone used in the encoder is Resnet50 [11]. Then a cross-modal fusion module is used to fuse the features of the two encoders, and finally the predicted image is obtained through an attention-gated decoder. We will elaborate the proposed network as follows.

3.1 S-Encoder

In general, CNNs tend to learn features related to image content [30]. Compared with the features introduced by image inpainting, they tend to focus on some statistical features, such as the distribution features of DCT coefficient correlations. In the existing forensics algorithms, there are already many methods to enhance tampering traces, among which the more common one is the Steganalysis Rich Model (SRM) [59], which is widely used in the field of image tampering forensics. The Bayar Constrained Convolution (Bayar) proposed by Bayar et al. [3] has a positive effect on obtaining tampering traces. The spatial domain high-pass filtering (PF) module proposed by Li et al. [21] can find the difference between

$$\frac{1}{4}\begin{bmatrix}0&0&0&0&0\\0&-1&2&-1&0\\0&2&-4&2&0\\0&-1&2&-1&0\\0&0&0&0&0\end{bmatrix}\quad \frac{1}{12}\begin{bmatrix}-1&2&-2&2&-1\\2&-6&8&-6&2\\-2&8&-12&8&-2\\2&-6&8&-6&2\\-1&2&-2&2&-1\end{bmatrix}\quad \frac{1}{2}\begin{bmatrix}0&0&0&0&0\\0&0&0&0&0\\0&1&-2&1&0\\0&0&0&0&0\\0&0&0&0&0\end{bmatrix}$$

(a)

$$\begin{bmatrix}0&0&0\\0&-1&0\\0&1&0\end{bmatrix}\quad \begin{bmatrix}0&0&0\\0&-1&1\\0&0&0\end{bmatrix}\quad \begin{bmatrix}0&0&0\\0&-1&0\\0&0&1\end{bmatrix}$$

(b)

Fig. 3. (a) The initialization weight of the SRM convolution kernel; (b) The initialization weight of the Bayar convolution kernel.

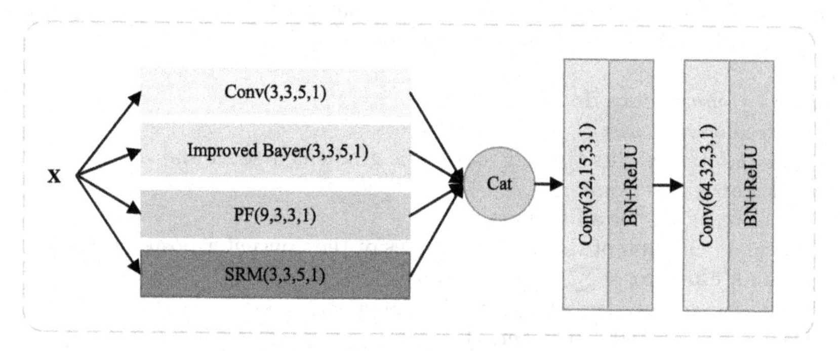

Fig. 4. The enhancement block. Conv+Improved Bayar+PF+SRM is the best combination, then the features are concatenated and further refined by two convolutions.

the inpainting area and the original area in the residual domain. However, it is difficult to learn multiple features of the data using the above-mentioned single enhanced trace method, resulting in unsatisfactory performance of the forensic network. Therefore, we first design a enhancement block in S-Encoder, including a parallel combination of the above methods, to suppress the content of the input image and enhance the inpainting traces, and then send the enhanced features to backbone network to extract features.

The feature map after SRM filtering emphasizes local noise features. It helps the network pay more attention to the noise information left by the tampering operation. As shown in Fig. 3 (a), the SRM filter kernel is used to initialize the convolution kernel, and the parameters of the convolution kernel are fixed. For a 3-channel input $\mathbf{X} \in \mathbb{R}^{3\times H\times W}$, a $5 \times 5 \times 3$ convolution kernel is used for noise feature extraction to obtain a 3-channel noise feature map $\mathbf{X_1} \in \mathbb{R}^{3\times H\times W}$. where H and W are the height and width of the feature map respectively.

Bayar constrained convolution can adaptively learn low-level predictive residual features by adding specific constraints to standard convolution kernels. The constraint rules are as follows:

$$\begin{cases} w_k(c,c) = -1 \\ \sum_{m,n\neq c} w_k(m,n) = 1 \end{cases} \tag{1}$$

where w_k represents the kth convolution kernel, (c, c) is the center position coordinate of wk, and (m, n) is the non-center position coordinate. After the constraints, the weight of the central position of the convolution kernel is −1, and the sum of the weights of other positions is 1. Constrained convolution solves the problem that SRM cannot be trained, but it is unstable in actual training, resulting in poor training results. Therefore, the improved constrained convolution (Improved Bayar) [56] is used, which maps the input to a 3-channel feature map $\mathbf{X_2} \in \mathbb{R}^{3\times H\times W}$. The training process is shown in Algorithm 1.

Algorithm 1. Training Algorithm of Improved Bayar

1: Initialize wk with the Laplacian-like weight
2: i=1
3: **while** i $\leq max\_iters$ **do**
4: Do feedforward pass
5: Update filter weights through stochastic gradient descent and backpropagate errors
6: Set $w_k(c,c) = 0$ for all k filters
7: Calculate the sum of the absolute values of the noncenter position weights for all k filters $S_k = \sum_{m,n\neq c} |w_k(m,n)|$
8: Let $w_k(m,n) = w_k(m,n) \div S_k$ to regularize w_k
9: Set $w_k(m,n) = 0.001$ if $w_k(m,n) \leq 0.001$
10: Set $w_k(c,c) = -S_k$ for all k filters
11: i = i+1
12: **end while**

PF is designed to obtain filtered residuals because in the residual domain, the inpainted regions are very different from the original ones. Therefore, the PF module is used to extract high-frequency features, which are designed as a $3\times 3\times 3$ convolution kernel, initialized with the first-order derivative high-pass filter kernel shown in Fig. 3 (b), and set to be learnable, which is convenient in fine-tuning during training. The PF module maps the input $\mathbf{X}$ to a 9-channel residual feature map $\mathbf{X_3} \in \mathbb{R}^{9\times H\times W}$.

To find a suitable combination of these special layers, we conducted experiments on the detection performance of different combinations. While combining special blocks, we also introduce ordinary convolution (Conv). Because all these special layers are equivalent to high-pass filters of different convolution kernels, if only the feature maps after these high-pass filters are used as input, the damage of compression and blurring to noise information is greater than that of semantic information, then the network's robustness will be worse. The final experiment found that the combination of Conv+Improved Bayar+PF+SRM has the best detection performance, and the feature map obtained by Conv is $\mathbf{X_4} \in \mathbb{R}^{3\times H\times W}$. The specific structure of the enhancement block is shown in Fig. 4. Therefore, we use this combination as the first layer in the enhancement block. Next, the feature maps $[\mathbf{X_1}, \mathbf{X_2}, \mathbf{X_3}, \mathbf{X_4}]$ obtained by this combination are stitched together in the channel dimension. Two standard convolutions are used to initially process the enhanced features and then sent to the backbone network Resnet50 for Spatial features $\mathbf{F_s}$, where $\mathbf{F_s} \in \mathbb{R}^{2048\times \frac{H}{16}\times \frac{W}{16}}$.

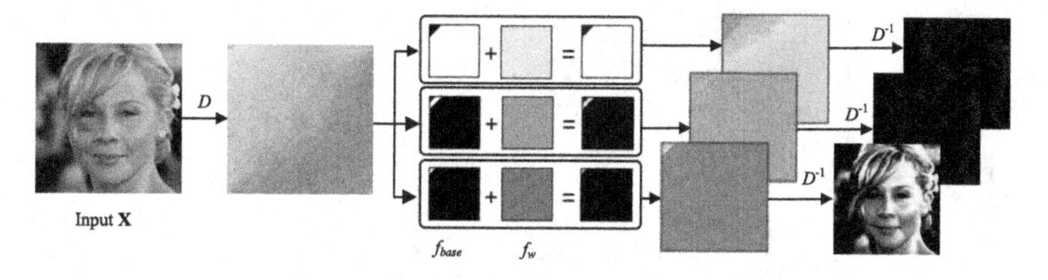

Fig. 5. The frequency block. A learnable combination filter adaptively extracts frequency features of interest.

3.2 F-Encoder

Most existing image inpainting methods use upsampling, and usually the upsampling operation will lead to abnormal frequency statistics of inpainted regions [55]. To this end, we introduce frequency information to help the network mine the essential differences between original and inpainted regions. For frequency-aware image decomposition, previous studies usually employ hand-crafted filters [6] on the spatial domain, thus failing to cover the complete frequency domain. Meanwhile, it is difficult to capture inpainting artifacts with fixed filtering settings adaptively. We propose the F-Encoder to solve this problem. A frequency block is designed, as shown in Fig. 5. The input image is adaptively decomposed according to a set of learnable frequency filters, and the decomposed frequency components can be inversely transformed to the spatial domain to obtain a series of image features with frequency awareness [35]. These features are concatenated along the channel axis and then fed into the backbone to mine the tampering traces left by inpainting comprehensively. Specifically, we manually designed three basic binary filters $\left\{f_{base}^{i}\right\}_{i=1}^{3}$ to divide the frequency domain into low, medium, and high-frequency bands. The low-frequency band is the first $1/16$ of the entire spectrum, the middle-frequency band is between $1/16$ and $1/8$ of the spectrum, and the high-frequency band is the last $7/8$. Then we add three learnable filters $\left\{f_{w}^{i}\right\}_{i=1}^{3}$ to these basic filters. The final combined filter is $f_{base}^{i}+\sigma\left(f_{w}^{i}\right)$, $i=\{1,2,3\}$, where $\sigma(x)=\frac{1-\exp(-x)}{1+\exp(-x)}$ is used to compress x in [-1,1]. Therefore, for an input image $\mathbf{X}$, the image frequency-domain perceptual feature $\mathbf{y}_i$ obtained through the frequency block can be expressed by:

$$\mathbf{y}_i=D^{-1}\left\{D(X)\odot\left[f_{base}^{i}+\sigma\left(f_{w}^{i}\right)\right]\right\}, i=\{1,2,3\} \tag{2}$$

where $\odot$ is the element-wise dot product and D refers to the discrete cosine transform. Typically most of the amplitude is concentrated in the low-frequency region, and we apply basic filters to divide the spectrum into bands of approximately equal energy from low to high frequency. The added learnable filter provides more adaptability to select the frequency of interest and can more comprehensively capture the subtle artifacts introduced by the inpainting. The image frequency domain perception features $\mathbf{y}_i$ is sent to the backbone to obtain the frequency domain feature $\mathbf{F_{fq}}$, where $\mathbf{F_{fq}}\in\mathbb{R}^{2048\times\frac{H}{16}\times\frac{W}{16}}$.

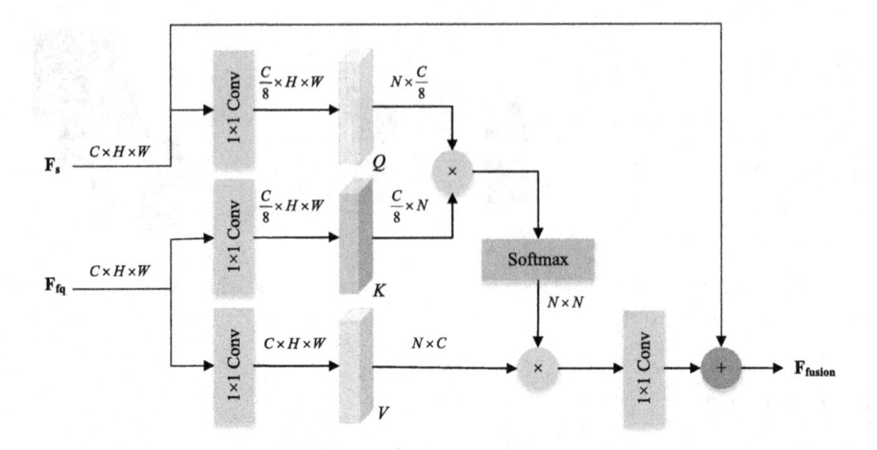

Fig. 6. The proposed cross-modal attention fusion module (CMAF) to fuse the features of the two encoders. ⊗ denotes matrix multiplication, ⊕ denotes element-wise sum.

3.3 CMAF

In order to better utilize frequency domain features to enhance spatial domain features, we use the query-key-value [38] mechanism to design a cross-modal attention fusion module to achieve feature interaction. Compared with simple feature concatenate and element-wise addition, this module performs better, while only a few parameters and calculations are added. As shown in Fig. 6. For the spatial feature $\mathbf{F_s}$ and the frequency-domain perceptual feature $\mathbf{F_{fq}}$, firstly using 1×1 convolution $Conv_q$, $Conv_k$, $Conv_v$ to encode them into Q, K, V respectively. The formula is as follows:

$$Q = Conv_q(\mathbf{F_s}), K = Conv_k(\mathbf{F_{fq}}), V = Conv_v(\mathbf{F_{fq}}) \tag{3}$$

In the above formula $\{Q, K, V\} \in \mathbb{R}^{256\times\frac{H}{16}\times\frac{W}{16}}$, they are then reconstructed as $\{Q, K, V\} \in \mathbb{R}^{256\times N}$. Then we perform similarity calculation on Q and K to obtain weights. Finally, using the softmax function to normalize the weights to obtain the attention map $A \in \mathbb{R}^{N\times N}$, whose calculation formula is:

$$A = softmax(\frac{Q^T K}{scale}) \tag{4}$$

where scale is $\sqrt{2048 \times N}$, and Q^T is the transpose of Q. Then the attention map A is multiplied by the transpose of V, the obtained result $\mathbf{F_a}$ is reconstructed as $\mathbb{R}^{2048\times\frac{H}{16}\times\frac{W}{16}}$, and then added to the original spatial feature $\mathbf{F_s}$ to obtain the final fusion feature $\mathbf{F_{fusion}}$, so that the feature is fully enhanced.

$$\mathbf{F_{fusion}} = \mathbf{F_a} + \mathbf{F_s} \tag{5}$$

where $\mathbf{F_{fusion}} \in \mathbb{R}^{2048\times\frac{H}{16}\times\frac{W}{16}}$, the obtained fusion features are sent to the decoder network.

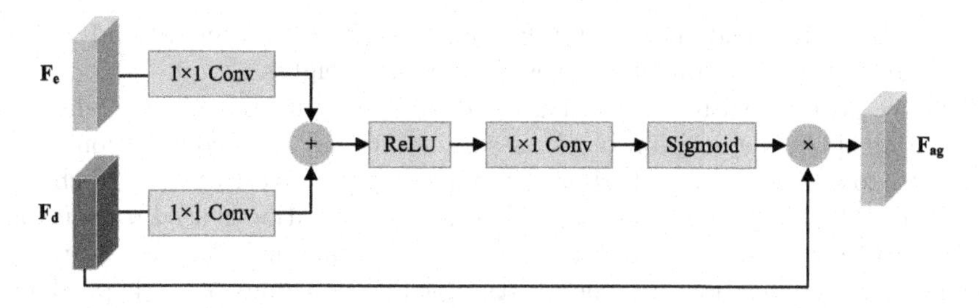

Fig. 7. The attention gate (AG) module. To suppress irrelevant regions in spatial features while highlighting salient features in specific local regions.

3.4 Attention-Gated Decoder Network

Continuous convolution operations in the encoder linearly increase the network's receptive field's size, enabling the network to learn shallower and rougher context and semantic representations gradually. Learning high-level semantic features makes the network slowly lose the ability to extract local features. However, this aspect is crucial for segmentation localization results [1]. In general, local information of features in the decoder can be augmented by skip connections from the encoder path to the decoder path at the same scale. However, this method treats the features of the encoder equally and introduces some useless information.

Based on the above considerations, We employ skip connections to exploit feature information in S-Encoder, including features after the enhancement block. To suppress irrelevant regions in spatial features while highlighting salient features in specific local regions, we propose an attention-gated decoder network. As shown in Fig. 2, we only add the feature information in S-Encoder to the decoder because the inpainting artifacts in the F-Encoder are relatively weak, and some features will be lost during the extraction process, so we do not add semantic information in the frequency domain in the decoder. The decoder network mainly consists of the upsampling module, the AG module and the decision block. We use transposed convolutions with a stride of 2 for upsampling, and each transposed convolution doubles the size of the input feature map and halves the number of channels. The AG module uses the features of each upsampling and the features of the same scale in S-Encoder to obtain the gating features. As shown in Fig. 7. Firstly, we use 1×1 point convolution to extract features from $\mathbf{F_e}$ and $\mathbf{F_d}$, then add them together, redistribute the weight of feature information, and use the ReLU function to remove redundant information to achieve the role of gating, where $\mathbf{F_e}$ is the spatial encoder features, $\mathbf{F_d}$ is upsampled to the same feature corresponding to the encoder feature resolution. Then use the 1×1 point convolution and the Sigmoid function to generate the attention feature map, and finally use the attention feature map to adjust Fd to obtain the attention-gated feature map $\mathbf{F_{ag}}$:

$$\mathbf{F_{ag}} = \mathbf{F_d} \otimes \sigma_1 \left(c_1 \left(\sigma_2 \left(c_2 \left(\mathbf{F_e} \right) + c_3 \left(\mathbf{F_d} \right) \right) \right) \right) \quad (6)$$

where σ_1 is the Sigmoid activation function, σ_2 is the ReLU activation function, c_1, c_2 and c_3 are 1×1 convolutions with different weights.

Finally, to deal with the checkerboard artifacts introduced by transposed convolutions [32], the decision block uses an additional 3×3 convolution with a stride of 1 to weaken the checkerboard artifacts while converting the 128-channel output into 32-channel features map. Then 1×1 convolution is used to transform 32 channels into a 1-channel feature map, which is then fed to a global Sigmoid function for classification to generate the localization map $\mathbf{Y}$ with pixel-level prediction.

3.5 Loss Function

We first use a binary cross-entropy (BCE) loss to supervise the training of DDEDNet, since the forensics localization problem is essentially a segmentation problem, i.e., to segment out the original region and the inpainting region. The final output of the network is $\mathbf{Y}$, and the standard BCE loss of the training data $(\mathbf{X}, \mathbf{Z})$ is:

$$\mathcal{L}_{bce} = \frac{-1}{HW}\sum_{i=1}^{H}\sum_{j=1}^{W}\left(\mathbf{Z}_{ij}\log\widehat{\mathbf{Y}}_{ij} + (1-\mathbf{Z}_{ij})\log(1-\widehat{\mathbf{Y}}_{ij})\right) \tag{7}$$

where $\widehat{\mathbf{Y}}_{ij}$ represents the (i, j)th element in the prediction probability matrix $\widehat{\mathbf{Y}}$, and $\mathbf{Z}_{ij}$ is the corresponding element in the label matrix. However, the proportion of tampered pixels in the inpainted image is relatively small, leading to the class imbalance problem. The standard BCE loss treats positive and negative samples equally and cannot fully explore the important information in positive samples. Therefore, to mitigate the effect of class imbalance, we adopt the focal loss [26]. It assigns a modulation factor α to the standard cross-entropy term, which reduces the weight of dominant and well-classified negative samples in the total loss. In this way, the network will pay more attention to the classification of positive samples.

$$\mathcal{L}_{focal} = \frac{-1}{HW}\sum_{i=1}^{H}\sum_{j=1}^{W}\begin{pmatrix}\alpha(1-\widehat{\mathbf{Y}}_{ij})^{\gamma}\mathbf{Z}_{ij}\log\widehat{\mathbf{Y}}_{ij} \\ +(1-\alpha)\widehat{\mathbf{Y}}_{ij}^{\ \gamma}(1-\mathbf{Z}_{ij})\log(1-\widehat{\mathbf{Y}}_{ij})\end{pmatrix} \tag{8}$$

γ is a focusing parameter that reduces the rate of weighting. In this paper $\alpha = 0.75$, $\gamma = 2$. So the overall loss is defined as follows:

$$\mathcal{L} = \mathcal{L}_{bce} + \mathcal{L}_{focal} \tag{9}$$

Using $\mathcal{L}$ to supervise the training of the network can not only alleviate the impact of category imbalance, but also increase the training speed of the network.

4 Experiment

To evaluate the performance of the proposed network in this paper, we test not only on the datasets of various deep inpainting methods but also on the datasets inpainted by various traditional methods. And compared with the state-of-the-art forensics methods, subjective and objective analysis is carried out. Meanwhile, ablation experiments are set up to demonstrate the effectiveness of different components of the network.

4.1 Experimental Setup

Datasets. This paper uses the dataset in [46] for training and testing. The training set randomly samples 24,000 images from the Places [58] and Dresden [9] datasets as basic images, and randomly samples masks from the irregular mask dataset [27], using the method of GC [53] to inpaint a dataset containing 48000 images. 38400 images are randomly selected as our training set. To increase the data diversity, we introduce additional datasets CelebA [17] and ImageNet [8] in the test set. Besides, a randomly generated series of basic shapes, such as rectangles, circles, ellipses, and polylines, are added as additional test masks, which can be located anywhere. These additional masks account for roughly the same proportion as the masks generated from [27]. A total of 10 representative methods are considered, 5 of which are deep learning-based inpainting methods proposed in recent years, namely GC [53], CA [52], SH [51], LB [47], and RN [54]. The remaining 5 methods are based on traditional inpainting methods, including TE [37], NS [4], LR [10], PM [13], and SG [15]. The test set of each method is 1000, a total of 10000.

Evaluation Metrics. We apply the Area Under the Receiver Operating Characteristic Curve(AUC) and F1 score as our evaluation metrics.The metrics are calculated on each image independently, and the mean values obtained over all images are reported in the following experiments.

Implementation Details. We implement our model using the Pytorch framework. The width and height of the input image are resized to 256 in the training phase. RandomHorizontalFlip is used to perform simple data augment. We set the batch size to 2 and have 40 Epochs. The learning rate is initially set to 0.0005, and it decays to 85% of the previous value after each Epoch. The model was optimized using the Adam optimizer [18] with momentum decay exponents $\beta_1 = 0.9$, $\beta_2 = 0.999$. All experiments are performed on a single NVIDIA 3060 GPU with 12GB memory.

4.2 Comparing with Previous Methods

For comparison, we selected 5 state-of-the-art methods, including LDI [22], MT-Net [49], HPFCN [21], IID-Net [46] and PSCC-Net [28], all using their public

Table 1. Quantitative comparisons by using AUC and F1 as metrics. These models are all retrained on our dataset, and Mean in the table represents the average result on 10 test sets. These models are all re-trained on our data set. Mean represents the average result on 10 test sets. The experimental results are unified using a percentage system, and the best results are indicated in bold.

Method	Metric	Test Dataset										Mean
		GC	CA	SH	LB	RN	TE	NS	LR	PM	SG	
LDI	AUC	47.63	40.91	49.20	44.70	46.23	56.88	71.75	47.97	49.93	52.17	49.94
MT-Net	AUC	91.26	91.09	95.13	95.31	90.36	81.62	79.97	91.25	89.65	98.62	90.42
HPFCN	AUC	98.48	88.93	99.24	98.06	98.87	95.76	97.09	99.68	99.45	99.91	97.55
PSCC-Net	AUC	98.17	**95.68**	99.56	99.30	99.14	95.59	96.93	98.74	99.04	99.85	98.20
IID-Net	AUC	96.77	95.39	99.67	99.80	**99.71**	96.12	97.65	**99.79**	99.54	**99.94**	98.44
DDEDNet	AUC	**99.33**	94.27	**99.79**	**99.81**	99.57	**97.77**	**98.04**	99.75	**99.58**	99.92	**98.78**
LDI	F1	15.08	2.30	6.26	7.83	14.17	16.96	25.02	3.03	2.65	15.92	10.26
MT-Net	F1	59.17	60.72	70.95	75.69	52.53	40.31	35.93	29.29	26.35	67.12	51.81
HPFCN	F1	86.43	52.32	88.14	60.69	82.82	76.57	78.56	83.84	58.65	91.71	75.97
PSCC-Net	F1	89.18	**84.12**	94.04	92.90	90.66	83.14	86.18	85.35	72.77	94.18	87.25
IID-Net	F1	83.10	80.80	**94.13**	**96.14**	**94.41**	82.05	84.98	87.28	71.43	**94.79**	86.91
DDEDNet	F1	**90.48**	78.24	93.92	95.35	92.24	**86.03**	**86.34**	**88.75**	**75.09**	94.68	**88.11**

code to retrain on our datasets and strictly follows the training procedure and parameter settings proposed in their paper. As shown in Table 1, the best results are marked in bold. It can be observed that the detection performance of LDI for traditional inpainting methods is relatively better than that of deep learning-based methods. The average AUC value is 49.97%, which is close to random guessing. This phenomenon may be because hand-engineered features are unreliable, especially for unseen inpainting methods. MT-Net achieves poor AUC/F1 performance, which may be because the network architecture of MT-Net is specially designed according to its original training dataset. The performance of HPFCN is average, its AUC is 97.55%, and its F1 is 75.97%, mainly because the model uses a specific repair method and a fixed repair mask, and a single PF layer cannot find the common characteristics of different inpainting methods. Since IID-Net uses NAS, its model has performed relatively well, but the search process is complicated and time-consuming. The AUC of PSCC-Net is 98.20%, and the F1 is 87.25%, which may be attributed to its use of the characteristics of dense cross-connections of different scales and the design of spatial channel correlation modules to achieve progressive mask prediction better. Due to the dual-domain encoder network, the hierarchical combination of different special layers in the spatial domain, and the subtle traces in the frequency domain, DDEDNet obtains 98.78% mean AUC value and 88.11% F1 score, both of which are significantly better than the test results of other comparison schemes.

In order to feel the detection effect of the method proposed in this paper more intuitively, a total of 6 images were randomly selected from the two test sets of GC and LR, and the qualitative results of different methods were shown. As shown in Fig. 8. It can be found that the localization effect of MT-Net and

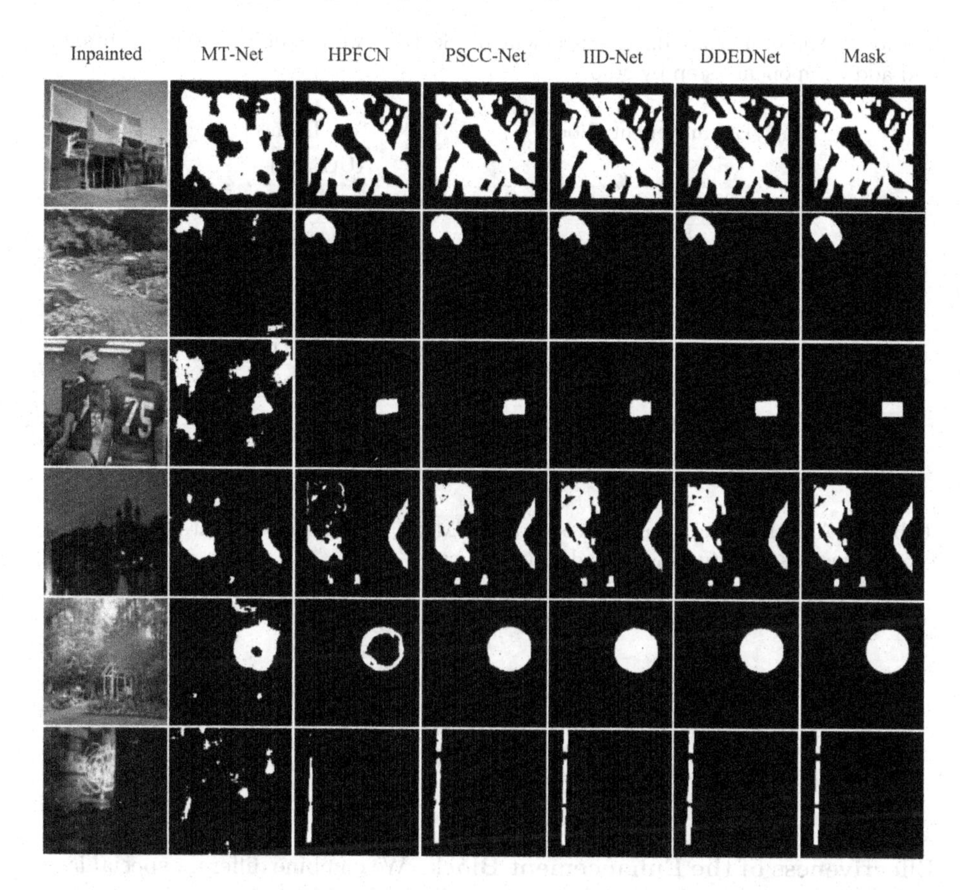

Fig. 8. The visualization of prediction results. For each column, the images from left to right are inpainted, detection result generated by MT-Net, HPFCN, PSCC-Net, IID-Net and our DDEDNet, the last column is ground truth mask.

HPFCN is relatively poor. For masks of different shapes, PSCC-Net and IID-Net can reasonably locate the repaired area, but DDEDNet can detect the complete outline of the inpainting area. The edge part of the area is predicted more accurately, even with some small isolated points. At the same time, maintaining a low false alarm for the original area.

4.3 Abation Study

We design three sets of ablation experiments to verify the effectiveness of some key components in our model, and in each ablation experiment, only one component is modified while others remain unchanged. Then, we retrain with the same training protocol as before, and report the average results of AUC and F1 score on the 10 test sets.

Table 2. Ablation study on enhancement block. Here we use only F-Encoder branch and add components step by step.

Enhancement Block	Test Dataset	
	AUC	F1
Conv	97.40	81.58
Bayar	97.96	83.40
Improved Bayar	98.20	84.06
PF	98.38	85.38
SRM	97.99	81.80
Conv+Improved Bayar	98.03	84.56
Conv+Improved Bayar+PF	98.34	85.73
Conv+Improved Bayar+PF+SRM	**98.42**	**85.82**

Table 3. Ablation study on F-Encoder and CMAF. We mainly compare CMAF with two common feature fusion methods add and cat.

F-Encoder&Fusion	Test Dataset	
	AUC	F1
Add	98.33	84.45
Cat	98.41	84.58
CMAF	**98.49**	**86.18**

Effectiveness of the Enhancement Block. We combine different special layers in Sect. 3.1, and only conduct experiments on the S-Encoder branch, and the experimental results are shown in Table 2. The effect of a single special convolution is better than Conv. Improved Bayar has improved performance compared to Bayar due to the new weight update method. Therefore, Improved Bayar is used in subsequent experiments. Conv+Improved Bayar+PF+SRM achieves the best performance due to its ability to capture various trace features and better generalization to different test sets.

Effectiveness of F-Encoder and CMAF. As shown in Table 3, using CMAF to fuse spatial and frequency domain encoders achieves the best performance. Using the two methods of Add and Cat to fuse, F1 decreased by 1.37% and 1.24%, respectively. This may be because the direct addition or concatenation of features of different modalities will destroy the learned inpainting traces or even cause feature redundancy. CMAF uses self-attention to extract more relevant features in the F-Encoder, then supplements features in S-Encoder to form more useful feature information.

Table 4. Ablation study on AG. This is done on top of the full model. Skip Connection brings a certain performance gain, and it increases significantly after joining AG.

Decoder	Test Dataset	
	AUC	F1
Skip Connection	98.72	87.12
Skip Connection+AG	**98.78**	**88.11**

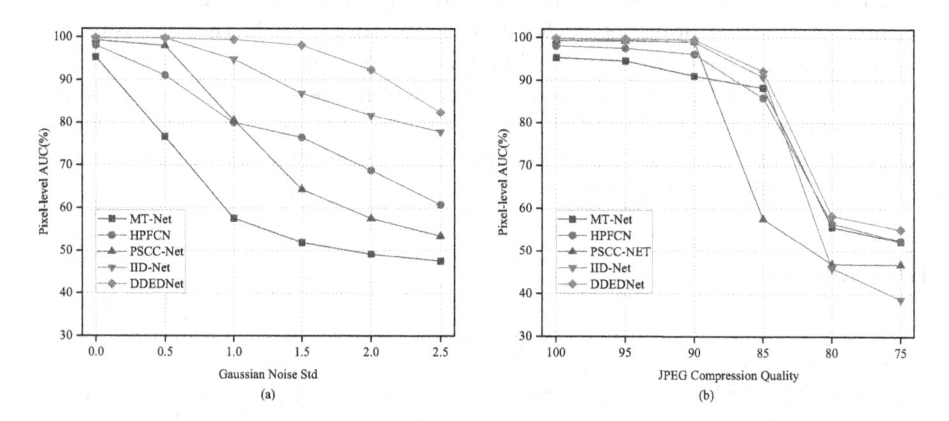

Fig. 9. Robustness evaluation on LB. (a) Performance curves w.r.t JPEG compression; (b) Performance curves w.r.t. Gaussian Noise.

Effectiveness of AG. As shown in Table 4, introducing skip connections from the S-Encoder can lead to significant performance gain, with 0.23% and 0.94% improvement in AUC and F1, respectively. To further optimize the performance, we add the AG to skip connections, which achieves the best results. This shows that AG can further optimize features in the S-Encoder and improve localization performance.

4.4 Robustness Evaluation

To analyze the robustness of DDEDNet, we use two common post-processing methods JPEG compression and Gaussian noise to process the LB test set, and report the test results on LB. The JPEG quality factor ranges from 75 to 100 with a step size of 5; the standard deviation of Gaussian noise is set from 0 to 2.5 with a step size of 0.5. The robustness analysis of different models under pixel-level AUC is shown in Fig. 9. It can be observed that with the increase of interference intensity, all schemes have different degrees of performance degradation. When the JPEG quality factor is lower than 85, the performance of all models drops faster, but DDEDNet can still outperform other schemes. With increasing Gaussian noise, the proposed method significantly outperforms other methods at different standard deviations.

5 Conclusions

In this paper, a two-branch encoder-decoder network based on spatial domain and frequency domain, called DDEDNet, has been proposed. The spatial domain branch uses the combination of different special layers to extract more general features, and the frequency domain branch designs an adaptive combination filter, aiming at capturing and repairing the artifact features left in the frequency domain. In addition, a cross-modal attention fusion module is used to fuse the features of the two branches to form richer features. To further improve the localization performance, an attention-gated skip connection is introduced in the decoder to properly incorporate features from the spatial encoder. Many experiments show that DDEDNet has better performance than state-of-the-art inpainting forensics methods and has strong robustness against JPEG compression and Gaussian noise attacks.

References

1. Azad, R., et al.: Medical image segmentation review: The success of u-net. arXiv preprint arXiv:2211.14830 (2022)
2. Barnes, C., Shechtman, E., Finkelstein, A., Goldman, D.B.: Patchmatch: a randomized correspondence algorithm for structural image editing. ACM Trans. Graph. **28**(3), 24 (2009)
3. Bayar, B., Stamm, M.C.: Constrained convolutional neural networks: a new approach towards general purpose image manipulation detection. IEEE Trans. Inf. Forensics Secur. **13**(11), 2691–2706 (2018)
4. Bertalmio, M., Bertozzi, A.L., Sapiro, G.: Navier-stokes, fluid dynamics, and image and video inpainting. In: Proceedings of the 2001 IEEE Computer Society Conference on Computer Vision and Pattern Recognition, CVPR 2001. IEEE (2001)
5. Chan, T.F., Shen, J.: Nontexture inpainting by curvature-driven diffusions. J. Vis. Commun. Image Represent. **12**(4), 436–449 (2001)
6. Chen, M., Sedighi, V., Boroumand, M., Fridrich, J.: Jpeg-phase-aware convolutional neural network for steganalysis of jpeg images. In: Proceedings of the 5th ACM Workshop on Information Hiding and Multimedia Security, pp. 75–84 (2017)
7. Criminisi, A., Pérez, P., Toyama, K.: Region filling and object removal by exemplar-based image inpainting. IEEE Trans. Image Process. **13**(9), 1200–1212 (2004)
8. Deng, J., Dong, W., Socher, R., Li, L.J., Li, K., Fei-Fei, L.: Imagenet: a large-scale hierarchical image database. In: 2009 IEEE Conference on Computer Vision and Pattern Recognition, pp. 248–255. IEEE (2009)
9. Gloe, T., Böhme, R.: The'dresden image database'for benchmarking digital image forensics. In: Proceedings of the 2010 ACM Symposium on Applied Computing, pp. 1584–1590 (2010)
10. Guo, Q., Gao, S., Zhang, X., Yin, Y., Zhang, C.: Patch-based image inpainting via two-stage low rank approximation. IEEE Trans. Visual Comput. Graph. **24**(6), 2023–2036 (2017)
11. He, K., Zhang, X., Ren, S., Sun, J.: Deep residual learning for image recognition. In: Proceedings of the IEEE Conference on Computer Vision and Pattern Recognition, pp. 770–778 (2016)

12. He, X., Li, W., Zhang, S., Li, K.: Efficient control of unscheduled packets for credit-based proactive transport. In: 2022 IEEE 28th International Conference on Parallel and Distributed Systems (ICPADS), pp. 593–600. IEEE (2023)
13. Herling, J., Broll, W.: High-quality real-time video in painting with pixmix. IEEE Trans. Visual Comput. Graph. **20**(6), 866–879 (2014)
14. Hu, J., Shen, L., Sun, G.: Squeeze-and-excitation networks. In: Proceedings of the IEEE Conference on Computer Vision and Pattern Recognition, pp. 7132–7141 (2018)
15. Huang, J.B., Kang, S.B., Ahuja, N., Kopf, J.: Image completion using planar structure guidance. ACM Trans. Graph. (TOG) **33**(4), 1–10 (2014)
16. Iizuka, S., Simo-Serra, E., Ishikawa, H.: Globally and locally consistent image completion. ACM Trans. Graph. (ToG) **36**(4), 1–14 (2017)
17. Karras, T., Aila, T., Laine, S., Lehtinen, J.: Progressive growing of gans for improved quality, stability, and variation. arXiv preprint arXiv:1710.10196 (2017)
18. Kingma, D.P., Ba, J.: Adam: a method for stochastic optimization. arXiv preprint arXiv:1412.6980 (2014)
19. Levin, A., Zomet, A., Weiss, Y.: Learning how to inpaint from global image statistics. In: ICCV, vol. 1, pp. 305–312 (2003)
20. Li, A., et al.: Noise doesn't lie: towards universal detection of deep inpainting. arXiv preprint arXiv:2106.01532 (2021)
21. Li, H., Huang, J.: Localization of deep inpainting using high-pass fully convolutional network. In: Proceedings of the IEEE/CVF International Conference on Computer Vision, pp. 8301–8310 (2019)
22. Li, H., Luo, W., Huang, J.: Localization of diffusion-based inpainting in digital images. IEEE Trans. Inf. Forensics Secur. **12**(12), 3050–3064 (2017)
23. Li, W., Chen, S., Li, K., Qi, H., Xu, R., Zhang, S.: Efficient online scheduling for coflow-aware machine learning clusters. IEEE Trans. Cloud Comput. **10**(4), 2564–2579 (2020)
24. Li, W., Yuan, X., Li, K., Qi, H., Zhou, X.: Leveraging endpoint flexibility when scheduling coflows across geo-distributed datacenters. In: IEEE INFOCOM 2018-IEEE Conference on Computer Communications, pp. 873–881. IEEE (2018)
25. Liang, Z., Yang, G., Ding, X., Li, L.: An efficient forgery detection algorithm for object removal by exemplar-based image inpainting. J. Vis. Commun. Image Represent. **30**, 75–85 (2015)
26. Lin, T.Y., Goyal, P., Girshick, R., He, K., Dollár, P.: Focal loss for dense object detection. In: Proceedings of the IEEE International Conference on Computer Vision, pp. 2980–2988 (2017)
27. Liu, G., Reda, F.A., Shih, K.J., Wang, T.-C., Tao, A., Catanzaro, B.: Image inpainting for irregular holes using partial convolutions. In: Ferrari, V., Hebert, M., Sminchisescu, C., Weiss, Y. (eds.) ECCV 2018. LNCS, vol. 11215, pp. 89–105. Springer, Cham (2018). https://doi.org/10.1007/978-3-030-01252-6_6
28. Liu, X., Liu, Y., Chen, J., Liu, X.: Pscc-net: progressive spatio-channel correlation network for image manipulation detection and localization. IEEE Trans. Circuits Syst. Video Technol. **32**(11), 7505–7517 (2022)
29. Liu, Y., Li, W., Qu, W., Qi, H.: Bulb: lightweight and automated load balancing for fast datacenter networks. In: Proceedings of the 51st International Conference on Parallel Processing, pp. 1–11 (2022)
30. Lu, M., Niu, S.: A detection approach using lstm-cnn for object removal caused by exemplar-based image inpainting. Electronics **9**(5), 858 (2020)

31. Nazeri, K., Ng, E., Joseph, T., Qureshi, F.Z., Ebrahimi, M.: Edgeconnect: generative image inpainting with adversarial edge learning. arXiv preprint arXiv:1901.00212 (2019)
32. Odena, A., Dumoulin, V., Olah, C.: Deconvolution and checkerboard artifacts. Distill **1**(10), e3 (2016)
33. Oktay, O., et al.: Attention u-net: Learning where to look for the pancreas. arXiv preprint arXiv:1804.03999 (2018)
34. Pathak, D., Krahenbuhl, P., Donahue, J., Darrell, T., Efros, A.A.: Context encoders: Feature learning by inpainting. In: Proceedings of the IEEE Conference on Computer Vision and Pattern Recognition, pp. 2536–2544 (2016)
35. Qian, Y., Yin, G., Sheng, L., Chen, Z., Shao, J.: Thinking in frequency: face forgery detection by mining frequency-aware clues. In: Vedaldi, A., Bischof, H., Brox, T., Frahm, J.-M. (eds.) ECCV 2020. LNCS, vol. 12357, pp. 86–103. Springer, Cham (2020). https://doi.org/10.1007/978-3-030-58610-2_6
36. Roy, A.G., Navab, N., Wachinger, C.: concurrent spatial and channel 'Squeeze & excitation' in fully convolutional networks. In: Frangi, A.F., Schnabel, J.A., Davatzikos, C., Alberola-López, C., Fichtinger, G. (eds.) MICCAI 2018. LNCS, vol. 11070, pp. 421–429. Springer, Cham (2018). https://doi.org/10.1007/978-3-030-00928-1_48
37. Telea, A.: An image inpainting technique based on the fast marching method. J. Graph. Tools **9**(1), 23–34 (2004)
38. Vaswani, A., et al.: Attention is all you need. Advances in neural information processing systems 30 (2017)
39. Wang, J., Liu, Y., Rao, S., Sherratt, R.S., Hu, J.: Enhancing security by using gift and ecc encryption method in multi-tenant datacenters. Comput. Mater. Continua **75**(2), 3849–3865 (2023)
40. Wang, J., Liu, Y., Rao, S., Zhou, X., Hu, J.: A novel self-adaptive multi-strategy artificial bee colony algorithm for coverage optimization in wireless sensor networks. Ad Hoc Netw. 103284 (2023)
41. Wang, J., Rao, S., Liu, Y., Sharma, P.K., Hu, J.: Load balancing for heterogeneous traffic in datacenter networks. J. Netw. Comput. Appl. **217**, 103692 (2023)
42. Wang, J., Yuan, D., Luo, W., Rao, S., Sherratt, R.S., Hu, J.: Congestion control using in-network telemetry for lossless datacenters. Comput. Mater. Continua **75**(1), 1195–1212 (2023)
43. Wei, W., Gu, H., Deng, W., Xiao, Z., Ren, X.: Abl-tc: a lightweight design for network traffic classification empowered by deep learning. Neurocomputing **489**, 333–344 (2022)
44. Wei, W., Gu, H., Wang, K., Li, J., Zhang, X., Wang, N.: Multi-dimensional resource allocation in distributed data centers using deep reinforcement learning. IEEE Trans. Netw. Serv. Manag. (2022)
45. Woo, S., Park, J., Lee, J.-Y., Kweon, I.S.: CBAM: convolutional block attention module. In: Ferrari, V., Hebert, M., Sminchisescu, C., Weiss, Y. (eds.) ECCV 2018. LNCS, vol. 11211, pp. 3–19. Springer, Cham (2018). https://doi.org/10.1007/978-3-030-01234-2_1
46. Wu, H., Zhou, J.: Iid-net: image inpainting detection network via neural architecture search and attention. IEEE Trans. Circ. Syst. Video Technol. **32**(3), 1172–1185 (2021)
47. Wu, H., Zhou, J., Li, Y.: Deep generative model for image inpainting with local binary pattern learning and spatial attention. IEEE Trans. Multimedia **24**, 4016–4027 (2021)

48. Wu, Q., Sun, S.J., Zhu, W., Li, G.H., Tu, D.: Detection of digital doctoring in exemplar-based inpainted images. In: 2008 International Conference on Machine Learning and Cybernetics, vol. 3, pp. 1222–1226. IEEE (2008)
49. Wu, Y., AbdAlmageed, W., Natarajan, P.: Mantra-net: manipulation tracing network for detection and localization of image forgeries with anomalous features. In: Proceedings of the IEEE/CVF Conference on Computer Vision and Pattern Recognition, pp. 9543–9552 (2019)
50. Xu, R., Li, W., Li, K., Zhou, X., Qi, H.: Darkte: towards dark traffic engineering in data center networks with ensemble learning. In: 2021 IEEE/ACM 29th International Symposium on Quality of Service (IWQOS), pp. 1–10. IEEE (2021)
51. Yan, Z., Li, X., Li, M., Zuo, W., Shan, S.: Shift-Net: image inpainting via deep feature rearrangement. In: Ferrari, V., Hebert, M., Sminchisescu, C., Weiss, Y. (eds.) Computer Vision – ECCV 2018. LNCS, vol. 11218, pp. 3–19. Springer, Cham (2018). https://doi.org/10.1007/978-3-030-01264-9_1
52. Yu, J., Lin, Z., Yang, J., Shen, X., Lu, X., Huang, T.S.: Generative image inpainting with contextual attention. In: Proceedings of the IEEE Conference on Computer Vision and Pattern Recognition, pp. 5505–5514 (2018)
53. Yu, J., Lin, Z., Yang, J., Shen, X., Lu, X., Huang, T.S.: Free-form image inpainting with gated convolution. In: Proceedings of the IEEE/CVF International Conference on Computer Vision, pp. 4471–4480 (2019)
54. Yu, T., et al.: Region normalization for image inpainting. In: Proceedings of the AAAI Conference on Artificial Intelligence, vol. 34, pp. 12733–12740 (2020)
55. Zhang, X., Karaman, S., Chang, S.F.: Detecting and simulating artifacts in gan fake images. In: 2019 IEEE International Workshop on Information Forensics and Security (WIFS), pp. 1–6. IEEE (2019)
56. Zhang, Z., Qian, Y., Zhao, Y., Zhu, L., Wang, J.: Noise and edge based dual branch image manipulation detection. arXiv preprint arXiv:2207.00724 (2022)
57. Zheng, J., Du, Z., Zha, Z., Yang, Z., Gao, X., Chen, G.: Learning to configure converters in hybrid switching data center networks. IEEE/ACM Trans. Netw. (2023)
58. Zhou, B., Lapedriza, A., Khosla, A., Oliva, A., Torralba, A.: Places: a 10 million image database for scene recognition. IEEE Trans. Pattern Anal. Mach. Intell. **40**(6), 1452–1464 (2017)
59. Zhou, P., Han, X., Morariu, V.I., Davis, L.S.: Learning rich features for image manipulation detection. In: Proceedings of the IEEE Conference on computer vision and Pattern Recognition, pp. 1053–1061 (2018)

Robust Medical Data Sharing System Based on Blockchain and Threshold Rroxy Re-encryption

Wenxuan Cheng[1,2], Bo Zhang[1,2(✉)], and Zhongtao Li[1,2(✉)]

[1] Shandong Provincial Key Laboratory of Network Based Intelligent Computing, University of Jinan, Jinan 250022, China
{ise_zhangb,lizt}@ujn.edu.cn

[2] School of Information Science and Engineering, University of Jinan, Jinan 250022, China

Abstract. In recent years, data processing technology has been continuously developed, and the hospital information system (HIS) has become more and more perfect. However, the upgrading speed of security attack technology is far beyond people's expectations. HIS is facing considerable challenges in the safe sharing of medical privacy data. Therefore, A blockchain-based robust medical data sharing system (BMDSS) is proposed. BMDSS uses blockchain technology to store medical privacy data and data-sharing records, realizing the decentralization of the entire system and the non-tamperable modification of data-sharing documents. BMDSS adopts threshold proxy re-encryption technology in the privacy data sharing process and uses proxy server clusters for ciphertext conversion without revealing the patient's private key, reducing the risk of the original key being leaked and enhancing the robustness of the entire system to achieve secure sharing, stable sharing, and transparent sharing of private data. Finally, a robust medical information security sharing system is obtained.

Keywords: Hospital information system · Blockchain · Data sharing · Threshold proxy re-encryption

1 Introduction

Hospital Information System (HIS) [12] was formally proposed in 1979. Since then, he has gradually received widespread attention from medical organizations in various countries. After more than 40 years of rapid development [4,11,25], HIS has become an essential part of hospital management and a core tool for medical institutions to manage patient data. Domestic and foreign policies and regulations have gradually increased the requirements for electronic medical

This work was supported by the Shandong Provincial Natural Science Foundation under Project ZR2022MF264.

Z. Tari et al. (Eds.): ICA3PP 2023, LNCS 14491, pp. 112–131, 2024.
https://doi.org/10.1007/978-981-97-0808-6_7

records. More and more medical institutions have begun to establish and use medical electronic file systems [17]. The functions and characteristics of the electronic medical library have been continuously expanded and improved, covering the information construction of outpatient, inpatient, physical examination, imaging, and other hospital departments [18]. He also plays a vital role in the quality and efficiency of healthcare [24]. Through functions such as electronic medical records and prescriptions, the hospital information system can realize the entire recording and management of the medical process, improve the doctor's diagnosis level and treatment effect, and reduce the medical error rate. With the advancement of technology, the data analysis and mining functions of hospital information systems have been continuously strengthened, which can provide data support for hospital management. Through data analysis, we can better understand its operating conditions and patient needs and optimize medical resource allocation and service processes.

However, there are also some problems in the traditional HIS system:

1. Unauthorized disclosure of private patient data: The hospital will share patient personal data with relevant institutions. However, the patient needs to be made aware of this move, and requires the patient's consent. For example: In 2015, the Federal Labor Relations Board (NLRB) filed a complaint against a healthcare organization for violating the Health Insurance Portability and Accountability Act (HIPAA) by disclosing patient health information in employee work emails. In 2019, a nanny in Arizona was accused of posting photos of private patient data such as the patient's name, medical record number, and disease diagnosis on social media and disclosing the patient's personal health information without the patient's authorization. 2. Interception or wiretapping of medical data transmission: In 2015, the Anthem insurance company in the United States suffered a massive data breach involving the personally identifiable information of more than 78 million patients. The investigation found that the attacker obtained a large amount of patient privacy data by eavesdropping on data packets during the transmission of medical data. In 2019, the Ukraine Registry of Medicines and Medical Products suffered a data breach involving the personally identifiable information of more than 50,000 patients. The investigation found that attackers obtained private patient data by eavesdropping on packets during medical data transmission. 3. Unserviceable attacks on hospital information systems: In May 2020, French hospital systems were hit by a large-scale DDoS attack, causing some hospital systems to be paralyzed and unable to operate normally. In February 2021, Rhoen-Klinikum, a prominent German medical group, was hit by a DDoS attack, causing some of the group's medical services to fail. In March 2021, the Singapore Public Hospital Group (SingHealth) was attacked by a DDoS again, resulting in patients being unable to use the medical information system. The above cases are only part of the DDoS attacks on hospital information systems around the world in recent years. These attacks not only seriously affected the hospital business but also exposed the loopholes and deficiencies in the robustness of hospital information systems.

Most existing hospital information systems use semi-trusted cloud servers to store structured data, resulting in various data transmission risks. Therefore, a robust blockchain medical data sharing system (BMDSS) is proposed. This drives the design of the threshold proxy re-encryption data sharing scheme used in this paper, where the scheme is designed to support secure storage and sharing of data. Precisely, the blockchain medical information system based on threshold proxy re-encryption consists of a data storage center and a data sharing center. The data storage center provides secure data storage and insertion, and the data sharing center provides security and controllability to ensure data sharing functions.

To address the above challenges, we use a Threshold proxy-based re-encryption blockchain medical information-sharing scheme. Specifically, we believe that the contributions of this paper are as follows:

1. In the system design, a dual-channel-based blockchain storage structure is used to store data storage information and data sharing records on two chains, aiming to support the confidentiality and integrity of medical privacy data while providing data. The characteristic is that shared records cannot be tampered with. While assisting the data-sharing process, the blockchain can also accurately record data access history to ensure transparent storage of medical data-sharing records.
2. This system uses a data sharing scheme based on threshold proxy re-encryption as a building block for designing our data sharing center. Threshold proxy re-encryption, as a privacy computing technology, can use proxy server clusters for ciphertext conversion without revealing the patient's private key. It reduces the risk of the original key being leaked while enhancing the robustness of the entire system and providing scalable performance provisioning. Utilizing the blockchain for proxy re-encryption key data transmission can not only guarantee the security of the patient's private key but also offer a transparent and recordable authorization process.

We then analyze the relative performance of BMDSS to demonstrate that it can achieve data confidentiality storage and efficient privacy data sharing within a given time. We noticed that in our scheme, secure authorization and robustness guarantees in the data-sharing process are realized, and a transparent and recordable authorization process is also provided. Its flexibility, isolation, and performance scalability help meet the needs of different medical scenarios.

The remainder of this paper is organized as follows. The next two sections present relevant literature and background information. Section 4 describes the blockchain medical information-sharing system based on threshold proxy re-encryption. Section 5 presents the simulation results of this system. The comparative summary shows that our BMDSS has a practically usable private data-sharing process and high system robustness. Finally, we conclude this work in Sect. 6.

2 Related Work

Blockchain is a distributed digital ledger based on encryption technology [19], which is represented as a decentralized peer-to-peer network. Once data is confirmed and stored in the blockchain, tampering with it isn't very easy. The immutability, decentralization, and anonymity of blockchain ensure the safe storage of medical data. With the widespread application of blockchain technology, blockchain-based medical information sharing has become the focus of current [3,6,13,15] medical data-sharing research. In a related study, researchers made various improvements to the system proposed by Cao et al. [23]. Medical data is stored securely using blockchain. They focus on architectural design but lack system implementation details and data-sharing processes. In particular, Nguyen et al. [7]. A specific system workflow is designed to enable sharing or access control of medical data. However, the system is implemented based on Ethereum, which needs to be improved for the hospital information system.

Due to the wide range of uses and privacy of medical data, the privacy protection needs cannot be met by using blockchain technology alone. In blockchain systems, the transparency of transaction records greatly increases the risk of privacy leakage. For example, patient data can be easily leaked and the analysis of transaction records can obtain the transaction patterns of users, so it needs to be coupled with more prominent privacy protection technologies. Kassab et al. [14] and Abu-Elezz et al. [1] studied common data protection mechanisms and analyzed the advantages of blockchain in medical data security protection and sharing. Xu et al [26] designed a blockchain technology-based privacy protection scheme for large-scale health data with fine-grained access control of medical data. Patients can revoke or add authorization to doctors autonomously through key management.

The systems proposed in [3,6,13,15,19] only use blockchain for secure medical data storage. They focus on architectural design but lack system implementation details and data-sharing processes. In addition, Xu et al. [26] used the PBFT consensus algorithm, and LIU et al. [21] adopted the DPOS consensus algorithm in their system, which makes data security relatively complicated. Some researchers have adopted a more lightweight Hyperledger Fabric (HF) architecture to implement blockchain networks. They have developed intelligent contracts for system access control and data management, but more research needs to be done on data sharing. Proxy re-encryption algorithms have been applied in different scenarios, enabling secure data sharing on semi-trusted cloud servers. Li et al. [20] designed a privacy enhancing scheme using to manage attributes. A content publisher generates an access policy based on the attributes defined by the third party and uses a random symmetric key to encrypt the data. However, its application in blockchain-based healthcare systems could be more developed. In this article [2], they propose a proxy re-encryption approach to secure data sharing in cloud environments. Data owners can outsource their encrypted data to the cloud using identity-based encryption, while proxy re-encryption construction will grant legitimate users access to the data.

Zeng et al. [10] proposed a medical data information system model based on blockchain, the Internet of Things, cloud storage, and proxy re-encryption algorithm, which realized reliable data collection and secure data storage, but the proxy they used. There may be a risk of data leakage in the re-encryption data sharing process. Zeng et al use a semi-trusted proxy server to provide the proxy re-encryption service, which can quickly become a performance bottleneck. When requests for a service increase, the server can be stressed, resulting in poor performance or downtime.

Dinh C. Nguyen [22] designed a blockchain for the secure sharing of mobile cloud-based e-health systems, where mobile devices are integrated with cloud computing to facilitate the exchange of medical data between patients and healthcare providers. Lanxiang Chen [8] designed a blockchain-based searchable encryption for electronic health record sharing. The index of electronic medical records is constructed and stored in the blockchain through complex logical expressions so that data owners can fully Control who can see their EHR data. Yi Chen [9] et al. designed a storage scheme based on blockchain and cloud storage to manage personal medical data, described the service framework for sharing medical records, and demonstrated and analyzed the medical area by comparing it with traditional systems?features of blockchain. Benjamin Barth [5] proposed a catalog of web services that provides access control for publishing, subscribing, and discovering disaster information and handles sensitive data for responders. Asma Khatoon [16] designed a blockchain-based intelligent contract system for healthcare management dedicated to facilitating multiple stakeholders involved in the healthcare system to provide better healthcare services and optimize costs.

The above studies focus on data storage security and have made long-term progress in the secure storage and authorized query of medical data. Only Zeng et al. [10] made a contribution to the secure sharing of medical data. However, his design still has a lot of room for improvement in terms of the security of the data-sharing process and the robustness of the data-sharing system. Therefore, it is necessary to design a blockchain medical information system based on threshold proxy re-encryption.

3 Preliminaries

3.1 Blockchain

Blockchain is a term in the field of information technology. A blockchain is a self-growing list of transaction records (also known as blocks) that are cryptographically concatenated to secure content. Each block contains the hash value of the previous block, the timestamp of this block, and the transaction data (usually represented by the hash value of the Merkle tree structure). This design makes the content of the block challenging to tamper with characteristics. Using a blockchain allows transactions to be efficiently recorded by multiple parties and permanently verifiable.

In essence, it is a shared database. The data or information stored in it has the characteristics of "unforgeable", "retaining traces throughout the process", "traceable", "open and transparent", and "collective maintenance". Based on these characteristics, blockchain technology has laid a solid "trust" foundation, created a reliable "cooperation" mechanism, and has broad application prospects.

The blockchain is formed by connecting blocks one by one, in which a block is a storage unit one by one which records all the communication information of each block node within a certain period of time.

Blockchain is a chained data structure that combines data blocks in a sequential manner in chronological order and is a cryptographically guaranteed, non-tamperable, and unforgeable distributed ledger.

Each block consists of two parts, Head and Body. The Head records the meta information of the current block, and the Body stores the actual data. Blocks are linked by random hash (also known as hash algorithm), and the last block contains the hash value of the previous block. With the expansion of information exchange, one block is followed by another to form. The result is called the blockchain. The included data is shown in Fig. 1.

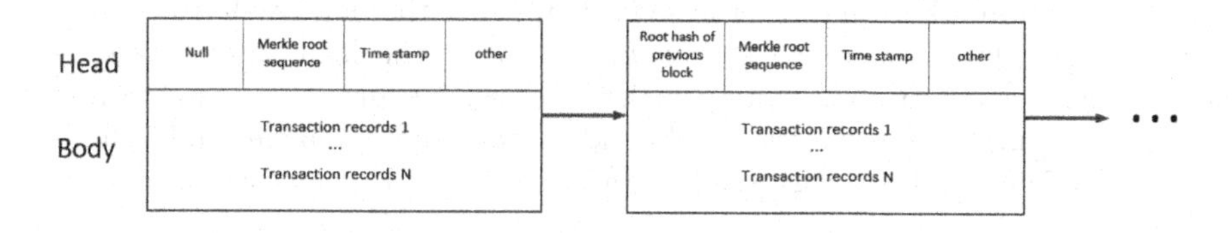

Fig. 1. Blockchain data structure.

To put it simply, the blockchain is a distributed shared ledger and database, which has the characteristics of decentralization, immutability, traceability throughout the process, traccability, collective maintenance, and openness and transparency. These characteristics ensure the "honesty" and "transparency" of the blockchain and lay the foundation for the creation of trust in the blockchain. The rich application scenarios of the blockchain are basically based on the fact that the blockchain can solve the problem of information asymmetry and realize collaborative trust and concerted action among multiple subjects. The blockchain is a distributed data storage, point-to-point transmission, consensus mechanism, and New application modes of computer technologies such as encryption algorithms.

Generally speaking, the blockchain system consists of a data layer, network layer, consensus layer, incentive layer, contract layer, and application layer.

The data layer encapsulates primary data and basic algorithms such as underlying data blocks and related data encryption and time stamps.

The network layer includes distributed networking mechanism, data dissemination mechanism, data verification mechanism, etc.

The consensus layer mainly encapsulates various consensus algorithms of network nodes.

The incentive layer integrates economic factors into the blockchain technology system, mainly including the issuance mechanism and distribution mechanism of financial incentives.

The contract layer mainly encapsulates various scripts, algorithms, and smart contracts, which is the basis of the programmable features of the blockchain.

The application layer encapsulates various application scenarios and cases of the blockchain. In this model, the chained block structure based on time stamps, the consensus mechanism of distributed nodes, the economic incentives based on consensus computing power, and the flexible and programmable smart contracts are the most representative innovations of blockchain technology.

The blockchain is encrypted using a hash algorithm, which is a one-way password mechanism in the blockchain to ensure that transaction information cannot be tampered with. Hash: y = hash(x), hashing x to get y can hide the original information x because there is no way to calculate x through y so as to achieve anonymity.

A "key pair" is used in the process of encryption and decryption. The two keys in the "key pair" have the characteristics of asymmetry. That is, in the process of sending information, the sender encrypts the data with a key. After receiving the data, the recipient can only decrypt the lead through another paired key. Asymmetric encryption makes it easier for any participant to reach a consensus, minimizes the friction boundary in value exchange, and can also achieve anonymity after transparent data and protect personal privacy.

3.2 Threshold Proxy Re-encryption

In proxy re-encryption, a semi-trusted proxy plays the role of ciphertext transformation. The text is converted to the ciphertext of the same plaintext encrypted by the delegator, who can then decrypt the converted ciphertext using his own private key. During the transformation process, the agent must have the ciphertext transformation key authorized by the entrusting party (that is, the agent re-encryption key), and at the same time, the agent cannot obtain any information about the corresponding plaintext in the ciphertext. Generally speaking, there are two classification methods for proxy re-encryption: one is based on the direction of ciphertext conversion, and proxy re-encryption can be divided into one-way and two-way. The former can only realize the ciphertext transformation from Alice to Bob, while the latter can not only recognize the ciphertext transformation from Alice to Bob but also vice versa; the other is based on the number of ciphertext transformations and can also re-encrypt the proxy Divided into one-time and multi-purpose, the former can only realize the ciphertext conversion from Alice to Bob and can convert the converted ciphertext again.

Additive notation is the norm that deals with elliptic curve cryptography, and this paper uses multiplication to represent operations on groups of elliptic curves, which is a common approach in the proxy re-encryption literature-a brief introduction to proxy server re-encryption. Proxy re-encryption is a particular type of public-key encryption that allows a proxy to convert one key into another without knowing the original message; to do this, the proxy must have an enabled re-encryption key. Therefore, it can be used to delegate decryption privileges, opening many applications that may charge for encrypted data. In the PRE literature, the parties involved are often referred to as the principal relationship:

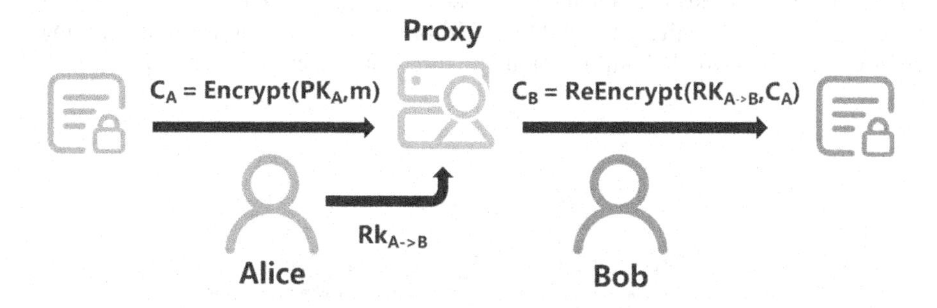

Fig. 2. Main actors and their interactions in the PRE environment.

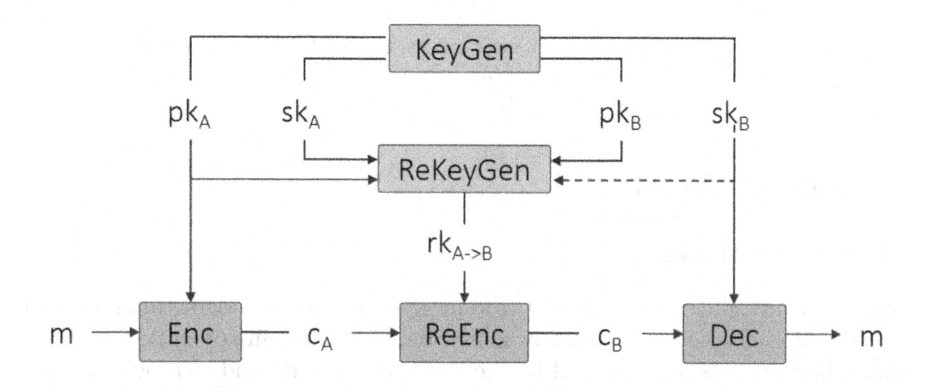

Fig. 3. Diagram of a proxy re-encryption scheme.

Figure 2. depicts the main actors and their interactions in the PRE environment. Since PRE is a particular type of PKE, the user also has a pair of public and private keys, as shown in the figure. Therefore, anyone who knows the public key can generate ciphertext for the corresponding recipient, which can only be decrypted with the corresponding decryption key. Figure 3. is a general diagram of a proxy re-encryption scheme.

The characteristic of Threshold Proxy Re-encryption lies in its split-key mechanism. Since it adopts the traditional proxy re-encryption scheme, the re-encryption process is carried out by a group of nodes, not just distributed by a single node. In order to perform the above tasks, threshold proxy re-encryption needs to specify a threshold as the minimum number of nodes that are allowed to perform re-encryption operations. In this way, the credentials are split among these nodes, very similar to the principle of Shamir's secret sharing, except that the re-encryption key is part of it instead of the shared private key. The name Threshold emphasizes the scheme's split-key property, which plays a central role in the distributed architecture of Threshold proxy re-encryption. Figure 4. Shows the main actors and their interactions in threshold proxy re-encryption.

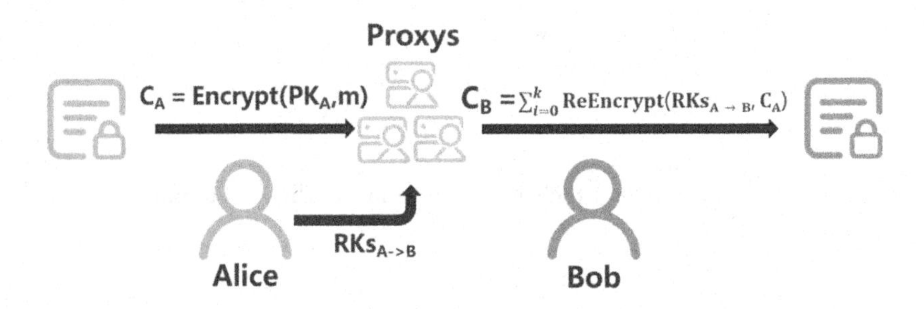

Fig. 4. Main actors and their interactions in the TPRE environment.

4 BMDSS Model

4.1 System Model

This paper introduces a threshold-based proxy re-encryption algorithm model to ensure that encrypted data is safely shared between two pairs of entities without valid information being obtained by the third-party entity and to provide secure storage of data in the cloud. Blockchain technology is introduced to ensure the consistency and immutability of data throughout the sharing process and to protect the integrity of cloud data. Based on blockchain and proxy re-encryption technology, this paper proposes A blockchain-based robust medical data-sharing system. (BMDSS). Figure 5 is the overall application architecture of the method in this paper, including Holder, Applicant, Proxys, Blockchain network, and Cloud storage.

According to the logical relationship between entities, the introduction is as follows:

Holder: The data holders in the TPR-BBHIS data-sharing process need to store the data safely and share the data with others safely. They are generally composed of patients from various hospitals, and they are managed by the hospital. Patients are mainly responsible for data collection, storage, and data

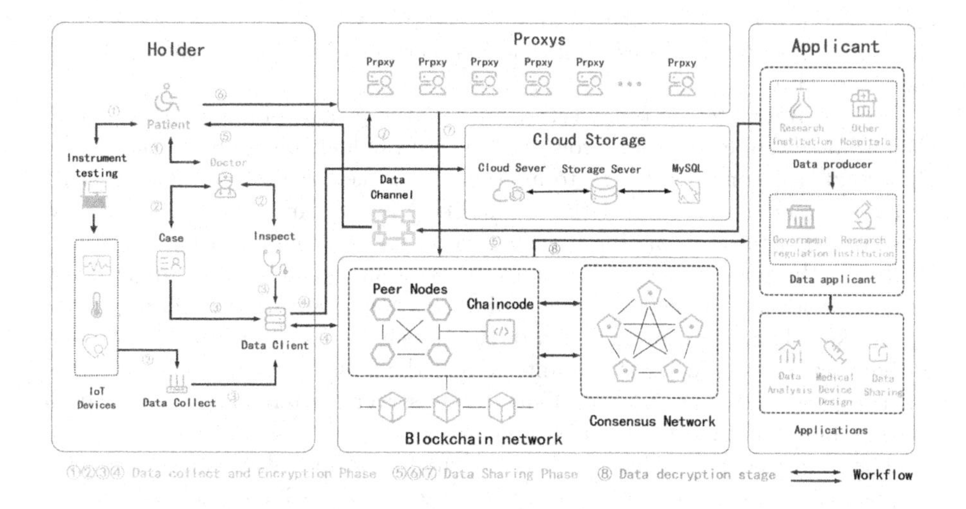

Fig. 5. System structure design.

sharing authorization. The first two are automatically assisted by the hospital's IoT devices and their data storage network. The patient's data-sharing authorization process is the main focus of this system.

Applicant: In this system, the requesting data holder needs to obtain the object of Internet of Things data for legal analysis and calculation. The data applicant is composed of medical institutions, government-related institutions, third-party research institutes, and other scientific researchers, mainly for data processing. Share the application, decrypt the encrypted data after obtaining the authorization, realize the analysis of the data, and support the release.

Blockchain network: The underlying blockchain network service can run on multiple nodes and achieve data consistency between nodes based on consensus mechanisms and P2P communication. The actual object of the node can be the two parties involved in the safe sharing of data, such as ordinary patients, doctors, medical equipment manufacturers, institutions, or universities for data analysis and research. Their common feature is that they hold data or need to mine the value of data. Out of distrust of third-party cloud service providers, they connect blockchain networks together to provide decentralized data recording services to ensure data consistency and integrity. The limitations of the blockchain itself make it impossible to store a large amount of file data, so it is only used to record the summary information of the shared data and verify the integrity of the data outside the chain. This system makes use of this feature of the blockchain, combined with the security mechanism of the proxy re-encryption algorithm, to form the security mechanism design of the system.

Cloud storage: Cloud storage resources provided by third-party cloud service providers mainly provide storage services for the entire data-sharing process. The cloud service stores the IoT data and the key data involved in the proxy re-encryption process in ciphertext, and the cloud service provider cannot decrypt the stored data to obtain user information, thus ensuring the confidentiality of the cloud data.

Proxys: Refers to the proxy server cluster, the primary function is to realize the authorization calculation process of the data holder, precisely to convert the ciphertext from one public key to another public key, realize safe data sharing and transmission, and protect the original key at the same time confidentiality. It is composed of multiple semi-trusted cloud servers connected to each other. It divides the patient's proxy key into multiple shares so that no single proxy server can obtain complete private key information, ensuring the security of the private key. Through the limitation of the threshold condition, the cooperation of the proxy server cluster is guaranteed, and the abuse or malicious behavior of a single proxy server is avoided. Thus, proxy server clusters enable secure data sharing and transfer using semi-trusted servers.

The above is the application entity structure proposed in this study. This chapter will introduce the design of the data security sharing model and its data sharing process in the BMDSS system.

4.2 Application

This section will take patient privacy data sharing as an example, describe the data sharing process based on the threshold proxy re-encryption algorithm, and focus on the formal operation process of relevant data. Assume that the data holder is a patient with physical examination data and related treatment data, and the data requester is a third-party medical research institution that requires the use of patient-related data due to scientific research needs. Data sharing is shown in Fig. 5., and the complete data security sharing process is formalized as follows:

Setup $(sec) \rightarrow (params)$:

1. The setup algorithm first determines a cyclic group $\mathbb{G}$ of prime order q, according to the security parameter *sec*. Let $g, U \in \mathbb{G}$ be generators. Let $H_2 : \mathbb{G}^2 \rightarrow \mathbb{Z}_q$, and $H_3 : \mathbb{G}^3 \rightarrow \mathbb{Z}_q$, and $H_4 : \mathbb{G}^3 \times Z_q \rightarrow \mathbb{Z}_q$ be hash functions that behave as random oracles. Let KDF: $\mathbb{G} \rightarrow \{0,1\}^{\updownarrow}$ be a key derivation function also modeled as a random oracle, where $\updownarrow$ is according to the security parameter *sec*. The global public parameters are represented by the tuple:

$$params = (\mathbb{G}, g, U, H_2, H_3, H_4, KDF) \tag{1}$$

2. Start the blockchain network, run the Fabric application, and achieve consistent state synchronization between nodes.
3. Output system public parameters to the blockchain for re-encryption calculation of proxy nodes.

KeyGen $(params) \rightarrow (pk, sk)$:
Patients and third-party medical institutions performer calculations separately to generate their own asymmetric keys:
Patients sample $a \in \mathbb{Z}_q$ uniformly at random, compute g^a and output the $keypair(pk_{P_i}, sk_{P_i}) = (g^a, a)$.
Third-party medical researcher sample $b \in \mathbb{Z}_q$ uniformly at random, compute g^b and output the $keypair(pk_{TH_i}, sk_{TH_i}) = (g^b, b)$.

Encrypt $(params, K, F) \rightarrow$
$(C_F, F_{hash}, F_{location}, C_K, K_{hash}, K_{location})$:

1. **FileEncrypt** $(params, K, F) \rightarrow (C_F, F_{hash}, F_{location})$:
   (a) The patient generates a symmetric key K and uses K to encrypt the data F to obtain the ciphertext C_F.
   (b) The patient calculates the hash F_{hash} of C_F using the hash function SHA-256. $F_{hash} = h_{256}(F)$
   (c) The patient uploads the ciphertext C_F to the cloud and obtains the ciphertext location information $F_{location}$ returned by the cloud.
2. **KeyEncrypt** $(params, K, F_{location}) \rightarrow (C_K, K_{hash}, K_{location})$:
   (a) Define the plaintext information M as follows:

$$M = K \parallel F_{location} \tag{2}$$

   (b) The patient encrypts the message M with his public key pk_{P_i} to obtain C_k.
   (c) The patient uses SHA-256 to calculate C_K to get K_{hash}.
   (d) The patient uploads the ciphertext C_K to the cloud and obtains the ciphertext location information $K_{location}$ returned by the cloud.

InfoUploade $(F_{hash}, F_{location}, K_{hash}, K_{location})$:
The patient calls the blockchain interface, puts F_{hash} and $F_{location}$ on the chain, then puts K_{hash} and $K_{location}$ winding.

KFragGen $(params, sk_{P_i}, pk_{TH_i}, C_F, th, sh) \rightarrow$
$(kFrags, kFrags_{hash}, kFrags_{location})$:

1. The third-party medical institution uses the channel to send its public key pk_{TH_i} to the patient as a data access request.
2. After receiving the request, the patient queries the authorized access list to determine whether there is authorization for pk_{TH_i}, if yes, proceed to step 3, otherwise terminate the process.
3. Query whether the $cFrags$ exists in the cloud through the blockchain and download it from the cloud if it exists. If they are exists, go to step 5. if there is no kF rags, go to step 4.
4. The patient calculates the re-encryption key rk offline according to pk_{TH_i} and sk_{P_i}, as follows:

(a) The patient generates the proxy secret key $RK_{A\rightarrow B}$based on his own secret key sk_{P_i} and Third-party medical institutions?s public key pk_{TH_i} and C_F.
(b) The patient calculates the N fragments of the re-encryption key between Alice and Bob according to the agent key $RK_{A\rightarrow B}$, the number of fragments N, threshold t, finally named $kFlags$.
(c) Calculate the hash $kFrags_{hash}$ of $kFrags$: $kFrags_{hash} = SHA_{256}(kFrags)$, upload $kFrags$ to the cloud location information $kFrags_{location}$, and upload $kFrags_{hash}$ and $kFrags_{location}$ to the blockchain together.
(d) The patient Send N $kFrag$ to N different proxy servers, ensuring that each server gets only one $kFrag$.
(e) The patient Send to the third-party medical institution and C_F to N proxy servers.

ReEncrypt $(params, C_F, kFrag) \rightarrow (cFrag, cFrag_{hash})$:

1. N proxy servers calculate their re-encryption key fragments $kFrag$ and C_K respectively, and reencapsuate the algorithm to reencapsulate the output C_K fragments $cFrag$ fragments.
2. Proxy calculates $cFrag_{hash} = H_{256}(cFrag)$, puts $cFrag_{hash}$ on the chain, and sends the return information of $cFrag$ and $cFrag_{hash}$ to the blockchain.

DencryptReEncrypt $(params, sk_{TH_i}, pk_{P_i},$
$Cfrags, C_F, kFrag,) \rightarrow (cFrag, cFrag_{hash})$:

1. The third-party medical institution receives the information returned from the blockchain and queries the blockchain according to the data returned from the blockchain to obtain $cFrag$ and the hash $cFrag_{hash}$ of $cFrag$.

The third-party medical institution receives $cFrag$ and the information returned from the blockchain and queries the blockchain according to the data returned from the blockchain to obtain the hash $cFrag_{hash}$ of $cFrag$.

2. Calculate $cFrag'_{hash} = H_{256}(cFrag)$ and compare whether $cFrag'_{hash}$ and $cFrag_{hash}$ are equal to check the integrity, if they are equal, proceed go to step 3, otherwise terminate the process.
3. The third-party medical institution decrypts $cFrag$ with the private key sk_{TH_i}, and obtains data M, including K,$F_{location}$,$cFrag_{hash}$.
4. The third-party medical institution downloads the ciphertext data C_F from the cloud according to $F_{location}$.
5. Calculate $F'_{hash} = H_{256}(C_F)$ and compare whether F'_{hash} and F_{hash} are equal to check the integrity, if they are equal, proceed go to step 6, otherwise terminate the process.
6. The third-party medical institution decrypts C_F with the symmetric key K to obtain the requested data.

5 System Evaluation

5.1 Environment

All experiments involved in this chapter are carried out on four high-performance CPU servers with the same configuration, and the specific design of each server is shown in Table 1 below.

Table 1. Hardware environment of the performance test.

Item	Specifications
CPU	Intel(R) Core(TM) i7-9700 CPU @ 3.00 GHz
RAM	16 GB
ROM	500 GB
Network Card	Intel,Red Hat, Inc. Virtio network device
OS	Ubuntu 20.04
Core Version	5.15.0-48-generic

For the software part, all the applications implemented in this experiment are deployed in the form of containers using Docker software, the development language includes Python and Golang, the Umbral algorithm is implemented based on the pyUmbral library of Python, the blockchain platform uses Hyperledger Fabric, and the functional testing software uses Postman is used, and the performance testing software uses the open source JMeter project of Apache Corporation. The specific version information of all the above software is shown in Table 2 below.

Table 2. Software environment of the performance test.

Item	Specifications
Docker	20.10.5
Python	3.9.4
Golang	1.15.1
JPBC	2.0.0
Hyperledger Fabric	1.4.0
Postman	8.0.10
Apache JMeter	5.4.1

5.2 Data Storage Performance

The Hyperledger Fabric platform used in this article is a blockchain consensus engine, which only provides the core functions of the blockchain, such as consensus mechanism, transaction ordering, state synchronization, upper-layer application programming interface (ABCI), etc. On this basis, this simulation experiment will be based on the blockchain officially provided by the hyperbook structure, and a simulation application program will be developed in Python language, including a user management module, proxy re-encryption module and IoT devices, cloud servers, proxy nodes, Blockchain data interaction module. The part of blockchain interaction uses gRPC to communicate with ABCI implemented in the Go language. gRPC is Google's open-source RPC framework that can be used to implement cross-language function communication. The blockchain part deploys the Fabric network to perform uplink operations and query operations, respectively.

The data interacting with the blockchain in this simulated application is mainly hash or file metainformation, which is small in scale, so the transaction start test in the performance test will construct some fixed-length data and upload it to the chain, while the query transaction test reads more The same blockchain data twice. The encryption and decryption steps in Umbral are performed offline, the re-encryption is performed by proxy nodes, and the storage of ciphertext data is performed by the cloud. All of the above have nothing to do with the blockchain, so this test does not include the overhead caused by the above steps but only provides for the overhead of on-chain test data and transaction queries.

In this experiment, Fabric is deployed in the form of Docker, including two organizations, a total of 4 peer nodes, an ordering node, and a ca node, to test the stability and response speed of the blockchain network. The response speed is measured using TPS (transactions per second). Regarding the environment deployment of this experiment, this performance test experiment is carried out on four high-performance servers, hereinafter referred to as servers 1–4, and the application services use Docker software to deploy container services. Among them, server one and Server 2 use multiple containers to simulate multiple node networks as a blockchain network cluster; Server 3 serves as a simulation application server and deploys simulation applications in a containerized manner; Server 4, as a client, deploys JMeter software for Initiate a request to the simulated application with a duration of 30 s. Experimentally tested, the transaction processing performance of Super Classified Fabric is up to 2400 TPS. When the number of concurrent requests increases, the performance of the blockchain network tends to decline, which is due to the impact of the network I/O rate. Generally speaking, when the concurrency increases, the performance of the blockchain network is stable without a timeout, which meets the requirements of the system.

5.3 Data Sharing Performance

The fuzzy algorithm used in this paper is a threshold-based proxy re-encryption algorithm, which divides a single proxy node into multiple proxy nodes in traditional proxy re-encryption. Therefore, when re-encrypting data, multiple proxy nodes are required to participate in collaborative computing to provide a secure data-sharing function. On this basis, this simulation experiment will use the official shadow Python library for simulation applications. The data shared in this simulation application is mainly symmetric key information, and the scale is small. Therefore, we will build and share some fixed-length data in the performance test and test the effect of data sharing by setting different thresholds and a total number of agents. Overhead. The experimental results are shown in Fig. 6.

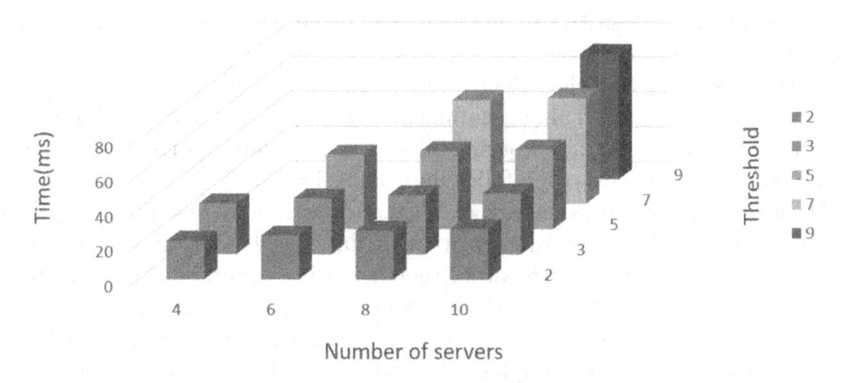

Fig. 6. Data Sharing Performance

According to the experimental test, the data-sharing performance of Umbra is at most within 70 milliseconds. When both the Threshold and the total number of agents increases, the blockchain network performance shows a downward trend, which is due to the influence of the network I/O rate. Generally speaking, when the threshold condition increases, the proxy re-encryption algorithm performs stably without a timeout, meeting the requirements of the system.

5.4 Security Analysis

BMDSS employs a new threshold proxy re-encryption method. The use of a new proxy re-encryption algorithm can ensure the security of the patient's private key and avoid the disclosure of patient privacy. But Zeng et al. adopted an insecure private key encapsulation method in the key conversion. They use the public key of the third-party medical institution to asymmetrically encrypt the patient's private key. This move significantly increases the risk of tampering and theft of patient privacy information. In the demo code, they store the converted proxy key in the transfer package but don't destroy it after use. Therefore, the existence of the above-mentioned risks can be confirmed.

Traditional proxy re-encryption schemes use a semi-trusted proxy server to provide proxy re-encryption services, which can quickly become a performance bottleneck. When requests for a service increase, the server can be stressed, resulting in poor performance or downtime. Once the server fails, all applications and services will stop running, which may cause business interruption and data loss. Individual servers are also difficult to scale because they require more resources. When more resources are required to support an application or service, you may need to replace it with a more powerful server, or increase the number of servers, adding cost and complexity. In BMDSS, the threshold proxy re-encryption algorithm solves this problem well. By constructing a proxy server cluster, the robustness of the entire system is guaranteed through hardware, and free choice of the authorization threshold can give users the right to set it. Moreover, authorization thresholds can be set, and the system can dynamically expand system performance.

In the design of the blockchain model, this paper uses a blockchain model with physically separated dual channels to ensure data security. However, we conducted multiple analysis experiments on the data sharing record data generation process, and came to the following conclusions:

Data-sharing authorization records should be stored on-chain by patients or proxy servers. The recording process of data applicants may result in the loss or falsification of usage records, thereby losing accuracy. As data owners, patients often authorize and record the devices used by patients to waste a lot of computing power and performance. The proxy server can be used as a scalable server cluster to complete this type of task very well. Therefore, we consider directly uploading the relevant keys to the blockchain network. This method completes the storage of authorization records while authorizing the applicant, which greatly improves the accuracy and security of usage records. However, if proxy key fragments are uploaded, all proxy servers can obtain all key fragments, resulting in one proxy server being able to complete all fragment proxy re-encryption processes, thereby losing the meaning of the threshold and making the system lose resistance.

In the end, we chose to store the authorization fragment re-encrypted by the proxy server on the blockchain, so that the applicant can obtain the authorization share while leaving the authorization record, and the anti-collusion still exists. In Zeng et al.'s system, data-sharing records are submitted by third-party medical institutions, real-time accuracy cannot be guaranteed, and it is challenging to track who has accessed which data. It can easily lead to data security issues, which can lead to data leakage, loss or misuse. This affects patients' trust in the system, leading to a crisis of faith. The relevant analysis results discussed in this chapter are shown in Table 3.

Table 3. Comparison of this system with other systems or schemes.

Character	Zeng et al	BMDSS
Security	N	Y
Robustness	N	Y
Authorization threshold	N	Y
Scalable	N	Y
Reliable records	N	Y
Collusion resistance	N	Y

6 Conclusion

In order to solve the problems existing in the process of medical privacy data sharing in the hospital information system, this paper proposes a privacy data sharing scheme based on threshold proxy encryption, which realizes the secure encapsulation and sharing of patient privacy data. Proxy re-encryption is used to secure data sharing, while blockchain is used to secure shared data. Aiming at the security problem and robustness of the semi-trusted proxy server of the traditional proxy encryption algorithm, we use the threshold value to divide the proxy server cluster from a single semi-trusted server to a decentralized authorization system, to ensure avoidance and strengthen the robustness and scalability of the system. In our design scheme, the security of data in the sharing process is tested and analyzed. The analysis results show that the security and system performance of this scheme meet the requirements of the system.

However, there are still some limitations in this study, and in future work, the following:

The threshold proxy re-encryption algorithm adopted in this paper is only suitable for data security sharing between one-to-one entities. It only supports one-to-many sharing, which is challenging to use at scale in dynamic and complex IoT environments. Future research will consider introducing an attribute-based proxy re-encryption mechanism to set different attribute policies for data requesters so that objects satisfying the specified attribute policies can access data and achieve fine-grained access control.

References

1. Abu-Elezz, I., Hassan, A., Nazeemudeen, A., Househ, M., Abd-Alrazaq, A.: The benefits and threats of blockchain technology in healthcare: a scoping review. Int. J. Med. Inform. **142**, 104246 (2020)
2. Agyekum, K.O.B.O., Xia, Q., Sifah, E.B., Cobblah, C.N.A., Xia, H., Gao, J.: A proxy re-encryption approach to secure data sharing in the internet of things based on blockchain. IEEE Syst. J. **16**(1), 1685–1696 (2021)
3. Ali, M.S., Vecchio, M., Pincheira, M., Dolui, K., Antonelli, F., Rehmani, M.H.: Applications of blockchains in the internet of things: a comprehensive survey. IEEE Commun. Surv. Tutorials **21**(2), 1676–1717 (2018)

4. Ball, M.J.: Hospital information systems: perspectives on problems and prospects, 1979 and 2002. Int. J. Med. Inform. **69**(2–3), 83–89 (2003)
5. Barth, B., Kabbinahithilu, G.C., de Cola, T., Bartzas, A., Pantazis, S.: A system for collaboration and information sharing in disaster management. In: Scholl, H.J., Holdeman, E.E., Boersma, F.K. (eds.) Disaster Management and Information Technology: Professional Response and Recovery Management in the Age of Disasters, pp. 53–64. Springer International Publishing, Cham (2023). https://doi.org/10.1007/978-3-031-20939-0_4
6. Cao, B., Li, Y., Zhang, L., Zhang, L., Mumtaz, S., Zhou, Z., Peng, M.: When internet of things meets blockchain: challenges in distributed consensus. IEEE Network **33**(6), 133–139 (2019)
7. Cao, S., Zhang, X., Xu, R.: Toward secure storage in cloud-based ehealth systems: a blockchain-assisted approach. IEEE Network **34**(2), 64–70 (2020)
8. Chen, L., Lee, W.K., Chang, C.C., Choo, K.K.R., Zhang, N.: Blockchain based searchable encryption for electronic health record sharing. Futur. Gener. Comput. Syst. **95**, 420–429 (2019)
9. Chen, Y., Ding, S., Xu, Z., Zheng, H., Yang, S.: Blockchain-based medical records secure storage and medical service framework. J. Med. Syst. **43**, 1–9 (2019)
10. Chen, Z., Xu, W., Wang, B., Yu, H.: A blockchain-based preserving and sharing system for medical data privacy. Futur. Gener. Comput. Syst. **124**, 338–350 (2021)
11. Farzandipur, M., Azimi, E., et al.: Factors affecting successful implementation of hospital information systems. Acta Informatica Medica **24**(1), 51 (2016)
12. internationale pour le traitement de l'information. Technical Committee 4" Information Processing in Health Care, F., Research", B., internationale pour le traitement de l'information. Working Conference on Hospital Information Systems (1979: Capetown), F., Shannon, R.H.: Hospital information systems: an international perspective on problems and prospects:[proceedings of IFIP Working Conference on Hospital Information Systems, Capetown, South Africa, 2–6 April 1979]. North Holland (1979)
13. Jin, H., Luo, Y., Li, P., Mathew, J.: A review of secure and privacy-preserving medical data sharing. IEEE Access **7**, 61656–61669 (2019)
14. Kassab, M., DeFranco, J., Malas, T., Laplante, P., Destefanis, G., Neto, V.V.G.: Exploring research in blockchain for healthcare and a roadmap for the future. IEEE Trans. Emerg. Top. Comput. **9**(4), 1835–1852 (2019)
15. Khan, M.A., Salah, K.: Iot security: Review, blockchain solutions, and open challenges. Futur. Gener. Comput. Syst. **82**, 395–411 (2018)
16. Khatoon, A.: A blockchain-based smart contract system for healthcare management. Electronics **9**(1), 94 (2020)
17. Lee, I., Sokolsky, O.: Medical cyber physical systems. In: Proceedings of the 47th Design Automation Conference, pp. 743–748 (2010)
18. Lee, I., et al.: Challenges and research directions in medical cyber-physical systems. Proc. IEEE **100**(1), 75–90 (2011)
19. Lee, T.F., Li, H.Z., Hsieh, Y.P.: A blockchain-based medical data preservation scheme for telecare medical information systems. Int. J. Inf. Secur. **20**, 589–601 (2021)
20. Li, B., Huang, D., Wang, Z., Zhu, Y.: Attribute-based access control for ICN naming scheme. IEEE Trans. Dependable Secure Comput. **15**(2), 194–206 (2016)
21. Liu, X., Wang, Z., Jin, C., Li, F., Li, G.: A blockchain-based medical data sharing and protection scheme. IEEE Access **7**, 118943–118953 (2019)

22. Nguyen, D.C., Pathirana, P.N., Ding, M., Seneviratne, A.: Blockchain for secure ehrs sharing of mobile cloud based e-health systems. IEEE access **7**, 66792–66806 (2019)
23. Pei, L., Legge, D., Stanton, P.: Hospital management in china in a time of change. Chin. Med. J. **115**(11), 1716–1726 (2002)
24. Qiu, H., Qiu, M., Liu, M., Memmi, G.: Secure health data sharing for medical cyber-physical systems for the healthcare 4.0. IEEE J. Biomed. Health Inform. **24**(9), 2499–2505 (2020)
25. Reichertz, P.L.: Hospital information systems-past, present, future. Int. J. Med. Inform. **75**(3–4), 282–299 (2006)
26. Xu, J., Xue, K., Li, S.: Tian: Healthchain: a blockchain-based privacy preserving scheme for large-scale health data. IEEE Internet Things J. **6**(5), 8770–8781 (2019)

FRAIM: A Feature Importance-Aware Incentive Mechanism for Vertical Federated Learning

Lei Tan[1,2], Yunchao Yang[1,2], Miao Hu[1,2], Yipeng Zhou[3], and Di Wu[1,2](✉)

[1] School of Computer Science and Engineering, Sun Yat-sen University, Guangzhou 510006, China
{tanlei6,yangych65}@mail2.sysu.edu.cn,
{humiao5,wudi27}@mail.sysu.edu.cn

[2] Guangdong Key Laboratory of Big Data Analysis and Processing, Guangzhou 510006, China

[3] School of Computing, Faculty of Science and Engineering, Macquarie University, Sydney, NSW 2109, Australia
yipeng.zhou@mq.edu.au

Abstract. Federated learning, a new distributed learning paradigm, has the advantage of sharing model information without revealing data privacy. However, considering the selfishness of organizations, they will not participate in federated learning without compensation. To address this problem, in this paper, we design a feature importance-aware vertical federated learning incentive mechanism. We first synthesize a small amount of data locally using the interpolation method at the organization and send it to the coordinator for evaluating the contribution of each feature to the learning task. Then, the coordinator calculates the importance value of each feature in the dataset for the current task using the Shapley value method according to the synthetic data. Next, we formulate the process of organization participation in the federation as a feature importance maximization problem based on reverse auction which is a knapsack auction problem. Finally, we design an approximate algorithm to solve the proposed optimization problem and the solution of the approximation algorithm is shown to be $\frac{1}{2}$-approximate to the optimal solution. Furthermore, we prove that the proposed mechanism is truthfulness, individual rationality, and computational efficiency. The superiority of our proposed mechanism is verified through experiments on real-world datasets.

Keywords: Vertical federated learning · Incentive mechanism · Feature importance · Reverse auction · Approximate algorithm

1 Introduction

With the rapid development of the Internet of Things (IOT), more and more devices of organizations are deployed in the network and a massive amount of data is generated from these network devices. Federated learning [1,13,16,20], as a new distributed computing paradigm, which advocates that organizations use their training data locally to collaboratively train a global model, avoiding the exposure of raw data to a central coordinator. In view of the distribution characteristics of data, federated learning is categorized as horizontal federated learning and vertical federated learning [31]. Horizontal

Z. Tari et al. (Eds.): ICA3PP 2023, LNCS 14491, pp. 132–150, 2024.
https://doi.org/10.1007/978-981-97-0808-6_8

federated learning is introduced when the data share the same feature space but the samples are different. In vertical federated learning, organizations may collect different features from the same set of users resulting in disjoint feature spaces on different organizations.

Although federated learning is excellent in enabling collaborative learning of global models while preserving data privacy, it still faces a critical challenge in practical applications, namely, incentivizing organizations to participate in federated learning through their contributions. Most current works on federated learning ideally assume that organizations altruistically contribute their data, computing, and communication resources to join in federated learning, which is unrealistic in the real world. Selfish organizations will be reluctant to spend their resources on federated learning without any reward.

In view of this challenge, many studies [4,9,36] applied traditional incentive mechanisms to federated learning, which are mainly applicable to horizontal federated learning. The incentive mechanism of vertical federated learning has not received much attention. For vertical federated learning, there are two new challenges in designing an incentive mechanism. First, the properties of disjoint feature spaces make it hard to motivate organizations to participate in vertical federated learning from the feature perspective. Existing federated learning incentive mechanisms motivate organizations assuming that each organization has a complete feature space which is invalid in vertical federated learning. However, in vertical federated learning, organizations may have the same sample IDs, the disjoint subsets of features, and each feature contributes differently to the learning performance. From an economic point of view, it is more efficient to select features with greater contributions to participate in federated learning. Second, through intermediate computation results, it is difficult to quantify the contribution of each feature to federated learning. The methods commonly used to measure the contribution of organizations in horizontal federated learning include loss reduction based on local models [4], organization reputation [9,36], and so on. In vertical federated learning, the outputs of local models are uploaded instead of local models or model parameters. As a result, these methods cannot be used in vertical federated learning directly.

To address the above challenges, in this paper, we propose a Feature impoRtance-Aware Incentive Mechanism for vertical federated learning, called *FRAIM*, which guarantees three economic properties, namely, truthfulness, individual rationality, and computational efficiency. To address the first challenge, FRAIM focuses on the feature perspective and motivates organizations to use features with high importance to participate in federated learning according to the importance of each feature to the learning task. Based on this idea, the mechanism models the federated learning process among each organization's features and the coordinator as a reverse auction. To address the second challenge, in FRAIM, a feature that intends to participate in vertical federated learning will use interpolation to synthesize a small amount of data locally, and then send synthetic data together with its bid to the coordinator through the organization where the feature is located. Then, the coordinator employs the Shapley value method to estimate the importance value of each feature to the learning task based on the received synthetic data so as to determine the contribution of each feature. Then we formulate the reverse auction problem as a knapsack problem and design an approximation algorithm to select participating features, we also provide a theoretical guarantee of $\frac{1}{2}$ approximation ratio for the solution of the approximation algorithm. Lastly, we conduct a series of experiments using real-world datasets to demonstrate the performance of FRAIM.

Overall, the novelty and technical contributions of this paper are summarized below.

- Different from existing works, we are the first to consider the data distribution characteristics of vertical federated learning and propose a feature importance-aware incentive mechanism, which satisfies truthfulness, individual rationality, and computational efficiency.
- We model the federated learning process among the organization's features and the coordinator as a reverse auction and propose an approximate algorithm with $\frac{1}{2}$ approximate ratio to solve the auction results. What's more, the auction only needs to be conducted once during the whole training process.
- We propose a method to estimate the contribution of features to the vertical federated learning task based on the Shapley value, and upload the data synthesized on the organization using the interpolation method to the coordinator to solve the problem that the feature contribution cannot be directly estimated on the coordinator side.
- We conduct a series of experiments using three real-world datasets to validate the effectiveness of our proposed mechanism. The experiment results show that FRAIM can perform better than baselines, and improve the test accuracy by 20.59% over the theoretically optimal organization selection mechanism.

2 Related Work

2.1 Federated Learning

With the generation of massive data, McMahan *et al.* [20] first proposed federated learning to relieve the pressure of communication and storage for centralized training and to protect data privacy. Because of its appealing property, federated learning has received tremendous attention in recent years.

Konečný *et al.* [11] designed structured and sketch updates to improve the communication efficiency of federated learning. Wang *et al.* [27] proposed a novel communication efficient adaptive federated learning method FedCAMS and provided theoretical convergence guarantees. In addition to horizontal federated learning, vertical federated learning has also attracted extensive interest. Considering the data characteristics of vertical federated learning, Luo *et al.* [19] proposed several feature inference attack methods to study the potential privacy leakage on model predictions in vertical federated learning. Wu *et al.* [29] provided a novel approach to preserve the privacy of vertical federated learning based on the tree model. In addition, to reduce communication and computational costs, Zhang *et al.* [37] also proposed AsySQN, an asynchronous stochastic quasi-Newton framework for vertical federated learning. Similarly, to improve the training speed of vertical federated learning, Fu *et al.* [5] propose a novel and efficient federated learning gradient boosting decision tree system and design a parallel training protocol to reduce idle periods.

The above works have studied federated learning from different perspectives, e.g., communication efficiency, statistical heterogeneity, and security. It is worth noting that these works are based on an ideal assumption that all organizations contribute data, communication, or computing resources to federated learning altruistically and submit honest updates. However, the reality is that without a reasonable incentive mechanism, organizations may be reluctant to contribute their resources to participate in federated learning or submit malicious updates.

2.2 Incentive Mechanism of Federated Learning

To motivate reliable organizations with high-quality data to participate in federated learning, Kang *et al.* [9] proposed an effective federated learning incentive mechanism based on reputation and contract theory. Similarly, Zhang *et al.* [36] designed a federated learning incentive mechanism based on reputation and reverse auction theory to select reliable organizations with high-quality data. Liu *et al.* [15] designed a privacy-preserving incentive mechanism for a federated cloud edge learning system, and formulated the incentive problem as a three-layer Stackelberg game. Zhan *et al.* [35] proposed an incentive mechanism based on deep reinforcement learning to determine the optimal pricing strategy for the coordinator and the optimal training strategy for organizations. To determine winners in auctions and maximize social welfare, Le *et al.* [12] introduced a primitive-dual greedy auction mechanism for federated learning. For multi-task federated learning, Deng *et al.* [4] proposed a quality-aware federated learning incentive mechanism based on a reverse auction to encourage the participation of high-quality organizations. Yu *et al.* [34] also designed an incentive mechanism for a federated learning and analysis system with multiple tasks based on a multi-leader-follower game. Ng *et al.* [22] introduced a two-level incentive mechanism to encourage data owners and organizations to participate in the training task of coding federated learning. Lim *et al.* [14] also proposed a two-layer incentive mechanism, the lower layer uses contract theory for cross-device federated learning, and the upper layer uses cooperative games for cross-silo to prevent free-rider attacks. Also for cross-silo federated learning, Tang *et al.* [25] introduced an incentive mechanism to address the heterogeneity of organizations and the characteristics of public goods.

These incentive mechanisms described above are all aimed at horizontal federated learning, whether in the cross-device scenario or the cross-silo scenario. They cannot be used for vertical federated learning because of the distribution characteristics of data. Lu *et al.* [17] proposed an incentive mechanism for vertical federated learning, but they assumed that each organization has label data, and the mechanism was not motivated from a feature perspective. There are few efforts dedicated to the incentive mechanism of vertical federated learning based on the feature perspective. To fill in the gap, our work designs a novel incentive mechanism that considers the feature importance of data distribution in vertical federated learning.

3 System Model and Problem Definition

In this section, We first describe the vertical federated learning and system model, then we define the feature importance-aware vertical federated learning problem.

Table 1. List of notations

Notation	Description
k, K	the index of organization, the total number of organizations
i, N	the index of data sample, the total number of data samples
x^k, y_i	the local dataset distribution of organization k, the label data
d, D	the index of feature, the total number of features
$\mathcal{F}, \mathcal{F}_k$	the set of all features, the feature set of organization k
d_k, ω^k	the feature number of local data in organization k, local model parameters of organization k
M^k, h_i^k	local model of organization k, the output of the local model M^k with samples x_i^k
b_d, r_d	the bid price of feature d, the reward feature d participating in the federated learning task
c_d	the cost of feature d participating in the federated learning task
u_d, U_k	the utility of feature d, the utility of organization k
FI_d	the Shapley value of feature d
Q_d, q_d	feature importance of feature d, the unit feature importance of feature d
α_d, S	binary variable indicating whether feature d is selected by the coordinator, the set of features selected by the coordinator
B	budget for the federated learning task provided by the initiator

3.1 Vertical Federated Learning

Vertical federated learning (VFL) is a federated learning system consisting of a set of organizations sharing the same sample IDs but disjoint feature data, and a trusted and impartial coordinator [31,37]. Consider a VFL scenario with K organizations and one coordinator with N data samples $\{x_i \in \mathbb{R}^D\}_{i=1}^N$, where D is the total number of all features, and the entire feature set can be expressed as $\mathcal{F} = \{1, ..., d, ..., D\}$. And each organization's local data set is denoted as $\{x_i^k \in \mathbb{R}^{d_k}\}_{k=1}^K$, where d_k is the feature number of local data in organization k. Let $\mathcal{F}_k$ denote the feature set corresponding to the data set x^k of organization k, obviously, $\mathcal{F} = \mathcal{F}_1 \cup ... \cup \mathcal{F}_k$. Among all the K organizations, only one organization has the label data $\{y_i \in \mathbb{N}\}_{i=1}^N$. Table 1 presents the main notations used in this paper and their meanings.

Like horizontal federated learning (HFL) [8,20], the VFL task is also an empirical risk minimization problem which can be formulated as:

$$\min_{\omega^1,...,\omega^K} \frac{1}{n} \sum_{i=1}^{n} \mathcal{L}(\omega^1, ...\omega^K; \{x_i, y_i\}), \tag{1}$$

where

$$\mathcal{L}(\omega^1, ...\omega^K; \{x_i, y_i\}) = \ell(\hat{y}_i; y_i), \tag{2}$$

$$\hat{y}_i = f(\sum_{k=1}^{K} h_i^k), \tag{3}$$

$$h_i^k = M^k(\omega^k; x_i^k). \tag{4}$$

Note that $\ell(\cdot)$ is the loss function, $f(\cdot)$ is the activation function, $\hat{y}$ is the predicted value, M^k and ω^k are the local model and model parameters of organization k, respectively. Naturally, h_i^k is the output of the local model M^k with samples x_i^k.

3.2 System Model

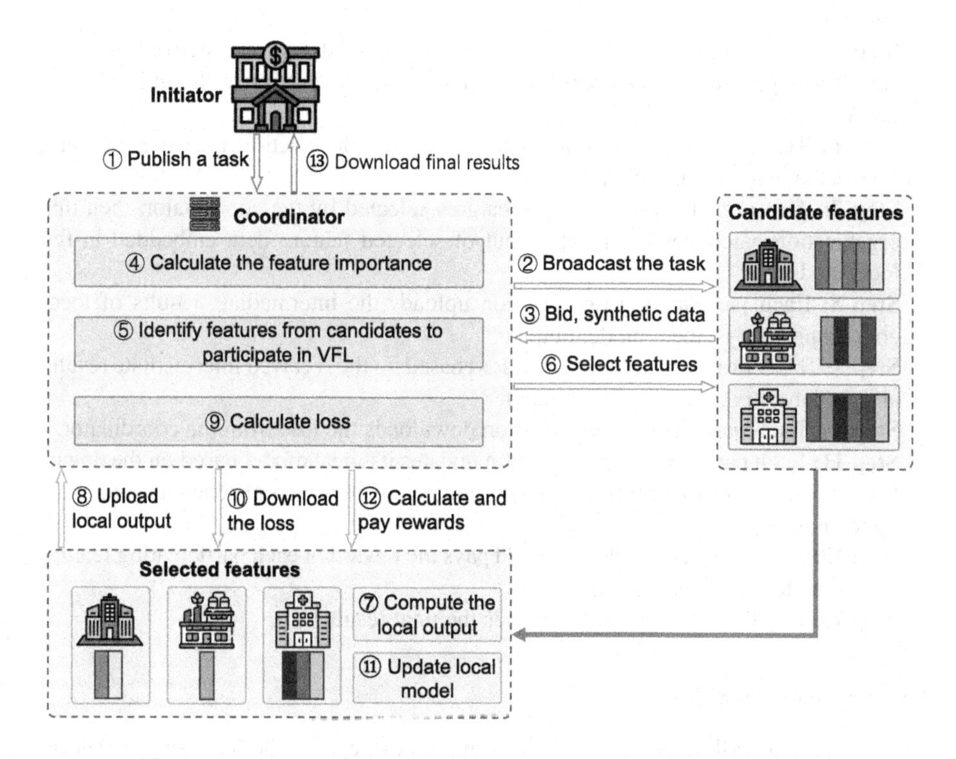

Fig. 1. The workflow of vertical federated learning incentive mechanism.

We consider vertical federated learning in the cross-silo scenario, where the participants are organizations. In such a community, the initiator releases a federated learning task with a limited budget to the community through the coordinator and hires other organizations to participate in the federated learning. The initiator is usually the label owner. In our design, those interested organizations consider the cost and expected remuneration of participating in federated learning, and decide whether to submit synthetic data and the corresponding bid for each feature to the coordinator. The coordinator utilizes the received synthetic data to calculate the importance value of the data features relative to the tasks submitted by the initiator, and then selects the participating features in combination with the collected bids. Participating organizations use their selected features to train a federated learning model. Finally, the initiator pays participants through the coordinator. The detailed workflow is shown in Fig. 1.

- **Step 1:** The initiator publishes a federated learning task to the coordinator and submits the task budget.
- **Step 2:** The coordinator broadcasts this federated learning task to all organizations.
- **Step 3:** Organizations interested in the task can submit bids for each feature of their local data and upload locally synthesized small amounts of data.
- **Step 4:** Based on the received synthetic data, the coordinator estimates the importance value of each feature to the federated learning task using the Shapley value method.
- **Step 5:** The coordinator identifies features from candidates to participate in federated learning based on bids and feature importance to maximize the utility of the initiator.
- **Step 6:** The coordinator announces the results of the auction to candidates and selects the features identified in Step 5.
- **Step 7:** If an organization has local features selected by the coordinator, then this organization calculates the local output of selected feature data embedded in the local model.
- **Step 8:** Each participating organization uploads the intermediate results of local computing, that is, local model output.
- **Step 9:** The coordinator calculates the loss based on the received intermediate results and the labels provided by the initiator.
- **Step 10:** Each participating organization downloads the loss from the coordinator.
- **Step 11:** Each participating organization updates its local model based on the downloaded loss. Return to **Step 7** until the number of iterations reaches the predetermined number.
- **Step 12:** The coordinator calculates and pays the reward of each participating feature according to the reward function.
- **Step 13:** Finally, the initiator downloads the final result.

3.3 Problem Definition

Following the workflow described in the previous subsection, the VFL system selects participating feature sets within a limited budget. Formally, the feature importance-aware vertical federated learning problem is defined below.

We introduce an auxiliary variable $\boldsymbol{\alpha} = \{\alpha_d\}_{d=1}^{D}$, where $\alpha_d \in \{0, 1\}$ is a binary variable to indicate whether the feature d is selected by the coordinator, which is equal to 1 if it is selected, otherwise $\alpha_d = 0$. Denote by c_d the training cost of the feature d, and its bid is b_d. Let $\boldsymbol{r} = \{r_d\}_{d=1}^{D}$, and r_d represent the reward for its participation in training, then the utility of feature d is u_d:

$$u_d = (r_d - c_d)\alpha_d. \tag{5}$$

Obviously, the utility U_k of the organization k is:

$$U_k = \sum_{d \in \mathcal{F}_k} u_d. \tag{6}$$

Next, we give a definition of feature importance value.

Definition 1. ***(Feature importance value).*** *The feature importance value refers to the contribution value of the feature to the task. FI_d is the Shapley value of the feature d, and $FI_d \geq 0$. The feature importance value $Q_d = \frac{1-e^{-FI_d}}{1+e^{-FI_d}}$ of feature d is represented by normalizing FI_d to the interval $[0, 1]$.*

Definition 2. ***(Initiator's utility).*** *The initiator's utility is the sum of the contributions of all features:*

$$\sum_{d=1}^{D} Q_d \alpha_d. \tag{7}$$

Therefore, the feature importance-aware vertical federated learning problem ($\mathcal{P}1$) can be formulated as:

$$\mathcal{P}1: \quad \max_{\boldsymbol{\alpha},\boldsymbol{r}} \sum_{d=1}^{D} Q_d \alpha_d, \tag{8a}$$

$$\text{s.t.} \quad \sum_{d=1}^{D} r_d \alpha_d \leq B, \tag{8b}$$

$$\alpha_d \in \{0, 1\}, \ \forall d, \tag{8c}$$

$$\alpha_d = 0, \ \forall d \notin S, \tag{8d}$$

where S is a set of selected features, and B is the available reward budget.

The feature importance-aware vertical federated learning problem not only needs to maximize the initiator's utility, but also guarantee three economic properties, namely, truthfulness, individual rationality, and computational efficiency [4,12,28,36].

Definition 3. ***(Truthfulness).*** *The organization can only maximize its utility by disclosing the true cost of each feature when bidding i.e., $b_d = c_d$. No organization can gain more utility by submitting an untruthful bid price.*

Definition 4. ***(Individual rationality).*** *The utility of each participant is non-negative, i.e., $r_d \geq c_d$.*

Definition 5. ***(Computational efficiency).*** *A mechanism is computationally efficient when feature selection and reward payments for participation can be computed in polynomial time.*

4 Incentive Mechanism Design of Vertical Federated Learning

4.1 Data Synthesis and Feature Importance Estimation

In the federated learning system, the data is stored locally on the organization, and the coordinator does not have any relevant data. In order to calculate the feature importance, we use the interpolation method to synthesize a small amount of data locally on the organization, and then send the synthesized data to the coordinator. Finally, the coordinator estimates the feature importance value of each feature based on these synthesized data.

Data Synthesis. In recent studies, data augmentation has been applied to federated learning to preserve raw data privacy [23,33]. In this paper, piecewise cubic Hermite interpolation is used to synthesize data and send it to the coordinator to preserve the raw data privacy. Compared with general interpolation methods, piecewise cubic Hermite interpolation [18] avoids Runge's phenomenon, provides sufficient interpolation accuracy and has low implementation complexity.

Assuming that the known function $f(x)$ is at $j+1$ mutually different nodes $x_i(i=0,1,...,j)$ on the interpolation interval $[A_1,A_2]$ satisfy $f(x_i)=f_i$ and $f'(x_i)=f_i'(i=0,1,...,j)$, if the function $H(x)$ exists, the following conditions are met:

- The polynomial degree of $H(x)$ in each interval $[x_i,x_{i+1}]$ is 3;
- $H(x_i)=f(x_i)$, $H'(x_i)=f'(x_i)$, $i=0,1,...,j$.

Then $H(x)$ is called the piecewise cubic Hermite interpolation polynomial of $f(x)$ at $j+1$ nodes x_i, and

$$H(x)=\beta x^3+\gamma x^2+\theta x+\epsilon. \tag{9}$$

Let us construct $H(x)$ using the basis function approach as in works [18,26]. Then we can get the expression of $H(x)$ is:

$$\begin{aligned}H(x)=&(1+2\frac{x-x_i}{x_{i+1}-x_i})(\frac{x-x_{i+1}}{x_i-x_{i+1}})^2 f(x_i)+(1+2\frac{x-x_{i+1}}{x_i-x_{i+1}})(\frac{x-x_i}{x_{i+1}-x_i})^2 f(x_{i+1})\\&+(x-x_i)(\frac{x-x_{i+1}}{x_i-x_{i+1}})^2 f'(x_i)+(x-x_{i+1})(\frac{x-x_i}{x_{i+1}-x_i})^2 f'(x_{i+1}).\end{aligned} \tag{10}$$

Note that the coordinator will not use synthetic data to train a model, the synthetic data is effective in evaluating feature importance, but is inefficient for model training because of its small amount of data. By the way, the coordinator is trusted and impartial, its responsibility is only to calculate the importance of features and the incentive mechanism implementation, and will not use synthetic data to train the model for the initiator.

Privacy Analysis: This interpolation method can still protect the original data from being leaked. Each piece of piecewise cubic Hermite interpolation is a cubic Hermite interpolation with two vertices. For each piece, the interpolation polynomial has 4 unknown parameters, and only 1 equation can be obtained according to an interpolated value. As we all know, the solution of a polynomial with 4 unknown parameters requires at least 4 equations, so the original data cannot be inferred from the interpolated values.

Feature Importance Value Calculation. The Shapley value, which satisfies the characteristics of symmetry, monotonicity, and linearity, is a classical method used to calculate the average marginal contribution of all players in cooperative game theory [6]. In this paper, the feature importance value is calculated by calculating the average marginal contribution of each feature based on the Shapley value method. Given a objective function $F(\cdot)$, the Shapley value FI_d for feature d is the weighted average change from adding d to subsets $A\subseteq S\setminus\{d\}$. The expression of the Shapley value is

$$FI_d=\sum_{A\subseteq S\setminus\{d\}}\frac{|A|!(|S|-|A|-1)!}{|S|!}(F(S\cup\{d\})-F(S)). \tag{11}$$

The computational complexity of the Shapley value is exponential, it cannot be directly used for feature importance estimation when the number of features is relatively large. Therefore, we utilize a Monte Carlo sampling-based Shapley value estimation algorithm [3,24] to estimate feature importance in vertical federated learning.

Assuming that the total number of features is d and the amount of generated data is G, an algorithm based on the Shapley value method is used to calculate the feature importance value of a single feature on the coordinator. Since the algorithm takes the contribution of all values within a feature as the important value of the feature, the algorithm can evaluate features with missing values. The specific details of the algorithm are shown in Algorithm 1. Where n and m are the numbers of sampling, and there is no uniform rule for their settings. From an empirical perspective, the larger the values, the higher the estimation accuracy, but the longer the calculation time.

Algorithm 1. Feature importance value calculation

Input: synthetic data $\{x_i, y_i\}_{i=1}^G$, objective function F, numbers of sampling n and m;
Output: $\{Q_d\}_{d=1}^D$: Feature importance value of each feature;
1: **for** $d = 1$ to D **do**
2: $FI_d = 0$;
3: **for** $i = 1$ to n **do**
4: Sample (x, y) from $\{x_i, y_i\}_{i=1}^G$;
5: $\delta = 0$;
6: **for** 1 to m **do**
7: Sample a random instance $z \in \{x_i, y_i\}_{i=1}^G$;
8: Construct two instances:
9: $x_{+d} = (\underbrace{x_1, ..., x_{d-1}, x_d,}_{values\ from\ x} \underbrace{z_{d+1}, ..., z_D}_{values\ from\ z});$
10: $x_{-d} = (\underbrace{x_1, ..., x_{d-1}}_{values\ from\ x}, \underbrace{z_d, z_{d+1}, ..., z_D}_{values\ from\ z});$
11: Compute $\delta = \delta + F(x_{+d}) - F(x_{-d})$;
12: **end for**
13: $FI_d = FI_d + \frac{\delta}{m}$;
14: **end for**
15: Compute $Q_d = \frac{1-e^{-FI_d}}{1+e^{-FI_d}}$;
16: **end for**
17: **return** $Q_1, Q_2, ..., Q_D$

4.2 Design of FRAIM

The problem $\mathcal{P}1$ can be formulated as the knapsack auction problem. We define the budget B as the seller's capacity, and the feature's bid price b_d as the bidder's capacity, assuming this is a variable. We define Q_d as the bidder's valuation, and it is fixed. In this way, the problem $\mathcal{P}1$ is transformed into a knapsack auction problem in which fixed-valuation bidders bid on knapsack capacity, and it is a 0–1 knapsack problem. As we all know, the knapsack problem is a classical NP-hard [7]. For the NP-hard problem of $\mathcal{P}1$, a simple heuristic idea is to greedily select features with greater unit feature importance. So we designed an approximation algorithm to solve the problem $\mathcal{P}1$ according to this idea, which can satisfy truthfulness, individual rationality, and computational efficiency while maximizing the objective. The design process of the approximation algorithm is described below.

First, we define the unit feature importance of each feature:

$$q_d = \frac{Q_d}{b_d}, \tag{12}$$

and we sort the sequence q_d in a descending order to get $q_1 \geq q_2 \geq ... \geq q_D$. Then, we find the smallest j, where j needs to satisfy:

$$\sum_{d=1}^{j} \frac{Q_d}{q_j} > B,\ 1 \leq j \leq D, \tag{13}$$

and we design the reward function:

$$r_d = \frac{Q_d}{q_j}. \tag{14}$$

The details of feature selection and reward payment are shown in Algorithm 2.

Algorithm 2. Approximation algorithm for solving the problem $\mathcal{P}1$

Input: available budget B, unit feature importance q_d;
Output: feature selection results $\boldsymbol{\alpha}$, reward set $\boldsymbol{r}$, selected feature set S;
1: Initialize $\alpha_d \leftarrow 0$ and $r_d \leftarrow 0$ for each d, $S \leftarrow \emptyset$;
2: Sort all unit feature importance $\{q_d\}_{d=1}^{D}$ in descending order, then get $q_1 \geq q_2 \geq ... \geq q_D$;
3: Find the smallest $j(1 \leq j \leq D)$ such that
4: $\sum_{d=1}^{j} \frac{Q_d}{q_j} > B$;
5: **for** $d = 1$ to $j - 1$ **do**
6: $\quad S \leftarrow S + \{d\}$;
7: $\quad r_d = \frac{Q_d}{q_j}$;
8: **end for**
9: **for** each $d \in S$ **do**
10: $\quad \alpha_d = 1$;
11: **end for**
12: **return** $(\boldsymbol{\alpha}, \boldsymbol{r}, S)$

The existing incentive mechanisms of HFL utilizing auction theory [4,12,36] need to perform the same number of auctions as iterations throughout the training process. In contrast, our designed mechanism is more lightweight and only needs to conduct an auction before training during the entire training process. The subset of features participating in vertical federated learning at each iteration is fixed, during the federated training process that occurs after the auction.

4.3 Theoretical Analysis

In this subsection, we will illustrate that the designed mechanism meets the requirements of truthfulness, individual rationality, and computational efficiency, and then discuss the performance of the algorithm 2.

Myerson's lemma. [21] An auction mechanism is truthful if and only if the selection strategy is monotone and the payment of each participant is the critical value:

- **Monotonicity.** If a feature d wins by the bid price b_d, it will win with bid price $\hat{b}_d < b_d$, too.
- **Critical value.** The feature will not win the bid if his bid price is greater than the critical value.

Theorem 1. *FRAIM is truthful.*

Proof. We first illustrate that the strategy for selecting features in the mechanism is monotone. Assuming that other features keep their bids unchanged, feature d wins with the bid b_d. When feature d submits a lower price $\hat{b}_d < b_d$, according to the sorting strategy in our mechanism, the new position of feature d will be ahead of the current position, and the worst also keeps its current position in the sorting, so it can still be selected by the coordinator. Therefore, monotonicity is proved.

Then, we show that a feature only gets its maximum utility if it submits a truthful bid $b_d = c_d$.

- Feature d wins with a truthful bid b_d, and utility of feature d is $u_d(b_d) = \frac{Q_d}{q_j} - b_d \geq 0$.
  - Untruthful bid $\hat{b}_d < b_d$, according to the above monotonicity proof, it can be seen that the feature d also wins with a bid $\hat{b}_d$, and the utility $u_d(\hat{b}_d) = \frac{Q_d}{q_j} - b_d = u_d(b_d)$.
  - Untruthful bid $\hat{b}_d > b_d$, based on the sorting strategy in our mechanism, the new position of feature d will fall behind the current position, and the best also keeps its current position in the sorting. For the former, the utility is $u_d(\hat{b}_d) = \frac{Q_d}{q_j} - b_d \leq 0$, and for the latter, the utility is $u_d(\hat{b}_d) = \frac{Q_d}{q_j} - b_d = u_d(b_d)$. We all have $u_d(b_d) \geq u_d(\hat{b}_d)$.
- Feature d loses with a truthful bid b_d, the utility of feature d is $u_d(b_d) = 0$ and $\frac{Q_d}{q_j} = \frac{b_j}{Q_j} Q_d \leq b_d$.
  - Untruthful bid $\hat{b}_d < b_d$, there are two situations for feature d at this time, lose or win. If feature d is still losing, then utility $u_d(\hat{b}_d) = 0$. And if feature d wins with the untruthful bid $\hat{b}_d$, the utility is $u_d(\hat{b}_d) = \frac{Q_d}{q_j} - b_d \leq 0$. We can always guarantee that $u_d(b_d) \geq u_d(\hat{b}_d)$.
  - Untruthful bid $\hat{b}_d > b_d$, because of the sorting strategy in our mechanism, feature d also loses with the untruthful bid $\hat{b}_d$, so the utility is $u_d(\hat{b}_d) = 0$.

Since $u_d(b_d) \geq u_d(\hat{b}_d)$ can be maintained in all cases, it indicates that the feature cannot obtain more utility by submitting an untruthful bid. Therefore, the truthful bid $b_d = c_d$ is the critical value.

According to Myerson's lemma, the mechanism FRAIM we proposed is truthful.

Theorem 2. *FRAIM is individually rational.*

Proof. If feature d is selected with truthful bid price $b_d = c_d$, since q_d is a descending sequence and j is the minimum value that satisfies Eq. (13), we can get

$$\frac{Q_d}{b_d} \geq \frac{Q_j}{b_j}, \tag{15}$$

then we have

$$\frac{b_j}{Q_j} Q_d \geq b_d. \tag{16}$$

Substituting Eq. (14) into Eq. (16), we can obtain:

$$r_d \geq b_d = c_d. \tag{17}$$

Therefore, the utility of the selected feature is $u_d \geq 0$. If feature d is not selected, its utility is $u_d = 0$. Finally, the above analysis, each feature d has utility $u_d \geq 0$.

Theorem 3. *FRAIM is computationally efficient.*

Proof. In Algorithm 2, the time complexity of sorting in line 2 is $O(NlogN)$, and the time complexity of finding the smallest j is $O(N)$ in lines 3–4. The time complexities of both the first loop (lines 5–8) and the second loop (lines 9–11) are $O(N)$. Above all, the total time complexity of Algorithm 2 is $O(NlogN)$.

Theorem 4. *The solution of Algorithm 2 is the $\frac{1}{2}$-approximate solution of the problem $\mathcal{P}1$.*

Proof. The fractional knapsack problem allows selecting capacities and assigning valuations by percentage. The bidders are sorted according to the sorting method in Algorithm 2, then winners are selected until the total capacity is filled, and at most only the last winner $j+1$ is selected according to the required fraction. The solution is the optimal solution to the fractional knapsack problem. The solution obtained by Algorithm 2 for the general 0–1 knapsack problem is the sum of the first j elements of the solution of the fractional knapsack problem, so the solution of Algorithm 2 is at least $\frac{1}{2}$-approximate solution of the optimal solution of the fractional knapsack problem.

5 Experimental Evaluation

5.1 Experiment Setup

Datasets. In this section, we evaluate our theoretical results on three real-world datasets that are also used in other VFL studies [19,37], namely, Credit card [32], MNIST [2], and Fashion MNIST [30]. The Credit card dataset contains data from 30,000 credit card users at a bank in Taiwan in 2005, and the goal is to predict whether an individual will default on a payment. The dataset consists of 23 features and 2 classes. Due to the imbalance of the two classes of data in the Credit card dataset, we adopt the method of undersampling to sample 6,000 samples in each class and combine them into a new dataset, and divide the training set and the test set according to the ratio of 5:1. The MNIST dataset contains 10 types of handwritten digits. Fashion MNIST is a dataset of clothing images with 10 classes. Both the MNIST and Fashion MNIST datasets have 784 features, each with 60,000 training samples and 10,000 test samples. We align the ID space of samples in all organizations by leveraging the private set intersection (PSI) technique [10].

Models. We use a simplified neural network (NN) model to classify the Credit card, MNIST, and Fashion MNIST datasets. The structure of the NN model used by the Credit card dataset has an input layer (The input size is equal to the number of features selected in each organization.), a hidden layer with 32 units, a dropout(p=0.2) unit, a ReLU activation, and an output layer with 2 units. For MNIST and Fashion MNIST datasets, the NN has an input layer (The input size is equal to the number of features selected in each organization.), a hidden layer with 64 units, a ReLU activation, and an output layer with 10 units. In addition, after aggregating the local output on the coordinator, sigmoid and softmax are respectively used as the activation functions for binary and multi-classification problems.

Table 2. Parameters settings

Setting	K	B	Bid price range	Number of features in each organization
Credit card	4	200	[20, 60]	[5, 6, 6, 6]
MNIST	7	200	[1, 1.5]	112
Fashion MNIST	7	200	[1, 1.5]	112

Simulation Settings. For all three datasets, the parameters n and m for estimating Shapley value in Algorithm 1 are set to 10 and 50, respectively. For the Credit card dataset, the learning rate η is set to 0.01, and the total number of iterations is set to 5000. The learning rate $\eta = 0.001$ and the total number of iterations is 1000 for the MNIST and Fashion MNIST datasets. All three datasets use full-batch training. Other parameter settings are shown in Table 2. Since the number of features in the Credit card dataset is 23, and both the MNIST and Fashion MNIST datasets have 784 features, the bid price range is not set to the same.

Baselines. Since we are the first to design an incentive mechanism for VFL from a feature perspective and cannot find more baselines, we cite the baseline from the work [4] and an optimal organization selection mechanism as the baselines.

- **Bid price first mechanism (BPF):** The coordinator selects features in increasing order of bid price from low to high. In this mechanism, the reward is equal to the bid, i.e., $r_d = b_d$. BPF cannot guarantee the truthfulness of the participants, that is, the organization can obtain more utility by submitting fake bids for features.

- **Theoretically optimal organization selection mechanism (O2S):** The sum of the importance of all features on the organization is regarded as the importance of the organization, and the coordinator then uses a depth-first search to find the optimal organization combination. O2S also cannot guarantee the truthfulness of participants.

5.2 Performance on Three Datasets

We evaluate our mechanism on three datasets, the budget size and feature number of each organization are set as shown in Table 2. Each figure shows the mean performance over three independent runs.

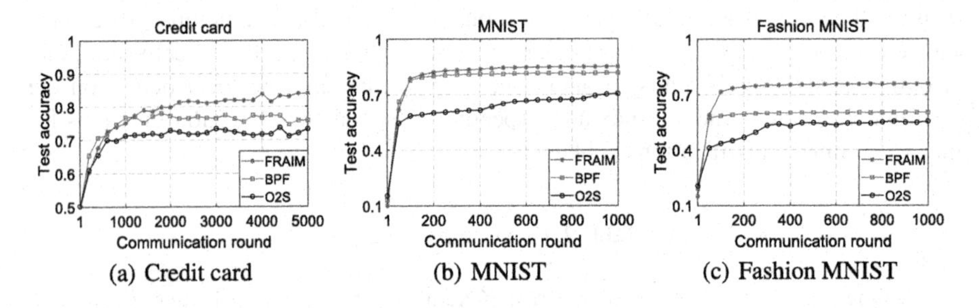

Fig. 2. The performance comparison of the three mechanisms on the Credit card, MNIST, and Fashion MNIST datasets.

As shown in Fig. 2, we can find that the BPF mechanism achieves 75.90%, 81.80%, and 60.48% accuracies on the Credit card, MNIST, and Fashion MNIST datasets, respectively. And O2S achieves the poorer performance (73.30%, 70.59%, and 55.44%) among the three mechanisms. In contrast, our proposed mechanism FRAIM achieves the highest performance (84.05%, 85.19%, and 76.03%). In summary, for these three datasets, FRAIM can improve the test accuracy by up to 15.55% compared with the BPF mechanism. And compared with the mechanism O2S based on organization-selection, FRAIM improves the test accuracy by up to 20.59%.

There are two main reasons why FRAIM shows the best performance. First, FRAIM can give priority to the features of high unit importance within a limited budget. Second, FRAIM is a feature selection mechanism, which is more flexible than O2S's organization selection method.

5.3 Performance with Different Budgets

To further investigate the effectiveness of the proposed mechanism, we compare the performance of the three mechanisms with different budget settings. We change the budget settings, and other parameter settings are consistent with those in Subsect. 5.2. For these three datasets, the budget is changed from 200 to 300 and 400. The performance comparison is shown in Fig. 3.

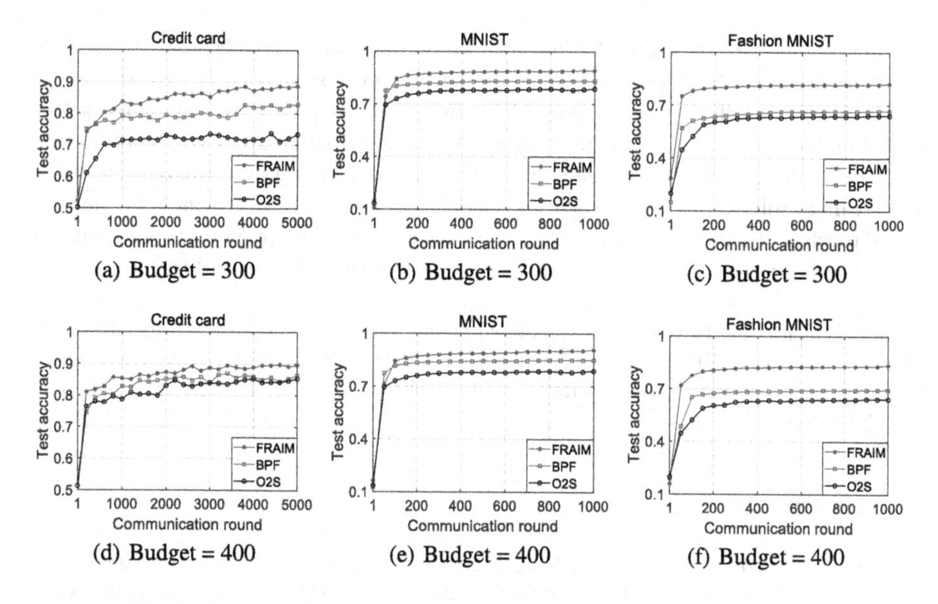

Fig. 3. The performance of the three mechanisms with different budgets.

It is obvious that our proposed mechanism FRAIM shows the best performance on all three datasets with different budgets. Specifically, FRAIM achieves 88.65% accuracy, while BPF and O2S only achieve 82.85% and 73.30% accuracies on the Credit card dataset with $B = 300$, respectively. And when $B = 400$, BPF and O2S only achieve 86.40% and 85.35% accuracies, while FRAIM achieves 89.60% accuracy on the Credit card dataset. Note that O2S has the same performance when the budget is 200 and 300, this is because O2S is based on organization selection. This selection method is less flexible, when the budget increases from 200 to 300, the additional budget does not allow O2S to select more organizations, so it can only select the same organizations as when the budget is 200. For the MNIST dataset, the proposed mechanism FRAIM achieves 89.25% and 90.85% accuracies respectively when the budget is 300 and 400. Correspondingly, the accuracies of BPF and O2S are 83.18%, 78.80% and 84.94%, 78.80%. And for Fashion MNIST dataset with $B = 300$ and $B = 400$, our mechanism FRAIM still shows the best performance (82.40% and 83.40%), O2S's performance (both 64.14%) is the worst, and the accuracy of BPF is 66.68% and 69.32%, respectively. It can also be found that whether on the MNIST or the Fashion MNIST dataset, when the budget is 300 and 400, the performance of O2S is the same, and the reason is still that the selection method of O2S is not flexible.

6 Conclusion

In this work, we considered the data distribution characteristics of vertical federated learning and proposed a feature importance-aware vertical federated learning incentive mechanism. The proposed mechanism can satisfy truthfulness, individual rationality, and computational efficiency. We first utilized a small dataset synthesized by interpolation to calculate the importance of features to the current task. Then we treated the behavior between the coordinator and the client as a knapsack auction problem that aims to maximize feature importance. Finally, we designed an approximate algorithm to solve the knapsack problem and gave a theoretical description of the quality of the approximate solution. Our empirical evaluation validated the effectiveness of the proposed mechanism. For our future works, we consider the problem of model heterogeneity and design reasonable incentives for personalized vertical federated learning.

Acknowledgements. This work was supported by the National Natural Science Foundation of China under Grant U1911201, U2001209, the Natural Science Foundation of Guangdong under Grant 2021A1515011369, and the Science and Technology Program of Guangzhou under Grant 2023A04J2029.

References

1. Cai, S., Zhao, Y., Liu, Z., Qiu, C., Wang, X., Hu, Q.: Mgfl: multi-granularity federated learning in edge computing systems. In: Algorithms and Architectures for Parallel Processing (ICA3PP), pp. 549–563. Springer International Publishing, Cham (2022). https://doi.org/10.1007/978-3-030-95384-3_34
2. Chen, F., Chen, N., Mao, H., Hu, H.: Assessing four neural networks on handwritten digit recognition dataset (MNIST). arXiv preprint arXiv:1811.08278 (2018). https://doi.org/10.48550/ARXIV.1811.08278
3. Covert, I., Lundberg, S.M., Lee, S.I.: Understanding global feature contributions with additive importance measures. Adv. Neural Inform. Process. Syst. (NeurIPS) **33**, 17212–17223 (2020)
4. Deng, Y., et al.: Fair: Quality-aware federated learning with precise user incentive and model aggregation. In: IEEE Conference on Computer Communications (INFOCOM), pp. 1–10. IEEE (2021). https://doi.org/10.1109/INFOCOM42981.2021.9488743
5. Fu, F., et al.: VF2Boost: very fast vertical federated gradient boosting for cross-enterprise learning. In: Proceedings of the 2021 International Conference on Management of Data, pp. 563–576 (2021)
6. Gul, F.: Bargaining foundations of shapley value. Econometrica: J. Econom. Society 81–95 (1989)
7. Horowitz, E., Sahni, S.: Computing partitions with applications to the knapsack problem. J. ACM **21**(2), 277–292 (1974)
8. Kairouz, P., et al.: Advances and open problems in federated learning. Found. Trends® in Mach. Learn. **14**(1–2), 1–210 (2021)
9. Kang, J., Xiong, Z., Niyato, D., Xie, S., Zhang, J.: Incentive mechanism for reliable federated learning: a joint optimization approach to combining reputation and contract theory. IEEE Internet Things J. **6**(6), 10700–10714 (2019). https://doi.org/10.1109/JIOT.2019.2940820
10. Kolesnikov, V., Kumaresan, R., Rosulek, M., Trieu, N.: Efficient batched oblivious prf with applications to private set intersection. In: Proceedings of the 2016 ACM Conference on Computer and Communications Security (CCS), pp. 818–829 (2016)
11. Konečnỳ, J., McMahan, H.B., Yu, F.X., Richtárik, P., Suresh, A.T., Bacon, D.: Federated learning: Strategies for improving communication efficiency. arXiv preprint arXiv:1610.05492 (2016)

12. Le, T.H.T., et al.: An incentive mechanism for federated learning in wireless cellular networks: an auction approach. IEEE Trans. Wireless Commun. **20**(8), 4874–4887 (2021). https://doi.org/10.1109/TWC.2021.3062708
13. Li, X., Chen, X., Wang, S., Ding, Y., Li, K.: Multi-initial-center federated learning with data distribution similarity-aware constraint. In: Algorithms and Architectures for Parallel Processing (ICA3PP), pp. 752–772. Springer Nature Switzerland, Cham (2023). https://doi.org/10.1007/978-3-031-22677-9_41
14. Lim, W.Y.B., et al.: Hierarchical incentive mechanism design for federated machine learning in mobile networks. IEEE Internet Things J. **7**(10), 9575–9588 (2020). https://doi.org/10.1109/JIOT.2020.2985694
15. Liu, T., Di, B., An, P., Song, L.: Privacy-preserving incentive mechanism design for federated cloud-edge learning. IEEE Trans. Network Sci. Eng. **8**(3), 2588–2600 (2021). https://doi.org/10.1109/TNSE.2021.3100096
16. Liu, Y., Wu, G., Zhang, W., Li, J.: Federated learning-based intrusion detection on non-iid data. In: Algorithms and Architectures for Parallel Processing (ICA3PP), pp. 313–329. Springer Nature Switzerland, Cham (2023). https://doi.org/10.1007/978-3-031-22677-9_17
17. Lu, J., Pan, B., Seid, A.M., Li, B., Hu, G., Wan, S.: Truthful incentive mechanism design via internalizing externalities and lp relaxation for vertical federated learning. IEEE Transactions on Computational Social Systems, pp. 1–15 (2022). https://doi.org/10.1109/TCSS.2022.3227270
18. Lu, S., Wang, Y., Wu, Y.: Novel high-precision simulation technology for high-dynamics signal simulators based on piecewise hermite cubic interpolation. IEEE Trans. Aerosp. Electron. Syst. **54**(5), 2304–2317 (2018). https://doi.org/10.1109/TAES.2018.2814278
19. Luo, X., Wu, Y., Xiao, X., Ooi, B.C.: Feature inference attack on model predictions in vertical federated learning. In: 2021 IEEE 37th International Conference on Data Engineering (ICDE), pp. 181–192. IEEE (2021)
20. McMahan, B., Moore, E., Ramage, D., Hampson, S., y Arcas, B.A.: Communication-efficient learning of deep networks from decentralized data. In: Artificial intelligence and statistics (AISTATS), pp. 1273–1282. PMLR (2017)
21. Myerson, R.B.: Optimal auction design. Math. Oper. Res. **6**(1), 58–73 (1981)
22. Ng, J.S., Lim, W.Y.B., Xiong, Z., Cao, X., Niyato, D., Leung, C., Kim, D.I.: A hierarchical incentive design toward motivating participation in coded federated learning. IEEE J. Sel. Areas Commun. **40**(1), 359–375 (2022). https://doi.org/10.1109/JSAC.2021.3126057
23. Shin, M., Hwang, C., Kim, J., Park, J., Bennis, M., Kim, S.L.: XOR Mixup: privacy-preserving data augmentation for one-shot federated learning. In: International Workshop on Federated Learning for User Privacy and Data Confidentiality in Conjunction with ICML (FL-ICML) (2020)
24. Štrumbelj, E., Kononenko, I.: Explaining prediction models and individual predictions with feature contributions. Knowl. Inf. Syst. **41**(3), 647–665 (2014)
25. Tang, M., Wong, V.W.: An incentive mechanism for cross-silo federated learning: a public goods perspective. In: IEEE Conference on Computer Communications (INFOCOM), pp. 1–10. IEEE (2021). https://doi.org/10.1109/INFOCOM42981.2021.9488705
26. Tseng, C.C., Lee, S.L.: Design of fractional delay filter using hermite interpolation method. IEEE Trans. Circuits Syst. I Regul. Pap. **59**(7), 1458–1471 (2012). https://doi.org/10.1109/TCSI.2011.2177136
27. Wang, Y., Lin, L., Chen, J.: Communication-efficient adaptive federated learning. In: Proceedings of the 39th International Conference on Machine Learning (ICML). Proceedings of Machine Learning Research, vol. 162, pp. 22802–22838 (17–23 Jul 2022)
28. Weng, J., Weng, J., Huang, H., Cai, C., Wang, C.: Fedserving: a federated prediction serving framework based on incentive mechanism. In: IEEE Conference on Computer Communications (INFOCOM), pp. 1–10. IEEE (2021)

29. Wu, Y., Cai, S., Xiao, X., Chen, G., Ooi, B.C.: Privacy preserving vertical federated learning for tree-based models. Proceedings of the VLDB Endowment 13(11) (2020)
30. Xiao, H., Rasul, K., Vollgraf, R.: Fashion-mnist: a novel image dataset for benchmarking machine learning algorithms. arXiv preprint arXiv: 1708.07747 (2017). https://doi.org/10.48550/arXiv.1708.07747
31. Yang, Q., Liu, Y., Chen, T., Tong, Y.: Federated machine learning: concept and applications. ACM Trans. Intell. Syst. Technol. **10**(2), 1–19 (2019)
32. Yeh, I.C., Lien, C.: The comparisons of data mining techniques for the predictive accuracy of probability of default of credit card clients. Expert Syst. Appl. **36**(2), 2473–2480 (2009)
33. Yoon, T., Shin, S., Hwang, S.J., Yang, E.: Fedmix: approximation of mixup under mean augmented federated learning. In: International Conference on Learning Representations (ICLR) (2021)
34. Yu, Y., Chen, D., Tang, X., Song, T., Hong, C.S., Han, Z.: Incentive framework for cross-device federated learning and analytics with multiple tasks based on a multi-leader-follower game. IEEE Trans. Netw. Sci. Eng. **9**(5), 3749–3761 (2022). https://doi.org/10.1109/TNSE.2022.3190377
35. Zhan, Y., Li, P., Qu, Z., Zeng, D., Guo, S.: A learning-based incentive mechanism for federated learning. IEEE Internet Things J. **7**(7), 6360–6368 (2020). https://doi.org/10.1109/JIOT.2020.2967772
36. Zhang, J., Wu, Y., Pan, R.: Incentive mechanism for horizontal federated learning based on reputation and reverse auction. In: Proceedings of the Web Conference 2021, pp. 947–956 (2021)
37. Zhang, Q., et al.: AsySQN: faster vertical federated learning algorithms with better computation resource utilization. In: Proceedings of the 27th ACM Conference on Knowledge Discovery & Data Mining (SIGKDD), pp. 3917–3927 (2021)

A Multi-objective Method for Energy-Efficient UAV Data Collection Communications

Jiajun Li, Jing Zhang, and Xin Feng(✉)

School of Computer Science and Technology, Changchun University of Science and Technology, changchun, China
fengxin@cust.edu.cn

Abstract. Unmanned aerial vehicles (UAVs) have been widely used to assist wireless sensor communications (WSN). However, due to the limited battery capacity of UAVs, their energy consumption as well as endurance issues are in need of additional attention. In this paper, a UAV-assisted data collection system is proposed to achieve efficient data collection from sensors by UAV. Specifically, a set of UAVs departs from the base station (BS) and collects data for each set of sensors, and returns to the BS. In this system, we formulate a data transmission multi-objective optimization problem (DSMOP) to maximize the transmission rate of UAVs and minimize the total energy consumption of UAVs, and we prove that the proposed DSMOP has trade-off. An enhanced multi-objective grasshopper optimization algorithm (EMOGOA) is proposed to better solve the DSMOP. Simulation results show that the results obtained by our proposed algorithm are superior to other comparative algorithms.

Keywords: UAV communications · wireless sensors · transmission rate · energy-efficient · multi-objective optimization

1 Introduction

With the rapid development of wireless technology and embedded electronics, wireless sensor networks (WSN) have been widely used in several fields [5,18]. The main goal of a sensor network is to sense the environment and send the information to a base station (BS) for further communication and processing [2]. However, the limited number of sensors may result in a wide range of sensor distribution. For some remote areas, the sensors are far away from the BS and cannot transmit data directly to the BS. In addition, for some systems that require real-time detection, such as weather and transportation systems, it is critical to obtain the latest data in a timely manner. Therefore, our main concern is to obtain the latest data collected by sensor transmission.

Unmanned aerial vehicles (UAVs) are widely used in mobile communications due to their advantages of high mobility, wide coverage, and easy deployment. Currently, the use of UAV-assisted WSN has become an effective way to

Z. Tari et al. (Eds.): ICA3PP 2023, LNCS 14491, pp. 151–162, 2024.
https://doi.org/10.1007/978-981-97-0808-6_9

achieve sensor data transmission [4], UAVs can provide efficient data transmission through the characteristics of real-time mobility and deployment anywhere. For example, RusyadiRamli et al. [12] formulate a hybrid medium access control protocol for data gathering in a WSN and UAV to address the problem of competing sensor nodes that need to transmit their data to UAV in a short time. Compared with ground-based sensor communications, UAVs can adaptively adjust their flight paths to obtain better channel conditions to provide on-demand communication coverage for sensor nodes and real-time data transmission. Moreover, UAVs have the flexibility to monitor the environment around sensor nodes from different angles and provide more data information.

However, the power consumption of UAVs is a huge challenge due to their limited battery capacity and heavy load. Therefore, it is of great importance to extend the flight time of energy-constrained UAVs in the UAV-assisted WSN. Wang et al. [17] consider a scenario with the objective of minimizing the overall time between all UAVs going from the data center and returning back after collecting data, as well as assigning each sensor a specific quantity of data and energy. Bouhamed et al. [1] propose an autonomous data-gathering mechanism utilizing self-trained UAVs as a flying mobile unit to collect data from ground sensor nodes that are spatially distributed within a designated geographic region during a predetermined time frame. However, the aforementioned works either focused on the energy consumption of the UAV or the sensor nodes, both of them have a significant impact on the endurance of the UAV and need to be considered at the same time. Therefore, there is an urgent need to come up with a solution that ensures effective data communications between UAV and sensor nodes while minimizing the energy consumption of the UAV.

In this paper, we design a UAV-assisted data collection system to implement that UAV obtains the data information from the sensor nodes. Specifically, a group of UAVs (N_U) departs from the BS and flies to different clusters of sensors separately for data collection, then returns to the BS. Moreover, a data transmission multi-objective optimization problem (DSMOP) is proposed to maximize the transmission rate of UAVs and minimize the total energy consumption of UAVs. The DSMOP is proven with trade-offs and it contains many variables, so we design an improved evolutionary computational method which is the enhanced multi-objective grasshopper optimization algorithm (EMOGOA) to solve the DSMOP. Note that the EMOGOA is used to improve the performance on the optimization objectives by introducing the chaotic solution initialization method, partially mapped crossover (PMX) algorithm and fused multi-objective salp swarm algorithm (MSSA) algorithm. In addition, the simulation results illustrate that our designed EMOGOA can achieve a superior data transmission effect and can effectively reduce the energy consumption of the UAVs compared with other algorithms.

The rest of this paper is organized as follows: Sect. 2 introduces the models and preliminaries. Section 3 formulates the DSMOP. Section 4 proposes the EMOGOA. Section 5 illustrates the simulation results and Sect. 6 summarizes the overall paper.

2 Models and Preliminaries

In this section, we first propose a UAV-based data collection system. Then, the corresponding network model is introduced. Finally, we present the energy consumption model of UAV.

2.1 System Model

we can see in Fig. 1, the UAV-based data collection system is designed. In this figure, a set of rotary-wing UAVs denoted as $\mathcal{U} = \{1, 2, ..., N_U\}$ intends to collect data information from different sensors denoted as $\mathcal{S} = \{1, 2, ..., N_S\}$ in the different sensor cluster $\mathcal{C} = \{1, 2, ..., N_C\}$ and transfers back to BS. Specifically, the UAVs over the position of BS, and they fly to the different sensor clusters, then each UAV flies to each sensor in turn in a fixed sensor cluster for information collection, after the data information of all the sensors in the cluster has been collected, the UAV returns to the BS for timely delivery of data information. Note that we set up that each of the UAVs is equipped with a single omnidirectional antenna to detect the positions of sensors. Moreover, to simplify the calculation, we assume the UAVs rise from their hovering positions to the vertical altitude, maintaining a fixed value (z^U). In addition, we employ the three-dimensional (3D) Cartesian coordinate system to manifest this work's general applicability in communications, the locations of the ith UAV, jth sensor, and BS are denoted as (x_i^U, y_i^U, z^U), (x_j^S, y_j^S, 0), (x^B, y^B, 0), respectively.

2.2 Network Model

In the designed system, we focus on the data link between the UAV and the sensor. The distance between the UAV and the sensor, as well as the obstacles and reflections in the signal propagation path of the sensor data received by the UAV, the above factors can have an impact on the quality of the signal propagation.

Firstly, the distance from the jth sensor to the ith UAV, $d_{j \to i}$, which are calculated by:

$$d_{j \to i} = \sqrt{(x_i^U - x_j^S)^2 + (y_i^U - y_j^S)^2 + (z^U)^2}. \tag{1}$$

Next, the received signal power of the receiver depends on the existence of a Line-of-Sight (LoS) link between the sender and the receiver. Moreover, the probability of the occurrence of LoS depends on the actual environment and the distance between the sender and the receiver. Considering the specific design scenario, the probability to have the LoS link from the jth sensor to the ith UAV is given by [13]:

$$\mathbb{P}_{j \to i} = \frac{1}{1 + v_1 e^{-v_2(\theta_{ji} - v_1)}} \tag{2}$$

where v_1 and v_2 are the Sigmoid function (S-curve) variables that depend on the communication environment. Moreover, θ_{ji} is the elevation angle of the jth sensor relative to the ith UAV, which can be calculated by:

$$\theta_{ji} = arctan\frac{z^U}{\sqrt{(x_i^U - x_j^S)^2 + (y_i^U - y_j^S)^2}} \tag{3}$$

Then, the path loss of the sensor-to-UAV (S2U) can be calculated by:

$$\beta_{j\to i} = 20\log(\frac{4\pi f_c(d_{j\to i})}{c}) + v_3\mathbb{P}_{j\to i} + v_4(1 - \mathbb{P}_{j\to i}) \tag{4}$$

where v_3 and v_4 are the attenuation factors that are determined by environmental factors. Moreover, c is the speed of light in m/s, and f_c is the carrier frequency in MHz.

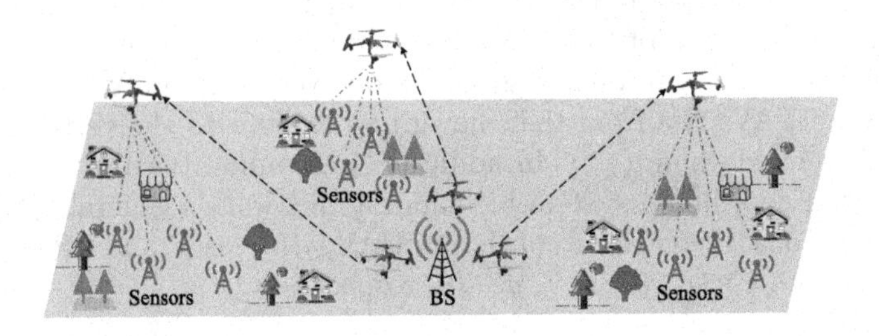

Fig. 1. A UAV-based data transmission system.

Finally, the transmission rate is employed to evaluate the transmission efficiency of a communication system. According to the transmission data link, the achievable transmission rate from the jth sensor to the ith UAV can be expressed by:

$$R_{ji} = B\log_2(1 + \frac{P_S\beta_{j\to i}}{\sigma^2}) \tag{5}$$

where B represents the channel bandwidth, P_S represents the transmission power of a sensor. Moreover, σ^2 denotes the additive white Gaussian noise.

As can be seen from Eqs. (1)-(5), the determining variable of the transmission rate is the positions of UAVs. Therefore, we have to pay attention to the key variable in our designed system.

2.3 Energy Consumption Model of the UAV

Both the communication energy consumption and propulsion energy consumption compose the overall energy consumption. In general, due to the smaller value of communication energy consumption, it is ignored in the calculation.

Thus, a UAV flights horizontally in the two-dimensional (2D) by v velocity, the power consumption of the UAV can be defined by [6]:

$$P(v) = P_B(1 + \frac{3v^2}{v_a^2}) + P_I(\sqrt{1 + \frac{v^4}{4v_t^4}} - \frac{v^2}{2v_t^2})^{\frac{1}{2}} + \frac{1}{2} d_f s_r \rho_a a_r v^3, \tag{6}$$

where P_B and P_I denote the blade profile power and induced power in the hovering conditions, respectively. Moreover, v_a and v_t denote the average induction velocity of the rotor and tip velocity of the rotor blade, and d_f and s_r represent the fuselage drag ratio and rotor solidity, respectively. In addition, ρ_a and a_r denote the air density and rotor disc area, respectively. Moreover, according to the propulsion energy consumption and flight and gravity energy consumption, the ascending or descending flight energy consumption of UAV in 3D can be calculated as follows:

$$E(T) \approx \int_0^T P(v(t))\, dt + \frac{1}{2} m_U \left(v(T)^2 - v(0)^2 \right) + m_U g (h(T) - h(0)), \tag{7}$$

where $v(t)$ is the instantaneous UAV speed of the UAV at moment t, and m_U denotes the aircraft mass of a UAV. Moreover, T and g represent the end time of flying and gravitational acceleration, respectively.

The energy consumption of UAVs, based on the energy consumption model, is mainly influenced by their positions. Additionally, in the network model, the location of the UAV is the decisive variable for communication quality. Consequently, it is crucial to deploy UAVs with a high degree of accuracy.

3 Problem Formulation and Analysis

In this section, we state the problem to be solved, and subsequently define the optimization objectives, then formulate the corresponding DSMOP.

3.1 Problem Statement

In the UAV-based data collection system, we utilize the UAVs to implement sensor data collection. In detail, we design a square area based on the location of each sensor for the UAV to move and receive the sensor's data content. It is clear from the network model that we need to optimize the transmission rate from the sensor to the UAV by regulating the positions of the UAVs. However, as the energy consumption of the UAV inevitably increases during constant movement, there is a need to design the UAV's position change carefully in order to minimize its additional energy consumption. Meanwhile, it is also necessary to consider how to reasonably balance the position and movement of the UAV to make the best use of its flight time and energy to improve the efficiency of the whole system and the quality of data acquisition. For this purpose, the trajectory of the UAV needs to be continuously designed and adjusted.

In general, our system is designed in such a way that a UAV needs to collect data information from all the sensors in a sensor cluster. However, since a UAV can only receive information from one sender at a time, it needs to re-position itself for the next reception after each UAV receives information from the sender. Therefore, the communication sequence between each UAV and the sensor also needs to be carefully designed, as it will directly affect the energy consumption of the UAV. Therefore, we need to optimize the movement order of UAV to minimize its energy consumption and improve the data collection efficiency in order to improve the stability and reliability of the designed system.

3.2 Optimization Objectives

In the UAV-based data collection system, we consider the following optimization objectives concurrently.

Optimization Objective 1: To ensure efficient transmission performance, the first optimization objective is to maximize the value of the transmission rate of the UAV from the sensor, which is represented as follows:

$$f_1(\mathbb{P}^U, \mathbb{D}) = \sum_{i=1}^{N_U} \sum_{j=1}^{N_S} R_{ji} \tag{8}$$

where $\mathbb{P}^U = (\mathbb{X}^U, \mathbb{Y}^U, Z^U) = \{\mathcal{P}_1^U, \mathcal{P}_2^U, \ldots, \mathcal{P}_i^U \mid i \in N_U\}$ denotes a set including the positions of all UAVs, in which $\mathcal{P}_i^U = \{x_i^U, y_i^U, z^U\}$ denotes the position of the ith UAV. Moreover, $\mathbb{D}$ denotes the order in which the UAV receives the different sensors of a sensor cluster.

Optimization Objective 2: In the designed system, each UAV needs to be constantly moved for position adjustment in order to achieve optimal transmission rate. Therefore, our second optimization objective is to minimize the overall energy consumption of all UAVs, which is expressed by:

$$f_2(\mathbb{P}^U, \mathbb{D}) = \sum_{i=1}^{N_U} \sum_{j=1}^{N_S} E_{i,j} \tag{9}$$

where $E_{i,j}$ denotes the flight energy consumption of the ith UAV, which can be calculated by Eqs. (6)-(7) according to the method in [6].

Consequently, the DSMOP can be formulated as follows:

$$\min F = \{-f_1, f_2\}, \tag{10a}$$

$$\text{s.t. } C1 : Px_{min} \leq x_i^U \leq Px_{max}, \forall i \in \mathcal{U}, \tag{10b}$$

$$C2 : Py_{min} \leq y_i^U \leq Py_{max}, \forall i \in \mathcal{U}, \tag{10c}$$

$$C3 : \mathbb{D} \in \mathcal{SD}, \tag{10d}$$

$$C4 : D_{m,n} \geq D_{min}, \forall m, n \in \mathcal{U}, \tag{10e}$$

where Px_{min} and Px_{max} indicate the minimum and maximum values of the range of UAV in the x-direction, and Py_{min} and Py_{max} indicate the minimum and maximum values of the range of UAV in the y-direction, respectively. Moreover, $\mathcal{SD} = \{\mathbb{D}_1, ..., \mathbb{D}_i | i \in N_C\}$ represents the set of communication sequences of a UAV and different sensors in a cluster. And $\mathbb{D}_i = \{3, 1, ..., 5\}$ denotes that the ith UAV will transfer data with sensor 3, sensor 1, ..., sensor 5 in a sensor cluster in turn. In addition, $D_{m,n}$ denotes the distance of the mth and nth UAVs, D_{min} is the minimum distance of two adjacent UAVs.

Lemma 1. The optimization objectives of the formulated DSMOP shown in Eq. (10) are trade-offs.

Proof: To achieve optimal information collection during the process of collecting sensor data information, the UAV requires constant movement. However, the movement leads to additional energy consumption. In addition, according to Eq. (6), the energy consumption will increase with the increase of the UAV speed. Conversely, when the UAV slows down, transmission time increases, leading to more hovering energy consumption. Therefore, the optimization objectives are trade-offs.

In this paper, we are faced with conflicting optimization objectives, and simply pursuing one of them may lead to adverse effects on the other objectives or failure to achieve them at all. Therefore, an efficient algorithm needs to be designed to balance the relationship between the optimization objectives in order to achieve the best overall performance.

4 Algorithm

For the DSMOP that the optimization objectives need to be balanced, a multi-objective optimization algorithm can be considered to solve the problem [7]. Multi-objective optimization algorithm refers to finding the best balance between multiple conflicting optimization objectives in order to obtain the optimal solutions while satisfying all objective conditions.

4.1 Conventional MOGOA

MOGOA [11] is a typical multi-objective optimization algorithm, which is originated from the grasshopper optimization algorithm (GOA) [14]. GOA can be used to simulate the swarming behaviors of grasshoppers in nature, and the position of the grasshopper in the swarm represents a possible solution of a given optimization problem. In this paper, the positions of UAVs in different dimensions can be updated as follows:

$$X_i^d = C_d \times \{ \sum_{j=1, j \neq i}^{N_U} C_d \times \frac{UB_d - LB_d}{2} q(|X_j^d - X_i^d|) \frac{X_j^d - X_i^d}{D_{i,j}} \} + T_d \quad (11)$$

where X_i^d and X_j^d denote the ith UAV and jth UAV solutions in the dth dimension, and UB_d and LB_d denote the upper and lower bounds in the dth dimension, respectively. Moreover, C_d is the corresponding coefficient which is calculated by Eq. (12), and q is the function that can be expressed by Eq. (13). In addition, $D_{i,j}$ denotes the distance between the ith and jth UAVs, and T_d is the value of the dth dimension in the target (best solution found so far).

Moreover, the coefficient C_d is a decreasing parameter to shrink the comfort area, repulsion area, and attraction area, which are expressed as follows:

$$C_d = c_{max} - t \times \frac{c_{max} - c_{min}}{t_{max}} \tag{12}$$

where c_{max} an c_{min} are the maximum and minimum value, respectively. Moreover, t indicates the current iteration, and t_{max} is the maximum number of iterations. In addition, the function q with u parameter can be calculated as follows:

$$q(u) = I_a \times e^{\frac{-u}{len}} - e^{-u} \tag{13}$$

where I_a denotes the intensity of attraction, and len is the attractive scale.

A multi-objective problem can have multiple solutions. In MOGOA, the Pareto optimal dominance method is utilized to compare the solutions, and the highest quality Pareto optimal solutions are saved in an archive. MOGOA can efficiently identify the Pareto optimal solutions, store them in an archive, and enhance their distribution. However, conventional MOGOA is not capable of solving the formulated DSMOP due to its inclusion of both continuous and discrete solution dimensions. To address this issue, we propose an EMOGOA, and the details are described below.

4.2 EMOGOA

In this section, we propose an EMOGOA, which is extended from MOGOA. We proposed EMOMA is improved from three aspects, which are the chaotic solution initialization, salp swarm algorithm (SSA)-based solution update strategy, and discrete solution update method. Moreover, the whole structure of EMOGOA is displayed in Algorithm 1 in which the marker ζ is a threshold ranges to restrict the computational conditions [15], and the details are shown as follows.

(i). Chaotic Solution Initialization: The conventional MOGOA typically produces initial solutions randomly, which can lead to a decrease in solution diversity and potentially result in a lack of direction in the search process. Moreover, the solution range for the formulated DSMOP discussed in Eq. (10) is large, poor performance of the initial solutions may cause solutions to become trapped in local optimal problems. Consequently, it becomes essential to enhance the performance of the initial solutions.

In this paper, the Logistic map is utilized to optimize the ith initial solutions in the dth dimension (X_i^d), and it is described as follows:

$$X_i^d = LB_d + m_d \times (UB_d - LB_d) \tag{14}$$

$$m_{d+1} = \mu m_d \times (1 - m_d) \tag{15}$$

Algorithm 1: EMOGOA

```
1  Set the parameters: population size N_pop, maximum iteration t_max and archive
   set Archive, etc.;
2  for i = 1 to N_pop do
3  |  Initialize the ith solution (X_i) by Eq. (14);
4  end
5  for t = 1 to t_max do
6  |  Calculate the objective function values and update values in the Archive;
7  |  for i = 1 to N_pop do
8  |  |  Update the continuous solutions (P^U) by Eq. (11);
9  |  |  if t < (t_max/2) and rand < ζ then
10 |  |  |  Update the continuous solutions (P^U) of X_i(X^U, Y^U) by Eq. (16);
11 |  |  end
12 |  |  Update the discrete solutions (D) of X_i by Algorithm 2;
13 |  end
14 end
15 Return Archive;
```

where m_d is the dth chaotic number of the Logistic sequence, and d is the index of the chaotic sequence. Moreover, μ is a parameter of the Logistic map.

(ii). SSA-based Solution Update Strategy: UAV consumes less energy when it is densely distributed during communications. Therefore, the algorithm can be improved by centralizing the horizontal positions of UAV ($\mathbb{X}^U$, $\mathbb{Y}^U$) in the earlier iterations would be beneficial. To achieve this objective, a SSA-based method inspired by SSA [21] is employed. In the initial iterations, the process of updating the horizontal positions of UAV elements by using SSA is defined as follows:

$$X_i^d(\mathbb{X}^U, \mathbb{Y}^U) = \begin{cases} T_d + c_1 c_2 \times ((UB_d - LB_d)) + LB_d & c_3 \geq 0.5 \\ T_d - c_1 c_2 \times ((UB_d - LB_d)) + LB_d & c_3 < 0.5 \end{cases} \tag{16}$$

where T_d represents the position of the target in the dth dimension, and c_1, c_2 and c_3 are rand numbers between 0 and 1.

(iii). Discrete Solution Update Method: The solutions for the formulated DSMOP contain two parts, one continuous ($\mathbb{P}^U$) and one discrete ($\mathbb{D}$) dimension. The continuous aspect of the solutions, the 3D coordinates ($\mathbb{P}^U$) of UAVs, can be obtained through SSA-based methods. On the other hand, the discrete aspect, cannot be obtained by using traditional swarm intelligence algorithms that were originally designed to solve problems with continuous solution spaces. In addition, when multiple sensors are present in an area, the communication sequence between the UAV and sensors has an important impact on the reliability of the transmission communication and the energy consumption of the UAV. To address this challenge, we present a method for updating the discrete solutions. The approach is implemented through partially mapped crossover (PMX) and discrete mutation operators [16]. Moreover, the specifics of the discrete solution update method are outlined in Algorithm 2.

Algorithm 2: Discrete solution update method

1 Determine the dimensionality of the discrete solution F ($\mathbb{D}_F$).
2 **if** $rand < 0.5$ **then**
3 | $X_i(\mathbb{D})$ intersects with $\mathbb{D}_F$ by the use of the PMX;
4 **else**
5 | According to the discrete mutation operator to update $X_i(\mathbb{D})$;
6 **end**
7 Return $X_i(\mathbb{D})$;

5 Simulation Results and Analysis

In this section, we verify that the performance of the proposed EMOGOA is excellent and remarkable by the simulation results.

5.1 Simulation Setups

During the simulation, we limit the quantity of UAVs (N_U) to 16, the numbers of sensor clusters (N_C) and sensors (N_S) in a cluster are 3 and 4. In addition to MOGOA, we also select several other multi-objective optimization algorithms to compare with EMOGOA. They are multi-objective dragonfly algorithm (MODA) [8], multi-objective multi-verse optimization (MOMVO) [9] and multi-objective ant lion optimizer (MALO) [10].

5.2 Simulation Results

Figure 2 shows the solutions distributions obtained by the different algorithms, where the solutions obtained by EMOGOA are on top performance as they are more concentrated and are closer to the Pareto front (PF). Moreover, Table 2

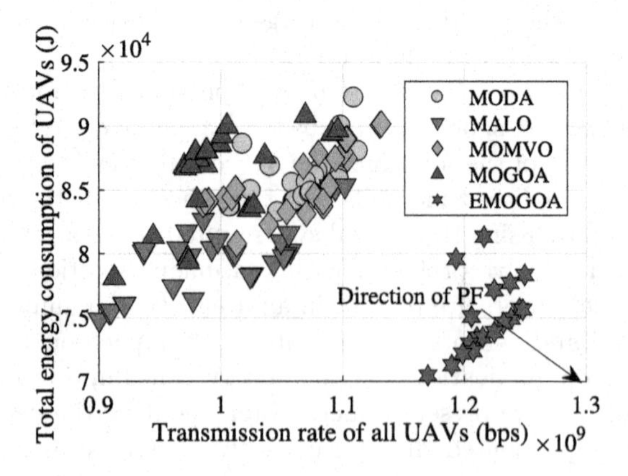

Fig. 2. Solutions distributions obtained by different algorithms.

Table 1. Numerical results obtained by different algorithms

Method	f_1 (bps)	f_2 (J)
MODA	1.07×10^9	8.86×10^4
MALO	1.05×10^9	7.98×10^4
MOMVO	1.10×10^9	8.75×10^4
MOGOA	9.12×10^8	7.81×10^4
EMOGOA	$\mathbf{1.25 \times 10^9}$	$\mathbf{7.57 \times 10^4}$

shows the results calculated by the above algorithms according to the transmission rate of UAVs (f_1) and the total energy consumption of UAVs (f_2). A larger value of f_1 and a smaller value of f_2 mean that the receiving signals of UAV are more and costuming energy consumption of UAV is smaller. As we can see, the proposed EMOGOA can achieve the best performance on maximizing f_1 and minimizing f_2 at the same time. In brief, the performance of EMOGOA is better than other algorithms, thus can better solve the formulated DSMOP (Table 1).

6 Conclusion

In this paper, the energy-efficient UAV-based data collection communications are confirmed. In detail, we design a UAV-based data collection system and utilize a set of UAVs to collect the data information gathered by the sensors. We formulate the DSMOP and prove it to be with trade-offs. Therefore, we design the EMOGOA to optimize the initial position of the UAVs, the updated position of the UAVs, and the communication sequence between the UAV and sensor simultaneously. Simulation results indicate that the present EMOGOA can achieve efficient data transmission and outperforms other comparative algorithms.

Acknowledgement. This study is supported in part by the Science and Technology Development Plan Project of Jilin Province (YDZJ202201ZYTS416). Xin Feng is the corresponding author.

References

1. Bouhamed, O., Ghazzai, H., Besbes, H., Massoud, Y.: A UAV-assisted data collection for wireless sensor networks: autonomous navigation and scheduling. IEEE Access **8**, 110446–110460 (2020). https://doi.org/10.1109/ACCESS.2020.3002538
2. Busi Reddy, V., Venkataraman, S., Negi, A.: Communication and data trust for wireless sensor networks using D-S theory. IEEE Sensors J. **17**(12), 3921–3929 (2017)
3. Hua, M., Wang, Y., Zhang, Z., Li, C., Huang, Y., Yang, L.: Power-efficient communication in UAV-aided wireless sensor networks. IEEE Commun. Lett. **22**(6), 1264–1267 (2018)

4. Jawhar, I., Mohamed, N., Al-Jaroodi, J.: UAV-based data communication in wireless sensor networks: Models and strategies. In: 2015 International Conference on Unmanned Aircraft Systems (ICUAS), pp. 687–694 (2015)
5. Ketshabetswe, L.K., Zungeru, A.M., Mangwala, M., Chuma, J.M., Sigweni, B.: Communication protocols for wireless sensor networks: a survey and comparison. Heliyon **5**(5), e01591 (2019)
6. Li, J., Kang, H., Sun, G., Liang, S., Liu, Y., Zhang, Y.: Physical layer secure communications based on collaborative beamforming for UAV networks: A multi-objective optimization approach. In: IEEE INFOCOM 2021, pp. 1–10
7. Li, J., et al.: Multi-objective optimization approaches for physical layer secure communications based on collaborative beamforming in UAV networks. IEEE/ACM Trans. Networking, pp. 1–16 (2023). https://doi.org/10.1109/TNET.2023.3234324
8. Mirjalili, S.: Dragonfly algorithm: a new meta-heuristic optimization technique for solving single-objective, discrete, and multi-objective problems. Neural Comput. Appl. **27**(4), 1053–1073 (2016)
9. Mirjalili, S., Jangir, P., Mirjalili, S.Z., Saremi, S., Trivedi, I.N.: Optimization of problems with multiple objectives using the multi-verse optimization algorithm. Knowl. Based Syst. **134**, 50–71 (2017). https://doi.org/10.1016/j.knosys.2017.07.018
10. Mirjalili, S., Jangir, P., Saremi, S.: Multi-objective ant lion optimizer: a multi-objective optimization algorithm for solving engineering problems. Appl. Intell. **46**(1), 79–95 (2017). https://doi.org/10.1007/s10489-016-0825-8
11. Mirjalili, S.Z., Mirjalili, S., Saremi, S., Faris, H., Aljarah, I.: Grasshopper optimization algorithm for multi-objective optimization problems. Appl. Intell. **48**(4), 805–820 (2018)
12. Rusyadi Ramli, M., Lee, J.M., Kim, D.S.: Hybrid MAC protocol for UAV-assisted data gathering in a wireless sensor network. Internet of Things **14**, 100088 (2021)
13. Samir, M., Assi, C., Sharafeddine, S., Ghrayeb, A.: Online altitude control and scheduling policy for minimizing AoI in UAV-assisted IoT wireless networks. IEEE Trans. Mobile Comput. **21**(7), 2493–2505 (2020)
14. Saremi, S., Mirjalili, S., Lewis, A.: Grasshopper optimisation algorithm: Theory and application. Adv. Eng. Softw. **105**, 30–47 (2017)
15. Sun, G., Li, J., Liu, Y., Liang, S., Kang, H.: Time and energy minimization communications based on collaborative beamforming for UAV networks: a multi-objective optimization method. IEEE J. Sel. Areas Commun. **39**(11), 3555–3572 (2021)
16. Sun, G., et al.: UAV-enabled secure communications via collaborative beamforming with imperfect eavesdropper information. IEEE Trans. Mob. Comput. 1–18 (2023). https://doi.org/10.1109/TMC.2023.3273293
17. Wang, Y., Hu, Z., Wen, X., Lu, Z., Miao, J.: Minimizing data collection time with collaborative UAVs in wireless sensor networks. IEEE Access **8**, 98659–98669 (2020). https://doi.org/10.1109/ACCESS.2020.2996665
18. Yin, R., Wang, D., Zhao, S., Lou, Z., Shen, G.: Wearable sensors-enabled human-machine interaction systems: from design to application. Adv. Funct. Mater. **31**(11), 2008936 (2021)

Black-Box Graph Backdoor Defense

Xiao Yang[1], Gaolei Li[1,2(✉)], Xiaoyi Tao[3], Chaofeng Zhang[4], and Jianhua Li[1,2]

[1] Shanghai Jiao Tong University, Shanghai 200240, China
gaolei_li@sjtu.du.cn
[2] Shanghai Key Laboratory of Integrated Administration Technologies for Information Security, Shanghai 200240, China
[3] Dalian Maritime University, Dalian 116026, China
[4] Advanced Institute of Industrial Technology, Tokyo 140-0011, Japan

Abstract. Recently, graph neural networks (GNNs) have been proven to be vulnerable to backdoor attacks, wherein the test prediction of the model is manipulated by poisoning the training dataset with trigger-embedded malicious samples during learning. Current defense methods against GNN backdoor are not practical due to their requirement for access to the GNN parameters and training samples. To address this issue, we present a **Bl**ack-box **G**NN **Ba**ckdoor **D**efense strategy, **BloGBaD**, that eliminates the backdoor without model parameter information and training dataset. Specifically, **BloGBaD** involves two primary phases: 1) test sample filtration, which identifies toxic graph nodes via the Gaussian mixture model and purifies their trigger features through clustering and filtration; and 2) model fine-tuning, which fine-tunes the model to a backdoor-free state by a loss function with a penalty regularization for poisoned features. We demonstrate the effectiveness of our method through extensive experiments on various datasets and attack algorithms under the assumption of black-box conditions.

Keywords: Graph neural networks · Backdoor attacks · Backdoor defense · Black-box defense

1 Introduction

Graph neural networks (GNNs) are designed to handle graph-structured data that can be represented by nodes and edges, such as social media, biological molecules, transportation networks, and knowledge graphs. GNN propagates information between nodes in the graph, and subsequently makes predictions based on the data contained within the graph. It has been successfully applied in a variety of domains, including recommendation systems, community detection, transportation demand forecasting, and biological analysis [2,23,32].

Despite the remarkable success of GNNs, they are still susceptible to one potential threat: backdoor attacks [14,16,17,27,28,31]. This attack is commonly implemented by poisoning the training set through the insertion of trigger nodes into the target graphs and modifying their labels as the premeditated class.

Z. Tari et al. (Eds.): ICA3PP 2023, LNCS 14491, pp. 163–180, 2024.
https://doi.org/10.1007/978-981-97-0808-6_10

Thereafter, the poisoned graphs will be employed by the model for learning along with the benign ones, and once the training is complete, trigger-embedded test graphs will be classified as premeditated class, since the model has learned the feature relation between the trigger and the premeditated class [15,29,34]. The general framework of the backdoor attack is shown in Fig. 1.

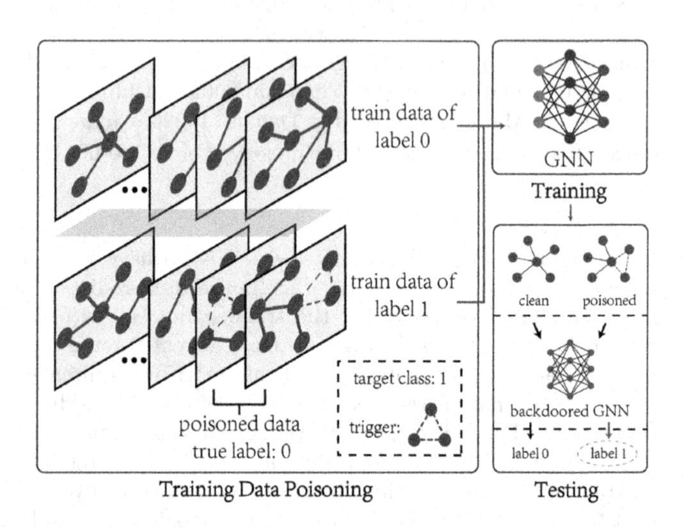

Fig. 1. Illustration for general GNN backdoor framework. In the training dataset, several graphs were poisoned by adding trigger nodes and modifying their true labels to the target class. This led to the model being backdoored during learning, causing it to output the target class when a trigger-embedded graph is inputted when testing.

The defense method against the GNN backdoor has not been extensively investigated. Several current studies primarily rely on either model parameters or explainability tools to identify suspect models or input samples, which is not applicable in real-world scenarios due to restrictions imposed by commercial property protection and user privacy policies that typically prohibit third-party access to the model parameters [12].

Our motivation to address this limitation derives from a statistical observation that for the trigger nodes in a poisoned graph, their feature distributions differ significantly from those of clean nodes, and if we could leverage the feature distribution to identify the poisoned nodes and analyze the trigger features, we could retrain the model to make it insensitive to these features, thereby mitigating the backdoor.

Based on the analysis above, in this paper, we present a novel **Bl**ack-box GNN **B**ackdoor **D**efense strategy: **BloGBaD**. It involves two main phases: 1) test sample filtration and 2) model fine-tuning. In phase 1), we first apply Gaussian mixture to model the graph data and detect potential suspect trigger nodes with anomalous distribution. Then, trigger features in detected nodes are identified through clustering algorithm, and filtered to clean state. While in phase 2), we fine-tune (retrain) the model by incorporating the trigger features detected in phase 1), and the loss function utilized for tuning will have a regularization penalty term associated with these features. Once the tuning is complete, the model will no longer be sensitive to the trigger, rendering the implanted backdoor ineffective. The entire defense process does not leverage any model parameter information, making it a black-box defense approach.

The contribution of this work can be summarized as follows:

- In this study, a novel two-phase GNN backdoor defense strategy, **BloGBaD**, is proposed. The proposed method consists of two phases: 1) test sample filtration and 2) model fine-tuning. In the first phase, we identify and remove trigger nodes and features that are responsible for activating the backdoor. In the second phase, the victim model is fine-tuned using trigger features to eliminate the backdoor. To the best of our knowledge, this is the first work that introduces black-box backdoor defense for GNN.
- We present a new method for identifying trigger nodes and features in poisoned graph that activate backdoors. It employs Gaussian mixture models to identify the trigger nodes and a clustering algorithm to compute the trigger features on these nodes. By clearing trigger features, the poisoned graph could be normally classified.
- Detailed experiments based on different real-world datasets and backdoor methods are conducted to evaluate the defense performances of **BloGBaD**, and it performs high detection success rate and significant attack success rate drop on malicious data without distinct model accuracy degradation.

The rest of this paper is organized as follows: Sect. 2 summarizes the related works; Sect. 3 provides the details of the proposed defense strategy, **BloGBaD**; Sect. 4 introduces the experimental settings and results; Sect. 5 concludes this paper and discusses the future trend.

2 Related Work

2.1 Graph Neural Network Backdoor

The vulnerability of GNNs to backdoors was initially exposed in [34]. The method utilizes a specific subgraph nodes as trigger to insert into graph training data and modifies its true label as target. When the model completes learning, it will predict the test graph sample with trigger as the target category. In a similar vein, [24] assesses backdoor feasibility through topology and employs specific sub-topology structures as trigger patterns to implant backdoors. While conventional GNN backdoors are implemented in a transductive training environment, [30] examines the transferability of GNNs to introduce a specific graph structure into the training set for inductive attacks. Federated learning has recently gained attention, and [25] investigates the susceptibility of federated GNNs to backdoor attacks: malicious attack that intends to alter the center's parameter update by introducing a specific trigger pattern into client training data.

The current attacks' trigger design is too apparent, so [7] proposes an adaptable trigger design to generate imperceptible poisoning data to implant a backdoor. Since GNN node data is aggregated from neighboring nodes, [5] introduces a neighbor-based attack technique to propagate the trigger's influence to nearby nodes. Additionally, there are some enhanced approaches based on triggers that have demonstrated increased success rates in attacks [4,6,26,35].

2.2 Black Box Backdoor Defenses

Study [3] first introduce a backdoor defense method for deep neural networks (DNNs) that operates under black box settings, meaning that it does not require access to the DNN's architecture or parameters for backdoor identification and mitigation. This approach utilizes the model parameters to recover the original training data and subsequently assesses the normality of the model. To make the defense available in a more strong assumption (without access to model parameters), [8] proposes a more detailed method based on analyzing the response of the network to recover a small set of synthetic inputs that are then employed to reveal the presence of a backdoor via inverse trigger construction. Guo et al. (2021) designed a backdoor defense method called Aeva, which works by analyzing the extreme values of the DNN's output when presented with different inputs and comparing them to a threshold that is learned during training. This approach can detect backdoors even under black-box conditions [9]. Aiming to make defense method applicable in the context of machine learning as a service, SCALE-UP method is presented. This approach could identify and filter out malicious testing samples by analyzing their prediction consistency during the pixel-wise amplification process [10]. Zhang et al. (2022) employ an auxiliary network to monitor the output of the target network and detect anomalies that may indicate the presence of a backdoor. This adaptive detection mechanism of the method allows it to adjust to new backdoor attacks and maintain high detection accuracy [33].

Most of the existing research on black-box defense methods focuses on computer vision domains, and to the best of our knowledge, there is currently no black-box defense method for GNN backdoor attacks.

3 Methodology

To defend against previous research on GNN backdoors, in this section, our defense strategy, **BloGBaD**, will be presented. **BloGBaD** comprises two phases. In the first phase, trigger nodes and features are identified through Gaussian mixture modeling and clustering algorithms. The second phase involves fine-tuning the model using a loss function that incorporates a penalty term related to the trigger features, aiming to eliminate backdoors. The general framework of **BloGBaD** is shown in Fig. 2.

We first outline our assumptions, objectives, and key intuition, and then provide a detailed description of **BloGBaD**.

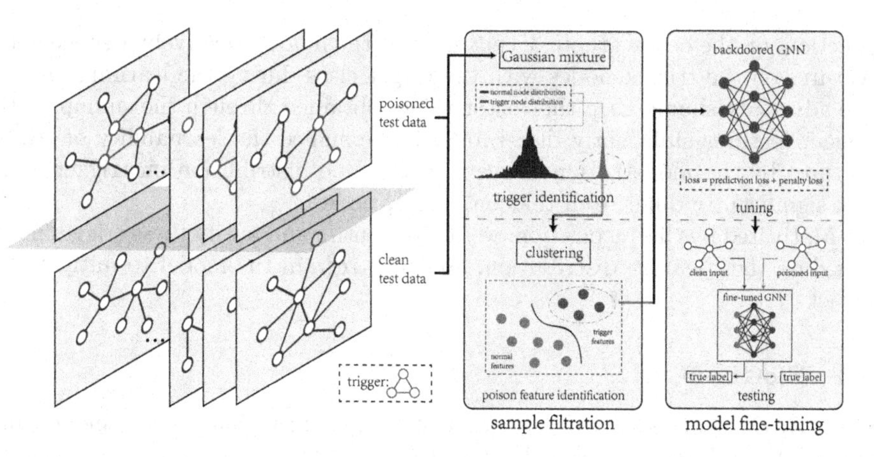

Fig. 2. Illustration for the proposed **BloGBaD** defense scheme. The test samples will undergo initial examination using a Gaussian mixture model to assess the presence of trigger nodes. In case trigger nodes are detected, clustering will be employed to extract the corresponding trigger features. Subsequently, a loss function incorporating penalty terms related to these features will be utilized for fine-tuning the original model, aiming to render it insensitive to triggers.

3.1 Defense Assumptions and Goals

We make some definitions of the capabilities and available resources of the defenders. It is first assumed that we have access to the output label and structure of the model, but not to the parameters. Additionally, we possess a set of samples (not the original training samples) that may potentially include poisoned samples.

Our goals include three specific ones:

- **Keep Model Performance**: For clean samples, it is essential to maximize the model's accuracy in classifying them.
- **Identify Malicious Samples**: Regarding poisoned graphs, we should accurately detect the infection status and perform purification procedures to enable correct classifications by the model. Meanwhile, the misdetection of clean samples as poisoned ones should be prevented.
- **Mitigate Model Backdoor**: When we detect poisoned graphs, we want to make the model immune to other trigger-embedded samples.

3.2 Key Intuition

Before **BloGBaD** is specified, we will explain what the intuition of the design is, and to simplify the narrative, we use the graph classification task of GNN as an example.

GNN operates by aggregating information from neighboring nodes within the graph and subsequently conducting a readout function, for instance, sumup function, on the entire graph. To ensure that the model effectively associates the features on the trigger nodes with the target class during the learning process, the adversary should keep the vector values obtained through the sumup on the trigger features sufficiently discernible (i.e., enhance the learnability of trigger features by the model). Hence, the feature value distribution on trigger nodes will significantly differ from that on normal nodes.

Motivated by this inspiration, we could identify and purify these trigger nodes based on their feature distribution, and then retrain the model to mitigate the attack.

3.3 BloGBaD

BloGBaD involves two phases: 1) test sample filtration and 2) model fine-tuning. We will sequentially introduce them.

1) Test Sample Filtration. Given a test graph sample $G = (V, E)$, our primary objective is to identify whether it is poisoned. If an infection is detected, our subsequent task is to ascertain the presence of trigger nodes and features within the graph. Based on the motivation discussed in Sect. 3.2, we employ Gaussian mixture to model all nodes V within a graph structure and estimate the distribution of trigger nodes [18]. By treating the feature vector of each node as individual data point, the formula for the posterior probability of one node belonging to specific distribution is:

$$\gamma_{ik} = \frac{\pi_k \cdot \mathcal{N}(\mathbf{x}_i|\boldsymbol{\mu}_k, \boldsymbol{\Sigma}_k)}{\sum_{j=1}^{K=2} \pi_j \cdot \mathcal{N}(\mathbf{x}_i|\boldsymbol{\mu}_j, \boldsymbol{\Sigma}_j)}, \tag{1}$$

where γ_{ik} represents the posterior probability of node i belonging to the kth distribution (we set K as 2 in accordance with our assumption of two node distributions, namely, a normal distribution $\mathcal{N}_{nor}$ and an infected node distribution $\mathcal{N}_{poi}$, and then our objective is to determine the trigger node distribution $\mathcal{N}_{poi}$), π_k is the mixing coefficient of the kth distribution, $\mathbf{x}_i$ is the observed node vector, $\boldsymbol{\mu}_k$ is the mean of the kth distribution, $\boldsymbol{\Sigma}_k$ is the covariance matrix of the kth distribution, and $\mathcal{N}(\mathbf{x}_i|\boldsymbol{\mu}_k, \boldsymbol{\Sigma}_k)$ denotes the probability density function of the multivariate normal distribution. π_k, $\boldsymbol{\mu}_k$, and $\boldsymbol{\Sigma}_k$ are iteratively computed using the expectation maximization (EM) method [19], the update equations for the model parameters are shown below:

$$\pi_i = \frac{N_i}{N}, \tag{2}$$

$$\boldsymbol{\mu}_k = \frac{1}{N_k} \sum_{i=1}^{N} \gamma_{ik} \mathbf{x}_i, \tag{3}$$

$$\boldsymbol{\Sigma}_k = \frac{1}{N_k} \sum_{i=1}^{N} \gamma_{ik} (\mathbf{x}_i - \boldsymbol{\mu}_k)(\mathbf{x}_i - \boldsymbol{\mu}_k)^\top, \tag{4}$$

where N_k is the node number in the kth distribution, while N is the total node number in the graph.

Through Gaussian mixture modeling, we obtain two distributions of a graph, and the smaller distribution is regarded as $\mathcal{N}_{poi}$ since the number of clean nodes generally exceeds that of the poisoned.

However, in the case of a clean graph, the node distribution is exclusively normal, without a poisoned node distribution, thus easily leading to misdetection. Consequently, to solve this problem, we employ the following formula to calculate a score value for graph and ascertain the presence of poisoning if it exceeds the valve:

$$S = \sum_{i=1}^{n} (\boldsymbol{\mu}_1^i - \boldsymbol{\mu}_2^i) \tag{5}$$

where $\boldsymbol{\mu}_k^i$ represents the ith value in the vector mean of the kth distributions obtained from Gaussian mixture. The idea of the equation design originates from a simple observation, where two different distributions exhibit significant disparities in their means, while two similar distributions demonstrate minimal differences in their means, which could be leveraged for distribution distinction.

Following the aforementioned procedure, poisoned nodes (trigger nodes, $\mathbf{x}_i \in \mathcal{N}_{poi}$) could be identified. Next, our aim is to detect the trigger features in these nodes.

In the current GNN backdoor attack settings, the trigger configuration typically assigns higher values to certain features, which are considerably larger in magnitude compared to the regular values. Consequently, we leverage a clustering algorithm for their identification and the cluster with a higher mean value will be recognized as trigger features.

For identified trigger node $\mathbf{x}_i = [f_1, f_2, ..., f_n] \in \mathcal{N}_{poi}$, we regard the values of all the features as independent data points (f_i) and then classify them in two categories (i.e., normal and trigger features) using a clustering algorithm. In this study, we employ the K-means algorithm for clustering [1]. The algorithm works by iteratively assigning data points to clusters and optimizing cluster centroids to minimize the overall intra-cluster variance, and the process is given by:

1. **Preparation**: Extract feature values as individual data points ($\{f_1, f_2, ..., f_n\}$).
2. **Setting**: Set the optimal number of clusters as two (i.e., normal and trigger features), and randomly initialize two cluster centroids (λ_1, λ_2) for cluster one C_1 and two C_2 respectively.
3. **Assignment** For each data point, calculate its distance to each cluster centroid and assign it to the cluster with the nearest centroid by

$$c_i = \arg\min_m \|f_i - \lambda_m\|^2, \tag{6}$$

   where λ_m is the centroid.
4. **Update**: Recalculate the centroids of each cluster by taking the mean of the feature values of all data points assigned to that cluster.

$$\lambda_m = \frac{1}{|C_m|} \sum_{f_i \in C_m} f_i, \tag{7}$$

where C_m is the cluster.

5. **Iteration**: Iterate the assignment and update steps until convergence, and the objective function is given by

$$J = \sum_{m=1}^{2} \sum_{i=1}^{n} \psi_{ij} \left\| f_i - \lambda_m \right\|^2, \tag{8}$$

where ψ_{ij} is an indicator function that takes the value of 1 if data point f_i belongs to cluster center $\lambda_m j$, and 0 otherwise.

Through the above process, two clusters will be found (C_1 and C_2), and we identify the smaller one as trigger feature cluster C_{poi}. By eliminating them, the nodes can revert to their clean state and thus be correctly classified by the victim model.

2) Model Fine-Tuning. By employing phase one, we can already achieve an input sample filtration-based backdoor protection in testing. But, to completely eliminate backdoors, the model should be fine-tuned so as to mitigate backdoor.

Utilizing the trigger features ($f_i \in C_{poi}$) obtained in the previous phase, we fine-tune (retrain) the model to keep it insensitive to malicious features, while preserving accuracy on normal sample testing. This is achieved by adding the loss function with a regularized penalty term and fine-tune the backdoored model [22]. The loss is provided as follows:

$$\begin{aligned} L =& \frac{1}{p} \sum_{i=1}^{p} (y_i - \hat{y}_i)^2 + R(\omega), \\ R(\omega) =& \alpha \sum_{i=1}^{l} \|\omega_i^1\|^2, \end{aligned} \tag{9}$$

where the first part of L is the normal loss calculation method (e.g., mean square error or cross entropy), $R(\omega)$ is the penalty regularization and ω_i^1 is the weight parameter related to trigger feature in the first layer of the model (we just need to specify the parameter position without knowing its specific value). Adding penalty item will minimize the attention given to these weights during parameter updates, and the final model activation will be less influenced by trigger features, thus invalidating the trigger and mitigating the backdoor impact on overall model performance.

In fact, directly setting ω_i^1 to 0 (also known as neuron pruning) could also reduce sensitivity to trigger features, but it may also lead to a loss of accuracy on clean samples. To circumvent this issue, incorporating a penalty term and subsequently fine-tuning the model enables the preservation of the model's normal performance while effectively reducing sensitivity to trigger features.

4 Experiment

In this section, we aim to evaluate the performance of our proposed algorithm: **BloGBaD**. We will first commence by introducing the experiment settings, followed by a presentation of the evaluation metrics employed in the experiment. Subsequently, experimental defense results will be analyzed. As **BloGBaD** is the first black-box backdoor defense in GNN, we mainly conduct ablation experiments in our study. To obtain a reliable estimate of the model's performance, all experiments will be repeated ten times, and the average values are computed.

4.1 Experiment Settings

To investigate the effectiveness of **BloGBaD**, we employ two classic backdoor attacks [24] (GTA attack) and [34] (GNN backdoor, we abbreviate it to GBA) will be utilized to test our defense approach, while GCN (which leverages convolution in GNN [13]), GAT (which employs attention mechanism in GNN [20]), GraphSAGE (which utilizes neighboring sampling and aggregating in GNN [11]), and GIN [21] will be set as victim models for experiments. As for datasets, 1) Bitcoin, 2) Twitter, 3) COLLAB, 4) AIDS, 5) ChEMBL, 6) Toxicant, and 7) MUTAG will be selected to analyze the defense performances under different tasks. Dataset Details are described in Table 1.

Table 1. Dataset information

Dataset	Graphs	Classes	Average Nodes	Task Type
Bitcoin	658	2	11.53	finance fraud detection
COLLAB	5,000	3	73.49	academic community detection
AIDS	2,000	2	15.69	molecule category prediction
Toxicant	10,315	2	18.67	chemical material classification
MUTAG	188	2	17.93	chemical molecule classification

GTA Attack Configuration. We trained both the corresponding topological and feature trigger generators and the backdoored models according to the method proposed in [24]. To maintain backdoor effectiveness and model performance, we adjust the poison rate (limited to 15%) to achieve an attack success rate exceeding 96% while ensuring an accuracy level above 93%. After the training is completed, the trigger generator is used to generate some poisoned graphs to test **BloGBaD**.

GBA Attack Configuration. We also poison the training dataset according to the method proposed in [34] and then leverage them model learning, thus acquiring a backdoored model. For the trigger pattern, we use Erdos Renyl and keep the poison rate within 20%. The backdoor attack achieves a success rate surpassing 75%, while the accuracy for normal samples remains above 74%. Subsequently, the backdoored model will be utilized for the generation of poisoned graphs to evaluate **BloGBaD**.

4.2 Evaluation Metrics

Several evaluation metrics are employed to assess the effectiveness of **BloGBaD**, including 1) attack success rate (ASR, successful attack trials to total trials), 2) model accuracy (Acc, the accuracy for clean samples), 3) detection rate (DR, detected poisoned samples to total attack trials), and 4) false detection rate (FDR, misdetected clean samples to all clean samples).

4.3 GTA Defense

We first present the experimental results of **BloGBaD** based on GTA, including results for various datasets, observations of decision boundary changes, results for different poison rates, and results for different trigger sizes.

Test Results Based on Different Datasets. We selected GCN, GAT, and GraphSAGE as victim models to conduct GTA attacks on the Bitcoin, COLLAB, AIDS, and Toxicant datasets, and utilized **BloGBaD** for defense. The defense performance is given in Table 2.

Table 2. Defense results against GTA under GCN, GAT, and GraphSAGE. The best performances are highlighted in this table.

Model	Dataset	ASR (backdoored)	ASR (purified)	Acc	DR	FDR
GCN	Bitcoin	91.4%	11.6%	**93.2%**	100%	0
	COLLAB	88.9%	**9.4%**	84.8%	100%	0
	AIDS	**96.9%**	19.7%	90.1%	99.7%	0.3%
	Toxicant	89.5%	17.6%	90.4%	99.9%	0.1%
GAT	Bitcoin	93.5%	14.4%	92.0%	100%	0
	COLLAB	89.5%	17.2%	87.56	100%	0
	AIDS	95.2%	13.7%	91.0%	99.9%	0.1%
	Toxicant	93.9%	20.3%	89.8%	99.7%	0.3%
GraphSAGE	Bitcoin	92.6%	9.8%	91.48%	100%	0
	COLLAB	86.1%	18.3%	88.6%	99.8%	0.2%
	AIDS	92.5%	20.8%	91.1%	99.9%	0.1%
	Toxicant	88.3%	14.7%	87.8%	99.8%	0.2%

For GCN, in the Bitcoin dataset, GTA achieves an ASR of 91.4% for the backdoored model, which significantly decreases to 11.6% after employing **BloGBaD** defense. The overall accuracy of the purified model reaches 93.2%, indicating its ability to maintain normal performance. **BloGBaD** exhibits a remarkable detection rate of 100% and no false discoveries. In the COLLAB dataset, the backdoored model has an ASR of 88.9%, reducing to 9.4% after purification. The accuracy of the purified model is 84.8%, and **BloGBaD** continues to demonstrate effective detection without any misdetection. In the AIDS dataset, the ASR for the backdoored model is 96.9%, which decreases to 19.7% after defense. The accuracy reaches 90.1%, and the defense showcases a high detection rate and a low false detection rate. In the Toxicant dataset, the ASR is 89.5%, which drops to 17.6% after employing **BloGBaD** defense. The accuracy remains relatively high at 90.4

Regarding GAT and GraphSAGE, the experimental results exhibited a consistent trend, albeit with slight differences in values.

Decison Boundary Changes. In order to further illustrate the change in the model fine-tuning on the decision boundary, we randomly select a poisoned graph sample and investigate the probability difference associated with its classification into the normal class versus the target class throughout the fine-tuning process. The GCN model and AIDS dataset will be selected as the target model, and target dataset for the analysis under 100 and 200 fine-tuning epochs. The results are shown in Fig. 3.

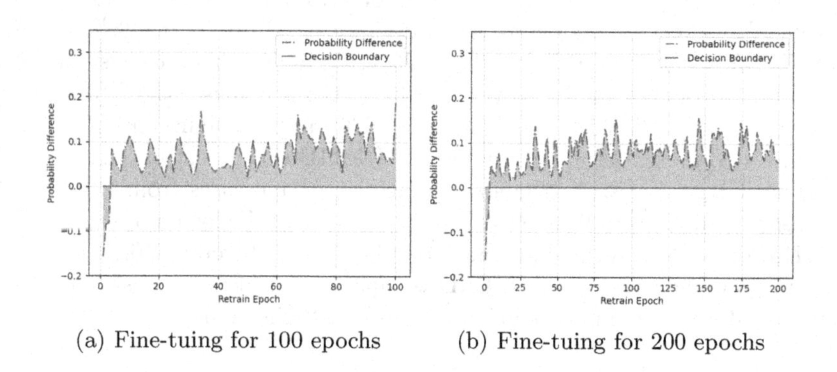

(a) Fine-tuing for 100 epochs (b) Fine-tuing for 200 epochs

Fig. 3. Illustration of the decision boundary changing process.

Figure 3(a) depicts the changing of the decision boundary under 100 fine-tuning epochs. Initially, the model predicts a higher probability of the poisoned sample belonging to the target class. However, as the fine-tuning progresses, the probability of classifying the poisoned sample to the normal class asymptotically surpasses the probability of classifying it to the target class. This suggests that the model is becoming more robust to triggers as it undergoes further fine-tuning.

Figure 3(b) shows the fine-tuning process for 200 epochs, which exhibits a similar trend to that of 100 epochs.

Poison Rate and Trigger Size. From empirical intuition, the increase in the poison rate and trigger size of a backdoor attack typically leads to a higher success rate. To assess the robustness of our defense method against such attack scenarios, we evaluate our defense scheme using different infection rates and trigger sizes. GCN and AIDS dataset are employed as the test model and data. The results are shown in Fig. 4.

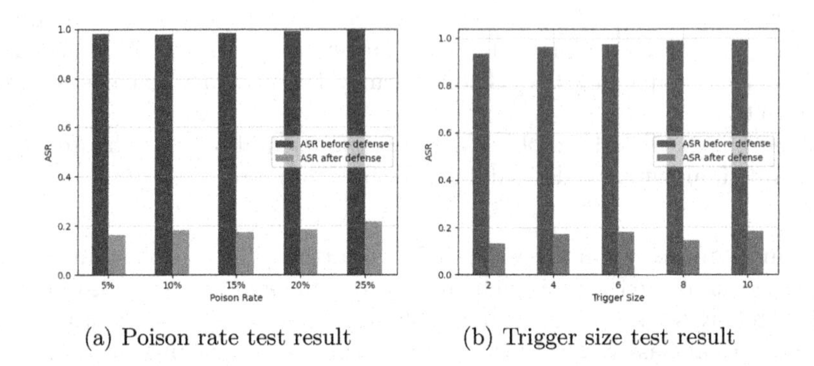

(a) Poison rate test result (b) Trigger size test result

Fig. 4. Illustration of the test results for various poison rates and trigger sizes.

Figure 4(a) illustrates the performance of **BloGBaD** against GTA attacks of varying poison rates. As the poison rate increases from 5% to 25%, the ASR gradually improves from 97% to 99%, while still maintaining a consistently high value (average of 98.4%). Moreover, purified by our defense, the ASRs remain stable at around 18%, with a minimum of 16.3% and a maximum of 21.8%.

Figure 4(b) presents the performance of **BloGBaD** against backdoor attacks of varying trigger sizes. As the size of the trigger increases from 2 to 10, the ASR also improves from 93% to 99% (average 96.9%). Following purification by **BloGBaD**, the ASR drops to a range of 13% to 18% (average 16.2%).

As Fig. 4 demonstrates, **BloGBaD** is capable of achieving consistently good defense results across different infection rates and trigger sizes.

4.4 GBA Defense

We further evaluate the effectiveness of our defense against GBA. Following the open source code in [34], GIN is employed as the victim model and Bitcoin, COLLAB, AIDS, and MUTAG datasets are leveraged for conducting the attacks. Then, **BloGBaD** is used for backdoor defense. The results under GBA attack are shown in Table 3.

In the Bitcoin dataset, GBA achieved an ASR of 89.3% for the backdoored model, which decreased to 11.7% after purification. The fine-tuned model achieved an accuracy of 82.9%. **BloGBaD** exhibited a perfect detection rate of 100% and a false detection rate of 0%, indicating accurate identification of poisoned graphs. In the COLLAB dataset, the ASR for the backdoored model was

Table 3. Defense results against GIN. The best performances are highlighted in this table.

Model	Dataset	ASR (backdoored)	ASR (purified)	Acc	DR	FDR
GIN	Bitcoin	**89.3%**	11.7%	**82.9%**	100%	0
	COLLAB	85.5%	9.9%	80.4%	100%	0
	AIDS	80.4%	13.1%	79.2%	99.8%	0.2%
	MUTAG	79.5%	**8.5%**	81.3%	99.7%	0.3%

85.5%, reducing to 9.9% after purification. The fine-tuned model achieved an accuracy of 80.4%. **BloGBaD** also demonstrated a high detection rate and a low false detection rate. For the AIDS dataset, GBA resulted in an ASR of 80.4% for the backdoored model and 13.1% for the purified model. The defended model achieved an accuracy of 79.2%. In the MUTAG dataset, GBA showed an ASR of 79.5% for the backdoored model, decreasing to 8.5% after purification. The accuracy of the fine-tuned model remained at 81.3%.

Decision Boundary Changes. In line with the GTA-based defense, we also witness the retraining procedure of a GBA attack defense. To better examine the change in decision boundaries, experimental evaluations will be performed at epochs 100 and 200. The test results are shown in Fig. 5.

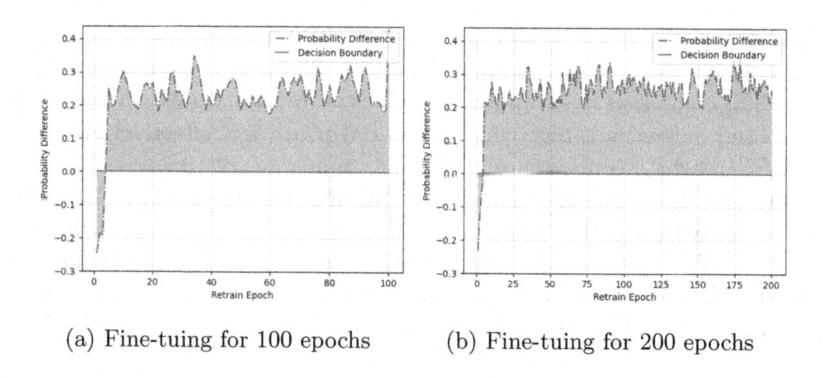

(a) Fine-tuing for 100 epochs (b) Fine-tuing for 200 epochs

Fig. 5. Illustration of the decision boundary changing process under GBA defense.

As depicted in Fig 5(a), for the experiment under 100 epochs, during the initial epochs, the model exhibits a higher probability of classifying the contaminated samples into the target class rather than the normal class (peaking at approximately 24%). Subsequently, as the fine-tuning process advances, the probability of accurate classification by the model gradually increases. Eventually, the probability of correct classification surpasses that of misclassification by an average margin of 25%.

For the experiment under 200 epochs, similar patterns are also observed in Fig. 5(b).

Poison Rate and Trigger Size. To assess the effectiveness of our defense approach against GBA attacks with varying infection rates and trigger sizes, we conducted additional defense experiments by systematically varying the poison rates and trigger sizes. The COLLAB dataset was selected as the test dataset for our experiments. The results are shown in Fig. 6.

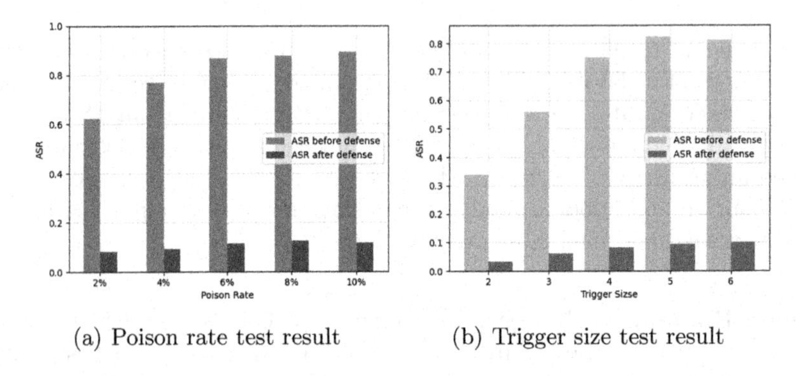

(a) Poison rate test result (b) Trigger size test result

Fig. 6. Illustration of the test results for various poison rates and trigger sizes under GBA defense.

Figure 6(a) gives the defense result against GBA under various poison rates. As the poison rate increased from 2% to 10%, the ASR increased to approximately 85%, compared to an initial success rate of around 62%. After applying our proposed defense strategy, the ASR success rate was effectively reduced to below 12% (reaching a minimum of 8% and an average of 10.66%).

Figure 6(b) illustrates the defensive performance of our **BloGBaD** approach against GBA with different trigger sizes. It can be observed that as the trigger size increased from 2 to 6, ASR gradually rose from 34% to 81% (with an average of 65.6%). When **BloGBaD** is applied, all ASR reduced to below 11% (ranging from a minimum of 3.4% to a maximum of 10.2%, with an average of 7.5%).

Phase 1 Performances. Based on Sects. 4.3 and 4.4, we have demonstrated the overall effectiveness of BloGBaD. In this section, we will further examine the phase 1 efficacy of BloGBaD.

We generated 50 test graphs via GBA on the COLLAB dataset to observe the average trigger identification rate (TIR, the ratio of correctly identified poisoned nodes to the total trigger nodes) and misidentification rate (MIR, misidentified nodes to all detected nodes). Trigger patterns were chosen as Erdős-Rényi subgraph (ER), Small World subgraph (SW), and Preferential Attachment subgraph (PA), with sizes set at 3, 5, 7, and 9 respectively. The test results are shown in Fig. 7.

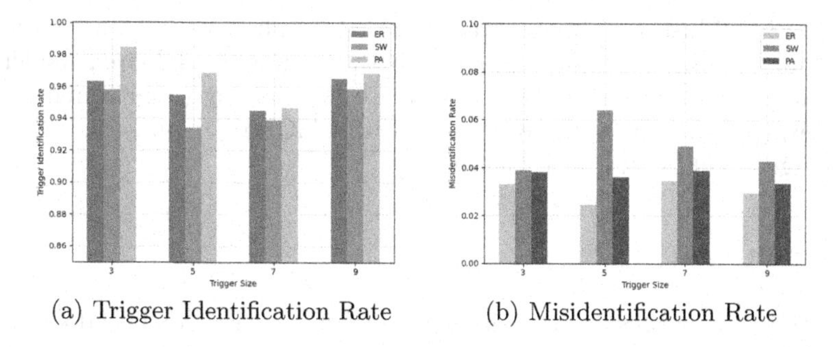

(a) Trigger Identification Rate (b) Misidentification Rate

Fig. 7. Test results of trigger identification rate and misidentification rate for phase 1.

Figure 7(a) gives the results about trigger identification rate. In the varying settings of different trigger modes and sizes, the TIR of **BloGBaD** in phase 1 consistently surpassed 90%, specifically, 95.69% for Erdõs-Rényi subgraph (ER), 94.73% for Small World subgraph (SW), and 96.70% for Preferential Attachment subgraph (PA).

Figure 7(b) shows misidentification rate results. Across different settings, **BloGBaD** maintained a misclassification rate of less than 7% in phase 1.

Phase 2 Performances. We further investigate the performance of phase 2 in BloGBaD. We compare our method with neuron pruning to observe the decrease in ASR and the preservation of Acc. We employ GBA attack under default setting in the code of [34] based on the COLLAB dataset as the target of defense. The test results are shown in Fig. 8.

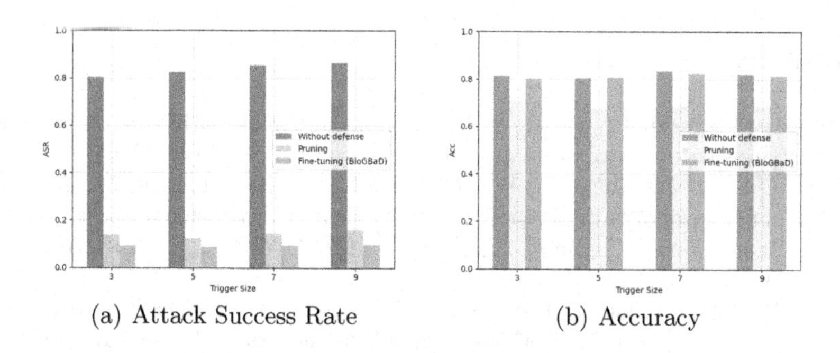

(a) Attack Success Rate (b) Accuracy

Fig. 8. Test results of ASR and Acc for phase 2.

As can be seen from Fig. 8(a), both neuron pruning and phase 2 in BloGBaD significantly reduce the Attack Success Rate, with BloGBaD demonstrating slightly better performance compared to pruning.

Figure 8(b) illustrates the accuracy performance of the model after processing. It can be observed that compared to the original backdoored model, the pruned model experiences a certain decrease in accuracy (approximately 11% reduction). However, after the processing with BloGBaD in phase 2, it does not exhibit significant accuracy degradation.

5 Conclusion

In this study, we first discuss the impracticality of current GNN backdoor defense methods: requiring access to model parameters, and then propose a black-box backdoor defense strategy in GNNs: **BloGBaD**. Our proposed approach utilizes a two-phase scheme to effectively eliminate backdoors. In the first phase, referred to as test sample filtration, Gaussian mixing and clustering algorithms are employed to detect and purify trigger nodes and features. The second phase involves model fine-tuning, wherein the trigger features obtained from the previous phase are utilized to retrain the model, ensuring that it becomes immune to the trigger and consequently eliminating the influence of backdoor. Experimental results across different datasets and utilizing two classical attack scenarios demonstrate the effectiveness of **BloGBaD** in reducing the attack success rate, detecting triggers, and maintaining model accuracy.

In future work, we plan to improve our approach by designing defense schemes that do not require any poison test samples or prior knowledge about the model.

Acknowledgement. This research is supported National Nature Science Foundation of China (No. 62202303, 62202302, U20B2048, and U2003206), Shanghai Sailing Program (No. 21YF1421700), Action Plan of Science and Technology Innovation of Science and Technology Commission of Shanghai Municipality (No. 22511101202), and JSPS KAKENHI (No. JP22K17884).

References

1. Ahmed, M., Seraj, R., Islam, S.M.S.: The k-means algorithm: a comprehensive survey and performance evaluation. Electronics **9**(8), 1295 (2020)
2. Chen, G., Wu, J., Yang, W., Bashir, A.K., Li, G., Hammoudeh, M.: Leveraging graph convolutional-lstm for energy-efficient caching in blockchain-based green iot. IEEE Trans. Green Commun. Netw. **5**(3), 1154–1164 (2021). https://doi.org/10.1109/TGCN.2021.3069395
3. Chen, H., Fu, C., Zhao, J., Koushanfar, F.: Deepinspect: a black-box trojan detection and mitigation framework for deep neural networks. In: IJCAI. vol. 2, p. 8 (2019)
4. Chen, J., Xiong, H., Zheng, H., Zhang, J., Jiang, G., Liu, Y.: Dyn-backdoor: backdoor attack on dynamic link prediction. arXiv preprint arXiv:2110.03875 (2021)
5. Chen, L., et al.: Neighboring backdoor attacks on graph convolutional network. arXiv preprint arXiv:2201.06202 (2022)
6. Chen, Y., Ye, Z., Zhao, H., Wang, Y., et al.: Feature-based graph backdoor attack in the node classification task. Int. J. Intell. Syst. 2023 (2023)

7. Dai, E., Lin, M., Zhang, X., Wang, S.: Unnoticeable backdoor attacks on graph neural networks. arXiv preprint arXiv:2303.01263 (2023)
8. Dong, Y., et al.: Black-box detection of backdoor attacks with limited information and data. In: Proceedings of the IEEE/CVF International Conference on Computer Vision, pp. 16482–16491 (2021)
9. Guo, J., Li, A., Liu, C.: Aeva: Black-box backdoor detection using adversarial extreme value analysis. arXiv preprint arXiv:2110.14880 (2021)
10. Guo, J., Li, Y., Chen, X., Guo, H., Sun, L., Liu, C.: Scale-up: an efficient black-box input-level backdoor detection via analyzing scaled prediction consistency. arXiv preprint arXiv:2302.03251 (2023)
11. Hamilton, W.L., Ying, Z., Leskovec, J.: Inductive representation learning on large graphs. In: Guyon, I., von Luxburg, U., Bengio, S., Wallach, H.M., Fergus, R., Vishwanathan, S.V.N., Garnett, R. (eds.) Advances in Neural Information Processing Systems 30: Annual Conference on Neural Information Processing Systems 2017(December), pp. 4–9, 2017. Long Beach, CA, USA, pp. 1024–1034 (2017), https://proceedings.neurips.cc/paper/2017/hash/5dd9db5e033da9c6fb5ba83c7a7ebea9-Abstract.html
12. Jiang, B., Li, Z.: Defending against backdoor attack on graph nerual network by explainability. arXiv preprint arXiv:2209.02902 (2022)
13. Kipf, T.N., Welling, M.: Semi-supervised classification with graph convolutional networks. In: 5th International Conference on Learning Representations, ICLR 2017, Toulon, France, April 24–26, 2017, Conference Track Proceedings. OpenReview.net (2017). https://openreview.net/forum?id=SJU4ayYgl
14. Li, Y., Wu, B., Jiang, Y., Li, Z., Xia, S.: Backdoor learning: a survey. CoRR abs/2007.08745 (2020). https://arxiv.org/abs/2007.08745
15. Li, Y., Li, Y., Wu, B., Li, L., He, R., Lyu, S.: Invisible backdoor attack with sample-specific triggers. In: 2021 IEEE/CVF International Conference on Computer Vision, ICCV 2021, Montreal, QC, Canada, October 10–17, 2021, pp. 16443–16452. IEEE (2021). https://doi.org/10.1109/ICCV48922.2021.01615
16. Liu, Y., et al.: Backdoor defense with machine unlearning. In: IEEE INFOCOM 2022 - IEEE Conference on Computer Communications, London, United Kingdom, May 2–5, 2022, pp. 280–289. IEEE (2022). https://doi.org/10.1109/INFOCOM48880.2022.9796974
17. Liu, Y., Ma, X., Bailey, J., Lu, F.: Reflection backdoor: a natural backdoor attack on deep neural networks. In: Vedaldi, A., Bischof, H., Brox, T., Frahm, J.-M. (eds.) ECCV 2020. LNCS, vol. 12355, pp. 182–199. Springer, Cham (2020). https://doi.org/10.1007/978-3-030-58607-2_11
18. Reynolds, D.A., et al.: Gaussian mixture models. Encycl. Biometrics **741**(659–663) (2009)
19. Sammaknejad, N., Zhao, Y., Huang, B.: A review of the expectation maximization algorithm in data-driven process identification. J. Process Control **73**, 123–136 (2019)
20. Velickovic, P., Cucurull, G., Casanova, A., Romero, A., Lio, P., Bengio, Y., et al.: Graph attention networks. stat **1050**(20), 10–48550 (2017)
21. Wang, X., Zhang, M.: How powerful are spectral graph neural networks. In: Chaudhuri, K., Jegelka, S., Song, L., Szepesvári, C., Niu, G., Sabato, S. (eds.) International Conference on Machine Learning, ICML 2022, 17–23 July 2022, Baltimore, Maryland, USA. Proceedings of Machine Learning Research, vol. 162, pp. 23341–23362. PMLR (2022). https://proceedings.mlr.press/v162/wang22am.html
22. Weng, C.H., Lee, Y.T., Wu, S.H.B.: On the trade-off between adversarial and backdoor robustness. Adv. Neural. Inf. Process. Syst. **33**, 11973–11983 (2020)

23. Wu, Z., Pan, S., Chen, F., Long, G., Zhang, C., Philip, S.Y.: A comprehensive survey on graph neural networks. IEEE Trans. Neural Netw. Learn. Syst. **32**(1), 4–24 (2020)
24. Xi, Z., Pang, R., Ji, S., Wang, T.: Graph backdoor. In: USENIX Security Symposium, pp. 1523–1540 (2021)
25. Xu, J., Wang, R., Liang, K., Picek, S.: More is better (mostly): On the backdoor attacks in federated graph neural networks. arXiv preprint arXiv:2202.03195 (2022)
26. Xu, J., Xue, M., Picek, S.: Explainability-based backdoor attacks against graph neural networks. In: Proceedings of the 3rd ACM Workshop on Wireless Security and Machine Learning, pp. 31–36 (2021)
27. Yan, Z., et al.: Dehib: Deep hidden backdoor attack on semi-supervised learning via adversarial perturbation. In: Thirty-Fifth AAAI Conference on Artificial Intelligence, AAAI 2021, Thirty-Third Conference on Innovative Applications of Artificial Intelligence, IAAI 2021, The Eleventh Symposium on Educational Advances in Artificial Intelligence, EAAI 2021, Virtual Event, February 2–9, 2021, pp. 10585–10593. AAAI Press (2021). https://doi.org/10.1609/aaai.v35i12.17266
28. Yan, Z., Li, S., Zhao, R., Tian, Y., Zhao, Y.: DHBE: data-free holistic backdoor erasing in deep neural networks via restricted adversarial distillation. In: Liu, J.K., Xiang, Y., Nepal, S., Tsudik, G. (eds.) Proceedings of the 2023 ACM Asia Conference on Computer and Communications Security, ASIA CCS 2023, Melbourne, VIC, Australia, July 10–14, 2023, pp. 731–745. ACM (2023). https://doi.org/10.1145/3579856.3582822
29. Yan, Z., Wu, J., Li, G., Li, S., Guizani, M.: Deep neural backdoor in semi-supervised learning: threats and countermeasures. IEEE Trans. Inf. Forensics Secur. **16**, 4827–4842 (2021). https://doi.org/10.1109/TIFS.2021.3116431
30. Yang, S., et al.: Transferable graph backdoor attack. In: Proceedings of the 25th International Symposium on Research in Attacks, Intrusions and Defenses, pp. 321–332 (2022)
31. Yao, Y., Li, H., Zheng, H., Zhao, B.Y.: Latent backdoor attacks on deep neural networks. In: Cavallaro, L., Kinder, J., Wang, X., Katz, J. (eds.) Proceedings of the 2019 ACM SIGSAC Conference on Computer and Communications Security, CCS 2019, London, UK, November 11–15, 2019, pp. 2041–2055. ACM (2019). https://doi.org/10.1145/3319535.3354209
32. Zhang, M., Cui, Z., Neumann, M., Chen, Y.: An end-to-end deep learning architecture for graph classification. In: McIlraith, S.A., Weinberger, K.Q. (eds.) Proceedings of the Thirty-Second AAAI Conference on Artificial Intelligence, (AAAI-18), the 30th innovative Applications of Artificial Intelligence (IAAI-18), and the 8th AAAI Symposium on Educational Advances in Artificial Intelligence (EAAI-18), New Orleans, Louisiana, USA, February 2–7, 2018, pp. 4438–4445. AAAI Press (2018). https://doi.org/10.1609/aaai.v32i1.11782
33. Zhang, X., Chen, H., Huang, K., Koushanfar, F.: An adaptive black-box backdoor detection method for deep neural networks (2022)
34. Zhang, Z., Jia, J., Wang, B., Gong, N.Z.: Backdoor attacks to graph neural networks. In: Proceedings of the 26th ACM Symposium on Access Control Models and Technologies, pp. 15–26. SACMAT '21, Association for Computing Machinery, New York, NY, USA (2021). https://doi.org/10.1145/3450569.3463560
35. Zheng, H., Xiong, H., Chen, J., Ma, H., Huang, G.: Motif-backdoor: rethinking the backdoor attack on graph neural networks via motifs. arXiv preprint arXiv:2210.13710 (2022)

DRL-Based Optimization Algorithm for Wireless Powered IoT Network

Mingjie Zhu, Shubin Zhang(✉), and Kaikai Chi

Zhejiang University of Technology, Hangzhou, China
zhangshubin@zjut.edu.cn

Abstract. The extensive applications of Internet of Things (IoT) bring development and prosperity to multiple industries and greatly influent all aspects of human lives, which rely on the timely computation and communication. However, the limited battery and computing capacity of the common IoT units cannot satisfy the need of processing computation intensive and time sensitive tasks. As a result, the combination of wireless power transmission (WPT) and mobile edge computing (MEC) provides a promising approach of dealing with aforementioned problems. The Hybrid Access Point (HAP) transmits radio frequency (RF) energy to provide power transmission for wireless devices (WDs), and processes tasks offloaded from WDs with the equipped edge computing servers (ECSs). In this paper, we consider a wireless powered mobile edge computing network containing an HAP, multiple WDs. By jointly optimizing energy transmission time duration, partial offloading decisions and transmission power allocation strategy, we obtain the maximum of the sum computation rate (SCR). Firstly, we formulate this as a non-convex problem which is hard to address. Secondly, we decompose the problem into a top-problem of optimizing the energy transmission time duration and a sub-problem of optimizing the partial offloading decisions and transmission power allocation strategy. Finally, we design a DRL-based offloading algorithm, which applies a DNN network with exploring and updating strategy, to address the top-problem and propose an effective optimizing algorithm to address the sub-problem. Extensive numerical results reveal that the proposed algorithm reaches near-optimal SCR and greatly reduces time latency and complexity compared to benchmark algorithms.

Keywords: Mobile-edge computing · wireless power transfer · reinforcement learning · partial offloading · resource allocation

1 Introduction

With the emergence of the increasing number of smart mobile devices, like smart phone, smart vehicle, and smart home, the Internet of Things (IoT) is becoming flourished and expanded, which prompts the extensive application of the IoT in

Supported by organization x.

Z. Tari et al. (Eds.): ICA3PP 2023, LNCS 14491, pp. 181–200, 2024.
https://doi.org/10.1007/978-981-97-0808-6_11

multiple industries, such as autonomous driving, remote surgery and smart city. Many applications rely on the communication instantaneity and computation rapidity. However, the rather limited battery capacity and low computation ability that IoT devices usually have can hardly sustain the large-scale high-speed computation tasks. As a result, solving the power and computation constraint of the IoT devices has become the key to the IoT application innovation. Radio frequency (RF) based wireless powered transmission (WPT) technology offers a promising manner of tackling the power constraint under long-term operation. Moreover, mobile edge computation (MEC) technology improves the computation capacity by deploying edge computing servers (ECSs) on the Hybrid Access Points (HAPs), and offloading the core data flux to edge server of network. Under the circumstances that the computing capability of wireless devices (WDs) is limited, they hold the option of offloading computation tasks to HAPs, where corresponding ECS performs high-speed computation. With the help of the WPT technology, WDs harvest energy broadcast by HAP and use it to drive CPU unit and offload tasks. The integration of WPT and MEC achieves continuable and power-free operations, and strengthens the carrying and computing capacity of WDs.

Although the combination of WPT and MEC empowers communication and computation improvements of the IoT network, it is still facing some challenges. Firstly, how to jointly optimize the great variety of computing resources becomes a difficult but vital problem. Secondly, due to the fast-fading quality of wireless channel, it is indispensable to devise a high-efficiency algorithm.

To tackle these problems, researchers have conducted many studies. [1] aimed at maximizing the total sum computation rate in a WP-MEC network based on NOMA and proposed a greedy algorithm, which jointly optimized the binary offloading decision of each WD, the duration for WPT, the time assigned for tasks offloading, along with the local computing rate for local computing WD and the transmission power level for offloading WD. Through jointly optimizing the time allocation of WPT, offloading time division of each WD along with the partial offloading scheme, [2] studied the maximization of computation sum rate and devised an online deep reinforcement learning algorithm, used for WPT duration optimization, while an optimal algorithm based on Lagrange duality is designed to optimize the offloading time allocation and offloading energy assignment. [3] applied an unmanned aerial vehicle (UAV) into the MEC network to extend service coverage. An iteration algorithm is proposed to maximize the minimum computation throughput of all WDs, by jointly optimizing the selection of offloading decision, the trajectory of UAV, and the computation frequency allocation.

In this paper, we focus on the WPT-MEC network adopting partial offloading, where HAP is equipped with multiple antennas. Our goal is maximizing the Sum Computation Rate (SCR) of the network by jointly optimizing the WET duration, offloading decision of each WD and transmission power allocation. However, this maximization is non-convex and hard to solve. To improve efficiency, a DRL-based algorithm is devised to optimize the energy transmission duration and resources allocation. The contributions are listed as follows:

1. The SCR maximization problem is formulated as a non-convex problem. We apply a decomposition strategy to break down it into a top-problem and a sub-problem. The top-problem is to optimize the wireless energy transmission time duration, while the sub-problem is to optimize the offloading decision and transmission power allocation of each WD under the given result from top-problem.
2. Unlike many existing deep learning methods that optimize all system parameters at the same time resulting infeasible solutions, in this paper, we decompose the original optimization problem into an offloading decision top-problem and a resource allocation sub-problem.
3. For the top-problem, we devise a DRL-based algorithm to generate WPT duration from the input of channel gain. Specifically, a deep neural network (DNN) together with a exploration strategy and a training strategy are designed to achieve self-learning from past experience and continue update the finest training samples. The algorithm shows great performance after convergence and outputs the near-optimal result with rather low computing latency. For the sub-problem, a joint optimization scheme is designed to use bi-section method and fmincon function to optimize the offloading decision and transmission power allocation, respectively.

The rest of this paper is organized as follows. Section 2 introduces the related work. The system model and problem formulation are illustrated in Sect. 3. In Sect. 4, we propose a DRL-based offloading algorithm. The numerical results are represented in Sect. 5. At last, Sect. 6 concludes this paper.

2 Related Work

In the field, there are two generally adopted offloading schemes called binary offloading and partial offloading for WP-MEC network. Binary offloading scheme represents a task is deemed to be a whole and each one of which is either computed locally or offloaded to the ECS on the HAP. Partial offloading scheme denotes a task is divisible and a certain proportion of the task is computed locally while the other part is offloaded.

2.1 Binary Offloading

Some works have been devoted to the application of binary offloading algorithms. Zeng *et al.* [1] aimed at maximizing the total sum computation rate in a WP-MEC network based on NOMA, and proposed a greedy algorithm, which jointly optimized the binary offloading choice of each WD, the duration for WPT, the time assigned for tasks offloading, along with the local computing rate for local computing WD and the transmission power level for offloading WD. To tackle the booming redundancy of computation and transmission the MEC system facing, together with the low sampling efficiency and slow convergence caused by popular DRL algorithm, [4] devised a cache-assisted collaborative task offloading and resource allocation strategy, which comprised a task caching method gaining the state of task caches and an efficient meta-reinforcement-learning-based computation offloading algorithm assisted by cache to obtain offloading decisions. Xu *et al.* [3] applied an unmanned aerial vehicle (UAV) into the MEC network to extend service coverage. Aiming to maximize the minimum computation throughput of all ground devices, an iteration algorithm employing the P-SCA method and D.C. framework was designed, which jointly optimized the selection of offloading decision, the trajectory of UAV, and the computation frequency allocation. Multiply simulations showed that the proposed algorithm outperformed the benchmark algorithms. In order to maximize the sum computation rate of all the wireless devices, [5] devised a decoupled optimization algorithm. By jointly optimizing the binary offloading selection decisions of all devices and the system transmission time allocation, the algorithm combined the bi-section search method with the alternating direction method of multipliers (ADMM) technique. Extensive simulations revealed that the proposed algorithm significantly outperformed the benchmark algorithms. On the basis of [5,6] proposed an effective online learning framework called deep reinforcement learning-based online offloading (DROO) algorithm to address the same objective optimization problem. The DROO algorithm approximately gain the sub-optimal results of the energy transmission duration time and helped obtain the maximal computation rate with lower complexity and time consumption. [7] investigated the backscatter-assisted data offloading scheme in OFDMA-based wireless-powered MEC system. Aiming at maximizing the sum computation rate, a fast-efficient algorithm was proposed by leveraging the block coordinate descent method to jointly optimize the transmit power at gateway, backscatter coefficient, time-splitting ratio, and binary decision-making matrices. The devised algorithm achieved a near-globally-optimal solution at a much lower complexity as compared to benchmark algorithm like the bisection-based algorithm.

2.2 Partial Offloading

Many efforts also devoted to the application of partial offloading algorithms. Mao *et al.* [8] investigated the total computation bits maximization problem for a WP-MEC network equipped with the intelligent reflecting surfaces (IRS) technique, which reconstituted the wireless communication environment by forming reflect beamforming. By jointly optimizing the downlink and uplink phase beamforming of IRS, transmission power and time block assigned for WPT and offloading, as well as the local computing frequencies of WDs, the maximum of computation bits is obtained. Wu *et al.* [9] considered an OFDMA-based WP-MEC network containing a hybrid access point and multiple users. By jointly optimizing the transmission power, the duration of WPT as well as the offloading task assignment, the authors pursued the maximization the weighted computation rate. Ren *et al.* [10] focused on the vehicular MEC. The vehicle uses a partial offloading policy to determine its computing tasks offloading assignment. By jointly optimizing offloading scheme and transmission power assignment, the minimization of the vehicular energy consumption is obtained, which is restricted by task-related response time. Zhang *et al.* [2] considered a WP-MEC network containing an edge computing server and multiple wireless devices. Through jointly optimize the time allocation of WPT, offloading time division of each WD along with the partial offloading scheme, the maximization of the proposed computation sum rate was obtained. An online deep reinforcement learning algorithm was used for WPT duration optimization while an optimal algorithm based on Lagrange duality was devised to optimize the offloading time allocation and offloading energy assignment. To tackle the contradiction between WDs' limited computing capability, power supply and growing needs on resources-relayed application and the high cost and large latency caused by MEC technique, [11] devised a hybrid metaheuristic algorithm named genetic simulated annealing-based particle swarm optimization that jointly optimized partial offloading assignment, WDs' CPU speed and transmission power and communication bandwidth to minimize the total energy the network consumed. Li *et al.* [12] focused on minimizing the total latency of the MEC system, where certain users' information was partly unknown. By devising a prediction-based information completion scheme and an adaptive-information-retransmission-and-predictive-completion-combined algorithm, impact of partial information and long-term system performance were signally improved. Saleem *et al.* [13] considered a device-to-device-enabled OFDMA-based MEC network and focused on the total minimization of the task execution latency sum in the shared range with interference. The task execution latency was reduced by jointly optimize the energy consumption, partial offloading data size, under the constraints of resource allocation. A novel algorithm, called Joint Partial Offloading and Resource Allocation (JPORA), was proposed. JPORA achieved the lowest latency as compared to the other schemes, which contained local computation, random offloading and total offloading and also limited the local energy consumption of WDs. Guo *et al.* [14] devised a Lyapunov-based partial computation offloading algorithm to minimize the total energy consumption in the MEC network. The authors considered the long-term average energy consumption and the discarding ratio of computation tasks as the quantitative indicators and compared the algorithm with several baselines and state-of-the-art algorithms, like local computing all, randomly partial computation offloading and so on, which showed the superiority of the proposed algorithm. Feng *et al.* [15] jointly optimized the RF power allocation of the base station, the offloaded data size and energy allocation of each WD. An iterative algorithm was devised to maximize the data utility and minimized the energy consumption of the operator under the offloaded delay constraint, which gained excellent performances in simulations compared with existing algorithms.

3 System Model and Problem Formulation

3.1 System Model

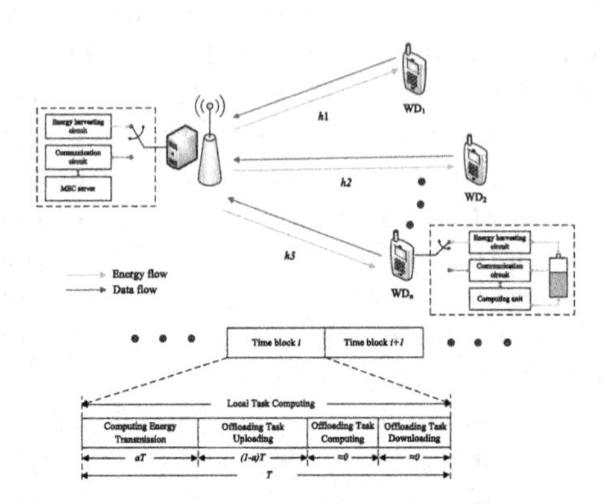

Fig. 1. Wireless powered MEC network.

As shown in Fig. 1, we consider a wireless powered mobile-edge computing network, which consists of an HAP and N wireless edge devices (WDs) denotes as $D_i(i \in \mathcal{N}, \mathcal{N} = \{1, 2, \ldots, N\})$. The HAP is equipped with an edge computing server (ECS), a power beacon and multiple antennas, which is greater in number than WDs. Both ECS and power beacon has stable energy sources, like cable power supply. With the ability of receiving computing tasks through antennas, the ECS provides HAP with considerable computing capability. Accordingly, we suppose the WDs may offload their computing tasks to the HAP. Power beacon can broadcast Radio-Frequency (RF) energy to the WDs, while each WD has a rechargeable battery that can obtain and store the harvested power to support its operations. Each WD contains an antenna and can capture RF energy provided by power beacon or transmit offloading tasks to ECS. Due to HAP's multi-antenna, HAP can receive offloading tasks transmitted by multiple WDs simultaneously. Moreover, we suppose the RF energy and task-offloading transmission are performed in the same frequency band. As a result, the Time division multiplexing (TDM) is applied to avoid interference between the charging process and the task-offloading process of WDs.

The system time is divided into successive time blocks with equal lengths T. Specifically, we further distribute each time block into wireless energy transmission (WET), offloading task uploading (OTU), offloading task computing (OTC), offloading task downloading (OTD) and local task computing (LTC), which is shown in Fig. 1. In WET, WDs capture and store the energy transmitted by power beacon. In OTU, WDs consume part of the captured energy to offload computing task to HAP. In OTC and OTD, HAP completes the computing process and transmits the result back to WDs. As the computation capability of HAP is much greater than WDs, the data size of the downloaded result is rather small and the transmission speed is very fast, the time occupied by OTC and OTD is so small that it is negligible when compared with others. Therefore, in our research, a time block is divided into the time for WET, OTU and LTC. The time of WET is denoted as aT, $a \in (0, 1)$, and the rest part of $(1-a)T$ is used for OUT, while the LTC can be conducted during the whole time block T.

However, the transmit power of HAP is not limitless. Let P_{max} denotes the total transmit power of HAP, and the transmit power of HAP to ith WD is denoted as p_i, which is fixed within a time block but may change between different time blocks. And the sum of each Pi must satisfy the restrain:

$$\sum_{N}^{i=1} p_i \leq P_{max}. \tag{1}$$

Both the amount of energy harvested by the WDs and the communication speed between HAP and WDs are related to wireless channel gain. In this paper, we consider a block-fading model. We assume that the wireless channel gain from HAP to a WD remains unchanged and the uplink and the downlink share the same channel gain in each time block, but the channel gain may vary from time blocks to time blocks. Let h_i denotes the wireless channel gain between the HAP and the ith WD. At the beginning of a time block, the energy captured by the process of WET is

$$E_i = \mu p_i h_i aT, \tag{2}$$

where μ denotes the energy conversion efficiency.

For each WD, we consider a partial offloading policy, which means that each WD can offload part of its computing task to HAP and compute the rest locally. We use $x_i \in [0, 1]$ as a indicator variable, that is Di can use $(1 - x_i)E_i$ amount of energy to compute locally and the leftover energy $x_i E_i$ is devoted to transmit task to the HAP.

3.2 Local Computing Mode

Each WD can compute its task throughout the entire time block even during the energy harvesting process. We denote the amount of processed bits in one time block as

$$\frac{f_i t_i}{\phi}, \tag{3}$$

where f_i denotes computing speed of the processor (cycles per second), $t_i \in [0, T]$ denotes the local computation time, and $\phi > 0$ denotes the number of cycles needed to process one bit of the task. Moreover, the constraint of the energy local computing process can consume is

$$K f_i^3 t_i \leq (1 - x_i)E_i, \tag{4}$$

where K represents the computation energy efficiency coefficient of the WDs. For the sake of computing the maximum amount of data, the WDs need to deplete the local computing energy $(1 - x_i)E_i$ and compute throughout the time block, that is, $t_i^* = T$. Accordingly, f_i^* is

$$f_i^* = \left(\frac{(1 - x_i)\, E_i}{KT}\right)^{\frac{1}{3}} = \left(\frac{(1 - x_i)\, \mu p_i a}{K}\right)^{\frac{1}{3}}. \tag{5}$$

As a result, the local computation rate (in bits per second) is

$$r_{l,i}^*(a, h_i, x_i, p_i) = \frac{f_i^* t_i^*}{\phi T} = \frac{(\mu)^{\frac{1}{3}}}{\phi}\left(\frac{(1 - x_i)p_i h_i a}{K}\right)^{\frac{1}{3}}. \tag{6}$$

3.3 Edge Computing Mode

Each WD can obtain the same width of communication bandwidth resource of the edge computing network denoted as B. In order to maximize the edge computing rate, each WD should deplete the energy $x_i E_i$, assigned to edge task offloading and communication between HAP and WDs, i.e. the transmission power of the ith WD is $p_i = \frac{x_i E_i}{(1-a)T}$. As we discussed above, we ignore the time of OTC and OTD. Accordingly, the edge computing rate equals to the data transmission rate over period of CTO, which can be denoted as

$$r_{o,i}^*(a, h_i, x_i, p_i) = \frac{BT\,(1 - a)}{v_u}\log_2\left(1 + \frac{\mu p_i x_i a h_i^2}{(1 - a)\, N_0}\right), \tag{7}$$

where N_0 represents noise power spectral density at HAP's receiver, and v_u represents the edge computing overhead.

3.4 Problem Formulation

Our goal is to seek the maximum of the sum computation rate of the wireless powered mobile-edge computing network. To achieve this goal, we have to jointly optimize the time of WET, the transmit power p_i distributed to ith WD, and the energy allocation of each WD within each time block. In addition, all the system parameters are fixed, except for the wireless channel gain $h = \{h_i, i \in \mathcal{N}\}$, which vary over time. The sum computation rate within one time block is

$$Q(\mathbf{h}, \mathbf{x}, \mathbf{p}, a) = \sum_N^{i=1}(r_{l,i}^*(a, h_i, x_i, p_i) + r_{o,i}^*(a, h_i, x_i, p_i)), \tag{8}$$

where $h = \{h_1, h_2, \ldots, h_N\}$, $x = \{x_1, x_2, \ldots, x_N\}$, $p = \{p_1, p_2, \ldots, p_N\}$, and $a \in (0, 1)$.

Our interest is in maximizing the sum computation rate of every time block with the wireless channel gain $\mathbf{h}$. It can be formulated as

$$(P1):\ Q^*(\mathbf{h}) = \underset{\mathbf{x},\mathbf{p},a}{\text{maximize}} \quad Q(\mathbf{h}, \mathbf{x}, \mathbf{p}, a) \tag{9a}$$

$$\text{subject to} \quad \sum_{i=1}^N p_i \le P_{max}, \tag{9b}$$

$$0 < a < 1, \tag{9c}$$

$$0 \le x_i \ge 1,\ \forall i \in \mathcal{N}. \tag{9d}$$

To tackle the fast-fading channel gain and optimize our network real-time, we propose a deep reinforcement learning based offloading algorithm to adjust the time of WET without being trained with former experience examples in advance.

4 Deep-Reinforcement-Learning-Based Offloading Algorithm

Notice that, as the objective function of the problem (P1) is not concave, it is hard to directly solve it by using the common optimization methods.

Theorem 1. *Once the computing energy transmission duration a is given, the problem (P1) will become problem (P2), and (P2) is concave with $\mathbf{x}$ and $\mathbf{p}$, separately.*

$$(P2): \; Q^*(\mathbf{h}, a) = \underset{\mathbf{x},\mathbf{p}}{maximize} \quad Q(\mathbf{h}, \mathbf{x}, \mathbf{p}, a) \tag{10a}$$

$$subject \ to \quad \textstyle\sum_{i=1}^{N} p_i \le P_{max}, \tag{10b}$$

$$0 \le x_i \ge 1, \ \forall i \in \mathcal{N}. \tag{10c}$$

Proof. As the value of a is given, let $g(x_i) = r_{(l,i)}(a, h_i, x_i, p_i)$, then the second derivative of $r_{(l,i)}(a, h_i, x_i, p_i)$ with respect to x_i can be expressed as

$$g''(x_i) = \frac{-Bp_i A(1 + Bp_i \ln 2)}{((1 + Bp_i x_i)) \ln 2)^2} \le 0, \tag{11}$$

where $A \triangleq \frac{(BT(1-a))}{v_u}$ and $B \triangleq \frac{\mu a h_i^2}{(1-a)N_0}$. For the other part of the $Q(\mathbf{h}, a, \mathbf{x}, \mathbf{p})$, let $h(x_i) = r_{(o,i)}(a, h_i, x_i, p_i)$,then the second derivative of $r_{(o,i)}(a, h_i, x_i, p_i)$ with respect to x_i can be expressed as

$$h''(x_i) = \frac{-2CD^2 p_i^2}{9} (D(1 - x_i)p_i)^{\frac{-5}{3}} \le 0, \tag{12}$$

where $C = \frac{\mu^{\frac{1}{3}}}{\phi}$ and $D = \frac{h_i a}{k}$. Therefore, (P2) is concave with $\mathbf{x}$. Similarly, (P2) can be proved to be concave with $\mathbf{p}$.

As a result followed by Theorem 1, we decompose the non-concave problem and devise a dual-layer method to solve Problem (P1):

- Top-problem: optimizing the WET duration a. A deep-reinforcement-learning-based DNN is applied to adaptively learn to generate the output of the near-optimal value a from the input channel gain $\mathbf{h}$.
- Sub-problem: optimizing the offloading decisions $\mathbf{x}$ and transmission power $\mathbf{P}$ of each WED under the given WET duration a solved in Top-problem. Aiming at obtaining the optimal solution of Problem (P2), we devise an efficient mixed algorithm based on bisection method and fmincon function to alternately optimize the variable $\mathbf{x}$ and $\mathbf{P}$.

4.1 Deep-Reinforcement-Learning-Based Method for Solving Top-Problem

To tackle the top problem, we utilize the DRL-based DNN to get the near-optimal WET duration a. Precisely, the DNN takes the channel gain h as the input and generates the value of a as the output, which can be expressed as

$$\hat{a} = \pi_{\theta}(\mathbf{h}), \tag{13}$$

where θ denotes the network parameters in DNN model. The DNN is equipped with self-enhancing process that is to train DNN with former experience of known optimal a from the past time blocks. The proposed DRL-based algorithm can obtain lower time latency and executive complexity compared with the baseline method (e.g. one-dimensional search method in this work). And the characteristic of low computing delay makes the DNN algorithm suitable for the situation with fast fading channel gains.

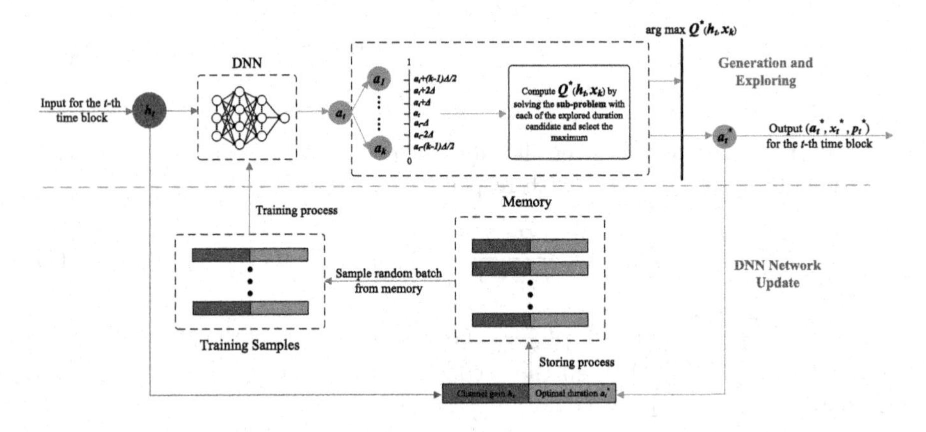

Fig. 2. The framework of proposed DRL-based offloading algorithm.

Figure 2 illustrates the structure of deep-reinforcement-learning-based method. In general, it contains two procedures, generating WET duration a and updating DNN network parameters θ. And the whole procedure contains the following steps:

1. Initially generate the duration $\hat{a}$ from input channel gain h with random network parameters θ.
2. Adopt an exploration policy to extract the set $\mathbf{a}$ of multiple alternative duration values, which may direct the optimal parameters of DNN.
3. Solve the sub-problem with each given value a in $\mathbf{a}$, and obtain the best value a that bring the maximal result of the sub-problem, which is denoted as the best-known solution a^* of the current time block.
4. Pair the channel gains $\mathbf{h}$ and the best solution a^* as $(\mathbf{h}, a^*)$, and save it in the memory component as a sample.

5. In each training interval, the network randomly selects a batch-size of group from samples in memory to train DNN and update its parameters θ.

Overall, the exploration policy and parameters updating scheme are two key parts of the DNN based deep reinforcement method.

Exploration Policy. As DNN network randomly generates parameters, the $\hat{a}$ obtained at preliminary stage may be rather random and poor. As a result, the exploration policy is proposed to indicate the parameter optimization direction. The exploration policy plays a role that help fast improve the quality of $\hat{a}$ and drive the value towards better fineness. Let Δ denotes the exploration length, and K denotes the total amount of candidates explored. Then, the set of multiple alternative duration values is formulated as

$$a(\hat{a}, K) = \{\hat{a} + \frac{K-1}{2}\Delta, \ldots, \hat{a} + \Delta, \hat{a} - \Delta, \ldots, \hat{a} - \frac{K-1}{2}\Delta\}.$$

For each a in $a(\hat{a}, K)$, solving the sub-problem of $\max_{\mathbf{x},\mathbf{p}} Q(\mathbf{h}, a)$ and obtaining the known optimal value, formulated as

$$a^* = \arg\max_{a \in a(\hat{a}, K)} Q(\mathbf{h}, a), \tag{14}$$

Which is further paired with corresponding $\mathbf{h}$ as $(\mathbf{h}, a^*)$ and stored in the memory component of the DNN network. Specifically, based on the optimization tendency, the longer a sample stay in the memory, the more likely it has a rather poor quality. Consequently, when the samples reach the maximum limitation storage of memory, a storage replacement policy of First-In-First-Out (FIFO) is applied to the memory, that is to replace the oldest sample in the memory.

Parameters Updating Scheme. At each training interval, a batch-size (a fixed number) amount of samples will be randomly extracted from the memory to train DNN and update network parameters. This work adopts Adam Optimization Algorithm to update DNN parameters. The Mean Square Error (MSE) loss is applied and the loss function denotes as

$$L(\theta) = \frac{1}{|\Gamma|} \sum_{\gamma=1}^{|\Gamma|} \left[\pi_\theta(h_\theta - a_\theta^*)\right]^2, \tag{15}$$

where Γ denotes the set of training samples and $|\Gamma|$ denotes the total amount of the training samples.

4.2 Mixed Optimization Method for Solving Sub-problem

As a is given as the result of top-problem, the sub-problem is aimed at finding the optimal offloading proportion set $\mathbf{x} = \{x_i \mid i \in \mathcal{N}\}$, denoted the task proportion each WD offloads to HAP, and transmission power set $\mathbf{p} = \{p_i \mid i \in \mathcal{N}\}$, denoted the energy transmission power HAP allocates to each WD, which cannot be optimized simultaneously. Therefore, an alternating optimization scheme is proposed to respectively optimize one variable while the other is assumed to be fixed.

Algorithm 1: A DRL based method to solve the offloading decision problem.

input : For tth time block, wireless channel gain $\mathbf{h}_t$ of N WDs
output: For tth time block, the WET duration a, the transmission power set $\{p_i \mid i \in \mathcal{N}\}$ and the offloading decision set $\{x_i \mid i \in \mathcal{N}\}$ of N WDs;

1 Initialize the DNN with random parameters θ_1 and empty memory; Set training interval λ and batch size β; **for** $t = 1, 2, \ldots, N$ **do**
2 Generate WET duration $\hat{a}_t = \pi_{\theta_t}(\mathbf{h}_t)$; Generate the set of multiple alternative duration values $a_t(\hat{a}_t, K)$ according to the explore policy; Solve the sub-problem for each a in $a_t(\hat{a}_t, K)$ using the mixed optimization method to get the optimal x_t^* and p_t^*; Select the best duration: $\mathbf{a}_t^* = \arg \max_{a \in a_t(\hat{a}_t, K)} Q^*(\mathbf{h}_t, a);$
3 Paired the a^* with corresponding h as (h_t, a_t^*) and update the memory by storing the (h_t, a_t^*) set as sample; **if** $t \bmod \lambda = 0$ **then**
4 Sample a batch of data sets $(h_\gamma, a_\gamma^*), \gamma \in \Gamma$ from memory; Train the DNN network with $(h_\gamma, a_\gamma^*), \gamma \in \Gamma$ and update parameter θ_t using the Adam algorithm;
5 **end**
6 **end**

Bisection Method for Optimization of Variable x.

Theorem 2. *Under the set parameters, the first-order derivative of $Q(\mathbf{h}, a)$ with respect to x_i,$\forall i \in \mathcal{N}$, contains one zero point.*

Proof. Theorem 1 proves that Problem (P2) is a concave problem and the second-order derivative of $Q(\mathbf{h}, a)$ is eternally not positive with respect to $x = \{x_i \mid x_i \in [0,1], i \in \mathcal{N}\}$. As a result, the first-order derivative of $Q(\mathbf{h}, a)$ monotonically decreases concerning any $x_i \in [0,1], i \in \mathcal{N}\}$. And the first-order derivative of $Q(\mathbf{h}, a)$ can be denoted as

$$g'(x_i) = \frac{\partial Q(x_i)}{\partial x_i} = \frac{-CD^{\frac{1}{3}}}{3}(1 - x_i)^{\frac{-2}{3}} + \frac{EF}{(1 + Fx_i)\ln 2}, \tag{16}$$

where $C = \frac{\mu^{\frac{1}{3}}}{\phi}$, $D = \frac{p_i h_i a}{K}$, $E = \frac{BT(1-a)}{V_u}$ and $F = \frac{\mu p_i a h_i^2}{(1-a)N_0}$. Let $h(x_i)$ denote $\frac{-CD^{\frac{1}{3}}}{3}(1 - x_i)^{\frac{-2}{3}}$ and $g(x_i)$ denote $\frac{EF}{(1+Fx_i)\ln 2}$. Then, $Q'(x_i) = h(x_i) + g(x_i)$.

Furthermore, for $\forall i \in \mathcal{N}$ value $x_i = 1$, $Q'(x_i) = Q'(1) = h(1) + g(1)$. We have $g(1) > 0$ but $g(1) \ll \infty$ and $h(1) \longrightarrow -\infty$. As the sum of $h(1)$ and $g(1)$, it is obvious that $Q'(1) \longrightarrow -\infty$. Then for $\forall i \in \mathcal{N}$ value $x_i = 0$, $Q'(x_i) = Q'(0) = h(0) + g(0) = \frac{-CD^{\frac{1}{3}}}{3} + fracEF\ln 2$. It is true that it can not be proved to be permanently positive. However, under the set values of parameters mentioned in Sect. 5, the value of $Q'(0)$ remains positive. Accordingly, the first-order derivative of $Q(h, a)$ with respect to x_i,$\forall i \in \mathcal{N}$, contains one zero point is proved.

As a result, following Theorem 2, we propose a bisection method of obtaining the near-optimal value of $x = \{x_i \mid x_i \in [0,1], i \in \mathcal{N}\}$. The bisection method is summarized in Algorithm 2.

Algorithm 2: A bisection method for optimizing variable x.

input : For tth time block, the WET duration a, wireless channel gain $\mathbf{h}_t$, and the transmission power set $\{p_i, i \in \mathcal{N}\}$ of N WDs
output: For tth time block, the offloading decision set $\{x_i \mid i \in \mathcal{N}\}$ of N WDs;

```
1  Fix the transmission power set; for t = 1, 2, ..., N do
2      set head = 0 and end = 1; calculate the first-order derivative Q'(x_i); while
       |Q'(x_i)| ≥ 0.0005 and x_i ≥ 0.00001 and |x_i − 1| ≥ 0.00001 do
3          if Q'(x_i) < 0 then
4              end = x_i; x_i = (head+end)/2;
5          end
6          if Q'(x_i) < 0 then
7              head = x_i; x_i = (head+end)/2;
8          end
9          calculate the first-order derivative Q'(x_i);
10     end
11 end
```

Scheme for Optimization of Variable p. Although the object function of $Q(\mathbf{h}, a)$ is concave with respect to variable $\mathbf{p}$, the constraint of $\sum_{i=1}^{N} p_i \le P_{max}$ makes the Lagragian method hard to apply. We adopt the optimization solver in MATLAB to solve this problem.

5 Numerical Results

In this section, our work evaluates the performance of our algorithm through simulations. The Rayleigh fading channel model, expressed as $h_i = \hat{h}_i a_i^t$, is applied to generate the wireless channel gain h_i of the ith WED. α_i^t denotes the random channel fading factor satisfing an exponentially distributed random variable with unit mean, and $\hat{h}_i$ denotes the average channel gain and follows the free-space path loss model, i.e.

$$\hat{h}_i = A_d \left(\frac{3 \times 10^8}{4\pi f_c d_i}\right)^{d_e}. \tag{17}$$

The antenna gain $A_d = 4.11$, the path loss exponent $d_e = 2.8$, and the frequency $f_c = 780MHz$. Moreover, d_i denotes the distance between HAP and the ith WED, which is randomly selected in $[10, 15]$ meters. In the simulations, the parameter setting is as follows. The energy harvesting efficiency $\mu = 0.51$ and the total transmission power of HAP to all WED is $P_{max} = 3Watts$ for energy transmission period. In addition, we assume equal computing energy efficiency $K = 10^{(-31)}$ and $\phi = 100$ for all WED. The data offloading bandwidth is $B = 2MHz$, the receiver noise power $N_0 = 10^{(-15)}$, the communication overhead $V_u = 1.1$, and the length of time block $T = 1$ second.

In our work, the proposed DRL-based algorithm uses a fully connected neural network DNN. DNN contains an input layer, two hidden layer and an output layer. The input layer includes N Neurons, and there are respectively 120 and 80 neurons in the two hidden layers, and the output layer contains one neuron. Each neuron applies sigmoid function as the activation function. We set the amount K of candidate that explore policy generates as 15 and the exploring step length δ as 0.005. The other parameter setting of neural network will be discussed in the subsequent performance evaluation.

5.1 Convergence Performance

Figure 3 shows the convergence performance within 5000 time blocks with the training rate set to be 0.0001. Specifically, as the parameters are randomly initialized by the DNN network at first, the quality of samples generated and the ability of the network are obviously poor. As a result, the training loss $L(\theta)$ is rather large during the primeval 500 time blocks. However, it decreases fast as the explore policy continuously ameliorates the quality of samples and consequently improves the ability of the DNN network. Then it converges to 0.0010 after nearly 2000 time blocks and stabilizes ever since.

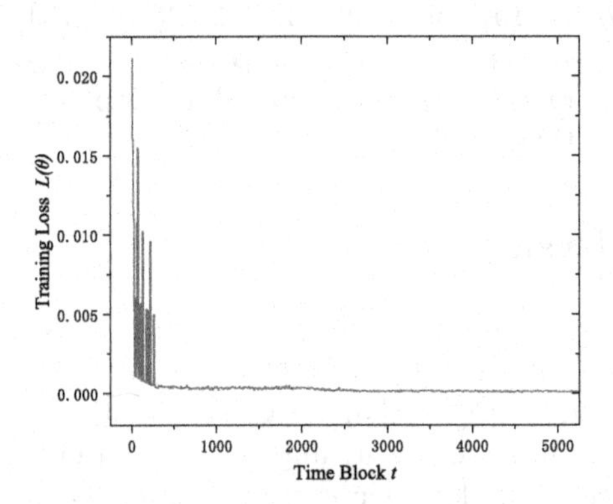

Fig. 3. Training loss of the DNN.

To evaluate the performance of our algorithm, we propose a normalized computation rate $Q(\mathbf{h}, a)$, which compare the result acquired by our algorithm and the benchmark algorithm. In particular, for benchmark algorithm, the one-dimensional search method is applied to obtain the optimal solution of the top problem. In order to acquire the corresponding optimal computation rate, the benchmark algorithm searches the range of $(0, 1)$ with step of 0.01 to find the best WET value a and use fmincon function to solve sub problem with each searched a value. The normalized computation rate $Q(\mathbf{h}, a)$ is defined as

$$\hat{Q}(\mathbf{h}, a) = \frac{Q(\mathbf{h}, a)}{\max_{a^* \in [0,1]} Q_{fmincon}(\mathbf{h}, a^*)}, \tag{18}$$

where $Q(\mathbf{h}, a)$ represents the computation rate obtained by our algorithm, and $\max_{a^* \in [0,1]} Q_{fmincon}(\mathbf{h}, a^*)$ denotes the best computation rate acquired by using the benchmark algorithm of one-dimensional search method and fmincon function.

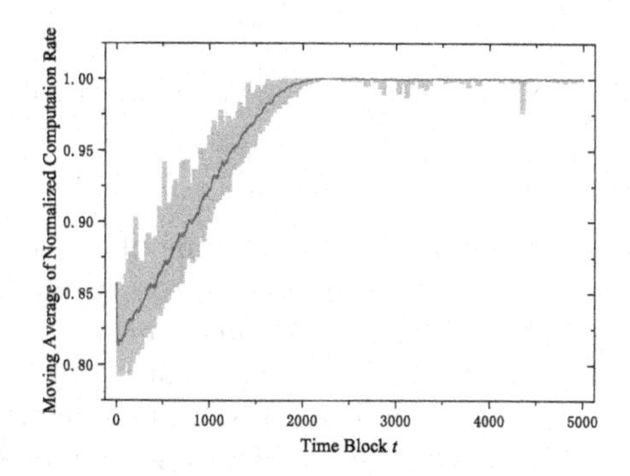

Fig. 4. The normalized computation rate of our algorithm.

In Fig. 4, each data point of the red curve represents the normalized computation rate of the past 50 time blocks, and purple shading represents the maximum and minimum rate in the past 50 time blocks. $Q(\mathbf{h}, a)$ exceeds 0.95 after about 600 time blocks and nearly converges after about 1500 time blocks. After the algorithm reaches its converge, the value of $Q(\mathbf{h}, a)$ is approximately close to 1 in most of the time blocks, and the average normalized computation rate is above 0.999 in the last 2000 time blocks. This shows that our algorithm obtains is very close to the benchmark algorithm.

5.2 Influence of Different Parameters on Result

From Fig. 5 to Fig. 8, we investigate the impact of different parameters on the convergence performance of the DNN model in our proposed algorithm, including the size of memory, the size of training samples batch, the explore step length, the number of explore candidates, the learning rate, and the training interval. Y-axis represents the moving average of normalized computation rate over the past 50 time blocks.

Figure 5 shows the impact of the memory size M on the convergence performance of the algorithm. When memory size is large (like 1600 or 800), the convergence speed is relatively slow. And the algorithm converges after about 1000 time blocks and holds a relatively fast convergence speed when $M = 300$ or

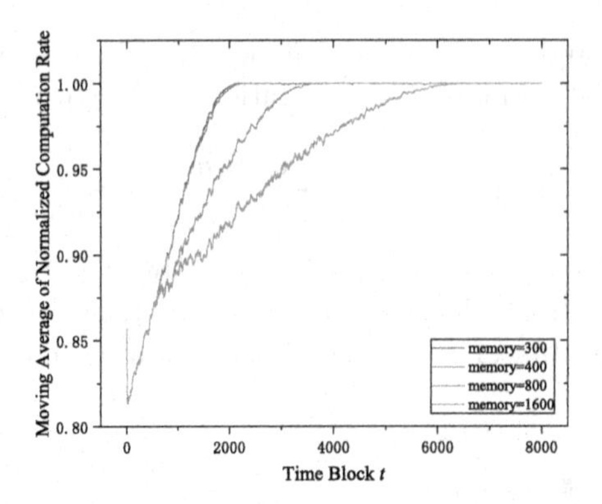

Fig. 5. Moving Average of Normalized Computation Rate with different memory sizes

$M = 400$. The reasons are as follows. In each time block, our algorithm applies explore policy to obtain the optimal sample and store it into the replay memory, which has a storage of M samples. After M time blocks, the memory reaches its upmost storage limitation and a FIFO replacement policy is applied to replace the oldest sample with the new one. For every training interval τ (like every 10 time blocks), the DNN-based algorithm randomly select a batch of samples from the memory to train the DNN model. That is, between each two training processes, there are only τ samples updated in the memory. In this way, there are some old samples from long ago being selected from memory. As a result, under the given batch size, the larger memory size is, the more similar the two batch of samples selected are, which further results to the poor diversity of training samples. On the contrary, the rather small memory size gives the DNN model fast updating ability with the memory always containing the newest samples. In the following evaluation, the memory size is set to 300.

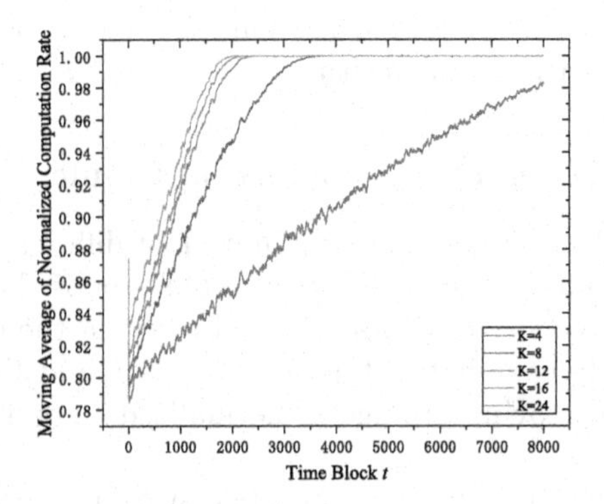

Fig. 6. Moving Average of Normalized Computation Rate with different K

Figure 6 shows the impact of different numbers of explore candidates on convergence performance. It is clear that the convergence speed of normalized computation rate increases as the number of explore candidates increases. With more explore candidates generated for each time block, the algorithm shows better performance. However, after the number of explore candidates reaches a certain amount, the improvement of the convergence speed are unconspicuous, but the cost of time and computation capacity considerably increase. As a result, we set the number of explore candidates K as 16 below to balance the performance and the consumption.

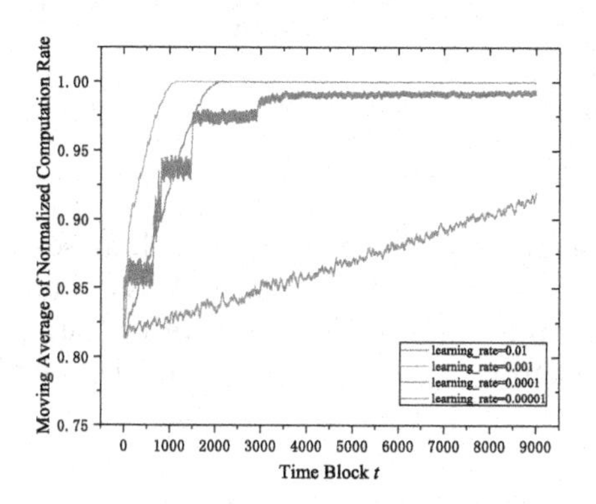

Fig. 7. Moving Average of Normalized Computation Rate with learning rates

As shown in Fig. 7, different learning rates influences the convergence performance greatly. If the learning rate is rather large (like 0.01) or rather small (like 0.00001), the algorithm can hardly converge. The reason is that the update of the network parameter is too oversize to find the optimal under the learning rate of 0.01 while the update is too undersize to find the optimal under the learning rate of 0.00001. Below we set the learning rate as 0.0001.

Figure 8 shows the convergence performance under different training intervals. We can see that the proposed algorithm converges slower as training interval increases. However, the total performance is always rather great with training interval under the considered range. Below we set the training interval as 5.

5.3 Numerical Evaluation

To evaluate the performance on computation rate maximization, we compare our algorithm with three benchmark algorithms.

- Fully local computing. All WEDs consume all harvested energy to perform computation locally, i.e., $x_i = 0, \forall i = 1, 2, \ldots, N$.
- Fully edge computing. HAP evenly allocate the transmission power resource and all WEDs consume all harvested energy to offload computation tasks, i.e., $x_i = 1, \forall i = 1, 2, \ldots, N$.

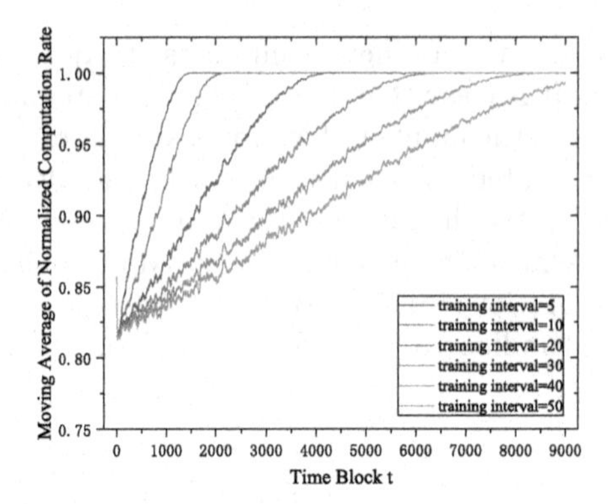

Fig. 8. Moving Average of Normalized Computation Rate with different training intervals.

- One-dimensional search + fmincon. This algorithm uses a search method with precision of 0.001 to find the optimal value of WET time a and apply a convex optimization solver fmincon function to solve the sub-problem of obtaining the optimal $\mathbf{x}$ and $\mathbf{p}$ group. This algorithm gains the solution with the optimal performance.

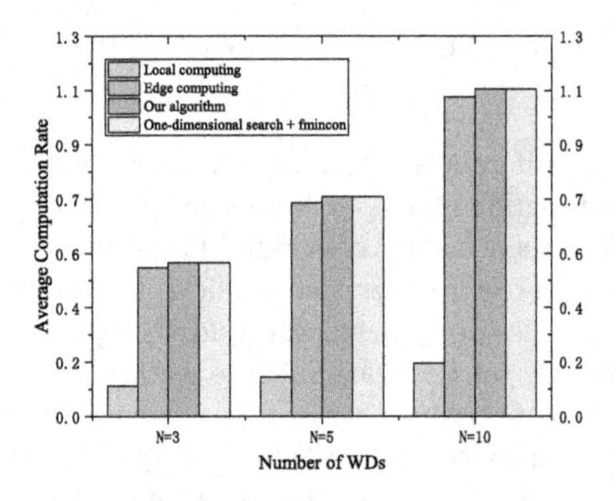

Fig. 9. Comparisons of average computation rate for different offloading algorithms.

As shown in Fig. 9, the average computation rates of our algorithm and three benchmark algorithms under different numbers of WDs. The local computing algorithm shows the worst performance, which is due to the poor computation capacity each WED holds. One-dimensional search + fmincon algorithm obtains the optimal result. Our proposed algorithm shows better performance over the local computing and edge computing, and both gain similar computing rate, which is close to the optimal computation rate.

Table. 1 demonstrates the time latency of the execution of the proposed algorithm and the benchmark algorithm of one-dimensional search+fmincon under

Table 1. Comparison of Execution Latency

Number of WDs	Our algorithm	One-Dimensional search + fmincon
$N = 3$	0.329 s	81.782 s
$N = 5$	0.698 s	382.536 s
$N = 10$	1.183 s	890.839 s

different amounts of WDs. Our algorithm significantly outperforms the benchmark algorithm with hundreds of times smaller time latency. Due to the fast-fading characteristic of the channel quality, our algorithm is much more applicative in the real situation, which can ensure the effectiveness of optimization strategy and data computed. To sum up, the proposed the algorithm obtains the near upmost computation rate and guarantees the efficiency and greatly decreases complexity.

6 Conclusion

In this paper, we consider the problem of maximizing the sum computation rate by jointly optimizing energy transmission time duration, partial offloading decisions and transmission power allocation strategy in wireless powered mobile edge computing network containing a HAP, multiple WDs. We decompose the non-convex problem into two parts of a top-problem and a sub-problem. For the top-problem, a DRL-based offloading problem is proposed to output the optimal energy transmission time duration, which contains a DNN network, an exploring strategy and a training strategy. For the sub-problem, an effective algorithm is devised to alternately optimize partial offloading decisions and transmission power allocation strategy under the given time duration. Numerical results reveal the effectiveness of the proposed algorithm in fast-fading channel circumstance and its low latency and complexity advantages compared with benchmark algorithms.

References

1. Zeng, M., Du, R., Fodor, V., Fischione, C.: Computation rate maximization for wireless powered mobile edge computing with Noma. In: IEEE 20th International Symposium on" A World of Wireless, Mobile and Multimedia Networks (WoWMoM), pp. 1–9. IEEE (2019)
2. Zhang, S. Gu, H., Chi, K., Huang, L., Yu, K., Mumtaz, S.: DRL-based partial offloading for maximizing sum computation rate of wireless powered mobile edge computing network. IEEE Trans. Wireless Commun. **21**(12), 10:934–10:948 (2022)
3. Xu, C., Zhan, C., Liao, J., Gong, J.: Computation throughput maximization for UAV-enabled MEC with binary computation offloading. In: ICC 2022-IEEE International Conference on Communications, pp. 4348–4353. IEEE (2022)

4. Chen, S., Rui, L., Gao, Z., Li, W., Qiu, X.: Cache-assisted collaborative task offloading and resource allocation strategy: a metareinforcement learning approach. IEEE Internet of Things J. **9**(20), 19:823–19:842 (2022)
5. Bi, S., Zhang, Y.J.: Computation rate maximization for wireless powered mobile-edge computing with binary computation offloading. IEEE Trans. Wireless Commun. **17**(6), 4177–4190 (2018)
6. Huang, L., Bi, S., Zhang, Y.-J.A.: Deep reinforcement learning for online computation offloading in wireless powered mobile-edge computing networks. IEEE Trans. Mob. Comput. **19**(11), 2581–2593 (2019)
7. Nguyen, P.X., et al.: Backscatter-assisted data offloading in OFDMA-based wireless-powered mobile edge computing for IoT networks. IEEE Internet Things J. **8**(11), 9233–9243 (2021)
8. Mao, S., et al.: Computation rate maximization for intelligent reflecting surface enhanced wireless powered mobile edge computing networks. IEEE Trans. Vehic. Technol. **70**(10), 10:820–10:831 (2021)
9. Wu, X., He, Y., Saleem, A.: Computation rate maximization in multi-user cooperation-assisted wireless-powered mobile edge computing with OFDMA. China Commun. **20**(1), 218–229 (2023)
10. Ren, C., Zhang, G., Gu, X., Li, Y.: Computing offloading in vehicular edge computing networks: Full or partial offloading? In: IEEE 6th Information Technology and Mechatronics Engineering Conference (ITOEC), vol. 6, pp. 693–698. IEEE (2022)
11. Bi, J., Yuan, H., Duanmu, S., Zhou, M., Abusorrah, A.: Energy-optimized partial computation offloading in mobile-edge computing with genetic simulated-annealing-based particle swarm optimization. IEEE Internet Things J. **8**(5), 3774–3785 (2020)
12. Li, Y., Zhang, X., Sun, Y., Liu, J., Lei, B., Wang, W.: Joint offloading and resource allocation with partial information for multi-user edge computing. In: IEEE Globecom Workshops (GC Wkshps), pp. 1736–1741. IEEE (2022)
13. Saleem, U., Liu, Y., Jangsher, S., Tao, X., Li, Y.: Latency minimization for D2D-enabled partial computation offloading in mobile edge computing. IEEE Trans. Veh. Technol. **69**(4), 4472–4486 (2020)
14. Guo, M., Wang, W., Huang, X., Chen, Y., Zhang, L., Chen, L.: Lyapunov-based partial computation offloading for multiple mobile devices enabled by harvested energy in MEC. IEEE Internet Things J. **9**(11), 9025–9035 (2021)
15. Feng, J., Pei, Q., Yu, F.R., Chu, X., Shang, B.: Computation offloading and resource allocation for wireless powered mobile edge computing with latency constraint. IEEE Wireless Commun. Lett. **8**(5), 1320–1323 (2019)

Efficiently Running SpMV on Multi-core DSPs for Banded Matrix

Deshun Bi, Shengguo Li(✉), Yichen Zhang, Xiaojian Yang, and Dezun Dong(✉)

National University of Defense Technology, Changsha, China
{bds123,nudtlsg,zhangyichen,yangxj,dong}@nudt.edu.cn

Abstract. Sparse matrix-vector multiplication (SpMV) plays a pivotal role in large-scale scientific computing. Despite the increasing use of low-power multicore digital signal processors (DSPs) in high performance computing (HPC) systems, optimizing SpMV on these platforms has been largely overlooked. This paper introduces the FT-M7032, a new CPU-DSP heterogeneous processor multi-core platform for high-performance computing. The FT-M7032 provides programmable memory units at multiple levels, but effectively utilizing these units poses a challenge. To address this, we evaluate the transfer capability between different units to map matrix elements to storage units. Based on our evaluation, we propose an efficient parallel implementation, SpMV_Band, specifically designed for banded matrices. Furthermore, we devise a computation pipeline that optimizes memory access overhead by overlapping data transfers and computations. To evaluate our approach, we compare its performance with a baseline executed on the general-purpose CPU cores of the FT-M7032 heterogeneous platform. Experimental results demonstrate that our techniques achieve a significant speedup of 2.0× compared to the competing baselines.

Keywords: SpMV · Multi-Core DSP · Performance Optimization · Double Buffering

1 Introduction

Sparse matrix-vector multiplication (SpMV), as the key program of the Basic Linear Algebra Subroutine (BLAS) library, has been widely used in scientific simulation, data analysis, deep learning and other fields [12,22,39]. Due to the wide application of SpMV, research on accelerated SpMV in various high-performance architectures such as multi-core CPU [34], GPGPU [15,41] and MIC [40] has been constantly emerging. Due to energy efficiency and power constraints, low-power embedded architectures are being introduced into heterogeneous high performance computing domains, such as digital signal processors (DSPs) [11,32,33]. Compared to CPUs and GPUs, DSPs typically have very long instruction words (VLIWs) or vector cores without out-of-order execution [38]. DSP cores typically operate on software-managed on-chip memory and incorporate a direct memory access (DMA) engine. Therefore, existing optimizations

Z. Tari et al. (Eds.): ICA3PP 2023, LNCS 14491, pp. 201–220, 2024.
https://doi.org/10.1007/978-981-97-0808-6_12

for SpMV on CPUs and GPUs cannot be directly applied to SpMV on DSPs. Although some studies have optimized general matrix multiplication [23,38], there is no consensus on how best to optimize SpMV on these emerging multi-core DSP platforms.

To improve the performance of SpMV, it is essential to optimize the irregular memory access pattern [21]. A naive implementation would simply store the matrix A and the vector **x** in main memory (e.g., DDR RAM) and access x irregularly. However, a more effective approach is to use on-chip programmable memory units, such as the scalar memory (SM), array memory (AM) and global shared memory (GSM) provided by the FT-M7032 platform [38], to cache the data. These memory units have different capabilities in terms of memory space, access latency and bandwidth. Therefore, it is crucial to strategically distribute data among these units to maximise their utilisation and improve performance. We evaluate the data transfer between different storage units. Our results show that GSM has a particularly significant impact on SpMV. By exploiting the caching capabilities of GSM for the vector **x**, significant improvements in SpMV performance can be achieved.

Banded matrices are a common type of matrix with a wide range of applications, and many sparse matrices in SuiteSparse [6] can be converted into banded matrices by using the Reverse Cuthill McKee (RCM) ordering [5,16]. To optimize the use of onchip storage for large-scale SpMV, we implemented a horizontal matrix partitioning approach. This approach partitions the banded matrix into row blocks, each of which corresponds to a vector **x** segment, which promotes good data locality for x. The row blocks are then computed sequentially using GSM. We refer to this new SpMV algorithm as SpMV_Band and storage format based the SELL format. We evaluate our approach on FT-M7032 and compare it with the SpMV implmenetaion on the general-purpose CPU cores. Our experimental results demonstrate that SpMV_Band outperforms other competing methods on the FT-M7032, achieving speedups of up to 2.0×, when DSP is used to accelerate SpMV.

The main contributions of this paper are as follows:

- We make an effective analysis to guide SpMV optimization on DSPs.
- We propose a new SpMV algorithm for banded matrices, which partitions the sparse matrix horizontally to achieve a better data locality of x.
- We design a computation pipeline that overlaps data transfers and computation to minimize memory access overhead.

2 Background

2.1 FT-M7032 Heterogeneous Processor

The FT-M7032 platform combines CPU and DSP capabilities for heterogeneous computing. It features a 16-core ARMv8 CPU that runs the operating system and four DSP clusters for general-purpose computing (GPDSP).

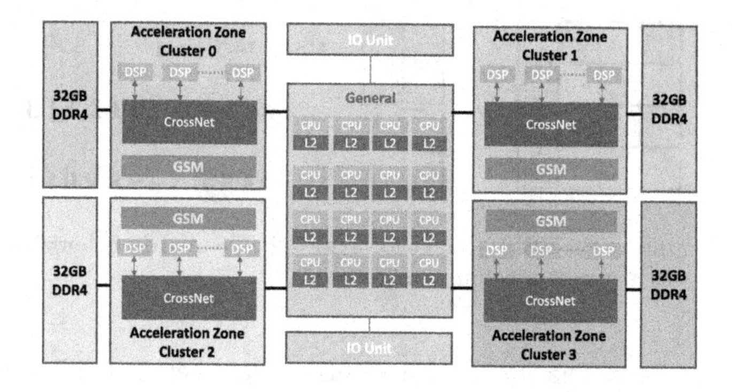

Fig. 1. Overview of the FT-M7032 architecture

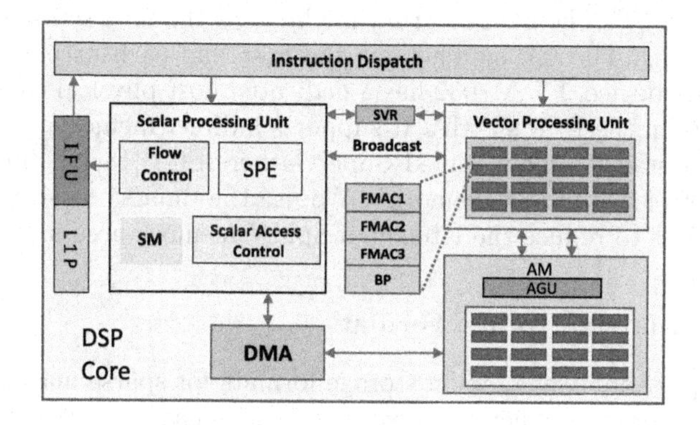

Fig. 2. Micro-Architecture of DSP Core in FT-M7032

Figure 1 shows the FT-M7032 platform's architecture. It consists of two areas: the general-purpose area (GP) with CPU cores and the acceleration area (ACC) with DSP cores. The ACC area has four clusters, each with eight DSP cores, a global shared memory (GSM), and an off-chip DDR memory. The CPU cores run at 2.0 GHz and the DSP cores run at 1.8 GHz. Each cluster in the ACC area can work independently and access its own GSM. The CPU cores manage the data transfers between different DSP clusters.

The DSP core is a VLIW-based architecture, as shown in Fig. 2. It has an instruction function unit (IFU), a scalar processing unit (SPU), a vector processing unit (VPU), and a DMA engine. The IFU can issue up to 11 instructions per cycle, including five scalar and six vector instructions. The SPU handles instruction flow control and scalar computations. It has an SPE (Scalar Processing Element) and an SM (Scalar Memory) of 64KB. The VPU performs vector computations in a SIMD manner. It has 16 VPEs (Vector Processing Elements), each with 64-bit registers and various units for arithmetic and memory operations. The VPU can process 1024 bits at once. It also has 768KB Array Memory (AM). The SPU and VPU communicate through broadcast instructions

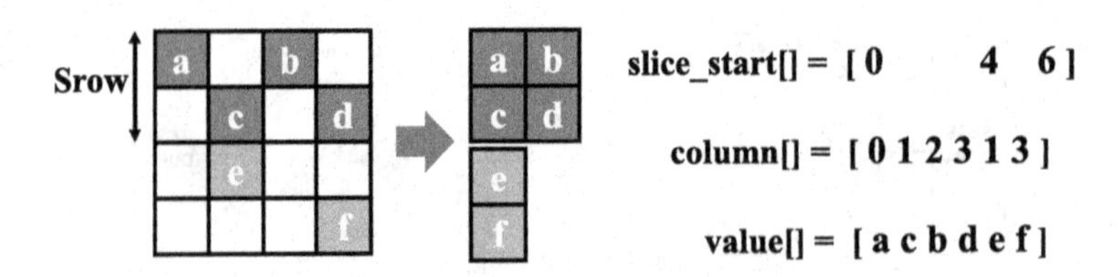

Fig. 3. An example of the SELL format for `Srow` = 2. The matrix is horizontally sliced into two slices, each stored in column-major order.

and shared registers. The DMA engine transfers data between different memory levels, such as main memory, GSM, and SM/AM.

On the software, `hthread` is provided to improve the programmability of multi-core DSPs [7]. The `hthread` runtime handles the interaction between GP and ACC regions. The GP side acts as the host and each acceleration cluster as a computing device. FT-M7032 has a dedicated host physical communication channel, APip, in the DSP's DMA. It supports indirect memory access by using a special data transfer mode called SuperGather (SG) [18] or dma_sg. It can move data between different memories. We use the dma_sg to support indirect memory accesses to reduce the latency of SpMV memory access.

2.2 Sparse Matrix Storage Format

This section presents two common storage formats for sparse matrices that suit SIMD-based processors. They compress the sparse matrix to save space.

ELLPACK. The ELLPACK [13] format is for GPUs and vector architectures. It stores nonzero elements in column order. Each row has K elements, where K is the largest number of nonzero elements in any row of the Sparse Matrix. Rows with fewer nonzero elements are padded with zeros. The drawback of ELL is that it wastes space with padding. It works well for matrices with similar numbers of nonzero elements per row.

Sliced ELLpack. The SELL (Sliced ELLpack) [27] format is a modified version of the ELL format. It is more efficient than ELL for storing and computing sparse matrices. It splits the sparse matrix into row slices. Each slice is stored in the ELL format. The slice size `Srow` is a parameter of this format. It determines how many adjacent rows are in each row slice.

The SELL format has three data structures: `slice_start`, `column`, and `value`. Figure 3 shows how they store the sparse matrix. `column` and `value` store the column indexes and values of the nonzero elements in column-major order. `slice_start` stores the index of the first element in each slice. The last number in `slice_start` is the total number of elements. For example, Fig. 3 has 6 elements. We can calculate the number of elements per row of a slice i, including the padding elements, by $(slice\_start[i+1] - slice\_start[i])/\texttt{Srow}$. When `Srow`

Algorithm 1: SpMV pseudocode for the SELL format

Input: Matrix A, vector x and $Srow$.
Output: Vector y.

```
1 // ns: number of slices
2 for k = 0 to ns do
3     for i = 0 to (A.slice_start[k + 1] − A.slice_start[k])/Srow do
4         for j = 0 to Srow do
5             idx = A.slice_start[k] + j + i × Srow
6             y[k × Srow + j]+ = x[A.column[idx]] × A.value[idx]
7         end
8     end
9 end
```

= 1, SELL is the same as the CSR format and does not need padding. When Srow = n (the dimension of the matrix), SELL is the same as the ELL format and has the most padding. A small Srow saves space and improves bandwidth usage. A large Srow is better for vectorization, which suits the DSP architecture.

Algorithm 1 shows the SELL-based SpMV algorithm. It has three loops. The outer loop goes through each slice. The middle loop goes through each column of the current slice. The inner loop multiplies and adds the nonzero element values of each column and the vector **x** to get the result vector y. Section 3 explains how the SELL format enables parallelization and vectorization.

2.3 Banded Matrix

If the nonzero elements of a matrix are distributed only along the main diagonal and its adjacent diagonal lines, it is called a banded matrix [10,30]. Banded matrices have a wide range of applications and many matrices can be converted to banded matrices by reordering. Since the nonzero elements of banded matrices have certain distribution characteristics, some specific storage and computation methods can be used to improve the efficiency of operations.

The bandwidth of a banded matrix refers to the maximum distance that nonzero element deviates from the main diagonal. In a banded matrix, the nonzero elements can be divided into three parts: the upper triangle region, the main diagonal and the lower triangle region. The upper triangle region contains all nonzero elements above the main diagonal, and its bandwidth is called the upper bandwidth. The lower triangle region contains all nonzero elements below the main diagonal, and its bandwidth is called the lower bandwidth. The upper bandwidth and the lower bandwidth can be different. As shown in the band matrix in the Fig. 7, we use SBL and SBU to denote the lower bandwidth and upper bandwidth respectively.

Table 1. Measured bandwidth between storage layers

Function	Direction	Bandwidth
dma_p2p	DDR↔SM	39 GB/s
dma_p2p	DDR↔AM	30 GB/s
dma_sg	DDR↔AM	4 GB/s
dma_p2p	GSM↔AM	206 GB/s
dma_sg	GSM↔AM	28 GB/s

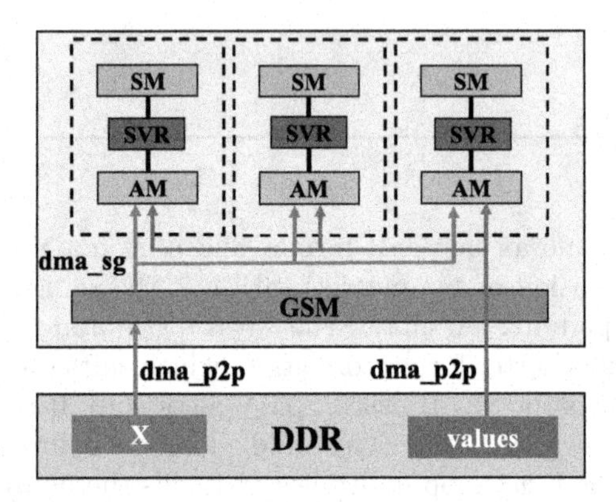

Fig. 4. The FT-M7032 storage hierarchy is abstracted into a three-level memory structure, with an example of three DSP cores in a cluster.

3 Our Approach

3.1 Architecture Analysis

We evaluate the memory access bandwidth between the levels of the FT-M7032 memory hierarchy. Based on the evaluation, we perform a data transfer analysis and predict the peak performance of SpMV. We then propose that a horizontal partitioning approach can guarantee data locality and efficiently execute SpMV on FT-M7032.

Performance Evaluation. The DSP memory architecture has three main levels, as shown in Fig. 4: an off-chip DDR, an on-chip GSM shared by DSP cores, and AM/SM private to a DSP core. AM is the storage space for VPUs. It needs to load the matrix nonzeros and the vector **x** before doing multiplications and additions of SpMV. SM is for scalar processing, such as scalar multiplication and addition, and vector reduction. GSM is a high-speed on-chip memory that acts as a shared L2 cache. Both SPUs and VPUs can access it.

The DMA engines on FT-M7032 move data between memory storages using two modes: the peer-to-peer mode (dma_p2p) and the super-gather mode

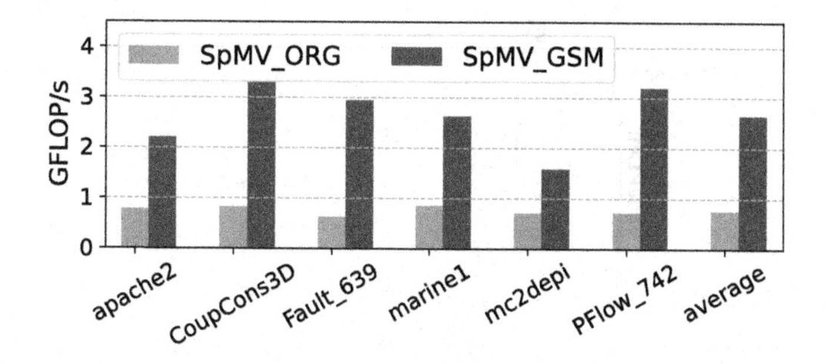

Fig. 5. Performance comparison of SpMV_ORG vs SpMV_GSM. SpMV_GSM, which uses GSM to cache x, but SpMV_ORG, which stores x in DDR without caching.

(dma_sg). The former is a non-blocking point-to-point transfer function. The latter is for gathering operations on **x**. Table 1 shows the measured bandwidth between storage layers. This is an average value selected after several tests, and the actual values may fluctuate around the table values. We can see that dma_sg has a much lower bandwidth than dma_p2p, especially when moving data from DDR to AM. Therefore, effective use of dma_sg is important for SpMV performance optimization on FT-M7032.

For the SpMV operation, the nonzero elements of matrix A are stored continuously while the elements of vector **x** are accessed irregularly. Therefore, improving data locality through caching or prefetching **x** can enhance SpMV performance. However, the bandwidth of the DMA gathering operation (i.e., dma_sg) for transferring data from DDR to AM is low (around 4 GB/s in Table 1), so we should avoid using dma_sg to transfer vector **x** from DDR to AM. Instead, we use GSM to cache as many elements of x as possible, which allows for faster data access and better bandwidth utilization. Figure 4 illustrates how we move matrix values and vector **x** among different storage units. First, we transfer the nonzero values of the sparse matrix from DDR to AM directly via dma_p2p. Second, we transfer vector **x** from DDR to GSM via dma_p2p and then gather it to AM from GSM via dma_sg. We can use SM to cache intermediate arrays used in SpMV, such as the `slice_start` in SELL format or the `row_pointer` [39] in CSR format, to reduce access latency. We transfer the intermediate arrays from DDR to SM via dma_p2p.

The peak bandwidth of dma_sg from GSM to AM is about 28 GB per second. This means that we can transfer 28/8=3.5G elements per second using double precision (FP64) for SpMV. Therefore, the peak throughput is 3.5* 2=7 GFLOPs, which is the maximum SpMV performance we can achieve using dma_sg on FT-M7032. This analysis provides a basis for our implementation and performance estimation of SpMV on DSP, regardless of the sparse matrix storage format we use. In this work, we choose the SELL format to implement. In the following subsections, we describe our design based on above analysis.

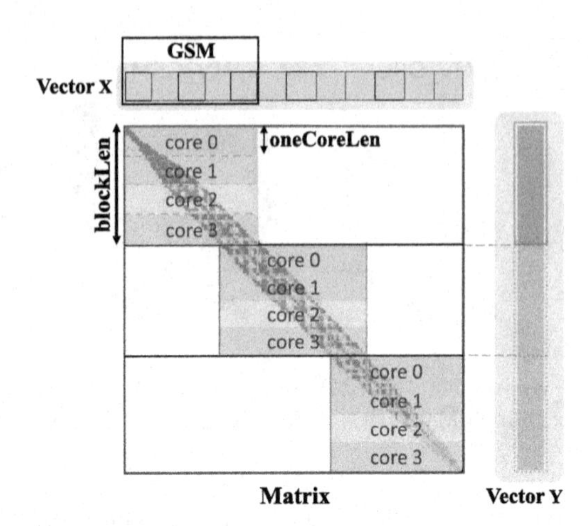

Fig. 6. Band matrix for horizontal division. SpMV_Band divides the matrix into row blocks. Each block is equally allocated to each DSP core for parallel computation. When a row block is to be computed, the corresponding segments of vector x are moved into GSM.

Strengthen Data Locality. In the previous section, we showed that caching the vector **x** in the on-chip memory GSM can reduce the memory access latency for SpMV. To evaluate the performance benefit of this optimization, we compared SpMV using GSM with SpMV_ORG, which stores **x** in DDR without caching. Figure 5 illustrates the speedup of SpMV using GSM over SpMV_ORG for different matrices. We can see that GSM improves the SpMV performance for all cases. Therefore, we conclude that storing **x** in GSM for indirect memory access is a general and effective optimization strategy on FT-M7032.

Although SpMV using GSM can perform well on the FT-M7032, it should be noted that the memory capacity of GSM is limited. If the size of the vector **x** is larger than `GSMlenMax` (6MB or a double precision vector of about 780,000 elements), GSM cannot hold the entire vector **x**. To solve this problem, we develop the SpMV_Band algorithm, which partitions the matrix in the horizontal direction, each row block is calculated one by one using GSM, and can handle arbitrarily size banded matrices.

As shown in the Fig. 6. Because the nonzero elements of each row block in the banded matrix are centrally distributed, good data locality is ensured and irregular memory accesses to the vector **x** are avoided.

3.2 SpMV_Band Design and Implementation

Loading Vector x to GSM. To move vector **x** into GSM space, the vector **x** range is calculated firstly based on the row block length `blockLen`, upper bandwidth `UBL`, and lower bandwidth `SBL`. As shown in the Fig. 7, according to the start row number `StartRow` and end row number `EndRow` (`StartRow` + `blockLen`) of the current row block, the moving range of **x** can be calculated

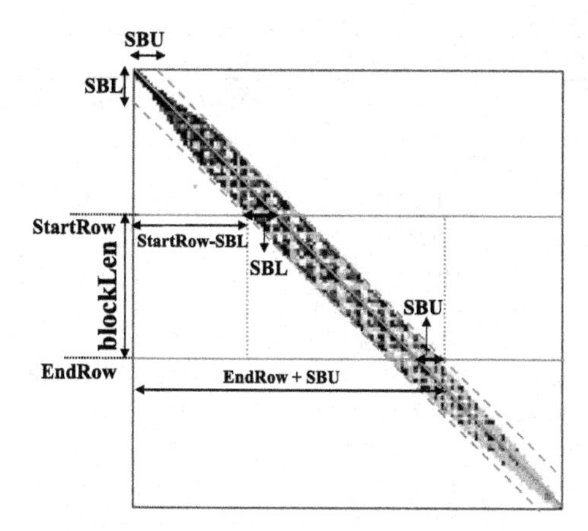

Fig. 7. Loading X to GSM in segments. According to the start row number StartRow and the end row number EndRow of the current row block, the corresponding movement range of x can be calculated.

as (`StartRow - SBL`, `EndRow` + `SBU`). Note that the **x** fragment corresponding to each row block has the overlap of `SBU` + `SBL` length.

SpMV_Band partitions the matrix into row blocks. When a row block is to be computed, the corresponding segments of vector **x** are moved into GSM. Since the segments of **x** are moved to GSM one by one, new segment will overwrite the old segment of **x** in GSM. So, the index of current segment of **x** in GSM is changed. Therefore, the column indexes of sparse matrix should be updated when accessing the elements of **x** stored in GSM via dma_sg. We can preprocess the `column` index array on the host side to ensure the subsequent matches is correct, and the preprocessing is shown in Algorithm 2.

`alpha` is the number of row blocks divided, it can be calculated from the matrix dimension `n` and `blockLen`. First, we determine whether there are more than one row block divided, and if only one row block is divided, no preprocessing is required (Line 1). Otherwise, the `column` array from the second row block to the penultimate row block is index shifted (Lines 4–11). The column index offset of the i-th row block is: $column[i] - col\_shift$, where $col_shift = i * \texttt{blockLen} - $ `SBL`. The last row block is the boundary value and needs to be handled separately (Lines 13–17). Before updating the `column` array, the indexes of the first and last slices in the row block are calculated (Lines 5–6). The column index of the element in the current row block can then be updated (Lines 7–9).

Multilevel Partitioning. Since AM is often too small to fit all matrix and vector elements, we designed a multilevel partitioning mechanism for SpMV to take full advantage of the memory hierarchy on FT-M7032. As shown in the Fig. 6, we partition the input matrix, A, into row blocks of length `blockLen`, where `blockLen` is equal to `coreNum*Srow*beta`. Here `beta` is a hyperparameter indicating that in each row block, each DSP core is responsible for processing

Algorithm 2: Column index preprocessing

```
   Input: Banded Matrix A, SBL, beta, coreNum and blockLen.
   Output: A.column.
1  if alpha > 1 then
2      // The second to last-1 BLOCK.
3      col_shift = blockLen − SBL
4      for i = 1 to alpha − 1 do
5          StartSlice = beta * coreNum * i
6          EndSlice = StartSlice + beta * coreNum
7          for j=A.slice_start[StartSlice] to A.slice_start[EndSlice] do
8              A.column[j] = (A.column[j] − col_shift)
9          end
10         col_shift+ = blockLen
11     end
12     // The last BLOCK.
13     StartSlice = beta * coreNum * (alpha − 1)
14     col_shift = (alpha − 1) * blockLen − SBL
15     for j=A.slice_start[StartSlice] to A.slice_start[end] do
16         A.column[j] = (A.column[j] − col_shift)
17     end
18 end
```

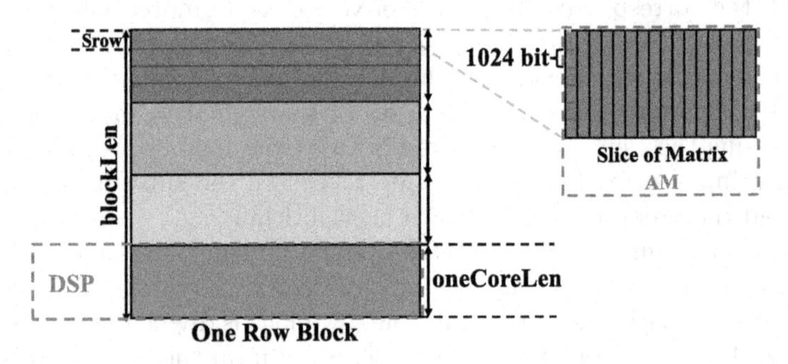

Fig. 8. Multilevel partitioning. For each row block, it is equally assigned to all DSP cores for processing. Four cores are used here as an example, and there are eight DSP cores in the actual experiment.

`beta` slices of length `Srow`. For each row block, multiple cores compute in parallel. The single row block partitioning process is illustrated in Fig. 8. Due to the capacity of AM, we use `AMlenMax` to denote the maximum size of matrix values and vector **x** that can be moved into AM at one time. The length of `Srow` cannot exceed `AMlenMax`. Each time, matrix elements of length `Srow` are processed. The computation is performed in the column-major order, and each column is divided into cells with a length of 16, and is calculated cell by cell. If the matrix cannot

Algorithm 3: SpMV_Band Algorithm

```
   Input: matrix A, vector xv, Srow, coreNum and Beta.
   Output: vector yv.
1  Function SpMV_Band(A, xv, Srow, coreNum, Beta)
2      // Get the ID of each running thread.
3      coreID = get_thread_id()
4      oneCoreLen = Beta * Srow
5      blockLen = coreNum * oneCoreLen
6      for blockCycle = 0 to alpha do
7          // X_Start = StartRow − SBL and X_End = EndRow + SBU.
8          X_Len = X_End − X_Start + 1
9          // Moving x from DDR to GSM via dma_p2p.
10         dma_p2p(xv_ddr + X_Start, X_Len * sizeof(double), xv_gsm)
11         group_barrier()
12         // Where blockoffset is equal to blockCycle * blockLen and coreOffset is
             equal to coreID * oneCoreLen. coreID ∈ (0,...,coreNum-1).
13         offset = blockOffset + coreOffset
14         sj0 = offset/Srow
15         // oneCoreLen is divided into Beta * Srow.
16         for sliceCycle = 0 to Beta do
17             yv_ddr = &yv[offset + sliceCycle * Srow]
18             sj = sj0 + sliceCycle
19             colwid = (A.slice_start[sj + 1] − A.slice_start[sj])/Srow
20             // Iterate across the column width of each slice.
21             for colCycle = 0 to colwid do
22                 sliceOffset = A.slice_start[sj] + colCycle * Srow
23                 val_ddr = &A.value[sliceOffset]
24                 col_ddr = &A.column[sliceOffset]
25                 dma_p2p(val_ddr, Srow * sizeof(double), val_am)
26                 dma_sg(xv_gsm, col_ddr, Srow * sizeof(double), xv_am)
27                 // Srow is divided into vpuTimes*16, and 16 doubles are
                     calculated each time.
28                 for vpuCycle = 0 to vpuTimes do
29                     yv_am[vpuCycle]+ = val_am[vpuCycle] * xv_am[vpuCycle]
30                 end
31             end
32             dma_p2p(yv_am, Srow * sizeof(double), yv_ddr)
33         end
34         group_barrier()
35     end
36 end
```

be divided equally to multiple DSPs, the remaining elements can be computed by the CPU cores within the GP region.

SpMV_Band. According to the multi-core feature of FT-M7032, we design a multi-level parallel SpMV algorithm based on the SELL format for banded matrices, namely SpMV_Band. SpMV_Band is implemented in `hthread`, which consists of host side and kernel side.

The host is responsible for allocating buffers, loading matrices and vectors, and managing the kernel side. The kernel side divides the sparse matrix into multiple row blocks, each with a length of `blockLen`. When performing SpMV operations, each row block corresponds to a part of the result vector y (Fig. 6), and the computations of row blocks are independent of each other. Each row block is divided into `coreNum` blocks of `oneCoreLen` length. Because it is stored in SELL format, each subblock of length `oneCoreLen`, consist of `beta` slices, and each slices consists of `Srow` rows.

Algorithm 3 describes our SpMV_Band algorithm. The operations between head and tail are thread-parallel SpMV (Lines 1–36). Before execution, get the id of each thread (Line 3). The matrix is divided into `alpha` row blocks according to the matrix dimension and `blockLen`, and Lines 6–35 are calculated one by one for each row block. Line 8 calculate the length of the x segment corresponding to each row block (Fig. 7). The start and end points of each segment must be greater than or equal to 0 and less than the matrix dimension. The vector **x** segment is moved from DDR memory to the GSM via dma_p2p (Line 10). Each row block is executed in parallel by multiple cores. As the matrix is stored in the SELL format, each row block consists of `coreNum`$*$`beta` slices. Each core is responsible for executing `beta` slices. After moving the vector **x**, each slice is traversed for calculation (Lines 16 and 33). To get the index of each slice in advance (Line 13, 14, 18), the column width of each slice can be calculated later (Line 19). Lines 21 and 31 are the core of the algorithm. For the current slice, it iterates through each column and load `Srow` nonzero elements into AM at a time via dma_p2p (Line 25). For the matched `Srow` length element in vector **x**, we load them into AM space via dma_sg (Line 26). Lines 28–30 are the specific calculation process, where 16 elements of the double type can be calculated each time. After the current slice has been calculated, the result vector is stored back in the DDR and the next slice is calculated (Line 32). It is noted that barrier synchronization should be set to ensure that data movement and SpMV calculation for each row block is correct (Line 11, 34).

SpMV_Band divides the matrix into multiple row blocks, which are processed simultaneously by different DSP cores. Each core computes different elements simultaneously, realizing thread-level parallelism and data-level parallelism. The VPU in the DSP core provides powerful vector computation capability and consists of 16 VPEs. The VPU can process 1024 bits in parallel, which means 16 double-precision operations at once, further developing data-level parallelism.

Multi-core Migration As mentioned earlier, data in GSM can be shared by all threads within the same DSP cluster. If only core 0 is responsible for moving the data, other cores cannot access the data in GSM until core 0 has completed the transfer of x to GSM. To improve performance, multiple cores can simultaneously

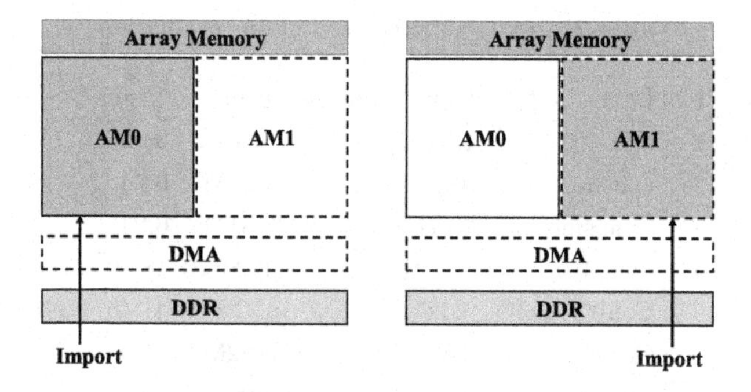

Fig. 9. Double buffer mechanism. We divide the AM space into two parts, one for data import and one for data computation and export, in alternating order.

move x from DDR to GSM via dma_p2p. Barrier synchronization should be used to ensure successful data movement before SpMV computation begins (Line 11). This strategy improves SpMV performance as shown in Fig. 10.

Double Buffering. We propose a two-level DMA-based double buffering strategy to optimize data transfer between multi-level memory. This strategy overlaps DMA data movement and computation time to further improve SpMV performance. As shown in Fig. 9, AM is equally divided into two parts: one for data import and the other for data computation and export, in alternating order. Except for the last part of the data, computations for all parts can be overlapped with data transfers to improve computational efficiency. The results are shown in Fig. 10.

4 Performance Results

4.1 Experimental Setup

Table 2 lists the banded matrices from the SuiteSparse Collection [6] used in our evaluation. These matrices vary in size and sparsity, which were extracted from different application domains. Therefore, they represent a wide range of inputs for SpMV in HPC applications. We compare the performance of SpMV_Band with the classical SpMV on the CPU cores in the GP area. The CPU cores run the Linux kernel version 5.4.0. The SpMV kernels are compiled with GCC version 9.3 and "-O2 -lpthread -fopenmp" option. Note that we use four threads for CPU execution.

4.2 Performance Evaluation

In this section, we evaluate the performance of SpMV_Band by using the matrices in Table 2. We compare SpMV_Band with highly optimized SpMV kernel that runs on CPU cores with different storage formats such ELL, SELL and

Table 2. Input matrices used in our evaluation

ID	Input	Dimensions(N)	#nnz	#nnz/N
1	afshell10	1.51M	52.67M	34.65
2	apache2	0.72M	4.82M	6.74
3	atmosmodd	1.27M	8.81M	6.94
4	bone010	0.99M	47.85M	48.50
5	CoupCons3D	0.42M	17.28M	41.45
6	CurlCurl_2	0.81M	8.92M	11.06
7	ecology1	1.00M	5.00M	5.00
8	ecology2	1.00M	5.00M	5.00
9	Emilia_923	0.92M	40.37M	43.74
10	Fault_639	0.64M	27.25M	42.65
11	marine1	0.40M	6.23M	15.55
12	mc2depi	0.53M	2.10M	3.99
13	PFlow_742	0.74M	37.14M	50.00
14	t2em	0.92M	4.59M	4.98
15	tmt_unsym	0.92M	4.58M	5.00

CSR, which are denoted by ELL_CPU, SELL_CPU and CSR_CPU in Fig. 10, respectively. The CPU side codes run on four CPU cores in the GP area of FT-M7032 and SpMV_Band uses eight DSP cores, and the results are shown in Fig. 10.

We implemented three versions of SpMV_Band:

- Band_SIN uses a single DSP core to move x from DDR to GSM via dma_p2p;
- Band_MUL uses multiple DSP cores to move x from DDR to GSM;
- Band_MUL_DB uses multiple DSP cores to move x and further uses double buffer movement strategy to overlap data transfers with computations.

The performances of Band_SIN and Band_MUL are compared to show the impact of multi-core migration optimization. The only difference between these two versions is the use of multi-core migration optimization. On average, the overall performance gains from multi-core migration are limited. This is because SpMV_Band moves only one segment of x to GSM at a time. To evaluate the benefits of double buffering, Band_MUL is compared with Band_MUL_DB. According to the results in Fig. 10, Band_MUL_DB is on average 9% faster than Band_MUL.

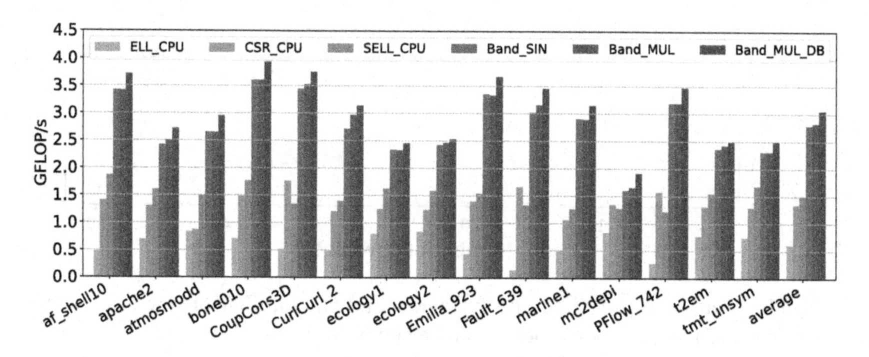

Fig. 10. Performance comparison of CPU-based SpMV vs SpMV_Band

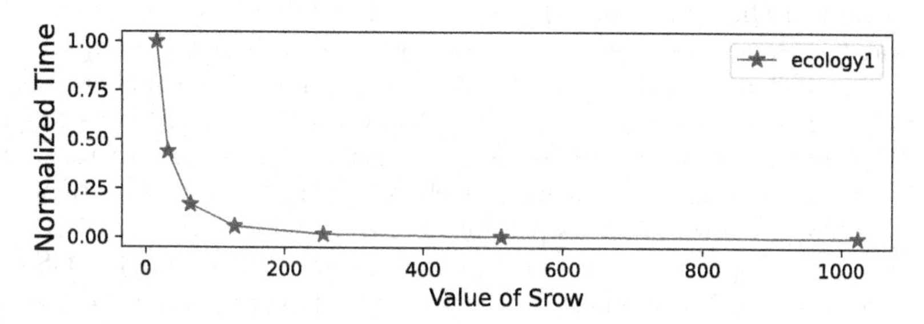

Fig. 11. The relationship between performance and the value of `Srow`. The vertical axis represents the normalized time. We normalize the time of ecology1 to a time range of 0 to 1 to make it easier to observe trends. Normalization formula is $time\_normalized = (time - time\_min)/(time\_max - time\_min)$.

SELL_CPU and CSR_CPU are generally superior to ELL_CPU for running CPU code in the GP area, with SELL_CPU typically showing the highest performance on average. The results in Fig. 10 show that the Band_MUL_DB implementation achieves an average speedup of 2.0× compared to the SELL_CPU implementation when running on the four CPU cores accessing the same DDR in a DSP cluster. And it is used as the final implementation of SpMV_Band.

Evaluation of the Parameter `Srow`. We evaluated the influence of the SELL format parameter `Srow` on the performance of the SpMV_Band algorithm. When `Srow` is small, the matrix is divided into many slices, resulting in higher latency costs (see Lines 25 and 26 of Algorithm 3). As `Srow` increases, more zero elements are filled, leading to higher bandwidth and computation costs. Thus, the time cost of the SpMV tends to decrease initially and then increase as `Srow` increases. However, the time cost is closely related to the sparse structure of the matrix, and different matrices may have different trends. The NNZ distribution of banded matrices is regular, such as matrix ecology1, with very few zero elements padding, the time cost generally decreases and then stabilizes as `Srow` increases (Fig. 11). In this work, we empirically set `Srow` to 512 as it gives good performance in our pilot studies.

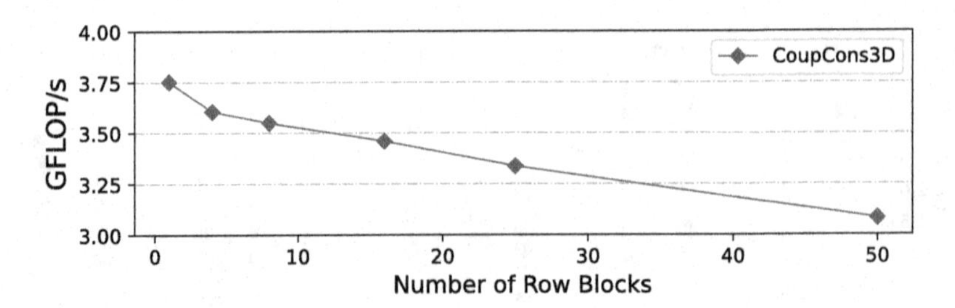

Fig. 12. The relationship between performance and the number of row blocks. As the number of row blocks increases, performance tends to decrease.

Evaluation of the Parameter Beta. SpMV_Band divides the matrix into row blocks and processes each row block one by one. We evaluated the impact of the number of row blocks on performance. We have conducted a number of experiments, which are illustrated here by the CoupCons3D matrix. Figure 12 shows that there is no performance benefit to increasing the number of partitioned row blocks. Therefore, our principle for choosing `blockLen` is to partition the matrix into as few row blocks as possible and to partition the matrix as evenly as possible, as long as the corresponding **x** segments can be stored in GSM. As can be seen in Fig. 7, the length of the **x** fragment corresponding to each row block is $\texttt{blockLen} + \texttt{SBL} + \texttt{SBU}$. The formula for choosing `blockLen` is as follows:

$$blockLenMax = \texttt{GSMlenMax} - SBL - SBU \tag{1}$$

$$num = \frac{n}{blockLenMax} \tag{2}$$

$$\texttt{blockLen} = \frac{n + num}{num + 1} \tag{3}$$

where n is the row dimension of sparse matrix. Since `blockLen` is equal to $\texttt{Srow} * \texttt{beta} * \texttt{coreNum}$, we can solve for `beta` as follows:

$$\texttt{beta} = \frac{\texttt{blockLen}}{\texttt{Srow} * \texttt{coreNum}}.$$

Bandwidth Utilization Analysis. Bandwidth utilization is an important evaluation criterion for memory-bounded operations and is calculated as:

$$bandwidth\ utilization = \frac{Actual\_bandwidth}{Stream\_bandwidth} \times 100\%,$$

where $Actual\_bandwidth$ is the actual bandwidth cost and $Stream\_bandwidth$ is the bandwidth measured by using the STREAM benchmark [24], which is modified by using `hthread` to run on FT-M7032. The total access memory of SpMV_Band equals to

$$slice\_start[end] * (4 + 8) + n * 8 + m * 8,$$

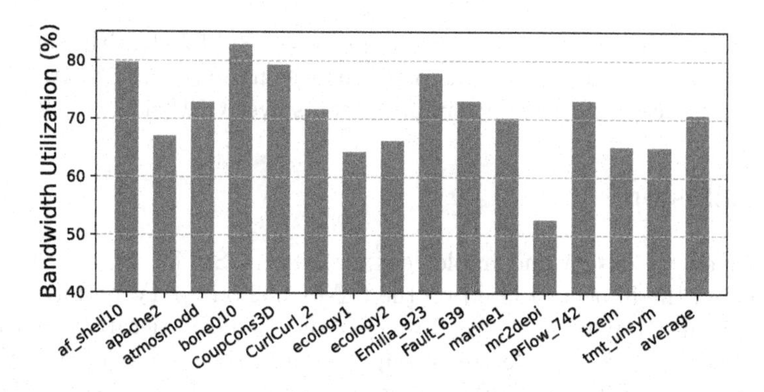

Fig. 13. Bandwidth utilization of SpMV_Band.

where *slice_start*[*end*] is the length of arrays `column` and `value` of SELL format (see Fig. 3), n and m are the row and column dimensions of the matrix, respectively.

Figure 13 shows the bandwidth utilization of the test matrices in Table 2 (Stream_bandwidth is measured by using the `dma_p2p` from DDR to AM, i.e., 30GB/s). The maximum and minimum bandwidth utilization ratio is 83% and 53%, respectively, with an average bandwidth utilization ratio of 71%.

5 Related Work

In this section, we present some of the current optimization methods for DSPs or multi-core platforms and some of the popular SpMV kernels. Prior works have optimized SpMV for CPUs [21,34] and GPUs [19,20,26,29] using storage formats like SELL-C-σ [14], SpV8 [17], CSR5 [19], ESB [21], HYB [2], HYB5 [4], and etc. Vectorization [3,14,35] and parallelization [1,25,28] approaches have also been proposed for SpMV.

Gao [8] et al. conducted a case study of SpMV on the TI C6678 DSP and found that DSP can achieve higher power efficiency than CPU and comparable power efficiency to GPU. Gao [9] et al. designed an on-chip scratchpad-based SpMV approach for the TI Keystone II DSP, improving performance by about 50%. Liu [23] et al. proposed a dense matrix multiplication vectorization method for a multicore long vector processor (MATRIX2). Sun [31] et al. introduced the SWCSR-SpMV, a method based on the CSR format, which effectively reduces the parallel SpMV bandwidth cost on the SW26010 platform. Xu [36] et al. proposed an adaptive sparse Matrix partitioning method based on the storage hierarchy of Matrix 2000b, which effectively improved the data reusability of SpMM.

Banded matrix has good data locality and can develop efficient SpMV operation with its own data locality. However, no one has yet optimized SpMV specifically for banded matrices. Yang [37] et al. proposed a parallel multi-core CSB format that has high scalability for banded matrices, but did not provide a specific optimization algorithm. In this paper, we propose a SpMV algorithm specifically for banded matrices and evaluate its performance on the FT-M7032 architecture, achieving significant performance improvements over CPU-side SpMV.

6 Conclusion

In this paper, we design and implement an efficient SpMV on the FT-M7032. We evaluated the impact of GSM on the FT-M7032 on SpMV performance and provided a quantitative analysis. Based on the memory hierarchy of the FT-M7032, SpMV_Band can effectively improve the performance of SpMV operations through multi-layer partitioning and multi-level parallelism. We have designed a double buffering strategy on the FT-M7032 that overlaps data transfers and computation to minimize memory access overhead. Our experimental results show that our optimization method significantly improves the SpMV performance for banded matrices on the FT-M7032 platform. Compared to CPU-based SpMV implementations, an acceleration ratio of 2.0x can be achieved with an average bandwidth utilisation of 71%.

Acknowledgement. This work was supported in part by the National Key R&D Program of China under grant agreement 2021YFB0300101, the National Science Foundation of China (NSFC) under grant agreements 61902411, 62032023, 12002382, 11275269, 42104078 and 62073333, the Excellent Youth Foundation of Hunan Province under grant agreement 2021JJ10050.

References

1. Alappat, C., et al.: Performance modeling of streaming kernels and sparse matrix-vector multiplication on A64FX. In: IEEE/ACM PMBS, pp. 1–7. IEEE (2020)
2. Bell, N., Garland, M.: Implementing sparse matrix-vector multiplication on throughput-oriented processors. In: Proceedings of the Conference on High Performance Computing Networking, Storage and Analysis, pp. 1–11 (2009)
3. Chen, L., Jiang, P., Agrawal, G.: Exploiting recent SIMD architectural advances for irregular applications. In: IEEE/ACM CGO, pp. 47–58. IEEE (2016)
4. Chen, S., Fang, J., Xu, C., Wang, Z.: Adaptive hybrid storage format for sparse matrix-vector multiplication on multi-core SIMD CPUs. Appl. Sci. **12**(19), 9812 (2022)
5. Crane, H., Jr., Gibbs, N.E., Poole, W.G., Jr., Stockmeyer, P.K.: Algorithm 508: Matrix bandwidth and profile reduction. ACM Trans. Mathematical Softw. (TOMS) **2**(4), 375–377 (1976)
6. Davis, T., Hu, Y.: The University of Florida sparse matrix collection. ACM Trans. Math. Softw. **38**(1), 1:1–1:25 (2011)
7. Fang, J., Zhang, P., Huang, C., Tang, T., Lu, K., Wang, R., Wang, Z.: Programming bare-metal accelerators with heterogeneous threading models: a case study of matrix-3000. Front. Inf. Technol. Electron. Eng. **24**(4), 509–520 (2023)
8. Gao, Y., Bakos, J.D.: Sparse matrix-vector multiply on the texas instruments c6678 digital signal processor. In: 2013 IEEE 24th ASAP, pp. 168–174. IEEE (2013)

9. Gao, Y., Zhang, F., Bakos, J.D.: Sparse matrix-vector multiply on the keystone ii digital signal processor. In: 2014 IEEE High Performance Extreme Computing Conference (HPEC), pp. 1–6. IEEE (2014)
10. Golub, G.H., Loan, C.F.V.: Matrix Computations, 3rd edn. The Johns Hopkins University Press, Baltimore, MD (1996)
11. Igual, F.D., Ali, M., Friedmann, A., Stotzer, E., Wentz, T., van de Geijn, R.A.: Unleashing the high-performance and low-power of multi-core DSPs for general-purpose HPC. In: SC'12: Proceedings of the International Conference on High Performance Computing, Networking, Storage and Analysis, pp. 1–11. IEEE (2012)
12. Im, E.J., Yelick, K.: Optimization of sparse matrix kernels for data mining. In: Submitted to First SIAM Conference on Data Mining (2000)
13. Kincaid, D.R., Oppe, T.C., Young, D.M.: ITPACKV 2D user's guide. Technical Report, Texas University, Austin, TX (USA). Center for Numerical Analysis (1989)
14. Kreutzer, M., Hager, G., Wellein, G., Fehske, H., Bishop, A.R.: A unified sparse matrix data format for efficient general sparse matrix-vector multiplication on modern processors with wide SIMD units. SIAM J. Sci. Comput. **36**(5), C401–C423 (2014)
15. Kubota, Y., Takahashi, D.: Optimization of sparse matrix-vector multiplication by auto selecting storage schemes on GPU. In: Murgante, B., Gervasi, O., Iglesias, A., Taniar, D., Apduhan, B.O. (eds.) ICCSA 2011. LNCS, vol. 6783, pp. 547–561. Springer, Heidelberg (2011). https://doi.org/10.1007/978-3-642-21887-3_42
16. Lewis, J.G.: Algorithm 582: The Gibbs-Poole-Stockmeyer and Gibbs-king algorithms for reordering sparse matrices. ACM Trans. Math. Softw. (TOMS) **8**(2), 190–194 (1982)
17. Li, C., Xia, T., Zhao, W., Zheng, N., Ren, P.: SpV8: Pursuing optimal vectorization and regular computation pattern in SpMV. In: 2021 58th ACM/IEEE Design Automation Conference (DAC), pp. 661–666. IEEE (2021)
18. Liu, S., Cao, Y., Sun, S.: Mapping and optimization method of SpMV on Multi-DSP accelerator. Electronics **11**(22), 3699 (2022)
19. Liu, W., Vinter, B.: CSR5: An efficient storage format for cross-platform sparse matrix-vector multiplication. In: 29th ACM ICS'15, pp. 339–350. ACM, New York (2015)
20. Liu, W., Vinter, B.: Speculative segmented sum for sparse matrix-vector multiplication on heterogeneous processors. Parallel Comput. **49**, 179–193 (2015)
21. Liu, X., Smelyanskiy, M., Chow, E., Dubey, P.: Efficient sparse matrix-vector multiplication on X86-based many-core processors. In: ICS'13, pp. 273–282. ACM, New York (2013)
22. Liu, Y., Schmidt, B.: LightSpMV: Faster CSR-based sparse matrix-vector multiplication on CUDA-enabled GPUs. In: 2015 IEEE 26th International Conference on Application-specific Systems, Architectures and Processors (ASAP), pp. 82–89. IEEE (2015)
23. Liu, Z., Tian, X.: Vectorization of matrix multiplication for multi-core vector processors. Chin. J. Comput. **41**(10), 2251–2264 (2018)
24. McCalpin, J.D.: Memory bandwidth and machine balance in current high performance computers. In: IEEE Computer Society Technical Committee on Computer Architecture (TCCA) Newsletter, pp. 19–25, December 1995
25. Merrill, D., Garland, M.: Merge-based parallel sparse matrix-vector multiplication. In: SC'16. Salt Lake (2016)
26. Mironowicz, P., Dziekonski, A., Mrozowski, M.: A task-scheduling approach for efficient sparse symmetric matrix-vector multiplication on a GPU. SIAM J. Sci. Comput. **37**(6), C643–C666 (2015)

27. Monakov, A., Lokhmotov, A., Avetisyan, A.: Automatically tuning sparse matrix-vector multiplication for GPU architectures. In: Patt, Y.N., Foglia, P., Duesterwald, E., Faraboschi, P., Martorell, X. (eds.) HiPEAC 2010. LNCS, vol. 5952, pp. 111–125. Springer, Heidelberg (2010). https://doi.org/10.1007/978-3-642-11515-8_10
28. Namashivayam, N., Mehta, S., Yew, P.C.: Variable-sized blocks for locality-aware SpMV. In: IEEE/ACM CGO, IEEE (2021)
29. Niu, Y., Zhengyang, L., Dong, M., Jin, Z., Liu, W., Tan, G.: TileSpMV: a tiled algorithm for sparse matrix-vector multiplication on GPUs. In: 35th IPDPS, pp. 68–78. IEEE (2021)
30. Saad, Y.: Iterative methods for sparse linear systems. In: SIAM (2003)
31. Sun, Q., Zhang, C., Wu, C., Zhang, J., Li, L.: Bandwidth reduced parallel SpMV on the SW26010 many-core platform. In: Proceedings of the 47th International Conference on Parallel Processing, pp. 1–10 (2018)
32. Tiwari, A., Kumar, V., Mitra, G.: High performance and energy optimal parallel programming on CPU and DSP based MPSOC. Ph.D. thesis, Ph. D. dissertation, IIIT-Delhi (2018)
33. Wang, Y., et al.: Advancing DSP into HPC, AI, and beyond: challenges, mechanisms, and future directions. CCF Trans. High Perform. Comput. **3**, 114–125 (2021)
34. Williams, S., Oliker, L., Vuduc, R., Shalf, J., Yelick, K., Demmel, J.: Optimization of sparse matrix-vector multiplication on emerging multicore platforms. In: Proceedings of the 2007 ACM/IEEE Conference on Supercomputing, pp. 1–12 (2007)
35. Xie, B., Zhan, J., Liu, X., Gao, W., Jia, Z., He, X., Zhang, L.: CVR: efficient vectorization of SPMV on x86 processors. In: IEEE CGO (2018)
36. Xu, H., Zhu, X., Wang, Q., Liu, J.: Efficiently executing sparse matrix-matrix multiplication on general purpose digital single processor. In: 2022 IEEE 24th International Conferenct on High Performance Computing & Communications, pp. 1–8. IEEE (2022)
37. Yang, B., Gu, S., Gu, T.X., Zheng, C., Liu, X.P.: Parallel multicore CSB format and its sparse matrix vector multiplication. In: Advances in Linear Algebra & Matrix Theory, vol. 2014 (2014)
38. Yin, S., Wang, Q., Hao, R., Zhou, T., Mei, S., Liu, J.: Optimizing irregular-shaped matrix-matrix multiplication on multi-core DSPs. In: 2022 IEEE International Conference on Cluster Computing (CLUSTER), pp. 451–461. IEEE (2022)
39. Zhang, Y., et al.: Memory-aware optimization for sequences of sparse matrix-vector multiplications. In: 37th IEEE International Parallel & Distributed Processing Symposium (IPDPS), IEEE (2023)
40. Zhang, Y., Li, S., Yan, S., Zhou, H.: A cross-platform SpMV framework on many-core architectures. ACM Trans. Archit. Code Optim. (TACO) **13**(4), 1–25 (2016)
41. Zhou, H., Fan, X., Zhao, L.: Optimizations on sparse matrix-vector multiplication based on CUDA. Comput. Meas. Control **18**(8) (2010)

SR-KGELS: Social Recommendation Based on Knowledge Graph Embedding Method and Long-Short-Term Representation

Xuechang Zhao[1,2] and Qing Yu[1,2](✉)

[1] Tianjin University of Technology (TUT), Tianjin 300382, China
xczhao@stud.tjut.edu.cn

[2] Tianjin Key Laboratory of Intelligence Computing and Novel Software Technology, Tianjin, China
yq08982023@163.com

Abstract. Data from user-item interactions and social data can be integrated to improve the effectiveness of social recommendations. Graph neural networks (GNNs) have gained popularity in social-based recommender systems due to their inherent integration of node information and topology. However, most research has focused on how to deeply model users using various datasets, with less emphasis on item relationships. Furthermore, users' changing interests over time, preferences in long-term patterns, and differences in rating behavior across perspectives have received little attention. In this work, we propose a social recommendation based on knowledge graph embedding method and long-short-term representation (SR-KGELS). The SR-KGELS learns user and item features by combining long- and short-term representations and using attention mechanisms to discern the strength of heterogeneity associated with social and relevant relationships. In addition, we treat the differences in users' scoring behaviors as a relative location difference problem in the embedding space, and model it with a knowledge graph embedding method called TransH to improve the generalization ability of the main rating model. Experiments on two real-world recommender system datasets validate the effectiveness of SR-KGELS.

Keywords: Social recommendations · Graph neural networks · Knowledge graph embedding · Attention mechanisms

1 Introduction

Recommender systems are an excellent technique to reduce information overload, particularly in online services. For example, information searches, product recommendations, and social media websites [1]. Therefore, social network-based recommendation systems can use information such as friendships and social behaviors in social networks to predict users' interests and behaviors. Thus, it may offer more personalized recommendations. Early studies mainly used matrix

Z. Tari et al. (Eds.): ICA3PP 2023, LNCS 14491, pp. 221–237, 2024.
https://doi.org/10.1007/978-981-97-0808-6_13

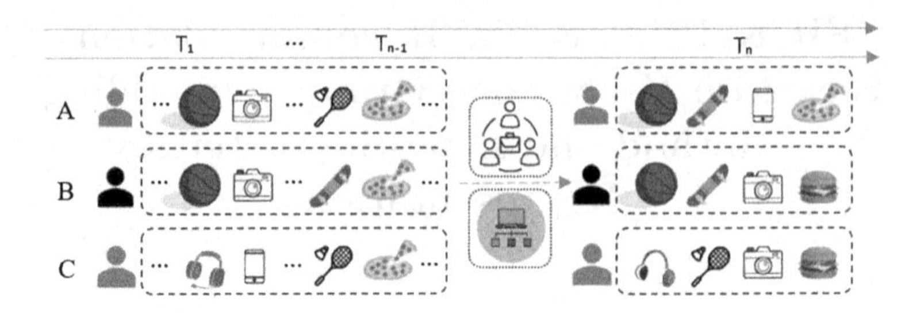

Fig. 1. Example of the user's changes over time under the influence of social friends.

factorization techniques to understand the latent factors of users and items by decomposing the user-item interaction matrix and the social graph adjacency matrix. Using the factor decomposition method of [2], just the first-order relationship of social method is addressed. With the great success of deep neural networks when applied to graph data, newer graph neural networks obtain meaningful representations by learning node features and graph structure data. When incorporating social information, graph neural networks combine user-item interaction data with observed social relationship graphs [3], and connect them to learn representations from users and items.

Although existing social recommendation systems perform well, they still have limitations. The first is data sparsity and unequal distribution. To solve this issue, some researchers have expanded the data by mining the implicit links between users and items [4]. However, the low cost of forming social relationships and the openness of social networks may generate significant noise that affects recommendation performance. Furthermore, different levels of rating indicate differences in the relative positions of users and items in the embedding space. In other words, a highly rated item should be more appealing to the user than a low-rated item. This difference in the model is not explicitly retained and generalized to other similar users and items, hampering the further development of recommendation systems. The relative location differences, the user's interest over time, and the attractiveness of the item over time should also be considered. In Fig. 1. For user *A* who prefers basketball sports, he will pay attention to basketball dynamics for a long time. And may be interested in skateboarding at some time by the influence of friend *B*. User *A* is likely to be interested in new cell phones because of similar interactions with friend *C*. Similarly, items show different attractiveness over time. Thus the user's social relationship and the attractiveness of the item have a great impact on the recommendation results.

In this paper, we use Graph Neural Networks (GNNs) to learn the implicit features of users and items. We propose a social recommendation based on knowledge graph embedding method and long-short-term representation (SR-KGELS) with three components: user modeling, item modeling, and relative location difference modeling. (1) User modeling: long-term and short-term representations of the user are modeled using the item-space. The influence of friends in the social graph is aggregated using the method of GraphRec [5]. Integration of item-space potential factors and social influence to get the final potential factor of the user. (2) Item modeling: long- and short-term attractiveness representations of the item are modeled using the user-space; the impact of candidate items on similar items in the correlation graph is aggregated. Integration of user-space potential factors and related item impacts to obtain the final potential factor of the item. (3) Modeling relative location differences by TransH [6] to constrain the relative locations of the user and item. The main contributions of this paper are summarized as follows:

- We propose an effective social recommendation model (SR-KGELS) that learns representations from both the user and item perspectives, while taking into account the effects of social graphs and related graphs.
- Modeling short-term interests by the bidirectional long-short-term memory (BiLSTM) [7] layer to obtain short-term representations of the user. Similarly, the short-term attractiveness of the item is captured.
- We model the relative position differences that high-rated items are closer to users' preferences than low-rated items. This is accomplished via the knowledge graph embedding method, which improves overall generalization ability.
- We conducted extensive experiments to verify the efficiency of the SR-KGELS. Experimental results show that our method is not only effective, but also outperforms the SOTA models.

2 Related Work

Deep Learning-Based Social Recommendation. With the development of deep learning techniques, scholars have found that graph neural networks can more efficiently reflect the attribute characteristics of the entity itself and the connection between the user and the item. The key to GNNs is to learn the representation of nodes by aggregating the feature information of the neighborhood, which is in line with the essence of collaborative filtering. NGCF [8] integrated user-item interactions into the embedding process and introduced Knowledge Graph Attention Networks (KGAT) [9] to explicitly model higher-order connectivity in knowledge graphs. GraphRec [5] added an attention mechanism to aggregate both user and item opinions and ratings in interaction graphs and social networks, resulting in good experimental results. LightGCN [10] simplifies NGCF [8] by eliminating nonlinear activation and feature transformations in graph convolution networks to improve recommendation performance. DGRec [11] is a state-of-the-art approach for session social recommendation that uses RNNs and graph attention networks (GATs) [12] to capture the dynamic interests and context-dependent social influences of users.

Knowledge Graph Embedding. The KGE maps entities and relations in the knowledge graph to points in a low-dimensional, dense vector space to automate the analysis and inference of the knowledge graph. TransH [6] works by adding a hyperplane to restrict the projection of entities and relations. The vector space between different relations is made independent, and the vector relations are computed on that hyperplane. For this paper, we transform the user rating behavior into a triplet of the form (user, rating, item). A knowledge graph can be obtained by considering the user, item and rating as head and tail entities and relations, respectively.

The KGE methods are: TransE [13] is remarkable for modeling simple relationships, but very unsatisfactory for modeling complex relationships. It is not applicable to the study of this paper. TransR [14] model entities and relationships through entity space and relationship space to solve multiple semantic problems in relationships. TransG [15] uses a Gaussian distribution to portray each semantics to solve multiple semantic problems. TransR [14] and TransG [15] are mainly used to solve multi-semantic problems, which need to be implemented with the help of more matrices. If applied to our model, the corresponding parameters will increase, and the computational complexity will be greatly increased. Based on the following two points: (1) The size of the adopted datasets and the characteristics of its data items. (2) The flexibility and diversity of embedding are increased due to the hyperplane of TransH [6] specific to multiple relations. It is to model different users giving the same rating to items to achieve the mapping of multiple relationships. Therefore, among the many knowledge graph embedding methods, we chose TransH to learn the relative location differences in the embedding space. The previous S4Rec [16] implements the modeling of relative location differences through ch1TransH, which is another reason why we do not use other knowledge graph embedding methods.

While existing models have good performance. None of them observe the dynamic preferences of users and ignore the relative location differences hidden in the embedding space in user ratings. Therefore, we used BiLSTM [7] to model users' short-term interests. For rating differences, it can be assumed that users are closer to highly rated items in the embedding space than to low-rated items. The relative location differences are preserved as constraints on the main rating model using the knowledge embedding method to improve the overall generalization ability.

3 Problem Definition

Let $U = \{u_1, u_2, \cdots, u_m\}$ and $V = \{v_1, v_2, \cdots, v_n\}$ denote the set of users and items, respectively, where m is the number of users and n is the number of items. $\boldsymbol{R} \in R^{m\times n}$ is the user-item interaction graph. r_{ij} represents the rating value given by user u_i to item v_j. $R(u_i)$ represents the set of items that interact with user u_i, and $R(v_j)$ denotes the set of users that interact with the item v_j.

In the user's social graph $\boldsymbol{S} \in S^{m\times m}$, $s_{ik} = 1$ if user i trusts another user k, otherwise $s_{ik} = 0$. $F_U(i)$ is used to denote the set of social friends of user i in the social graph: $F_U(i) = \{k \mid s_{ik} = 1\}$. In the item-related graph $\boldsymbol{I} \in I^{n\times n}$. We use $F_V(j)$ to denote the set of similar items of item j in the related graph $\boldsymbol{I}$. Each interaction is recorded as a time triplet (u_i, v_j, t) and t is a timestamp. Let $T_v(u_i)$ be the ascending sequence of all timestamps of the user u_i interactions, then the sequence of items that user u_i has interacted with can be written

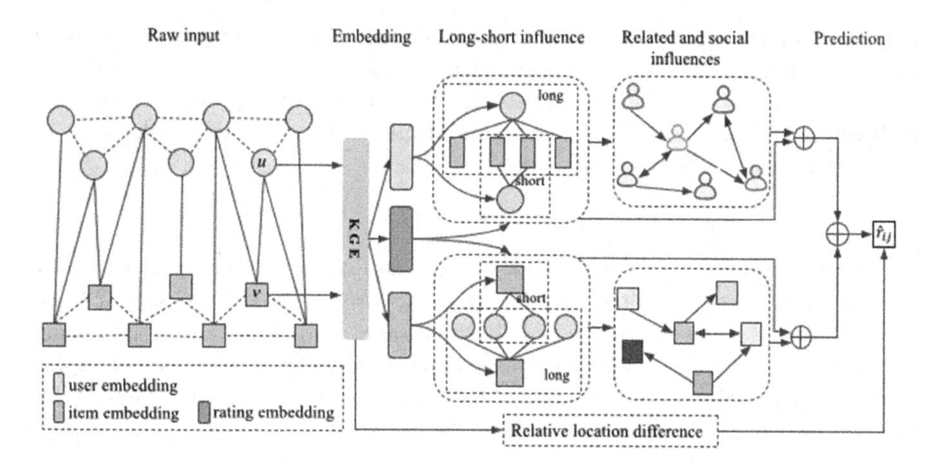

Fig. 2. A schematic diagram of our proposed SR-KGELS model.

as $S_{v_{j(t)}}(u_i) = \{v_{j(t)} \mid j(t) \in T_v(u_i)\}$. Similarly, if $T_u(v_j)$ represents the ascending sequence of item v_j interactions, then the corresponding user sequence is $S_{u_{i(t)}}(v_j) = \{u_{i(t)} \mid i(t) \in T_u(v_j)\}$.

Therefore, we define the social recommendation problem as follows: Given the observed the user-item interaction graph $\boldsymbol{R}$, the social graph $\boldsymbol{S}$, and the item-related graph $\boldsymbol{I}$. Our goal is to predict unknown rating values in $\boldsymbol{R}$ and recommend items to users who are predicted to have higher ratings.

4 The Proposed Model

The SR-KGELS framework is shown in Fig. 2. The data is first pre-processed in the input layer, and all rating relationship triplets are used to train the TransH model. Then, the embedding parameters of the obtained user, item and rating are initialized to the model's relevant parameters. The embeddings of the user and item are learned using long-term and short-term representations. The overall impact of user-item interactions, social connections, and item correlations was calculated. Predicting ratings by combining user and item potential factors under the influence of TransH. The components of the model are described next.

4.1 Embedding Layer and Input Data Pre-processing

Embedding Layer. We define the user u_i rating for item v_j to be denoted as r_{ij}, and the rating r_{ij} usually takes discrete values. In our work, each rating is 1, 2, 3, 4, or 5. Each rating is embedded into a d dimensional vector. Let $\mathbf{p_i}, \mathbf{q_j} \in \mathbb{R}^d$ be the embedding of user u_i and item v_j, respectively.

Input Data Pre-processing. We ignore the effect of interaction time on the modeling of relative location differences, since the source of relative location differences is primarily ratings. And still consider the time factor when representing long- and short-term. First, the user's rating behavior on items is transformed into a relational triplet to build a knowledge graph. TransH is used to constrain the relative positions of users and items in the embedding space. To learn the embedding of head and tail entities and relations of the relational triplet (u, r, v), we optimize the translation principle:

$$\mathbf{p}_i^r + \mathbf{e}_r \approx \mathbf{q}_j^r, \tag{1}$$

The representation of the projection of $\mathbf{p}_i$ and $\mathbf{q}_j$ on the relation hyperplane can be called $\mathbf{p}_i^r$, $\mathbf{q}_j^r$, denoted as: $\mathbf{p}_i^r = \mathbf{p}_i - \mathbf{w}_r^\top \mathbf{p}_i \mathbf{w}_r$, $\mathbf{q}_j^r = \mathbf{q}_j - \mathbf{w}_r^\top \mathbf{q}_j \mathbf{w}_r$, where $\mathbf{e}_r \in \mathbb{R}^d$ denotes the embedding of the relation r, and $\mathbf{w}_r \in \mathbb{R}^d$ is the normal vector. In addition, we first pre-train the TransH model with all relational triplets to obtain embedding parameters for the user, item and rating. We then initialize the obtained parameters into the model to get the new interaction embedding.

$$\mathbf{x}_{ij} = L_U([\mathbf{q}_v \oplus \mathbf{e}_r]), \ \mathbf{y}_{ji} = L_V([\mathbf{p}_u \oplus \mathbf{e}_r]), \tag{2}$$

L_U and L_V are both two-layer perceptrons, and the outputs are the interactive embedding of v_j to u_i and u_i to v_j, respectively. $[\oplus]$ denotes a connection operation. It should be noted that $\mathbf{p}_u$ and $\mathbf{q}_v$ are obtained after pre-training initialization and are different from $\mathbf{p}_i$ and $\mathbf{q}_j$.

4.2 User Modeling

The purpose of user modeling is to discover the user's latent factors. The problem is how to naturally combine the interaction graph and the social network. To address this issue, we employ two types of aggregation methods to learn latent user features from both graphs. The first aggregation, known as item-space aggregation, is composed of two parts: short-term representation and long-term representation. The second is social aggregation, which uses the social graph to learn about possible user factors. The above two latent factors are combined to get the final latent factors for the user. Next, we will describe item-space aggregation, social aggregation, and how to combine the potential user factors.

Item-Space Aggregation. The item-space aggregate user potential factors are divided into two parts: modeling the user's short- and long-term representation.

Short-term Representation Modeling. This can be understood as the modeling of users' short-term preferences. After getting the user interaction embedding $\mathbf{x}_{ij}$, the item of u_i interaction in all sequences can be written as $\mathbf{X}(i) = \{\mathbf{x}_{ij(t)} | j(t) \in T_v(u_i)\}$. In order to obtain the short-term preference of the user, we need to model the sequences. Since RNN has powerful modeling and representation capabilities for sequence modeling, Bidirectional Long-Short-Term Memory (BiLSTM) [7] uses contextual information in both directions and has good performance in sequence modeling. Therefore, we use it to model the user's short-term preference representation $\mathbf{h}_u^S$.

$$\mathbf{h}_u^S = \mathbf{BiLSTM}(\mathbf{X}(i)), \tag{3}$$

Because the BiLSTM contains forward and backward hidden states, we have the output of the last hidden layer. And for longer item sequences, we intercept a fixed length to reduce the loss.

Long-Term Representation Modeling. This is the modeling of users' long-term preferences. For all user interactions, we embed each interaction with $\mathbf{x}_{ij}$ as an edge representation of the user-item graph. Consider that each interaction contributes differently to the user's latent factor. Inspired by the attention mechanism [17], We use the aggregation function G_v to obtain long-term user preferences.

$$G_v(\{\mathbf{x}_{ij}, \forall j \in R(u_i)\}) = \sum_{j \in R(u_i)} \eta_{ij}\mathbf{x}_{ij}, \tag{4}$$

The attention weight of the interaction between user u_i and item v_j is η_{ij}. Different from the self-attention in graph attention networks that ignores the edge features, we adopt the following attention network to incorporate edge features. η_{ij} is the contribution learned at the central node representation $\mathbf{p}_u$ and the edge representation $\mathbf{x}_{ij}$:

$$\eta_{ij} = \frac{\exp(\mathbf{W}_2 \cdot \sigma(\mathbf{W}_1 \cdot [\mathbf{p}_u \oplus \mathbf{x}_{ij}] + \mathbf{b}_1) + \mathbf{b}_2)}{\sum_{j \in R(u_i)} \exp(\mathbf{W}_2 \cdot \sigma(\mathbf{W}_1 \cdot [\mathbf{p}_u \oplus \mathbf{x}_{ij}] + \mathbf{b}_1) + \mathbf{b}_2)}, \tag{5}$$

σ denotes the nonlinear activation function. $(\mathbf{W}_1, \mathbf{b}_1)$ and $(\mathbf{W}_2, \mathbf{b}_2)$ are the weights and biases of the first and second layers of the two-layer neural network, respectively. Equation (4) and (5) can obtain the long-term preference of the user $\mathbf{h}_u^L$.

$$\mathbf{h}_u^L = g^L(\mathbf{p}_u, \{\mathbf{x}_{ij} : j \in R(u_i)\}), \tag{6}$$

We combine the short-term representation and the long-term representation to obtain the user's potential factor $\mathbf{h}_u^i$ in the item-space via Hadamard.

$$\mathbf{h}_u^i = \mathbf{h}_u^L \odot \mathbf{h}_u^S, \tag{7}$$

Social Aggregation. Friends frequently impact users' behavior. So, we should model the latent factors for the user in combination with social information. Since users are more likely to have higher interests in strong relationships than in weak relationships. We apply the attention mechanism of the graph to aggregate representative friends to obtain the user's social influence $\mathbf{h}_u^{So}$.

$$\mathbf{h}_u^{So} = g^S(\mathbf{p}_u, \{\mathbf{h}_o^i : o \in F_U(i)\}), \tag{8}$$

$F_U(i)$ denotes the set of friends of user u_i. In order to better model the potential factor of the user, we combine long- and short-term representations and friends' representations. The final potential factor representation of the user u_i is $\mathbf{h}_{u_i}$.

$$\mathbf{h}_{u_i} = L_u^i([\mathbf{h}_u^i \oplus \mathbf{h}_u^{So}]), \tag{9}$$

4.3 Item Modeling

Item modeling is used to study an item's latent factors. A critical issue is how to intrinsically combine the interaction graph and the correlation graph. As a result, we use two forms of aggregation to learn two different latent factors from two graphs. The first is user-space aggregation, which is used to learn item latent factor from user-item interaction graph. The second is related items aggregation, which learns the item's latent factor from the correlation graph. The final latent factor of the item is obtained by combining two factors.

User-Space Aggregation. It consists of two parts: Modeling the short-term and long-term representations of the item. It should be noted that the short-term and long-term representations here refer to the item's short-term and long-term attraction.

Short-Term Representation Modeling. It can be viewed as a representation of the short-term attractiveness of the item. After obtaining the item interaction embedding $\mathbf{y}_{ji}$, the user of v_j interaction in all sequences can be written as $\mathbf{Y}(j) = \{\mathbf{y}_{ji(t)} | i(t) \in T_u(v_j)\}$. We use the same approach as modeling the short-term representation of the user to aggregate the item's short-term attractiveness representation $\mathbf{h}_v^S$.

$$\mathbf{h}_v^S = \mathbf{BiLSTM}(\mathbf{Y}(j)), \tag{10}$$

We output the last hidden layer of BiLSTM and use a fixed length for longer user sequences to reduce computing costs.

Long-Term Representation Modeling. It can be viewed as a model of the item's long-term attractiveness. For all user interactions, we embed each interaction with $\mathbf{y}_{ji}$ as an edge representation of the interaction graph. Considering that each interaction contributes differently to the item's latent factor, we use the attention mechanism to calculate the contribution of each interaction to the item's potential factor. We use the aggregation function G_U to obtain long-term item attractiveness.

$$G_U(\{\mathbf{y}_{ji}, \forall i \in R(v_j)\}) = \sum_{i \in R(v_j)} \xi_{ji} \mathbf{y}_{ji}, \tag{11}$$

We use attention networks to calculate the weight ξ_{ji}.

$$\xi_{ji} = \frac{\exp(\mathbf{W}_2 \cdot \sigma(\mathbf{W}_1 \cdot [\mathbf{q}_v \oplus \mathbf{y}_{ji}] + \mathbf{b}_1) + \mathbf{b}_2)}{\sum_{i \in R(v_j)} \exp(\mathbf{W}_2 \cdot \sigma(\mathbf{W}_1 \cdot [\mathbf{q}_v \oplus \mathbf{y}_{ji}] + \mathbf{b}_1) + \mathbf{b}_2)}, \tag{12}$$

σ is the nonlinear activation function, $\mathbf{W}_1$ and $\mathbf{b}_1$ are the weights and biases of the first layer of the attention network, and $\mathbf{W}_2$ and $\mathbf{b}_2$ are the weights and biases of the second layer of the attention network. At this point, Eq. (11) and (12) can be used to obtain the latent factor $\mathbf{h}_v^L$ of the item v_j in user-space.

$$\mathbf{h}_v^L = g^L(\mathbf{q}_v, \{\mathbf{y}_{ji} : i \in R(v_j)\}), \tag{13}$$

We combine the short-term representation and the long-term representation to obtain the item's latent factors $\mathbf{h}_v^j$ in the user-space via Hadamard.

$$\mathbf{h}_v^j = \mathbf{h}_v^L \odot \mathbf{h}_v^S, \tag{14}$$

Related Items Aggregation. Since items are not independent and may have related or similar items. It is necessary to further combine related graphs to enrich the potential factors of the item. We use the attention mechanism of the graph to aggregate the effects of related graphs, known as $\mathbf{h}_v^{Re}$.

$$\mathbf{h}_v^{Re} = g^I(\mathbf{q}_v, \{\mathbf{h}_r^j : r \in F_V(j)\}), \tag{15}$$

$F_V(j)$ denotes the set of related items of item v_j in the related graph. In order to better model the final potential factor of the item, we need to consider the long- and short-term representations and the representation of the correlations. The final potential factor of item v_j is defined as $\mathbf{h}_{v_j}$.

$$\mathbf{h}_{v_j} = L_v^j([\mathbf{h}_v^j \oplus \mathbf{h}_v^{Re}]), \tag{16}$$

Rating Prediction. Our method is mostly used for rating prediction recommendations, and it calculates the predicted rating $\hat{r}_{ij}$ as follows:

$$\hat{r}_{ij} = L_{\text{out}}([\mathbf{h}_{u_i} \oplus \mathbf{h}_{v_j}]), \tag{17}$$

4.4 Relative Location Difference Modeling

TransH has two roles in this paper, the first one is to pre-train TransH to obtain parameters initialized to the corresponding parameters of the model, which has been introduced in Sect. 4.1. The second is to maintain the relative position of the user and the item through the loss function. Next, the second function is described in detail.

The relative positional differences are kept as constraints on the main scoring model using the knowledge graph embedding method: TransH. By optimizing the translation principle $\mathbf{p}_u^r + \mathbf{e}_r \approx \mathbf{q}_v^r$. The scoring function of the TransH is $f(u, r, v)$.

$$f(u, r, v) = ||\mathbf{p}_u^r + \mathbf{e}_r - \mathbf{q}_v^r||_2^2, \tag{18}$$

For this scoring function, we adopt the L_{KG} as the loss function of the TransH model, which requires the real triplet to produce lower scores than the generated triplet to maintain the relative position between users and items, thus solving the problem of scoring differences in the embedding space.

$$\begin{aligned} L_{KG} = & \sum_{(u,r,v,v')} [f(u,r,v) + \gamma - f(u,r,v')]_+ \\ & + \sum_{(u,r,v,u')} [f(u,r,v) + \gamma - f(u',r,v)]_+ , \end{aligned} \tag{19}$$

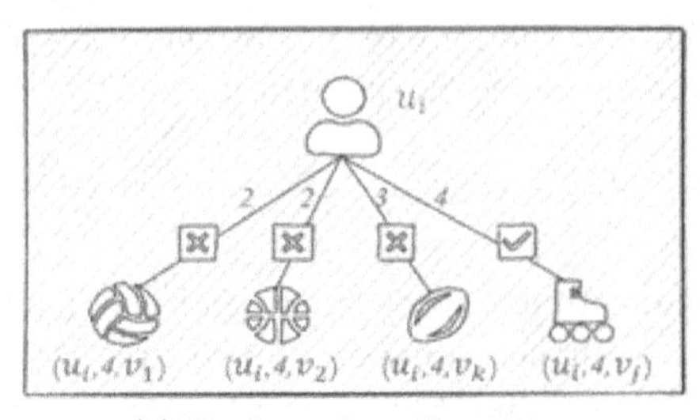

(a) Replace the tail entities.

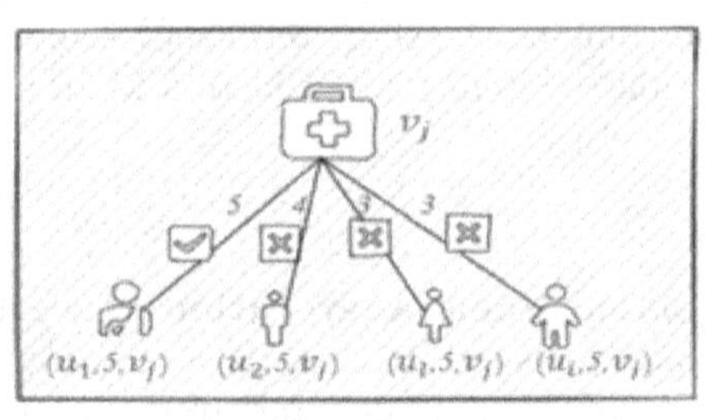

(b) Replace the head entities.

Fig. 3. Negative example of triplets replacement strategy.

Where the triplet (u, r, v') and (u', r, v) are formed from the real triplet (u, r, v) by replacing the tail and head entities, which are negative example triplets. The goal is to ensure that the training data is balanced. $[f(\cdot)]_+$ denotes $\max(0, [f(\cdot)])$. We apply a rating-based cross-sampling and replacement strategy to generate negative example relational triplets.

Specifically. All faked items in the negative example triplet should be drawn from among the scored items rated lower than the true triplet. As shown in Fig. 3(a), a user is interested in roller skates and scores 4. We select items from the set of items with user ratings lower than 4 to form the negative example triplet $(u_i, 4, v_1)$, etc. The user u_i actually rated Volleyball v_1 as 2. Using the same strategy as in Fig. 3(b), we select users from the set of items with v_j scores lower than 5 to form negative example triplets $(u_2, 5, v_j)$, etc. The item v_j was actually rated 4 by the user u_2. In this manner, we generate a list of negative example triplets for each true triplet. These fixed-size triplets are chosen at random to constrain the model. For example, $N_S = 30$, instead of using all triplets in the training process.

4.5 Model Training

To estimate the model performance of SR-KGELS for the task of score prediction, the common loss function.

$$L_1 = \frac{1}{2\langle \mathcal{O} \rangle} \sum_{(u_i, v_j) \in \mathcal{O}} (\hat{r}_{ij} - r_{ij})^2, \tag{20}$$

Where r_{ij} is the true score.The final loss function can be obtained as:

$$L = L_1 + \lambda L_{KG}, \tag{21}$$

λ is the regularization parameter, while SR-KGELS is optimized via gradient descent. To avoid overfitting effects on learning caused by too many interactions, we use the Dropout [18] approach in our work.

5 Experiments

5.1 Experimental Settings

Datasets. To evaluate the proposed approach, we conducted experiments on two real-world datasets. They both have social connections and rating information.

Table 1. Statistics of the two datasets.

Feature	Ciao	Epinions
Users	2,379	22,167
Items	16,862	296,278
Ratings	35,990	920,073
Social Relations	57,544	355,813

Ciao and Epinions[1] are datasets from the popular social networking sites Ciao and Epinions. Users can rate items, read and post reviews, and add friends to their "circle of trust" on each social networking account. They offer a plethora of ratings and social data. Table 1 displays the statistics for Ciao and Epinions, which filter away users with ratings less than 5 to reduce the dataset's sparsity.

Baselines. We compare SR-KGELS to the baseline to measure the effectiveness of our model: (1) Prediction using interaction data and considering social influence. SoReg [19] uses social network information in a traditional Matrix Factorization (MF) framework. SocialMF [2] combines trust propagation mechanisms with MF technique used for social recommendations. DeepSoR [1] uses neural networks to extract user complexity and intrinsic nonlinear features from social relationships. (2) Session-based recommender systems: NARM [20] applies attention mechanisms to capture users' sequential behavior. STAMP [21] captures users' interest from session context, long- and short-term memories. SSRM [22] uses MF-based attention models to understand the uncertainty of user behavior. (3) GNN-based social recommendation: DGRec [11] is a state-of-the-art approach to considering sessions and social influences. GraphRec [5] applies two attention networks (GATs) in interaction graphs and social graphs. Aggregating and combining user-embedded information for rating prediction. GraphRec+ [23] provides recommendations for related graphs to improve user and item representation. GILSR [24] is used to balance the short-term and long-term interests of users as well as capture the short-term and long-term changes in the item life cycle for prediction. It is the latest model that considers both long- and short-term impacts of users and items and has good performance.

[1] https://cse.msu.edu/$\sim$tangjili/trust.html

Item Correlation Graph Construction. The item correlation graph is constructed from items and their similar items. This is because neither dataset contains explicit information about the display between items. The natural way to connect related items is based on their similarity or relevance. Therefore, we use cosine similarity to calculate the similarity between items, and the higher the value, the more similar the two items are to each other. We extract the top-k items of each item to construct the item similarity graph $\boldsymbol{I}$.

Evaluation Metrics. We use the Mean Absolute Error (MAE) and Root Mean Square Error (RMSE) to evaluate the prediction accuracy of the recommendation algorithm. The lower the value of the MAE and the RMSE, the better. We repeat the experiment five times and report the average performance of the test dataset.

Table 2. Performance comparison of different recommender models.

Algorithms	Ciao		Epinions	
	RMSE	MAE	RMSE	MAE
SoReg	1.0848	0.8611	1.1703	0.9119
SocialMF	1.0501	0.8270	1.1328	0.8837
DeepSoR	1.0316	0.7739	1.0972	0.8383
NARM	1.0540	0.8349	1.1050	0.8648
STAMP	1.0827	0.9558	1.0829	0.8820
SSRM	1.0745	0.9211	1.0665	0.8800
DGRec	0.9943	0.8029	1.0684	0.8511
GraphRec	0.9894	0.7486	1.0673	0.8123
GraphRec+	0.9794	0.7387	<u>1.0631</u>	0.8168
GILSR	<u>0.9683</u>	<u>0.7134</u>	1.0679	<u>0.8056</u>
SR-KGELS	0.9579	0.6585	1.0590	0.7973

Parameter Settings. Our proposed model is built on PyTorch[2]. We randomly split the rating dataset with 80%, 10%, 10% for training, validation and testing set, respectively. Batch size $B = 256$. The embedding dimension $d = 128$, the BiLSTM hidden layer $L_l = 4$. The regularization parameter $\lambda = 2$, and the length of user and item sequences is intercepted by 30. The item correlation graph is built with $k = 100$, the Dropout $d_r = 0.5$. The sample neighborhood size is 30 for both the social graph and the correlation graph. The RMSprop optimizer selects the learning rate at $\{0.0001, 0.0005, 0.001, 0.005\}$. The parameters of the baseline algorithm are initialized in the corresponding paper and then carefully tuned to achieve optimal performance.

[2] https://pytorch.org.

5.2 Comparative Results

The performance of all models on the two datasets are shown in Table 2. The underlined values indicate the best performance between baselines. (1) ch1SoReg, SocialMF, and DeepSoR exploit rating and social information. These results demonstrate the need for neural network-based models to fully exploit social relationships for social recommendation. (2) The table data reveals that the session-based strategy used by ch1NARM, STAMP and SSRM outperforms the traditional social recommendation approach on the Epinions dataset overall. Because the Epinions dataset offers a amount of session data and the session-based technique captures changing interests better. The session-based technique is otherwise based on the traditional social method in the Ciao dataset. These comparison findings show that social data do carry useful information. (3) ch1DGRec, ch1GraphRec, and GraphRec+ utilize GNNs to combine with social information. This indicates that GNNs have good representational learning capabilities.

Table 3. Ablation Study on Ciao and Epinions datasets.

Algorithms	Ciao		Epinions	
	RMSE	MAE	RMSE	MAE
SR-KGELS-SN	0.9903	0.6936	1.0736	0.8029
SR-KGELS-IN	0.9707	0.6822	1.0774	0.8164
SR-KGELS-BI	1.0144	0.7240	1.0668	0.8179
SR-KGELS-LO	0.9659	0.6956	1.0682	0.8098
SR-KGELS-PR	0.9611	0.6919	1.0663	0.8112
SR-KGELS	0.9579	0.6585	1.0590	0.7973

We can see that our SR-KGELS outperforms all other baseline methods in Table 2. Compared to GNN-based methods, our approach provides information for integrating user-item interaction graphs, social graphs, and related graphs. This allows better integration of user or item information, since it improves the representation of the user and item and improves the performance of the recommender system. In Table 2, SR-KGELS shows improvements of 1.04% and 0.41% in RMSE and up to 5.49% and 0.83% in MAE on the Ciao and Eipnions datasets, respectively, when compared to the strongest baseline. Although the relative improvements in some metrics or datasets are small. Yet, previous studies [25] have indicated that small improvements in MAE or RMSE may have a significant impact on the quality of top-K recommendations.

5.3 Ablation Study

To validate the impact of the different components in our proposed SR-KGELS. We conducted the ablation experiment: *SR-KGELS-SN* removed social network information. This variable ignores the effect of social relationships on rating prediction. *SR-KGELS-IN* removes the related graph and does not consider the effect of related items on rating prediction. *SR-KGELS-BI* eliminates the effect of short-term representations and models only long-term representations of users and items. To better understand the role of relational triplet constraints in SR-KGELS. We use two variants: *SR-KGELS-LO* eliminates the triplet constraint in the loss function. *SR-KGELS-PR* removes the pre-training strategy and directly trains fixed-size triplets to train model.

The hyper-parameters were set identically in these variants. As Table 3 summarizes the performance of the *SR-KGELS* variant in both datasets, *SR-KGELS-SN* and *SR-KGELS-IN* differ in Ciao, but have similar performance in the Epinions Dataset. This indicates that social information and related item data have a significant impact on prediction. Similarly, it can be seen from Table 3 that *SR-KGELS-BI*, without considering short-term modeling, has higher values on both datasets. This indicates that short-term data are indispensable for model modeling. The *SR-KGELS-LO* and *SR-KGELS-PR* variants that validate the triplet constraint perform similarly to *SR-KGELS* on the Epinions dataset, but differ in Ciao. The pre-training and initialization strategies improve the performance of the model. Combining the experimental data from the multiple variants again validates the effectiveness of our SR-KGELS.

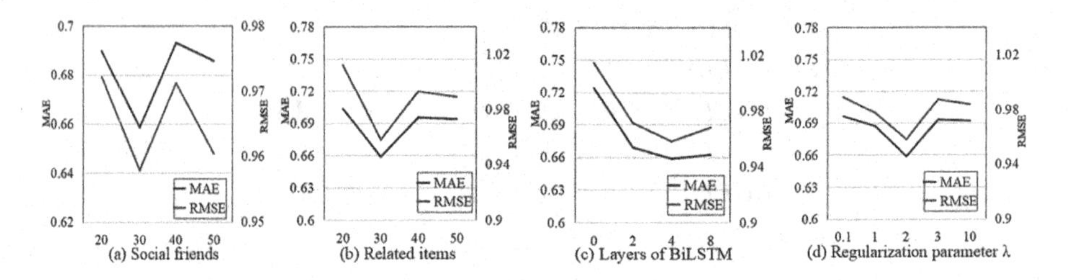

Fig. 4. Performance on Ciao different hyper-parameters.

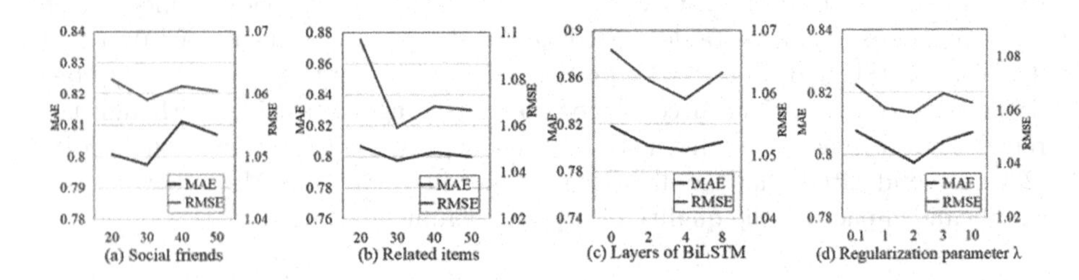

Fig. 5. Performance on Epinions different hyper-parameters.

5.4 Analysis of Parameters

To verify how the performance of our model changes with the hyper-parameters. We compare: (a) The number of social friends. (b) The number of relevant items in the correlation graph. (c) The number of BiLSTM layers. (d) The hyper-parametric analysis of the regularization parameter λ.

According to Fig. 4. The performance is better when the number of social friends is 30 or 50 on the Ciao dataset. Setting social friends to 50 will increase the computer's cost. According to the combined performance of MAE and RMSE, the number of user friends and similar items set to 30 is the optimal parameter. The model works best with layers $L_n = 4$ of BiLSTM, and the time complexity is lower. For the regularization parameter $\lambda = 2$ the experimental results are optimal. Figure 5 shows the performance on the Epinions dataset. The number of friends of the user and the similar items set to 30 are the optimal parameters. The number of BiLSTM layers and the regularization parameter have similar performance on MAE and RMSE as on Ciao. These experimental results show that a reasonable combination of long- and short-term information and the choice of regularization parameters can improve the performance of the model.

6 Conclusion

In this work, we propose a social recommendation based on knowledge graph embedding method and long-short-term representation (SR-KGELS). In particular, different aggregation operations are proposed to model graph data for better learning of the user and item representations. The good performance of the model can be attributed to: (1) The recommended model aggregates social network and related graph impacts with the use of graph attention mechanisms (GATs) to model the learning of user and item representations. (2) Combining user-item interaction graphs using the BiLSTM and attention methods to produce short-term and long-term representations, respectively. (3) The issue of relative location differences. The relational triplet (user, rating, item) that we created offers a good generalization technique to enhance overall performance. Comprehensive experiments on two datasets show that our model is effective.

For future work, since the model does well at modeling short-term representations. We will validate the effectiveness of the proposed method with social datasets that include timestamps. In addition, we intend to conduct more experiments in order to make it clear when and how the triplet constraint enhances the model's performance and whether it can be applied in other scenarios.

References

1. Fan, W., Li, Q., Cheng, M.: Deep modeling of social relations for recommendation. In: Proceedings of the AAAI Conference on Artificial Intelligence, vol. 32 (2018)
2. Jamali, M., Ester, M.: A matrix factorization technique with trust propagation for recommendation in social networks. In: Proceedings of the Fourth ACM Conference on Recommender Systems, pp. 135–142 (2010)

3. Yang, L., Liu, Z., Dou, Y., Ma, J., Yu, P.S.: ConsisRec: enhancing GNN for social recommendation via consistent neighbor aggregation. In: Proceedings of the 44th International ACM SIGIR Conference on Research and Development in Information Retrieval, pp. 2141–2145 (2021)
4. Yu, J., Gao, M., Li, J., Yin, H., Liu, H.: Adaptive implicit friends identification over heterogeneous network for social recommendation. In: Proceedings of the 27th ACM International Conference on Information and Knowledge Management, pp. 357–366 (2018)
5. Fan, W., et al.: Graph neural networks for social recommendation. In: The World Wide Web Conference, pp. 417–426 (2019)
6. Wang, Z., Zhang, J., Feng, J., Chen, Z.: Knowledge graph embedding by translating on hyperplanes. In: Proceedings of the AAAI Conference on Artificial Intelligence, vol. 28 (2014)
7. Huang, Z., Xu, W., Yu, K.: Bidirectional LSTM-CRF models for sequence tagging. arXiv preprint arXiv:1508.01991 (2015)
8. Wang, X., He, X., Wang, M., Feng, F., Chua, T.S.: Neural graph collaborative filtering. In: Proceedings of the 42nd International ACM SIGIR Conference on Research and Development in Information Retrieval, pp. 165–174 (2019)
9. Wang, X., He, X., Cao, Y., Liu, M., Chua, T.S.: KGAT: knowledge graph attention network for recommendation. In: Proceedings of the 25th ACM SIGKDD International Conference on Knowledge Discovery & Data Mining, pp. 950–958 (2019)
10. He, X., Deng, K., Wang, X., Li, Y., Zhang, Y., Wang, M.: LightGCN: simplifying and powering graph convolution network for recommendation. In: Proceedings of the 43rd International ACM SIGIR Conference on Research and Development in Information Retrieval, pp. 639–648 (2020)
11. Song, W., Xiao, Z., Wang, Y., Charlin, L., Zhang, M., Tang, J.: Session-based social recommendation via dynamic graph attention networks. In: Proceedings of the Twelfth ACM International Conference on Web Search and Data Mining, pp. 555–563 (2019)
12. Veličković, P., Cucurull, G., Casanova, A., Romero, A., Lio, P., Bengio, Y.: Graph attention networks. arXiv preprint arXiv:1710.10903 (2017)
13. Bordes, A., Usunier, N., Garcia-Duran, A., Weston, J., Yakhnenko, O.: Translating embeddings for modeling multi-relational data. In: Advances in Neural Information Processing Systems, vol. 26 (2013)
14. Lin, Y., Liu, Z., Sun, M., Liu, Y., Zhu, X.: Learning entity and relation embeddings for knowledge graph completion. In: Proceedings of the AAAI Conference on Artificial Intelligence, vol. 29 (2015)
15. Xiao, H., Huang, M., Hao, Y., Zhu, X.: TransG: a generative mixture model for knowledge graph embedding. arXiv preprint arXiv:1509.05488 (2015)
16. Yuan, K., Liu, G., Wu, J., Xiong, H.: Semantic and structural view fusion modeling for social recommendation. IEEE Trans. Knowl. Data Eng. **35**, 11872–11884 (2022)
17. Chen, C., Zhang, M., Liu, Y., Ma, S.: Neural attentional rating regression with review-level explanations. In: Proceedings of the 2018 World Wide Web Conference, pp. 1583–1592 (2018)
18. Chen, J., Xin, X., Liang, X., He, X., Liu, J.: GDSRec: graph-based decentralized collaborative filtering for social recommendation. IEEE Trans. Knowl. Data Eng. **35**, 4813–4824 (2022)
19. Ma, H., Zhou, D., Liu, C., Lyu, M.R., King, I.: Recommender systems with social regularization. In: Proceedings of the Fourth ACM International Conference on Web Search and Data Mining, pp. 287–296 (2011)

20. Li, J., Ren, P., Chen, Z., Ren, Z., Lian, T., Ma, J.: Neural attentive session-based recommendation. In: Proceedings of the 2017 ACM on Conference on Information and Knowledge Management, pp. 1419–1428 (2017)
21. Liu, Q., Zeng, Y., Mokhosi, R., Zhang, H.: STAMP: short-term attention/memory priority model for session-based recommendation. In: Proceedings of the 24th ACM SIGKDD International Conference on Knowledge Discovery & Data Mining, pp. 1831–1839 (2018)
22. Guo, L., Yin, H., Wang, Q., Chen, T., Zhou, A., Quoc Viet Hung, N.: Streaming session-based recommendation. In: Proceedings of the 25th ACM SIGKDD International Conference on Knowledge Discovery & Data Mining, pp. 1569–1577 (2019)
23. Fan, W., et al.: A graph neural network framework for social recommendations. IEEE Trans. Knowl. Data Eng. **34**(5), 2033–2047 (2020)
24. Wang, J., Zhang, Z.: Graph neural network with item life cycle for social recommendation. In: Proceedings of the 2022 6th International Conference on Computer Science and Artificial Intelligence, pp. 160–165 (2022)
25. Koren, Y.: Factorization meets the neighborhood: a multifaceted collaborative filtering model. In: Proceedings of the 14th ACM SIGKDD International Conference on Knowledge Discovery and Data Mining, pp. 426–434 (2008)

CMMR: A Composite Multidimensional Models Robustness Evaluation Framework for Deep Learning

Wanyi Liu[1], Shigeng Zhang[1,3](✉), Weiping Wang[1], Jian Zhang[1], and Xuan Liu[2]

[1] School of Computer Science, Central South University, Changsha 410083, China
{wyliu,sgzhang,wpwang,jianzhang}@csu.edu.cn

[2] College of Computer Science and Electronic Engineering, Hunan University, Changsha 410002, China
xuan_liu@hnu.edu.cn

[3] Science and Technology on Parallel and Distributed Processing Laboratory (PDL), Changsha 410073, China

Abstract. Accurately evaluating the defense models against adversarial examples has been proven to be a challenging task. We have recognized the limitations of mainstream evaluation standards, which fail to account for the discrepancies in evaluation results arising from different adversarial attack methods, experimental setups, and metrics sets. To address these disparities, we propose the Composite Multidimensional Model Robustness (CMMR) evaluation framework, which integrates three evaluation dimensions: attack methods, experimental settings, and metrics sets. By comprehensively evaluating the model's robustness across these dimensions, we aim to effectively mitigate the aforementioned variations. Furthermore, the CMMR framework allows evaluators to flexibly define their own options for each evaluation dimension to meet their specific requirements. We provide practical examples to demonstrate how the CMMR framework can be utilized to assess the performance of models in enhancing robustness through various approaches. The reliability of our methodology is assessed through both practical examinations and theoretical validations. The experimental results demonstrate the excellent reliability of the CMMR framework and its significant reduction of variations encountered in evaluating model robustness in practical scenarios.

Keywords: Robustness evaluation · Adversarial attacks · Adversarial machine learning

1 Introduction

In recent years, with the deepening of deep learning models in research and practical applications, there has been rapid development in the field of deep learning models. Although deep learning models have been shown to be vulnerable to adversarial attacks [1], defense methods against them have also emerged [2–4].

Z. Tari et al. (Eds.): ICA3PP 2023, LNCS 14491, pp. 238–256, 2024.
https://doi.org/10.1007/978-981-97-0808-6_14

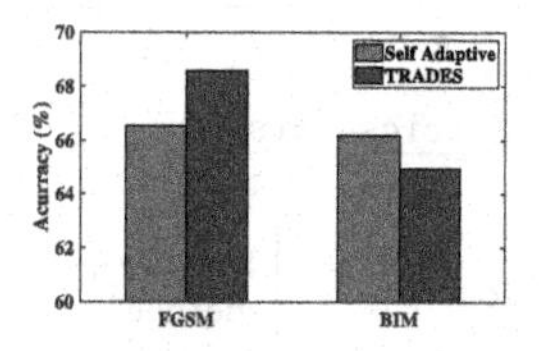

(a) The accuracy of Self Adaptive and TRADES under FGSM and BIM with $\epsilon = 0.03$

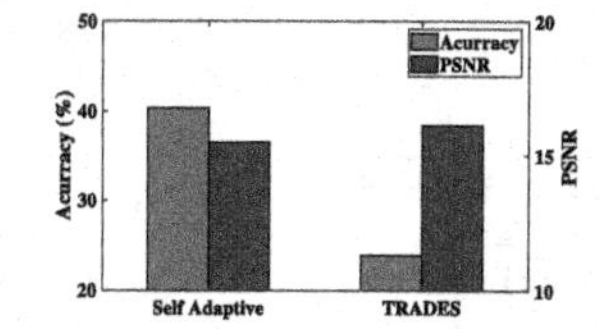

(b) The accuracy and PSNR of Self Adaptive and TRADES under FGSM with $\epsilon = 0.12$

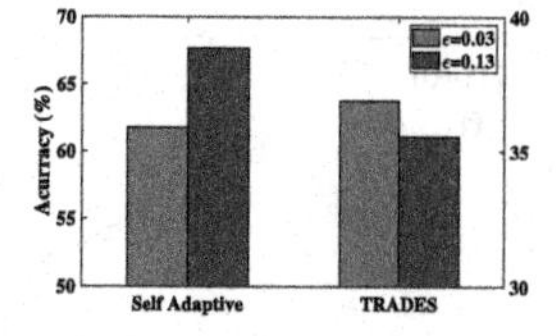

(c) The accuracy of Self Adaptive and TRADES under FGSM and BIM with $\epsilon = 0.03$ and $\epsilon = 0.13$

Fig. 1. Examples of evaluation under different adversarial attack methods, experimental settings, and metrics sets.

However, many proposed defense methods are quickly proven to have implemented incorrect or incomplete evaluations after their publication [5–7]. Therefore, we need a more comprehensive and accurate approach to evaluating the robustness of models.

In the current research on evaluating the robustness of deep learning models, several issues have been identified, including the following: inconsistent evaluation results due to the use of different attack methods, inconsistent evaluation results due to variations in the parameters set for the attack experiments, and inconsistent evaluation results due to the use of different evaluation metrics. We will now elaborate on each of these issues.

Attack Methods: It is common for researchers to propose new attack methods to circumvent newly developed defense methods, and subsequently, new defense methods are proposed to counter the previous attack methods, thus creating a continuous cycle. Therefore, evaluating the robustness of a deep learning model cannot rely solely on a single attack method. For instance, defense methods designed based on gradients can be easily defeated by attack methods that do not rely on gradient descent [8–10]. As shown in Fig. 1(a), models TRADES [4] and Self Adaptive [11] with the same perturbation budgets, the Self Adaptive achieves higher accuracy than the Trades model under the BIM, but its accuracy is lower than that of the TRADES model under the FGSM.

Experimental Parameters: When evaluating a model, it is generally assumed that the optimal parameters achieve the maximum attack success rate with the minimum attack cost, to observe the lower limit of model performance. However, for another attack method, the optimal experimental parameters may differ. Specifically, as shown in Fig. 1(c), at $\epsilon = 0.03$, TRADES exhibits higher classification accuracy than Self Adaptive under the FGSM attack method, but when $\epsilon = 0.13$, its performance is inferior to adaptive.

Evaluation Metrics: When assessing whether a model meets the user's requirements, it is necessary to select the metrics to be observed by the evaluated model and then assess the performance of these metrics. Currently, most evaluation metrics for model robustness only consider model classification accuracy, which is inadequate for defense methods. For instance, if a human observer can directly identify the differences between an adversarial sample and a normal sample, it

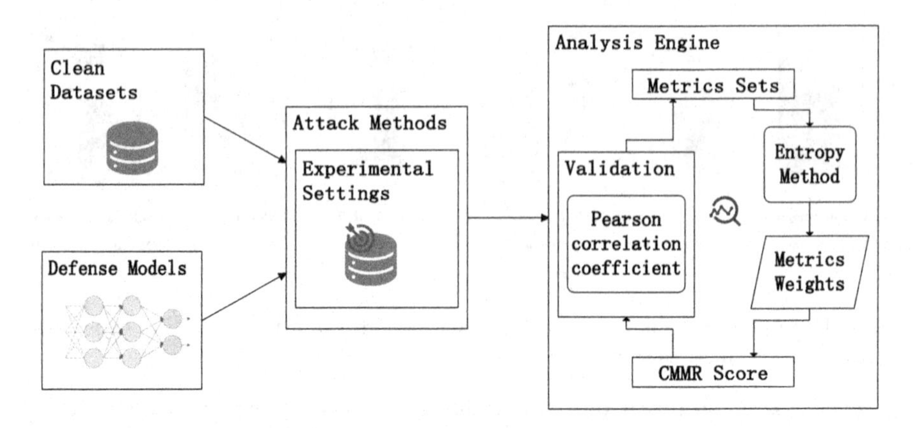

Fig. 2. Overview of CMMR. We confront clean data with defense models facing different types of adversarial attacks under different experimental parameter settings and finally obtain a multidimensional set of metric results. These metric result sets are fed into Analysis Engine to obtain our CMMR score.

can be avoided from being input into the model. Moreover, the choice of different evaluation metrics leads to different evaluation results. As shown in Fig. 1(b), The values of Acc and PSNR for both Self Adaptive and TRADES under FGSM attack with $\epsilon = 0.12$. It can be seen that Self Adaptive's accuracy is higher than TRADES' accuracy, but Self Adaptive's PSNR is lower than TRADES' PSNR.

Due to the diverse range of evaluation dimensions mentioned above, it is challenging to establish a definitive criterion for assessing model robustness. Without a unified standard, fair comparisons cannot be made. DEEPSEC, proposed by Ling et al. [12], evaluates the robustness of each defense as an average, rather than based on the most effective attack against that defense [13]. Dong et al. [7] introduced robustness curves, but they did not demonstrate the model's performance under multiple attacks. Wu et al. [14] proposed the PSC Framework to address the issue of result discrepancies caused by different experimental settings in model robustness evaluation. However, they also did not consider the scenario where the model is subjected to multiple adversarial sample attacks. To address these issues, we propose a multidimensional comprehensive robustness evaluation method to accurately, comprehensively, and holistically assess the robustness of models. Specifically, our method is shown in Fig. 2. The evaluation process includes the following four steps: In the first step, select the counterattack method, set the parameters of the attack method, and select the set of metrics for evaluation. The second step inputs the evaluated model. In the third step, the entropy weight method is designed to obtain the weight of each metric. The new metrics M are calculated based on the metric weights and metric values. In the fourth step, the metrics M under each attack method are calculated and equally weighted to obtain the final composite multidimensional model robustness evaluation score (CMMR). The CMMR is finally obtained to measure the robustness of the model under different attack methods, metrics, and experimental parameters.

Our contributions can be summarized as follows: We conducted extensive experiments to demonstrate the differences in model robustness under different attack methods, metrics, and experimental parameters. We analyze the factors that contribute to the discrepancies in model robustness evaluation and propose a Composite Multidimensional framework for evaluating Model Robustness (CMMR) in order to reduce the robustness evaluation discrepancies and provide a comprehensive and accurate assessment of model robustness. We selected two sets of adversarial training models with different perturbation budgets and observed and analyzed their robustness using our proposed method.

2 Attacks, Defense, and Metrics

In this section, we summarize typical adversarial attack methods, defense methods, and commonly used evaluation metrics.

2.1 Attack Methods

White-Box Attacks: White-box attack means that the attacker knows the parameters and structure of the models. Most white-box attack methods craft adversarial examples based on input gradients. The Fast Gradient Sign Method (FGSM) [15] is a classical single-step attack algorithm that calculates the perturbation value for adversarial attacks solely based on the sign of the gradient. The Basic Iterative Method (BIM) [16] is an iterative version built upon FGSM, also known as the Iterative Fast Gradient Sign Method (I-FGSM). In BIM, the approach involves taking multiple small steps instead of a single large step as in FGSM. Projected Gradient Descent (PGD) [17] is another extension of FGSM that replaces the single large step with multiple small steps. Carlini and Wagner [18] proposed a group of optimization-based adversarial attacks, known as C&W attacks, which can generate adversarial samples CW_0, CW_2, and CW_∞ under L_0, L_2, and L_∞ norm constraints, respectively. Deep Fool [19] is an attack method based on hyperplane classification, aiming to find the minimum perturbation that leads to misclassification. Momentum Iterative Attack (MIA) [20] integrates momentum into the BIM attack and derives a new attack iteration algorithm. Its essence lies in the fact that the current perturbation is not only dependent on the current gradient direction but also on previous gradient directions.

Black-Box Attacks: Transfer-based black-box attack: transfer-based attack craft adversarial examples against a substitute model against another unknown model with different parameters. The basic idea of SVRG [21] is to reduce the intrinsic variance of Stochastic Gradient Descent (SGD) using prediction variance reduction, while reducing the intrinsic gradient variance of multiple models. Object-based diverse input (ODI) method [22] is proposed, which expands objects to draw counter images on 3D objects and classifies rendered images as target classes. **Decision-based black-box attack**: in this setting, only the probabilities or logits of the model are provided. Boundary [23] is a method for decision-based black-box attacks that simulates local geometric shapes to search for directions, effectively reducing the dimensionality of the search space. **Score-based black-box attack**: refers to situations where the attacker has access to the predicted probabilities from the model's final layer. CG-attack [24], whose main idea is to develop a new adversarial transferable mechanism that is robust to agent bias. The "$\mathcal{N}$ attack" [25] method focuses on finding the probability density distribution within a small region centered around the input, allowing the sampling from this distribution to potentially yield adversarial examples without accessing the internal layers or weights of the DNN.

2.2 Defense Methods

The field of adversarial attacks and defenses can be seen as a game, where the continuous emergence of new attack methods leads to the development of corresponding defense techniques. In this section, we classify the defense techniques into two categories, including model and data perspectives. We will provide an overview of defense strategies from these two angles: the model and the data.

Model: The methods [26,27] were initially proposed to enhance the model's generalization ability and render it highly resilient to adversarial examples. It involves the utilization of defensive distillation to smooth the trained model during the training process. However, in 2017, Carlini and Wagner [18] declared the ineffectiveness of this method. Adversarial training [28,29] is another defense method in which noise is introduced and parameters are regularized to alter the model's parameters, thereby improving its robustness. However, Shafahi et al. [30] demonstrated that no matter how much adversarial training is performed, there will always exist adversarial examples capable of deceiving neural networks. APMSA [31], AID [32] as a model-assisted classifier that does not change the original model structure and assists in defending against adversarial attacks. In addition, there are adversarial examples detection methods [33] for detecting adversarial samples to avoid input models.

Data: Luo et al. [34] proposes a defense mechanism based on the foveation mechanism, which can defend against adversarial perturbations generated by L-BFGS and FGSM methods. The assumption behind this defense is that a CNN classifier trained on a large dataset is robust to image scaling and transformation variations. Xie et al. [35] discovered that introducing random resizing to training images can weaken the strength of adversarial attacks. Other methods include random padding and image augmentation during the training process.

2.3 Comparison Metrics

To ascertain whether a model satisfies the criteria set by evaluators, it is imperative to carefully select the metrics that will be monitored to evaluate the performance of the model under consideration. In the case of deep learning models, classification accuracy is undoubtedly a fundamental metric used for evaluating model performance. However, in practical applications, if an image undergoes substantial perturbations that render it easily identifiable to the human eye, subsequent evaluations become inconsequential. Therefore, supplementary metrics are employed to gauge the quality of images before and after they are subjected to adversarial attacks. This section presents an introduction to the chosen metrics.

Accuracy. If $\mathcal{A}_{\epsilon,p}$ represents an attack setting for generating adversarial examples with perturbation size ϵ under the ℓ_p, and $x^{adv} = \mathcal{A}_{\epsilon,p}(x)$ denotes the adversarial example generated from a clean sample x under this attack setting, C represents a model classifier with a defense method. Then, the classification accuracy of the model classifier C under adversarial attacks can be expressed as

$$ACC\left(C, \mathcal{A}_{,p}\right) = \frac{1}{N}\sum_{i=1}^{N} \mathbf{1}\left(C\left(\mathcal{A}_{,p}\left(\mathbf{x}_i\right)\right) = y_i\right), \tag{1}$$

where $\{\boldsymbol{x}_i, y_i\}_{i=1}^{N}$ is test set, $\mathbf{1}(\cdot)$ is the indicator function.

Average Structural Similarity (ASS). The change of ASS [36] before and after an image against attack is expressed as

$$ASS(\mathbf{x}, \mathcal{A}_{\epsilon,p}) = [l(\mathbf{x}, \mathbf{x^{adv}})]^{\alpha}[c(\mathbf{x}, \mathbf{x^{adv}}])]^{\beta}[s(\mathbf{x}, \mathbf{x^{adv}}])]^{\gamma}, \tag{2}$$

where $l()$ means luminance, $c()$ means contrast, and $s()$ means structure.

Mean Squared Error (MSE). x denotes the original image with dimensions $m * n$, and x^{adv}denotes the image obtained by subjecting it to an adversarial attack method. The average mean squared error (MSE) of a dataset can be mathematically expressed as

$$MSE\left(x, \mathcal{A}_{\epsilon,p}\right) = \frac{1}{N}\sum_{i=1}^{N}\left(\frac{1}{mn}\sum_{i=0}^{m-1}\sum_{j=0}^{n-1}[x(i,j) - x^{adv}(i,j)]^2\right). \tag{3}$$

Average L_2 Distortion (ALD_2) is used to measure the similarity of two images. The ALD_2 of the original dataset after the adversarial attack is expressed as

$$ALD_2(\mathbf{x}, \mathcal{A}_{\epsilon,p}) = \frac{1}{N}\sum_{i=1}^{N}(\|x_i - x_i^{adv}\|_2). \tag{4}$$

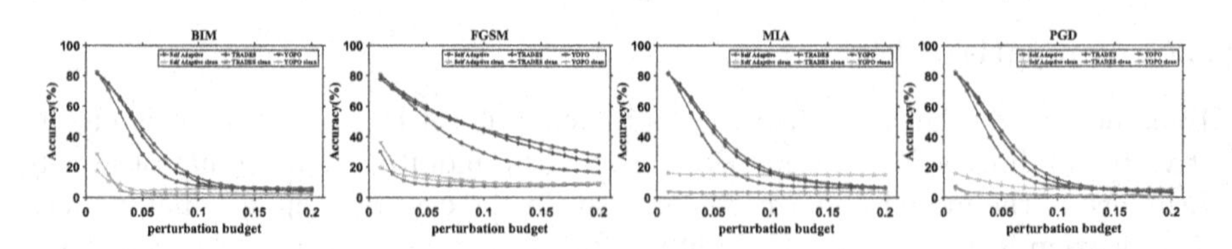

Fig. 3. The accuracy vs. perturbation budget curves of the 6 models on CIFAR-10 against untargeted white-box attacks under the ℓ_∞ norm.

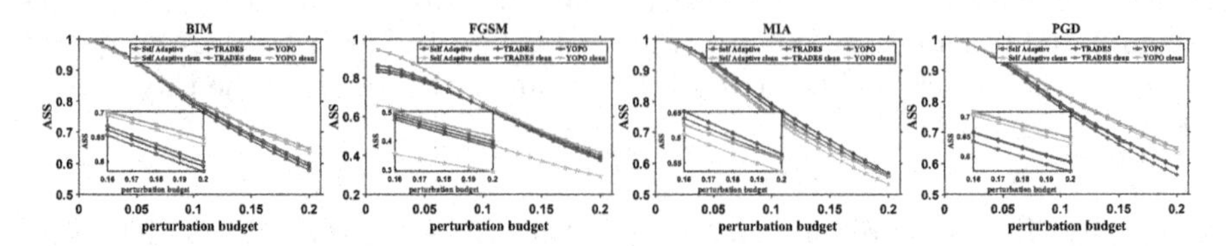

Fig. 4. The ASS vs. perturbation budget curves of the 6 models on CIFAR-10 against untargeted white-box attacks under the ℓ_∞ norm.

Peak Signal-to-Noise Ratio (PSNR). is one of the standards used to measure image quality. For color channels, the MSE values of the three RGB channels are calculated separately and then averaged to obtain the PSNR value, and the formula for calculating PSNR is

$$PSNR = 10 \cdot \log_{10}\left(\frac{MAX_I^2}{MSE}\right) = 20 \cdot \log_{10}\left(\frac{MAX_I}{\sqrt{MSE}}\right), \tag{5}$$

where MAX_I represents the maximum pixel value of the image.

3 Composite Multidimensional Model Robust Evaluation Method

The extensive deployment of deep learning models in practical applications underscores the critical importance of accurately evaluating the effectiveness of a classification model in the face of the continuous and profound development of adversarial attack methods. In this section, we will analyze how to evaluate the robustness of a model in three dimensions and introduce our Composite Multidimensional Model Robustness Evaluation Framework (CMMR).

3.1 Motivation

Some defense methods are specifically designed to counter a particular type of adversarial attack, but their defensive capabilities are greatly diminished against other attack methods. For instance, the original Fast Gradient Sign Method (FGSM) [15] generates adversarial samples based on gradient information. However, this method becomes ineffective when confronted with gradient masking

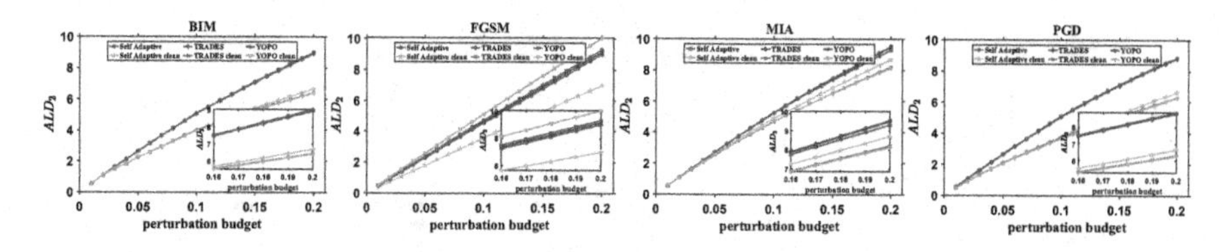

Fig. 5. The ALD_2 vs. perturbation budget curves of the 6 models on CIFAR-10 against untargeted white-box attacks under the ℓ_∞ norm.

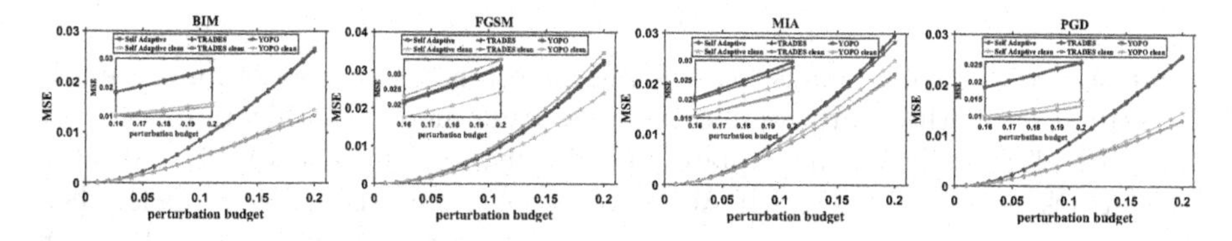

Fig. 6. The MSE vs. perturbation budget curves of the 6 models on CIFAR-10 against untargeted white-box attacks under the ℓ_∞ norm.

caused by simple adversarial training [37]. As indicated in Fig. 3 under FGSM attack, in terms of the achieved accuracy (Acc) in the experimental results, the Self Adaptive Robust model performs significantly better than the other three models under FGSM attacks. This suggests that the defense method employed by Self Adaptive effectively evades FGSM attacks. During the training process, Self Adaptive dynamically corrects mislabeled samples based on the model's predictions. Since FGSM is a simple single-step gradient-based attack, the defense mechanism of Self Adaptive can successfully evade such attacks and achieve a higher accuracy rate. Therefore, in order to comprehensively evaluate the robustness of a model, it is advisable to consider employing multiple adversarial attack methods for model evaluation.

Considering the second scenario, the experimental results vary even for models under the same attack method with different adversarial perturbation size settings. Specifically, as shown in Fig. 3 under FGSM attack, the TRADES model consistently outperforms the Self Adaptive model when $\epsilon < 0.09$. However, when $\epsilon > 0.09$, the performance gap between TRADES and Self Adaptive models increases with the increase of perturbation budget, and the Self Adaptive model obtains higher accuracy (Acc). In order to comprehensively evaluate the performance of the model under different perturbation strengths, the second feature of this study's methodology is to evaluate the model in multiple experimental environments. More precisely, we compare different adversaries in a specific comparison range $\epsilon = [0.01, 0.2]$ with an interval of 0.1, under the same attack method, resulting in a total of 20 parameter settings.

Furthermore, the concept of "robustness" refers to the capacity of a system to maintain specific performance characteristics when subjected to perturbations in certain parameters (such as structure or magnitude) [38]. Robustness plays a vital role in ensuring the system's survival in abnormal and hazardous

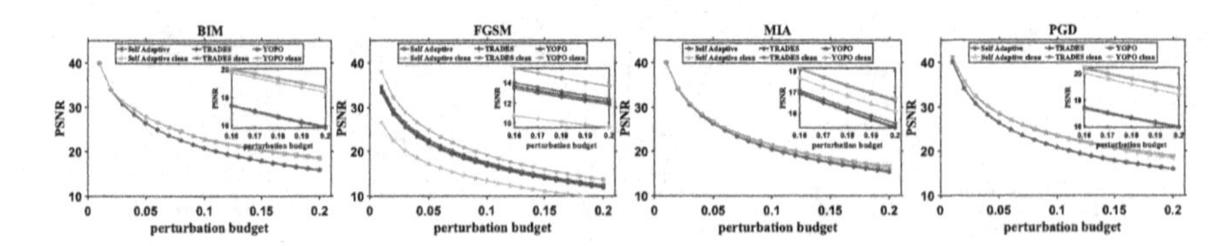

Fig. 7. The PSNR vs. perturbation budget curves of the 6 models on CIFAR-10 against untargeted white-box attacks under the ℓ_∞ norm.

circumstances. When it comes to assessing the robustness of models, many existing evaluation methods are confined to measuring model classification accuracy [7,39], which is evidently inadequate. Specifically, in Fig. 3 and Fig. 4, under the MIA attack with $\epsilon = 0.05$, Self Adaptive outperforms the other models when classification accuracy is used as the evaluation criterion, while the TRADES model outperforms the other models when ASS is used as the evaluation criterion. Therefore, in order to comprehensively evaluate the robustness of the model, we chose multiple evaluation metrics to assess the performance of the model under different attack methods.

In this study, we assume that all adversaries possess the capability to apply the maximum perturbation to attack the model, in order to observe the model's performance under the strongest attack method.

3.2 CMMR Framework

Getting Metrics Weights by Entropy Weight Method. For the selected multiple metrics, different evaluators find it challenging to make consistent judgments regarding the relative importance of each metric. Therefore, we employ the entropy weighting method to objectively assign weights to the selected metrics. The entropy weighting method was initially proposed within Shannon's formula [40]. In the field of statistics, it is widely acknowledged that as data becomes more dispersed, the entropy value decreases, indicating greater importance of the corresponding metric. This concept is also applicable in the context of adversarial attacks. For example, when evaluating models, if the variation of metric A is relatively small among all models under the same attack and at the same attack intensity, selecting it as part of the evaluation result would have minimal impact on the final evaluation outcome. Conversely, if metric B exhibits a significant variation, it would have a substantial influence on the final evaluation result. The following are the detailed steps of the improved entropy weighting method used in this approach.

Step 1: Define the evaluation target and establish the evaluation metric system. Construct the preference matrix R', where the horizontal vectors of R' represent the set of evaluation metrics. R'_1 to R'_5 represent $Acc, ASS, MSE, ALD_2, PSNR$, respectively. Additionally, an extra metric R'_6 is included to balance the importance of selecting metric weights. The column vector represents the values of these metrics under the setting $\epsilon = [0.01, 0.2]$. For example, R'_{11}represents the value of Acc under attack method A with $\epsilon = 0.01$.

Step 2: Normalize the preference matrix to obtain R.

Step 3: Calculate the entropy value for each matrix using the formula

$$H_j = -k \sum_{i=1}^{m} f_{ij} \cdot \ln f_{ij}, \tag{6}$$

where $f_{ij} = \frac{r_{ij}}{\sum_{i=1}^{m} r_{ij}}, k = \frac{1}{\ln m}$, f_{ij}represents the weight of the i parameter setting's metric value under the j metric.

Step 4: Determine the weight of each metric as

$$\lambda_j = \frac{\lambda'_j w_j}{\sum_{j=1}^{n} \lambda'_j w_j}, \tag{7}$$

where $w_j = \frac{1-H_j}{\sum_{j=1}^{n}(1-H_j)}$ denotes the entropy weight of the j metric.

Step 5: Based on the weights assigned to each indicator, a new evaluation indicator M,indicator m is calculated as

$$M = \sum_{j=1}^{n} R \cdot w_j. \tag{8}$$

Finally, employ Pearson's coefficients to analyze the reliability and consistency between M and R_i. Adjust the weights iteratively until the three coefficients reach their maximum values, indicating that the final evaluation metric is reliable.

Synthesizing Adversarial Attacks by Equivalent Weighting. Assuming a lack of prior knowledge regarding the evaluated model's defense mechanisms against specific types of attacks, the model's susceptibility to any attack is considered equally probable. Prior to this, we computed the value of M for each adversarial attack method using Eq. 8. Following that, equal weights are assigned to the integrated metrics, M, for each adversarial attack method. The resulting values are then used to calculate the Comprehensive Multidimensional Model Robustness (CMMR) Score, which serves as a combined assessment score for evaluating the robustness of the model. The formula for calculating the CMMR Score is as

$$CMMR = \sum_{j=1}^{N} M \cdot w_j, w_j = \frac{1}{N}, \tag{9}$$

where N represents the total number of adversarial attack methods employed against the model.

Table 1. We show the structure of the defense models and their clean models that were incorporated into our adversarial robustness evaluation framework. We also show the original threat models (i.e., the threat models in the original paper where the defense system was trained to be robust or evaluated;), and the accuracy (%) of each method on clean data. The accuracies are recalculated by ourselves.

Defense Method	Model	Intended Threat	Clean Acc.
TRADES	WRN-34-10	L(=0.031)	84.59
YOPO	WRN-34-10	L(=0.03)	86.8
Self-adaptive	WRN-34-10	L(=0.031)	83.48
TRADES Natural	WRN-34-10	-	94.93
YOPO Natural	WRN-34-10	-	95.05
Self-adaptive Natural	WRN-34-10	-	66.33
FAT_MART_62	WRN-28-10	L(=16/255)	80.64
FAT_TRADES_62	WRN-34-10	L(=16/255)	82.41
FAT_MART_031	WRN-28-10	L(=8/255)	90.56
FAT_TRADES_031	WRN-34-10	L(=8/255)	89.44

Table 2. The Summary of Notations

Attack Method	Description	Utility Metrics	Description
FGSM	Fast Gradient Sign Method	ASS	Average Structural Similarity
BIM	Basic Iterative Method	PSNR	Peak Signal to Noise Ratio
PGD	Projected L Gradient Descent attack	MSE	Mean Square Error
MIA	Momentum Iterative Attack	ALDp	Average Lp Distortion

Curving CMMR Scores. If the model consistently exhibits superior performance compared to the adversary across the entire spectrum of perturbations, the problem can be considered straightforward. However, evaluations frequently exhibit intersections at specific points. To achieve a more comprehensive assessment of the model's robustness performance, we utilize the dimensionality reduction technique mentioned earlier to reduce the evaluation results from three dimensions to a single dimension. Subsequently, we plot the aggregated scores of diverse models at varying levels of perturbation intensity to gain insights into a more comprehensive evaluation outcome.

4 Experimental Analysis for CMMR

In Sect. 3, we introduced the steps of the Composite Multidimensional Models Robustness Evaluation method (CMMR). In this section, we will employ the CMMR method to demonstrate the process of evaluating model performance from three dimensions to CMMR. Finally, we will show the validation of the CMMR method from practical and theoretical aspects respectively.

Dimension. Given that the evaluation of model robustness mentioned in the introduction requires multiple dimensions, including adversarial attack methods, evaluation metric sets, and experimental parameter settings, we will visually demonstrate the process of evaluating model robustness in three dimensions, as well as the CMMR evaluation process of our method, in order to demonstrate the intuitiveness, wholeness, and correctness of our method.

Setting. We selected five groups of classification models under the CIFAR10 dataset, three of which were used as baseline comparisons, including the original model and the model with the defense method applied. The other two groups consist of models trained under different levels of adversarial perturbations using each of the two defense methods. Table 1 provides details of the defense models, the structure of the models, the budget of the adversarially trained perturbations, and the clean accuracy.

Validation. The CMMR validation consists of two steps: Practical validation and Theoretical validation. In the practical validation step, we compare the results of models with applied defense methods and models without defense methods, using CMMR, to determine if the outcomes align with real-world scenarios. In the theoretical validation part, we assess the reliability of the reduced data through Kendall's coefficient of concordance, as well as examine the correlation between the reduced data and the original set of metrics using Pearson's coefficient.

Acronyms and Notations. For convenient reference, we summarize the acronyms and notations in Table 2.

4.1 3-Dimension Models Robustness Evaluation.

Figures 3, 4, 5, 6, and 7 presents line graphs that demonstrate the relationship between the evaluation metric sets Acc, ASS, ALD_2, MSE, and $PSNR$ and the perturbation budget curves for three comparative models. The graphs are arranged from top to bottom and represent the performance of the models under non-targeted attacks such as FGSM, BIM, PGD, and MIA.

Acc. As the perturbation budget increases, both the natural models and the defense models experience a gradual decrease in classification accuracy. However, there is a difference in behavior. The classification accuracy of the model without defense methods significantly drops to its lowest point when the perturbation is small, and then only slightly decreases as the perturbation size increases. In contrast, robust models with the same structure maintain relatively high classification accuracy even with small perturbation budgets. Interestingly, the selected defense methods are all effective in defending against the FGSM attack. Even at the maximum perturbation budget, the model's classification accuracy remains better than the corresponding natural model. However, it is worth noting that the FGSM attack was not specifically designed to evaluate model robustness against strong attacks [41]. Another interesting observation is that although the Self Adaptive natural model exhibits a significant decrease in classification accuracy at small perturbation budgets, there is no significant change in accuracy

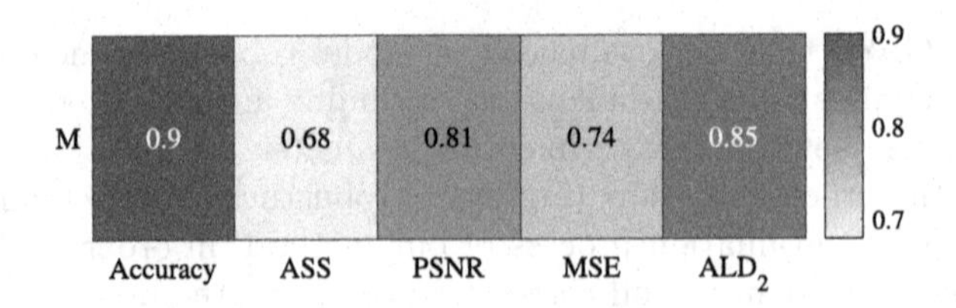

Fig. 8. The Pearson correlation coefficients of the metric M with the five metrics.

as the perturbation budget increases. In fact, Self Adaptive clean model outperforms even most defense models for $\epsilon > 0.11$ under FGSM attacks. Additionally, we can observe that for the BIM attack method, the YOPO model demonstrates the highest robustness at $\epsilon > 0.13$. However, within this perturbation budget range, the YOPO model may not be the most robust against the other attack methods.

ASS. Among the different adversarial attack methods, the ranking of the three robust classification models in terms of $SSIM$ under the FGSM attack differs. When the perturbation size $\epsilon < 0.1$, the model rankings, from best to worst, are as follows: Self Adaptive, Trades, YOPO. At $\epsilon = 0.1$, the rankings are nearly the same, but as the perturbation size increases, the rankings change to YOPO, Trades, Self Adaptive, moving in the opposite direction. For the other adversarial attack methods, the model rankings do not change with the perturbation size.

$PSNR$. We are aware that a higher PSNR indicates better image quality, and it aligns with prior knowledge that the $PSNR$ of all models decreases as the perturbation budget increases. However, as the perturbation size increases, there is typically an increasing gap in $PSNR$ between natural models and robust models, with the curves of the robust models positioned below those of the natural models. Does this imply that natural models are less vulnerable to attacks compared to robust models? Not necessarily. This observation suggests that adversarially trained models often require more substantial changes in the image to induce misclassifications when confronted with adversarial attacks of the same perturbation size. Similar to the classification accuracy findings, the Self Adaptive clean model demonstrates distinct behavior compared to other robust models under the FGSM attack.

ALD_2. Compared to the aforementioned three metrics, its variation with respect to the perturbation budget is more consistent, meaning the curve of this metric is closer to a straight line.

MSE. The difference in MSE between the model groups increases with the perturbation size. However, under the FGSM attack, the MSE of the Self Adaptive clean model is lower than that of all models. For the other adversarial attack methods, the MSE curve of the Self Adaptive model lies between the curves of the robust models and the clean model.

The previous section showed the values of the five metric sets for the three comparison models under four attacks, and in this section, we show the process

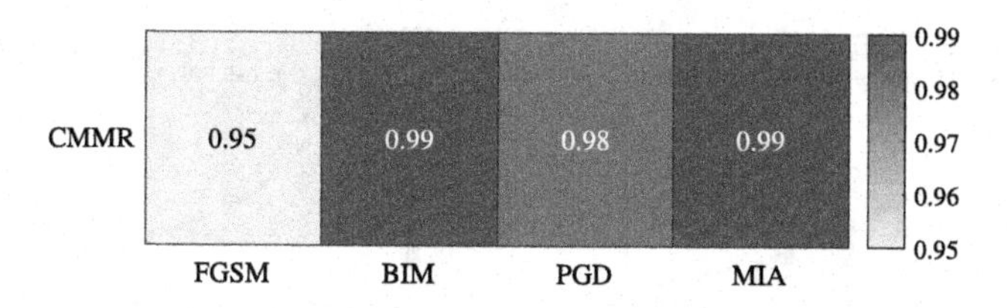

Fig. 9. The Pearson correlation coefficients of CMMR with each model indicator M under each attack method.

Table 3. The weights of each metric are obtained by the entropy weighting method

Metric	Weights Value
Accuracy	0.49
ASS	0.12
PSNR	0.12
MSE	0.12
ALD_2	0.07
w	0.08

of evaluating the three comparison models using the CMMR method based on the above test results. Based on the test results, the weights of each metric are obtained by the entropy weighting method, where ω is the hyperparameter used to balance the importance of each weight. We set $\omega = 0.08$, and the calculated weights of each index are shown in Table 3 to obtain the composite metric M. Then we calculate the composite metric M under FGSM, BIM, PGD, and MIA respectively, and assign them with equal weights to obtain CMMR.

Figure 10 shows the CMMR scores of TRADES, Self Adaptive, Yopo model, and their clean model. It is evident that the robust model consistently outperforms the clean model in terms of CMMR scores. Furthermore, as the perturbation magnitude increases, the disparity in CMMR scores between the clean and robust models diminishes, leading to score convergence. While the CMMR score of the Self Adaptive clean model surpasses that of the Self Adaptive robust model at $\epsilon = 0.13$, the difference is not statistically significant. Interestingly, empirical observations demonstrate that in the presence of attacks like BIM, PGD, and MIA, the accuracy of the Self Adaptive clean model at $\epsilon = 0.13$ exceeds that of the Self Adaptive robust model, thus validating the observations.

In a subsequent stage, we performed a theoretical validation. We used the Pearson correlation coefficient to test the correlation between the data since the metric M value and the five metrics two-by-two satisfy the following conditions:

- The relationship between the two variables is linear and both are continuous data.
- The overall distribution of the two variables is normal or near-normal with a single-peaked distribution.

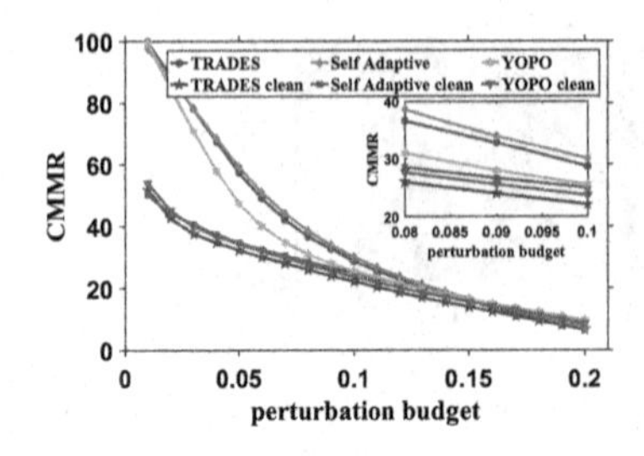

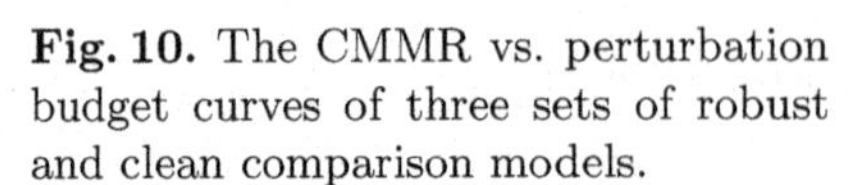
Fig. 10. The CMMR vs. perturbation budget curves of three sets of robust and clean comparison models.

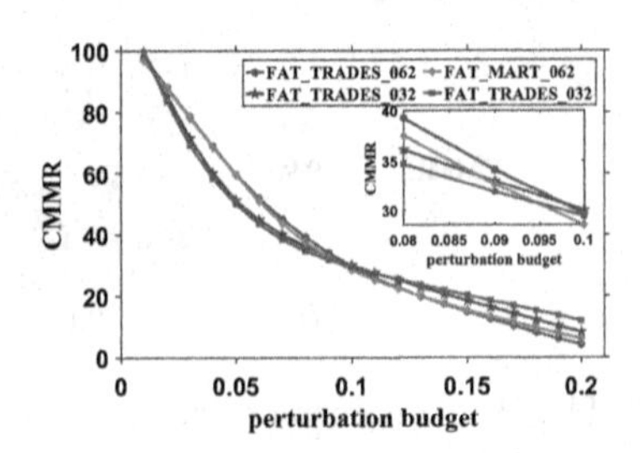

Fig. 11. The CMMR vs. perturbation budget curves of two sets of robust models trained by different perturbation budget.

- The observations of the two variables are paired, and each pair of observations is independent of each other.

Figure 8 shows the Pearson correlation coefficients of the metric M with the five metrics, all of which are greater than 0.6, indicating that metric M is strongly correlated with the five metrics of the original measure. Similarly, Fig. 9 shows the Pearson correlation coefficients of CMMR with each model indicator M under each attack method. Since the assumption is that the model suffers from the same probability of each type of attack, the difference between the metric M and the final CMMR under each attack method is not significant and all are greater than 0.9, which is strongly correlated. Therefore, we can consider that in theory the CMMR represents the performance of the model exhibited by the combined selected metrics.

4.2 Evaluating the Robustness of Two Sets of Models by CMMR

The robustness of the robust and clean models has been analyzed above. Now, let's examine the robustness performance of models trained with different perturbation magnitudes in adversarial training. We have selected two sets of models trained with different perturbation magnitudes for this study. The first set includes FAT for TRADES models trained with adversarial perturbations at $\epsilon = 8/255$ and $\epsilon = 16/255$, denoted as FAT_TRADES_031 and FAT_TRADES_062, respectively. The second set includes FAT for MART models trained with adversarial perturbations at $\epsilon = 8/255$ and $\epsilon = 16/255$, denoted as FAT_MART_031 and FAT_MART_062, respectively.

Figure 11 shows the relationship between the CMMR scores of the two groups of models and the perturbation budget. It is evident that the robustness relationships of the FAT for TRADES group and the FAT for MART group are not consistently aligned. The red curve corresponds to the FAT for MART group, which exhibits two inflection points in its robustness relationship. For $\epsilon < 0.01$, the CMMR of FAT_TRADES_031 is higher than that of FAT_TRADES_062. However, as epsilon increases,

FAT_TRADES_062 surpasses FAT_TRADES_031 until reaching $\epsilon = 0.95$. Subsequently, for epsilon values greater than 0.06, FAT_TRADES_031 outperforms FAT_TRADES_062 until $\epsilon = 0.2$. On the other hand, the blue curve represents the FAT for TRADES group, which also displays two inflection points in its robustness relationship. For $\epsilon < 0.01$, the CMMR of FAT_MART_031 is higher than that of FAT_MART_062. Similar to the previous group, as epsilon increases, FAT_MART_062 surpasses FAT_MART_031 until reaching $\epsilon = 0.1$. Once again, for $\epsilon > 0.098$, FAT_MART_031 outperforms FAT_MART_062 until $\epsilon = 0.2$, and this pattern persists. Therefore, we can draw the conclusion that for models trained through adversarial training to enhance robustness, the robustness of the models is not directly correlated with the perturbation magnitude used during adversarial training. It is not necessarily the case that larger perturbations lead to better robustness.

5 Conclusion

In this study, we propose a Composite Multi-dimensional Robustness (CMMR) score to evaluate the robustness of models from multiple dimensions, including adversarial attack methods, selected metrics, and experimental settings. Currently, there is no unified framework in the literature for comprehensive multi-dimensional evaluation. The computation of the CMMR score involves three main steps, all of which can be standardized. We provide examples of how to assess the performance of models that enhance robustness in different ways. To ensure its reliability, we employ three coefficients that measure the consistency and reliability of the evaluation data. Evaluators also have the flexibility to define their own options for each evaluation dimension to meet their specific requirements. We believe that this approach will standardize and expedite the equitable comparison of model robustness.

Acknowledgements. This work was supported in part by the National Key Research and Development Program of China under Grant 2022YFC3400404 and Grant 2021YFB3101201, the National Science Foundation of China under Grant 62172154 and 62372473. the Hunan Provincial Natural Science Foundation of China under Grant 2023JJ30702 and the Changsha Municipal Natural Science Foundation under Grant kq2208283. The authors are grateful for resources from the High Performance Computing Center of Central South University.

References

1. Szegedy, C., et al.: Intriguing properties of neural networks. arXiv preprint arXiv:1312.6199 (2013)
2. Jia, X., Zhang, Y., Wu, B., Ma, K., Wang, J., Cao, X.: LAS-AT: adversarial training with learnable attack strategy. In: Proceedings of the IEEE/CVF Conference on Computer Vision and Pattern Recognition, pp. 13398–13408 (2022)
3. Jin, G., Yi, X., Huang, W., Schewe, S., Huang, X.: Enhancing adversarial training with second-order statistics of weights. In: Proceedings of the IEEE/CVF Conference on Computer Vision and Pattern Recognition, pp. 15273–15283 (2022)

4. Zhang, H., Yu, Y., Jiao, J., Xing, E., El Ghaoui, L., Jordan, M.: Theoretically principled trade-off between robustness and accuracy. In: International Conference on Machine Learning, pp. 7472–7482. PMLR (2019)
5. Carlini, N., Wagner, D.: Defensive distillation is not robust to adversarial examples. arXiv preprint arXiv:1607.04311 (2016)
6. Cornelius, C., Das, N., Chen, S.-T., Chen, L., Kounavis, M.E., Chau, D.H.: The efficacy of shield under different threat models. arXiv preprint arXiv:1902.00541 (2019)
7. Dong, Y., et al.: Benchmarking adversarial robustness on image classification. In: Proceedings of the IEEE/CVF Conference on Computer Vision and Pattern Recognition, pp. 321–331 (2020)
8. Uesato, J., O'donoghue, B., Kohli, P., Oord, A.: Adversarial risk and the dangers of evaluating against weak attacks. In: International Conference on Machine Learning, pp. 5025–5034. PMLR (2018)
9. Mohammadian, H., Ghorbani, A.A., Lashkari, A.H.: A gradient-based approach for adversarial attack on deep learning-based network intrusion detection systems. Appl. Soft Comput. **137**, 110173 (2023)
10. Zhang, Y., Wang, C., Shi, Q., Feng, Y., Chen, C.: Adversarial gradient-based meta learning with metric-based test. Knowl. Based Syst. **263**, 110312 (2023)
11. Huang, L., Zhang, C., Zhang, H.: Self-adaptive training: beyond empirical risk minimization. Adv. Neural. Inf. Process. Syst. **33**, 19365–19376 (2020)
12. Ling, X., et al.: DEEPSEC: a uniform platform for security analysis of deep learning model. In: 2019 IEEE Symposium on Security and Privacy (SP), pp. 673–690. IEEE (2019)
13. Carlini, N.: A critique of the DeepSec platform for security analysis of deep learning models. arXiv preprint arXiv:1905.07112 (2019)
14. Wu, J., Zhou, M., Zhu, C., Liu, Y., Harandi, M., Li, L.: Performance evaluation of adversarial attacks: discrepancies and solutions. arXiv preprint arXiv:2104.11103 (2021)
15. Goodfellow, I.J., Shlens, J., Szegedy, C.: Explaining and harnessing adversarial examples. arXiv preprint arXiv:1412.6572 (2014)
16. Kurakin, A., Goodfellow, I.J., Bengio, S.: Adversarial examples in the physical world. In: Artificial Intelligence Safety and Security, pp. 99–112. Chapman and Hall/CRC (2018)
17. Madry, A., Makelov, A., Schmidt, L., Tsipras, D., Vladu, A.: Towards deep learning models resistant to adversarial attacks. arXiv preprint arXiv:1706.06083 (2017)
18. Carlini, N., Wagner, D: Towards evaluating the robustness of neural networks. In: 2017 IEEE Symposium on Security and Privacy (SP), pp. 39–57. IEEE (2017)
19. Moosavi-Dezfooli, S.-M., Fawzi, A., Frossard, P.: DeepFool: a simple and accurate method to fool deep neural networks. In: Proceedings of the IEEE Conference on Computer Vision and Pattern Recognition, pp. 2574–2582 (2016)
20. Dong, Y., et al.: Boosting adversarial attacks with momentum. In: Proceedings of the IEEE Conference on Computer Vision and Pattern Recognition, pp. 9185–9193 (2018)
21. Xiong, Y., Lin, J., Zhang, M., Hopcroft, J.E., He, K.: Stochastic variance reduced ensemble adversarial attack for boosting the adversarial transferability. In: Proceedings of the IEEE/CVF Conference on Computer Vision and Pattern Recognition, pp. 14983–14992 (2022)

22. Byun, J., Cho, S., Kwon, M.-J., Kim, H.-S., Kim, C.: Improving the transferability of targeted adversarial examples through object-based diverse input. In: Proceedings of the IEEE/CVF Conference on Computer Vision and Pattern Recognition, pp. 15244–15253 (2022)
23. Brendel, W., Rauber, J., Bethge, M.: Decision-based adversarial attacks: reliable attacks against black-box machine learning models. arXiv preprint arXiv:1712.04248 (2017)
24. Feng, Y., Wu, B., Fan, Y., Liu, L., Li, Z., Xia, S.-T.: Boosting black-box attack with partially transferred conditional adversarial distribution. In: Proceedings of the IEEE/CVF Conference on Computer Vision and Pattern Recognition, pp. 15095–15104 (2022)
25. Piplai, A., Chukkapalli, S.S.L., Joshi, A.: NAttack! Adversarial attacks to bypass a GAN based classifier trained to detect network intrusion. In: 2020 IEEE 6th International Conference on Big Data Security on Cloud (BigDataSecurity), IEEE International Conference on High Performance and Smart Computing, (HPSC) and IEEE International Conference on Intelligent Data and Security (IDS), pp. 49–54. IEEE (2020)
26. Ba, J., Caruana, R.: Do deep nets really need to be deep? In: Advances in Neural Information Processing Systems, vol. 27 (2014)
27. Hinton, G., Vinyals, O., Dean, J.: Distilling the knowledge in a neural network. arXiv preprint arXiv:1503.02531 (2015)
28. Zhou, J., et al.: Adversarial training with complementary labels: on the benefit of gradually informative attacks. arXiv preprint arXiv:2211.00269 (2022)
29. Niu, Z.-H., Yang, Y.-B.: Defense against adversarial attacks with efficient frequency-adaptive compression and reconstruction. Pattern Recogn. **138**, 109382 (2023)
30. Shafahi, A., Najibi, M., Xu, Z., Dickerson, J., Davis, L.S., Goldstein, T.: Universal adversarial training. In: Proceedings of the AAAI Conference on Artificial Intelligence, vol. 34, pp. 5636–5643 (2020)
31. Khan, M.A., et al.: A resource-friendly certificateless proxy signcryption scheme for drones in networks beyond 5G. Drones **7**(5), 321 (2023)
32. Hwang, D., Lee, E., Rhee, W.: AID-purifier: a light auxiliary network for boosting adversarial defense. Neurocomputing **541**, 126251 (2023)
33. Wei, J., Yao, L., Meng, Q.: Self-adaptive logit balancing for deep neural network robustness: defence and detection of adversarial attacks. Neurocomputing **531**, 180–194 (2023)
34. Luo, Y., Boix, X., Roig, G., Poggio, T., Zhao, Q.: Foveation-based mechanisms alleviate adversarial examples. arXiv preprint arXiv:1511.06292 (2015)
35. Xie, C., Wang, J., Zhang, Z., Zhou, Y., Xie, L., Yuille, A.: Adversarial examples for semantic segmentation and object detection. In: Proceedings of the IEEE International Conference on Computer Vision, pp. 1369–1378 (2017)
36. Wang, Z., Bovik, A.C., Sheikh, H.R., Simoncelli, E.P.: Image quality assessment: from error visibility to structural similarity. IEEE Trans. Image Process. **13**(4), 600–612 (2004)
37. Tramèr, F., Kurakin, A., Papernot, N., Goodfellow, I., Boneh, D., McDaniel, P.: Ensemble adversarial training: attacks and defenses. arXiv preprint arXiv:1705.07204 (2017)
38. Rousseeuw, P.J., Hampel, F.R., Ronchetti, E.M., Stahel, W.A.: Robust Statistics: The Approach Based on Influence Functions. Wiley (2011)

39. Liu, Y., Cheng, Y., Gao, L., Liu, X., Zhang, Q., Song, J.: Practical evaluation of adversarial robustness via adaptive auto attack. In: Proceedings of the IEEE/CVF Conference on Computer Vision and Pattern Recognition, pp. 15105–15114 (2022)
40. Shannon, C.E.: A mathematical theory of communication. ACM SIGMOBILE Mob. Comput. Commun. Rev. **5**(1), 3–55 (2001)
41. Carlini, N., et al.: On evaluating adversarial robustness. arXiv preprint arXiv:1902.06705 (2019)

Efficient Black-Box Adversarial Attacks with Training Surrogate Models Towards Speaker Recognition Systems

Fangwei Wang[1,2], Ruixin Song[2], Qingru Li[1,2], and Changguang Wang[1,2](✉)

[1] Key Laboratory of Network and Information Security of Hebei Province, Hebei Normal University, Shijiazhuang 050024, China
{qingruli,wangcg}@hebtu.edu.cn

[2] College of Computer and Cyberspace Security, Hebei Normal University, Shijiazhuang 050024, China

Abstract. Speaker Recognition Systems (SRSs) are gradually introducing Deep Neural Networks (DNNs) as their core architecture, while attackers exploit the weakness of DNNs to launch adversarial attacks. Previous studies generate adversarial examples by injecting the human-imperceptible noise into the gradients of audio data, which is termed as white-box attacks. However, these attacks are impractical in real-world scenarios because they have a high dependency on the internal information of the target classifier. To address this constraint, this study proposes a method applying in a black-box condition which only permits the attacker to estimate the internal information by interacting with the model through its inputs and outputs. We use the idea of the substitution-based method and transfer-based method to train various surrogate models for imitating the target models. Our methods combine the surrogate models with white-box methods like Momentum Iterative Fast Gradient Sign Method (MI-FGSM) and Enhanced Momentum Iterative Fast Gradient Sign Method (EMI-FGSM) to boost the performance of the adversarial attacks. Furthermore, a transferability analysis is conducted on multiple models under cross-architecture, cross-feature and cross-architecture-feature conditions. Additionally, frequency analysis also provides us with valuable findings about adjusting the parameters in attack algorithms. Massive experiments validate that our attack yields a prominent performance compared to previous studies.

Keywords: Speaker Recognition Systems · Adversarial Example · Adversarial Attack · Information Security

1 Introduction

Nowadays, speaker recognition has become one of the mainstream approaches for user verification. This technique uses the acoustic features extracted from

Z. Tari et al. (Eds.): ICA3PP 2023, LNCS 14491, pp. 257–276, 2024.
https://doi.org/10.1007/978-981-97-0808-6_15

the voices to distinguish different people, which have been widely adopted in a variety of scenarios, ranging from smart homes to automatic driving and commercial applications [1]. There is no doubt that SRSs are ubiquitous due to their advancements in identity authentication.

With the rapid development of DNNs, acoustic systems have begun to utilize DNNs-powered models as the most important internal module in the procedure of speaker recognition. But DNNs have been proved that they have vulnerabilities to tiny perturbations [2]. So it is undeniable that DNNs-based SRSs are fragile in the face of adversarial examples, which are constructed by injecting human-imperceptible perturbations into clean audio files. Attackers launch adversarial attacks with these polluted examples to manipulate the models into giving wrong predictions.

For adversarial attacks, previous studies divide them into three categories: white-box setting [3–5], gray-box setting [6,7] and black-box setting [8–12]. As the name implies, if the target model is completely transparent to the adversary, this kind of attack is called a white-box attack. Since the adversary can get access to full information about the target model, the success rate of this attack is significantly higher. However, this scenario is not very realistic because almost all companies do not publish the internal details of their systems. On the contrary, black-box setting is more challenging which the attackers cannot get any knowledge of the target models. The attacker can only query the model to obtain the prediction labels or confidence scores and analyze the information to adjust the parameters in the procedure of generating adversarial examples. If the attacker has limited partial knowledge about the victim systems, the attack can be performed in a gray-box environment. Furthermore, adversarial attacks can also be classified into targeted attack and untargeted attack. The untargeted attack only cares about whether the target model gives a false recognition result, while the targeted attacks require the output label of the target model to be the specified class.

There is no surprise that the black-box attack is much more difficult compared to other attack methods. Existing studies have proposed many methods for black-box attacks against SRSs, such as combining transfer attack and query attack [8], using optimization-based algorithms [9], designing an adversarial perturbation generator [12], etc. The experimental results have shown under the condition that critical information such as the internal architecture, parameters, and gradients of the target model is not available, the attack success rate is rather limited. Some attacks damage the audio files seriously to get a high success rate of attack. However, the quality of the audio adversarial examples may not be guaranteed, and people will notice the presence of malicious disturbances. In addition, attacks based on plenty of queries will make the attacks detected by victim models.

To overcome the above-mentioned drawbacks, this study proposes a novel black-box attack combining transfer-based methods and substitution-based methods. In order to improve the success rate of attack and the quality of speech, we are inspired by the idea of model-stealing attacks [13–16] in the image domain. Interaction with the target model involves obtaining its inputs and outputs. And using these information helps develop a surrogate model capable of imitating the real target model to some extent. If this surrogate model fits well the target model, better attack results will be obtained. In addition, excellent white-box methods such as MI-FGSM [17] and EMI-FGSM [18] are applied to our work which could considerably increase the success rate of attack. We briefly summarize the following contributions.

1. An idea of model-stealing attacks to train surrogate models for simulating target models is adapted. Combining these well-trained models with advanced gradient-based methods helps construct adversarial examples.
2. A transferability analysis of adversarial examples generated on four state-of-the-art SRSs under cross-architecture, cross-feature and cross-architecture-feature circumstances.
3. An investigation of the effect of the perturbations constrained in different frequency bands on attack results. The frequency analysis reveals that noise in low frequency has the greatest influence on the effectiveness of the attack. In addition, we prove that adding distinct levels of perturbation on different frequency bands may result in better attack performance.

The remainder of this paper is structured as follows. Section 2 introduces related studies about adversarial attacks against SRS. Section 3 presents detailed explanations of our proposed method. The experimental results are listed in Sect. 4. Defense methods are provided in Sect. 5. A summary of this paper is provided in Sect. 6.

2 Related Work

In this section, the basic knowledge as well as the existing approaches regarding the adversarial attacks against SRSs are discussed, which have been a hot topic in recent years. SRSs implement the analysis of the input speech and then give the label of the current speaker, which makes the identity authentication of several platforms much more convenient. We assume that $F(x) = y$ as the ordinary process of speaker recognition, where F indicates a classifier, x is a clean example and y denotes the ground-truth label of x. The adversarial attack uses tiny malicious noises injected into the benign speech to mislead the classifier to produce the wrong results. The formula $F(x + \delta) = y^*$ denotes the adversarial attack, where δ indicates a tiny perturbation as small as possible and y^* is an incorrect label that is distinct from the real class of speaker. Actually, such attacks have been introduced into the audio domain just lately, so many existing studies borrow the ideas from the image domain, such as Fast Gradient Sign Method (FGSM) [19], Carlini and Wagner Attacks (C & W) [20], etc. Depending on the attacker's mastery levels of the prior information about the victim model, the attack scenarios can be divided into white-box, gray-box and black-box.

2.1 White-Box and Gray-Box Attacks

In terms of white-box attacks, Zhang et al. [21] introduced an attack method called Adaptive Decay Attack (ADA), which obviously improved the imperceptibility of the adversarial examples. Luo et al. [22] attempted to attack d-vectors with a Generalized End to End (GE2E) loss function. This novel loss function helped the perturbation generator produce effective adversarial examples and restrict the perturbation magnitude which guaranteed the generated adversarial examples have a high similarity with the original data. Zhang et al. [23] designed a two-step algorithm to craft adversarial examples and introduced Room Impulse Response (RIR) in the experiment, which validated the robustness of adversarial examples when they were played over the air. Additionally, Shamsabadi et al. [24] inspired by speech steganography and proposed an approach called FoolHD which exploited a Gated Convolutional Autoencoder (GCA) to generate human-imperceptible perturbation. In addition, gray-box attacks allow the adversary to get limited access to the victim classifier. Zhang et al. [7] applied zeroth order optimization-based method to design a practical attack called VMask, utilizing the idea of psychoacoustic masking to inject noise into audio files. However, the above-mentioned studies both have limitations which they assumed that the target models are exposed to attackers. There is no doubt that this type of attack is impractical in real scenarios.

2.2 Black-Box Attacks

To break such a dilemma, researchers pay more attention to practical black-box attacks. The adversary can only get the classification result or the confidence scores in this scenario. Many existing studies consider adversarial perturbation generation as an optimization problem.

Occam was proposed in [9] which combined the idea of adversarial attacks with model inversion attacks to fool commercial applications in an extremely strict black-box setting. SirenAttack [10] applied gradient-free Particle Swarm Optimization (PSO) strategy to generate adversarial examples. Furthermore, Zhang et al. [11] successfully constructed imperceptible waveform-level adversarial examples using differential evolution algorithm. Chen et al. [25] proposed FakeBob, applying Natural Evolution Strategy (NES) as a gradient estimation technique, and this strategy was also applied in the SEC4SR platform [26]. Besides, studies [8,27] illustrated that depending on the similarity between internal structures among different models, adversarial examples generated by a certain model can also attack other models. Xie et al. [12] used universal audio adversarial perturbation generator (UAPG) to generate universal adversarial audios, which could be added into arbitrary benign examples. Most methods rely on massive queries on the target, but this may cause the target classifier to detect the attacks.

Since our work pays attention to black-box attacks, we summarize the detailed information about black-box studies in Table 1 to facilitate subsequent comparisons. They provided the experimental benchmark for our study.

Table 1. Related studies on black-box adversarial attacks against SRSs.

Year	Article	(Un)target	Strategy	Target Model	Corpus
2019	SirenAttack	Target	PSO	VGG19,DenseNet	VCTK
	FakeBob	Both	Evolutionary	i-vector,x-vector	LibriSpeech
2020	VMask	Target	Transferability	Commercial services	LibriSpeech
2021	Occam	Target	Optimization	Commercial services	LibriSpeech
	SEC4SR	Both	Evolutionary	AudioNet,x-vector	LibriSpeech
	UAPG	Both	Generative model	x-vector	VCTK
2022	NMI-FGSM-Tri	Target	Transferability	x-vector ECAPA-TDNN	LibriSpeech
	AS2T	Both	Transferability	AudioNet ECAPA-TDNN	LibriSpeech
2023	Zhang	Target	Optimization	x-vector,i-vector	LibriSpeech VoxCeleb

3 Methodology

3.1 Problem Description

Our goal is to design an attack algorithm that can inject adversarial perturbation into the benign data and these modified examples can manipulate the well-trained classifier into producing wrong recognition results. Suppose that we define F as a state-of-the-art target model, (x, y) as benign example-label pair. δ is adversarial perturbation that can be inserted into benign data to generate an adversarial example x^*, the formula is $x^* = x + \delta$. $F(\cdot)$ outputs the prediction label of a given example. Further, $\| \cdot \|_p$ denotes the L_p norm which can constrain the added perturbation in extremely small magnitude, where p can be 0, 1, 2, ∞. ϵ denotes the maximum range of added perturbation.

For untargeted attack, the equation can be described as:

$$argmax\ F(x) = y,\ and\ argmax\ F(x^*) = y^*, s.t. \parallel x - x^* \parallel_p \leq \epsilon \tag{1}$$

where $y*$ is the incorrect label not equal to the ground-truth label.

For targeted attack, the equation can be described as:

$$argmax\ F(x) = y,\ and\ argmax\ F(x^*) = t, s.t. \parallel x - x^* \parallel_p \leq \epsilon \tag{2}$$

where t denotes the target speaker class.

3.2 The Proposed Framework

In this section, the framework of our proposed method is presented. The whole adversarial attack contains two main modules: training process of surrogate models and adversarial examples generation. Due to the limited availability of the target model, attackers cannot obtain the crucial information, thus this study uses the surrogate model as a substitute for the target model. All the detailed information of the well-trained surrogate models is exposed to the adversary. Then the attacker can generate adversarial examples on surrogate models, using the advanced gradient-based algorithm. These adversarial examples are much more transferable if the substitutes fit well with the victim systems. Figure 1 shows the overall workflow of our approach, and the specific implementations are elaborated in the following subsections.

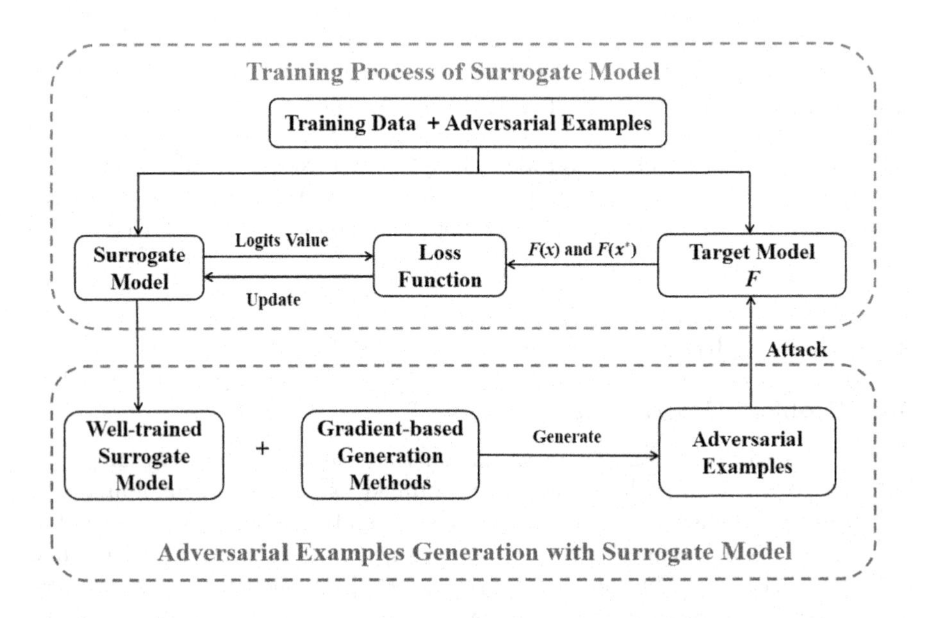

Fig. 1. The workflow of our framework.

3.3 Surrogate Models

Our approach assumes that the internal structures of different models are similar to each other to a certain extent, and these models have some commonalities in processing audio files. Additionally, depending on the feature distribution similarity theory in [8], the relative positions of the audio features in different distributions are largely similar, regardless of which feature extraction method the models use. Based on these assumptions, we consider that the adversarial examples have transferability. The adversarial examples generated by a certain model can lead another black-box classifier to produce incorrect predictions, even though these models are different in terms of architectures and feature types.

Based on such theory, the surrogate models are developed which can highly imitate the target models. The adversarial examples generated by surrogate models transfer to the target models, realizing the black-box attacks. Noted that the

Table 2. Details of four models.

Model	Architecture	Feature Type	Feature Dimension	Accuracy (%)
ECAPA-TDNN	TDNN	MFCC	80	97.99%
AudioNet	CNN	fBank	32	99.90%
x-vector	TDNN	fBank	32	99.62%
ResNet34	CNN	MFCC	80	98.00%

choice of substitute is important. The more commonalities SRSs have, the easier adversarial examples transfer to each other. However, due to the lack of knowledge of the victim model, it is almost impossible for us to adopt a model with the same configurations as the target model. To match a practical attack environment, this study attempts to leverage various models which differ from the target model in many aspects.

To illustrate that our proposed framework is applicable to multiple models, four different mainstream SRSs are utilized, namely ECAPA-TDNN [28], AudioNet [29], x-vector [30], and ResNet34 [31]. These models differ in model structure, input data type, and feature dimensions, which makes the experiments more challenging. They cover two architectures including DNNs and Time Delay Neural Network (TDNN), and they are also equipped with the commonly used feature extraction methods Mel-Frequency Cepstral Coefficients (MFCC) and fBank. The basic information of the four models is shown in Table 2.

As shown in Fig. 2, according to the configurations of the models, the transfer-based attack circumstances are classified into three categories: cross-feature, cross-architecture and cross-feature-architecture. A transferability analysis is conducted in Sect. 4.2 to investigate the impact of diverse factors on the attack performance.

3.4 The Training of Surrogate Model

In the training procedure of the surrogate model, a data augmentation method is applied for enhancing the fitting ability of the surrogate model. Existing study [16] proves that adding a set of perturbed data into the training dataset can make the surrogate model simulate the target model well. In our experiments, we construct adversarial examples x^* with MI-FGSM [17], which is proved to be the most effective algorithm to boost the transferability of adversarial attacks. The specific formulation of this algorithm is explained in Sect. 3.5.

Our study focuses on decision-based black-box attacks, where adversaries query the target model to acquire the classification label. In each iteration, the mixed dataset is fed into the target model and surrogate model. Then the target model provides the query feedback, and the surrogate model gives the logit value of all the speakers. Subsequently, the loss function calculates these results and updates the parameters, which leads the fitting ability of surrogate model to improve gradually. The loss function for training the surrogate models is

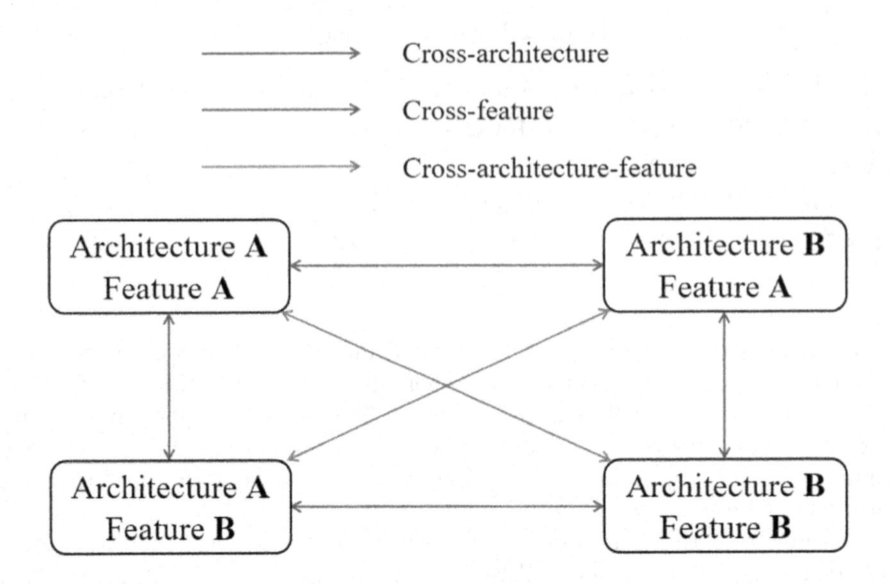

Fig. 2. The description of three transfer attack circumstances.

described as follows.

$$L_{total} = L(D_{logits}(x), F(x)) + L(D_{logits}(x^*), F(x^*)), \quad (3)$$

where L is the cross-entropy function. D is a surrogate model, and $D_{logits}(x)$ denotes the output probabilities of input data x. Adversaries query target classifier F to obtain the feedback F(x) and $F(x^*)$ which correspond to the input x and x^*, respectively.

Besides, our study performs L_∞ norm attack. Assuming that x' is the intermediate example in each iteration which is clipped into a specific range to guarantee the imperceptibility of the final example x^*. The clipping function is as follows.

$$x^* = clip_{x,\epsilon}(x'), = max(min(x', x + \epsilon), x - \epsilon) \quad (4)$$

where ϵ denotes the maximal bound of perturbation, x is an original example.

3.5 Adversarial Examples Generation

This part introduces the two advanced gradient-based adversarial examples generation methods employed in this paper: MI-FGSM [17] and EMI-FGSM [18]. These two attacks both come from the image domain. It is reasonable to assume that image processing and audio processing have high similarity in deep neural networks, so our method is adapted from the image domain. With the knowledge of the architecture and parameters of a well-trained surrogate model, numerous advanced white-box attack methods can be applied to get better attack results.

FGSM is the most classical gradient-based algorithm, which induces an iterative version of FGSM (I-FGSM) [32]. FGSM uses the sign function to calculate the gradient and applies the result to the real example in one step. However, the adversarial examples generated by FGSM have poor attack ability. So the multi-step algorithm like I-FGSM is proposed, which enhances the attack ability of the generated adversarial examples significantly. But it crafts the adversarial example by greedily moving the update direction along the sign of the gradient in each iteration. To escape from these defects, the following two methods integrate momentum into the iterative algorithm, yielding a widespread class of methods with higher attack ability.

Momentum Iterative Fast Gradient Sign Method (MI-FGSM) [17]. This method helps the update process barrel through the poor local maxima by recording and accumulating the velocity vector in the gradient direction.

$$g_{t+l} = \mu \cdot g_t + \frac{\nabla L((x_t)^*, y)}{\| \nabla_x L((x_t)^*, y) \|_1}, \tag{5}$$

$$x^*_{t+l} = x^*_t + \alpha \cdot sign(g_{t+1}) \tag{6}$$

Enhanced Momentum Iterative Fast Gradient Sign Method (EMI-FGSM) [18]. EMI-FGSM memorizes the past gradients of multiple data points which exist in the neighborhood of the current data point, instead of adopting the gradient accumulation of only one example.

$$\overline{x}^*_t[i] = x^*_t + a_i \cdot \overline{g}_{t-1}, \tag{7}$$

$$\overline{g}_t = \frac{1}{N}\sum_{i=1}^{N} \nabla_{\overline{x}^*_t[i]} L(\overline{x}^*_t[i], y), \tag{8}$$

$$g_t = \mu \cdot g_{t-1} + \frac{\overline{g}_t}{\| \overline{g}_t \|_1}, \tag{9}$$

$$x^*_{t+l} = x^*_t + \alpha \cdot sign(g_t), \tag{10}$$

where N denotes the sampling number which is bound in $[-\eta, \eta]$ and a_i indicates one of the N coefficients. Compute the average gradient of the N sample points which are located in the vicinity of the data point by Eq. (8). Then x^* is updated by Eq. (9) and Eq. (10).

4 Experiments

In this section, there are several ways to demonstrate the performance of our approach. To evaluate the effectiveness and generality of our method, comprehensive evaluations are conducted by incorporating multiple models as target and surrogate models. Further, the transferability analysis of adversarial examples by classifying the attacks into three circumstances: cross-architecture, cross-feature and cross-architecture-feature, which supplements the evaluation. For exhaustive comparison, a variety of indicators are adopted to reflect the excellent capability of our method. Besides, we explore the influence of adversarial perturbations constrained in different frequency bands on the attack results.

The experiments are carried out on a service with Pytorch 1.2, 2080 Ti-11G GPU, Intel Xeon E5-2696 v2 CPU.

4.1 Experimental Settings

Dataset. Open-access and widely used voice dataset Librispeech [8,11,23,25,27] is applied to our experiments. This dataset is very large containing 153516 utterances which come from approximately 9000 speakers. To improve the efficiency of the experiment, a portion of the whole dataset is used. We employ Spk251-train and Spk251-test datasets which are proposed in [26] for the training phases and testing phases respectively. They contain the same 251 individuals, but there are no overlapping sentences between the two datasets. For each speaker, 90% of the audio files are categorized into Spk251-train, and the remaining 10% are added into Spk251-test. Studies [8,27] choose ten individuals to form a speaker group for testing the attack method. For comparison with them, we organize a subset of Spk251-test called Spk10-attack for attack evaluations in our experiments. Spk10-attack covers 10 speakers which are randomly selected in Spk251-test.

Evaluation Metrics. Evaluation metrics for adversarial attacks are classified into two categories: practicability and imperceptibility. The generated adversarial perturbations should not only make the target model produce the wrong prediction, but also be inaudible to humans. To measure the practicability, we use Attack Success Rate (ASR). To validate the imperceptibility, we use Signal-to-Noise Ratio (SNR) [33], Perceptual Evaluation of Speech Quality ($PESQ$) [34], Short-Time Objective Intelligibility score ($STOI$) [35] and L_2 distance [20]. The definitions of them are briefly introduced as follows.

Attack Success Rate (ASR): It represents the proportion of successful attacks to total attacks. The formula is expressed as $ASR = M/N \times 100\%$. N refers to the total number of adversarial attacks. For untargeted attack, M is the number of examples which are classified into labels which are not equal to the ground-truth labels. For target attack, M is the number of examples that are predicted as the specific class.

Signal-to-Noise Ratio (SNR): It is the ratio of the original voice power to the noise power. Higher SNR means high perceptual quality of adversarial example, which guarantees the attack is imperceptible to humans.

Perceptual Evaluation of Speech Quality (PESQ): The calculation of $PESQ$ is more complex, and it is also used for perceptual evaluation of audio quality. The range of this indicator is from -0.5 to 4.5, higher score indicates better stealthiness.

Short-Time Objective Intelligibility Score (STOI): STOI is a critical measure of speech intelligibility. It varies in $[0, 1]$, and the value which is infinitely close to 1 indicates that the speech can be fully understood.

L_2 Distance: L_2 norm computes the distance between original and adversarial examples. Smaller L_2 means the adversarial perturbation is imperceptible enough.

4.2 Transferability Analysis

Evaluation Setup. In this experiment, four state-of-the-art models are adopted which are mentioned in Table 2. One of them is regarded as the target model and the remaining three are used as surrogate models. Based on this, we completed 12 experiments. Throughout the attack evaluation, we set the perturbation magnitude ϵ to 0.002 with step size $\alpha = \epsilon/5 = 0.0004$ for both MI-FGSM and EMI-FGSM. According to the results of [17], the decay factor μ is set to 1. In addition, the sampling interval $\eta = 10$ and the sampling number $N = 11$ are used for EMI-FGSM in this work. Notably, to simplify the description, we formulate a set of representations A-B to denote the surrogate model (A) and target model (B) respectively, which means that the adversarial examples generated by surrogate A transfer to black-box model B. Due to the space limitations, only the results of the targeted attacks are reported in Table 3 and Table 4. This type of attack is more difficult than other settings, which can better demonstrate the advantages of our method.

Analysis. We can see that the success rates of the attacks on four models can all reach more than 60%, proving the generality of our approach. Additionally, MI-GSM is quite effective in most cases, but for AudioNet-ECAPA-TDNN and ResNet34-x-vector pairs, EMI-GSM has more prominent effects which can slightly improve *ASR*.

For the attacks under cross-architecture circumstances, the transferability from a CNN model to TDNN model is fairly limited, especially the *ASR* of the targeted attack AudioNet-x-vector, even drops to only 0.86%. However, the transferability of these two models is not symmetric. The adversarial examples constructed by TDNN models transfer to CNN models quite well. This conclusion coincides with the previous results of the study [27].

In addition, the cross-feature attacks among TDNN models can achieve better attack performance, but CNN-based models do not transfer well to each other in this setting. Additionally, the surrogate models with fBank features transfer to target models with MFCC features quite well, such as the attack of x-vector-ECAPA-TDNN produces the *ASR* of 87.34%.

While the factors affecting the attacks under cross-feature-architecture conditions are relatively complicated. In this scenario, models have more transferability gaps due to the difference between both internal architectures and feature types. Indeed, TDNN models are more transferable to CNN models regardless

Table 3. Attack performance of different conditions using MI-FGSM.

Cross	Target Model	Surrogate Model	*ASR* (%)	*SNR* (dB)	*PESQ*	*STOI*	*L2*
Architecture	ResNet34	ECAPA-TDNN	67.23	31.10	2.69	0.98	0.64
	ECAPA-TDNN	ResNet34	28.45	30.64	2.91	0.99	0.87
	AudioNet	x-vector	79.67	30.78	3.06	0.98	0.63
	x-vector	AudioNet	0.86	30.76	3.05	0.99	0.86
Feature	ECAPA-TDNN	x-vector	87.34	31.36	2.59	0.99	0.59
	x-vector	ECAPA-TDNN	69.83	30.67	2.62	0.99	0.87
	AudioNet	ResNet34	8.13	30.66	3.11	0.98	0.64
	ResNet34	AudioNet	12.07	31.05	2.97	0.99	0.61
Architecture-Feature	ECAPA-TDNN	AudioNet	9.48	30.62	2.84	0.99	0.87
	AudioNet	ECAPA-TDNN	73.17	30.86	3.04	0.98	0.63
	x-vector	ResNet34	0.86	30.54	2.90	0.99	0.65
	ResNet34	x-vector	32.76	30.75	2.72	0.99	0.63

Table 4. Attack performance of different conditions using EMI-FGSM.

Cross	Target Model	Surrogate Model	*ASR* (%)	*SNR* (dB)	*PESQ*	*STOI*	*L2*
Architecture	ResNet34	ECAPA-TDNN	42.02	30.33	2.64	0.98	0.70
	ECAPA-TDNN	ResNet34	25.86	30.77	2.88	0.99	0.63
	AudioNet	x-vector	71.54	30.50	2.95	0.98	0.66
	x-vector	AudioNet	0.86	30.82	2.93	0.99	0.85
Feature	ECAPA-TDNN	x-vector	65.52	30.61	2.41	0.99	0.64
	x-vector	ECAPA-TDNN	43.10	30.73	2.65	0.99	0.64
	AudioNet	ResNet34	10.57	30.54	3.01	0.98	0.65
	ResNet34	AudioNet	11.21	31.09	2.81	0.99	0.61
Architecture-Feature	ECAPA-TDNN	AudioNet	10.34	30.94	2.64	0.99	0.85
	AudioNet	ECAPA-TDNN	34.15	29.85	2.94	0.97	0.71
	x-vector	ResNet34	1.72	30.45	2.86	0.99	0.66
	ResNet34	x-vector	25.86	30.42	2.61	0.99	0.66

of the feature types they use. This finding, to some extent, suggests that the model structure has more influence on the outcomes in transfer-based attacks.

4.3 Frequency Analysis

Evaluation Setup. According to the conclusions in previous studies [5,36], our study aims to investigate whether the perturbations constrained on different frequency regions cause different attack results. To achieve this, we limit the attack space to three subspaces. We first transform the input raw waveform to a spectrogram with a linear scale. Then a set of band-pass filters are applied to split the frequency regions to low-frequency bands $[0k, 2k]$, medium-frequency bands $[2k, 5k]$ and high-frequency bands $[5k, 8k]$. As indicated in Fig. 3, the attacks are categorized as low-frequency attack (LF attack), medium-frequency attack (MF attack) and high-frequency attack (HF attack) based on the frequency bands where the perturbations cover.

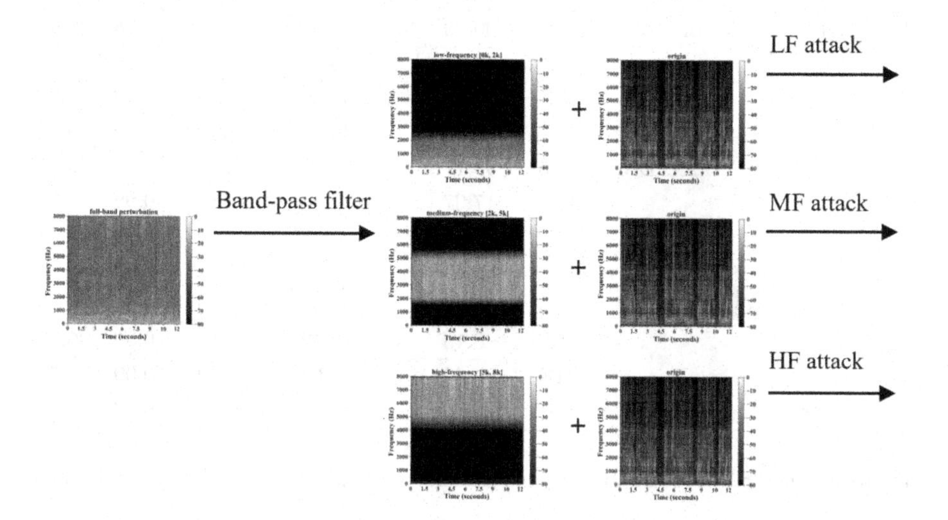

Fig. 3. Three types of attacks based on frequency properties.

Besides, our hypothesis is extended that applying distinct levels of noise to three frequency subspaces may provide us with interesting findings. Since the three bands respond differently to perturbation, the attackers do not have to change the size of the perturbation on the full band when adjusting the attack strength. Inspired by this idea, this study tries to tune the noise magnitude only in one specific band, which has more flexibility to balance the imperceptibility and effectiveness of the attack.

Analysis. In this experiment, AudioNet is considered as the target model and MI-FGSM is utilized as the adversarial example generation method. The results of untargeted and targeted attacks are reported in Table 5 and Table 6 respectively. Three frequency bands affect the performance of the attack differently. We can observe that perturbations in low-frequency bands are particularly effective in both untargeted and targeted attacks. However, if the perturbations are restricted to high-frequency subspaces, they perform worse in terms of the success rate of attack. The higher ASR is obtained at the cost of lower SNR. But the distance among the SNR values of the three attacks is not large. Compared with full-frequency attacks, LF attacks improve the imperceptibility of adversarial examples with a little sacrifice on ASR. For example, the adversarial examples generated by ECAPA-TDNN attack AudioNet, the SNR value of LF attacks is increased by 4. Remarkably, this LF attack which only has one-third of polluted regions can achieve the ASR of 82.93% when the ASR of full-frequency attack is

Table 5. The attack results of untargeted attacks w.r.t. perturbation frequency bands.

Target Model	Surrogate Model	Type	Practicability	Imperceptibility			
			ASR (%)	*SNR* (dB)	*PESQ*	*STOI*	*L2*
AudioNet	x-vector	Full	100.00	30.58	3.02	0.98	0.65
		HF	46.34	36.51	3.70	0.99	0.33
		MF	73.17	35.56	3.71	0.99	0.37
		LF	77.24	35.16	3.71	0.99	0.38
	ECAPA-TDNN	Full	99.17	30.55	2.97	0.98	0.66
		HF	17.07	37.28	3.68	0.99	0.30
		MF	58.54	35.58	3.71	0.99	0.37
		LF	82.93	34.83	3.62	0.98	0.40
	ResNet34	Full	34.60	30.68	3.11	0.98	0.65
		HF	8.13	39.12	3.67	0.99	0.25
		MF	24.39	39.68	4.01	0.99	0.23
		LF	26.83	33.34	3.68	0.98	0.47

Table 6. The attack results of targeted attacks w.r.t. perturbation frequency bands.

Target Model	Surrogate Model	Type	Practicability	Imperceptibility			
			ASR (%)	*SNR* (dB)	*PESQ*	*STOI*	*L2*
AudioNet	x-vector	Full	79.67	30.78	3.06	0.98	0.63
		HF	21.95	36.50	3.64	0.99	0.33
		MF	13.00	35.88	3.74	0.99	0.35
		LF	43.09	35.27	3.73	0.98	0.38
	ECAPA-TDNN	Full	73.17	30.86	3.04	0.98	0.63
		HF	6.50	37.61	3.77	0.99	0.29
		MF	13.82	35.97	3.73	0.99	0.35
		LF	35.77	34.95	3.67	0.98	0.39
	ResNet34	Full	8.13	30.66	3.11	0.98	0.64
		HF	1.62	39.44	3.70	0.99	0.24
		MF	4.07	39.82	4.02	0.99	0.23
		LF	12.20	33.31	3.69	0.98	0.47

99.17%. As illustrated in Table 6, sometimes LF attacks can even obtain better results compared to full-frequency attacks when the target model is AudioNet and the surrogate model is ResNet34.

Several metrics are used in order to test the imperceptibility of the adversarial examples. However, this is not comprehensive because the noise can be incorporated into the background sound. It is possible for the human ear to detect the noise even though a high *SNR* is produced. Thus, a human perception test is conducted to supplement the evaluation. To ensure the validity of our experiments, test voices are selected among the examples that can successfully attack the model. For each surrogate model, three types of attacks (LF, MF, HF) are used to generate adversarial examples. Additionally, we survey ten participants including four females and six males. The test voices are allowed to play through the speakers of the computers or mobile phones, which is more practical in real-world scenarios.

Each participant is asked to distinguish the degree of similarity between the adversarial audios and the original audios. There are two options for the results. "Noisy" means that the participant regards that the adversarial example has some noises. And another choice is "Clean", which means that the participant cannot feel any noise. In other words, the volunteer believes this adversarial audio is basically similar to its corresponding original audio. The results demonstrate that 100% of participants believe that the adversarial voices generated by HF attacks and MF attacks are clean enough, while only 20% of participants hear a little noise in the adversarial voices generated by LF attacks. Hence, this indicates our attack is human-imperceptible.

For extended experiments, we will verify the assumptions that adding different levels of perturbations on three frequency bands will bring helpful findings for adjusting attack parameters. MI-FGSM is used to generate adversarial examples on three substitute models against the target model AudioNet. For a clear and intuitive representation, we label the different configurations as shown in Table 7. The experimental results are reported in Fig. 4. Note that the symbols (A, B, C, ...) correspond to the values of horizontal coordinates in Fig. 4.

Table 7. Configurations of attacks. Index corresponds to the value of x-coordinate in the Fig. 4.

Perturbation Region	Perturbation Magnitude								
HF	0.001	0.001	0.002	0.002	0.002	0.002	0.003	0.003	0.003
MF	0.001	0.001	0.001	0.002	0.002	0.002	0.002	0.003	0.003
LF	0.001	0.002	0.001	0.001	0.002	0.003	0.002	0.002	0.003
Index	A	B	C	D	E	F	G	H	I

Obviously, as the intensity of the added perturbation increases, the perceptual quality of the generated adversarial examples is gradually decreased in most cases. The ASR of attack F is very close to that of attack I in some cases. Likewise, the attacks which correspond to Index B and Index E also have similar results. This finding inspires us that, if the attacker wants a higher success rate of attack, he can enlarge the intensity of noise only in the low-frequency band instead of covering the full bands with larger perturbations.

Besides, when the noise magnitude is relatively small, the low-frequency band is more sensitive to the change of perturbation. When the perturbation size in the low-frequency band is adjusted from 0.001 to 0.002, which are denoted by symbols A and B respectively, the changes between them are the most obvious compared to other settings. Just as shown in the subplots, the variation of the results between attacks A and B is greater than that of configurations E and F in most cases.

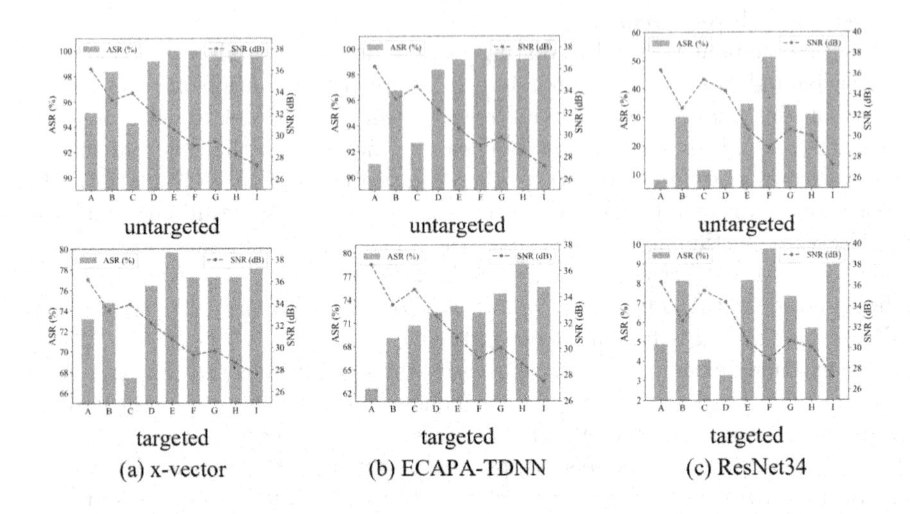

Fig. 4. The evaluation of effectiveness (ASR) and imperceptibility (SNR). The captions of the subfigures indicate the used surrogate model.

4.4 Performance Comparison

As illustrated in Table 8, compared with previous studies, our approach is more prominent in many aspects. This paper chooses the results of targeted attacks for comparisons since this attack setting is relatively difficult, which is more capable of demonstrating the superiority of our method.

In [27], researchers perform transfer attacks with AudioNet and ResNet34 as victim models in a strict black-box setting, where they even cannot query the target model. So we remark that the success rates of these attack experiments are extremely low. While our proposed method utilizes limited queries to train surrogate models, which brings explosive growth in success rates.

For ECAPA-TDNN, the study [8] combines an ensemble-based approach with Nesterov accelerated gradient (NAG) to generate adversarial examples. In addition, method [27] mount attacks using novel loss functions with FGSM and Project Gradient Descent (PGD) as the optimization approaches. The experimental results indicate our method produces stealthier adversarial examples and enhances the attack effectiveness than compared works.

Existing studies [11,26] refer that x-vector is a relatively impregnable system to be attacked, perhaps due to the massive trainable parameters which make

Table 8. Performance comparison of different methods for targeted attacks.

Target Model	Methods	*ASR* (%)	*SNR* (dB)	*PESQ*
AudioNet	AS2T [27]	5.60	-	-
	Our method	**79.67**	**30.78**	**3.06**
ECAPA-TDNN	AS2T [27]	86.50	31.14	2.74
	NMI-FGSM-Tri [8]	66.30	-	-
	Our method	**87.34**	**31.36**	**2.59**
x-vector	NMI-FGSM-Tri [8]	43.00	28.42	2.60
	Our method	**69.83**	**30.67**	**2.62**
ResNet34	AS2T [27]	14.30	-	-
	Our method	**67.23**	**31.10**	**2.69**

the model complicated. Recent work [8] shows that the *ASR* against this model is only 43%, our proposed attack yields over 69% success rate of attack, which is 26% higher than the results of the previous study. It proves that, even on the most difficult model, the adversarial examples crafted by our approach have superb performance in terms of practicability and imperceptibility.

5 Discussion of Potential Defenses

This section discusses the potential defense methods that could effectively defend against our proposed attack. The first method is adversarial training, which incorporates the generated adversarial examples into the original clean dataset to retrain the model. This helps enhance the robustness of the model against the adversarial attacks. However, it should be noted that this defense mechanism may not work well if the attacker changes the attack parameters frequently. The second one is to downsample the input audios. By reducing the sampling rates, there might be a loss of high-frequency information in the audio, leading to a partial reduction in perturbations present in the adversarial audios. The experimental results demonstrate that the lack of the high-frequency components in the perturbations affects the effectiveness of our proposed attack. Further, it is important to consider that this downsampling method may potentially cause distortion in the original audio.

In conclusion, a thorough examination of defense methods and mitigation strategies becomes crucial in safeguarding against adversarial attacks. This will be the primary focus of our future work.

6 Conclusions and Future Work

More researchers leverage the vulnerability of DNNs-based SRSs to construct adversarial examples. This paper proposes a framework for performing black-box adversarial attacks. The adversarial examples we crafted exhibit higher effectiveness, better transferability and good confidentiality. Previous studies emphasize the white-box attacks which allow the adversaries to obtain all the information of the target model. Our method overcomes this shortcoming, which is applicable to the case where all the prior knowledge is unknown. Sufficient experiments are conducted on four mainstream speaker recognition models, and the results illustrate the prominent ability of our method. Additionally, the analysis for perturbation frequency distribution is instructive for us to leverage the weakness of the SRSs.

Our works expose the vulnerability of SRSs, and this also reminds us that our future works should focus on exploring the defense methods which can protect the models from adversarial attacks.

Acknowledgements. This research was funded by NSFC under Grant 61572170, Natural Science Foundation of Hebei Province under Grant F2021205004, Science and Technology Foundation Project of Hebei Normal University under Grant L2021K06, Science Foundation of Returned Overseas of Hebei Province Under Grant C2020342, and Key Science Foundation of Hebei Education Department under Grant ZD2021062.

References

1. Hanifa, R.M., Isa, K., Mohamad, S.: A review on speaker recognition: technology and challenges. Comput. Elec. Eng. **90**(3), 107005 (2021)
2. Szegedy, C., Zaremba, W., Sutskever, I., Bruna, J., Erhan, D., Goodfellow, I.: Intriguing properties of neural networks. In: Proceedings of the 2nd International Conference on Learning Representations (ICLR). IEEE (2014)
3. Li, X., Zhong, J., Wu, X., Yu, J., Liu, X., Meng, H.: Adversarial attacks on GMM i-vector based speaker verification systems. In: IEEE International Conference on Acoustics, Speech and Signal Processing, ICASSP 2020, pp. 6579–6583. IEEE (2020)
4. Tan, H., Wang, L., Zhang, H., Zhang, J., Shafiq, M., Gu, Z.: Adversarial attack and defense strategies of speaker recognition systems: a survey. Electronics **11**(14), 2183 (2022)
5. Li, J., Zhang, X., Xu, J., Ma, S., Gao, W.: Learning to fool the speaker recognition. In: Proceedings of the IEEE International Conference on Acoustics, Speech and Signal Processing, ICASSP 2020, pp. 2937–2941. IEEE (2020)
6. Li, J., et al.: Universal adversarial perturbations generative network for speaker recognition. In: 2020 IEEE International Conference on Multimedia and Expo (ICME), pp. 1–6. IEEE (2020)
7. Zhang, L., Meng, Y., Yu, J., Xiang, C., Falk, B., Zhu, H.: Voiceprint mimicry attack towards speaker verification system in smart home. In: Proceedings of the 39th IEEE Conference on Computer Communications, INFOCOM 2020, pp. 377–386. IEEE (2020)
8. Zhang, J., et al.: NMI-FGSM-Tri: an efficient and targeted method for generating adversarial examples for speaker recognition. In: 2022 7th IEEE International Conference on Data Science in Cyberspace (DSC), pp. 167–174. IEEE (2022)
9. Zheng, B., et al.: Black-box adversarial attacks on commercial speech platforms with minimal information. In: Proceedings of the 2021 ACM SIGSAC Conference on Computer and Communications Security, pp. 86–107. ACM (2021)

10. Du, T., Ji, S., Li, J., Gu, Q., Wang, T., Beyah, R.: SirenAttack: generating adversarial audio for end-to-end acoustic systems. In: Proceedings of the 15th ACM Asia Conference on Computer and Communications Security, pp. 357–369. ACM (2020)
11. Zhang, X., Zhang, X., Sun, M., Zou, X., Chen, K., Yu, N.: Imperceptible black-box waveform-level adversarial attack towards automatic speaker recognition. Complex Intell. Syst. **9**(1), 65–79 (2023)
12. Xie, Y., Li, Z., Shi, C., Liu, J., Chen, Y., Yuan, B.: Enabling fast and universal audio adversarial attack using generative model. In: Proceedings of the AAAI Conference on Artificial Intelligence, AAAI 2021, pp. 14129–14137 (2021)
13. Kariyappa, S., Prakash, A., Qureshi, M.K.: MAZE: data-free model stealing attack using zeroth-order gradient estimation. In: Proceedings of the IEEE/CVF Conference on Computer Vision and Pattern Recognition (CVPR), pp. 13814–13823. IEEE (2021)
14. Wang, Y., et al.: Black-box dissector: towards erasing-based hard-label model stealing attack. In: Avidan, S., Brostow, G., Cissé, M., Farinella, G.M., Hassner, T. (eds.) 17th European Conference on Computer Vision, ECCV 2022. LNCS, Tel Aviv, Israel, 23–27 October 2022, Proceedings, Part V, pp. 192–208. Springer, Cham (2022). https://doi.org/10.1007/978-3-031-20065-6_12
15. Yuan, X., Ding, L., Zhang, L., Li, X., Wu, D.O.: ES attack: model stealing against deep neural networks without data hurdles. IEEE Trans. Emerg. Top. Comput. Intell. **6**(5), 1258–1270 (2022)
16. Wang, F., Ma, Z., Zhang, X., Li, Q., Wang, C.: DDSG-GAN: generative adversarial network with dual discriminators and single generator for black-box attacks. Mathematics. **11**(4), 1016 (2023)
17. Dong, Y., et al.: Boosting adversarial attacks with momentum. In: Proceedings of the IEEE Conference on Computer Vision and Pattern Recognition (CVPR), pp. 9185–9193. IEEE (2018)
18. Wang, X., Lin, J., Hu, H., Wang, J., He, K.: Boosting adversarial transferability through enhanced momentum. arXiv preprint arXiv: 2103.10609 (2021)
19. Goodfellow, I.J., Shlens, J., Szegedy C.: Explaining and harnessing adversarial examples. arXiv preprint arXiv:1412.6572 (2014)
20. Carlini, N., Wagner, D.: Towards evaluating the robustness of neural networks. In: 2017 IEEE Symposium on Security and Privacy (SP), pp. 39–57. IEEE (2017)
21. Zhang, X., Xu, Y., Zhang, S., Li, X.: A highly stealthy adaptive decay attack against speaker recognition. IEEE Access **10**(11), 118789–118805 (2022)
22. Luo, H., Shen, Y., Lin, F., Xu, G.: Spoofing speaker verification system by adversarial examples leveraging the generalized speaker difference. Secur. Commun. Netw. **2021**, 1–10 (2021)
23. Zhang, W., et al.: Attack on practical speaker verification system using universal adversarial perturbations. In: Proceedings of the IEEE International Conference on Acoustics, Speech and Signal Processing, ICASSP 2021, pp. 2575–2579. IEEE (2021)
24. Shamsabadi, A.S., Teixeira, F.S., Abad, A., Raj, B., Cavallaro, A., Trancoso, I.: FoolHD: fooling speaker identification by highly imperceptible adversarial disturbances. In: Proceedings of the IEEE International Conference on Acoustics, Speech and Signal Processing, ICASSP 2021, pp. 6159–6163. IEEE (2021)
25. Chen, G., et al.: Who is real bob? Adversarial attacks on speaker recognition systems. In: 2021 IEEE Symposium on Security and Privacy (SP), pp. 694–711. IEEE (2019)
26. Chen, G., Zhao, Z., Song, F., Chen, S., Fan, L., Liu, Y.: SEC4SR: a security analysis platform for speaker recognition. arXiv preprint arXiv:2109.01766 (2021)

27. Chen, G., Zhao, Z., Song, F., Chen, S., Fan, L., Liu, Y.: AS2T: arbitrary source-to-target adversarial attack on speaker recognition systems. arXiv preprint arXiv:2206.03351 (2022)
28. Desplanques, B., Thienpondt, J., Demuynck, K.: ECAPA-TDNN: emphasized channel attention, propagation and aggregation in TDNN based speaker verification. arXiv preprint arXiv:2005.07143 (2020)
29. Becker, S., Ackermann, M., Lapuschkin, S., Müller, K.R., Samek, W.: Interpreting and explaining deep neural networks for classification of audio signals. arXiv preprint arXiv:1807.03418 (2018)
30. Snyder, D., Garcia-Romero, D., Sell, G., McCree, A., Povey, D., Khudanpur, S.: Speaker recognition for multi-speaker conversations using x-vectors. In: IEEE International Conference on Acoustics, Speech and Signal Processing, pp. 5796–5800 (2019)
31. Son Chung, J., Nagrani, A., Zisserman, A.: VoxCeleb2: deep speaker recognition. arXiv preprint arXiv:1806.05622 (2018)
32. Kurakin, A., Goodfellow, I.J., Bengio, S.: Adversarial examples in the physical world. In: Proceedings of the Workshop of the 5th International Conference on Learning Representations, ICLR 2017, pp. 99–112. IEEE (2017)
33. Yuan, X., et al.: CommanderSong: a systematic approach for practical adversarial voice recognition. In: Proceedings of the 27th USENIX Security Symposium, pp. 49–64. IEEE (2018)
34. Rix, A.W., Beerends, J.G., Hollier, M.P., Hekstra, A.P.: Perceptual evaluation of speech quality (PESQ) - a new method for speech quality assessment of telephone networks and codecs. In: 2001 IEEE International Conference on Acoustics, Speech, and Signal Processing, pp. 749–752. IEEE (2001)
35. Taal, C.H., Hendriks, R.C., Heusdens, R., Jensen, J.: An algorithm for intelligibility prediction of time-frequency weighted noisy speech. IEEE Trans. Audio Speech Lang. Process. **19**(7), 2125–2136 (2011)
36. Sharma, Y., Ding, G.W., Brubaker, M.: On the effectiveness of low frequency perturbations. arXiv preprint arXiv:1903.00073 (2019)

SW-LeNet: Implementation and Optimization of LeNet-1 Algorithm on Sunway Bluelight II Supercomputer

Zenghui Ren, Tao Liu(✉), Zhaoyuan Liu, Min Tian, Ying Guo, and Jingshan Pan

Shandong Computer Science Center (National Supercomputer Center in Jinan), Qilu University of Technology (Shandong Academy of Sciences), Jinan, China
liutao@sdas.org

Abstract. Nowadays, convolutional neural networks are representative of deep learning algorithms. With the development of convolutional neural networks, their network structures become more complex, and the number of parameters for training becomes larger and larger. The parallelization of convolutional neural network algorithms on multicore or many-core processors is essential for training convolutional neural networks. In this paper, we propose a parallel algorithm of LeNet-1 based on the Sunway Bluelight II supercomputer, named SW-LeNet. Moreover, we propose a two-level parallelization scheme, including thread-level optimization and process-level optimization. In thread-level optimization, the following optimization methods are used, including CPEs parallelism, hybrid scheme and DMA optimization, register optimization, and SIMD data parallelism. Data parallelism optimization and parameter packing optimization are used in process-level optimization. Compared with the original LeNet, SW-LeNet can achieve 4.94x speedups in a single core group. Moreover, SW-LeNet can be scaled up to 2,048 processes, with 133,120 cores, and achieves 84.93% parallel efficiency.

Keywords: Sunway · Many-core processor · Parallel optimization · LeNet-1

1 Introduction

In recent years, deep learning (DL) has been widely used in many fields, such as natural language processing [1] and computer vision [2]. Convolutional Neural Network (CNN) is an essential foundation of deep learning. LeNet-1 [3] is one of the earliest published convolutional neural networks, which has attracted wide attention due to its advantages in image recognition and classification. After AlexNet [4] emerged, convolutional neural networks developed rapidly. GoogLeNet [5], ResNet [6], and many other powerful algorithms are in application and promotion in academia and industry. With the development of convolutional neural networks, their network structures become more complex, and

Z. Tari et al. (Eds.): ICA3PP 2023, LNCS 14491, pp. 277–298, 2024.
https://doi.org/10.1007/978-981-97-0808-6_16

the training datasets become larger and larger. Therefore, reducing the training time of convolutional neural network algorithms is critical. Compared with using the traditional CPUs, using the new heterogeneous processors such as Sunway many-core processor [7], GPU [8], and DCU [9] to train convolutional neural networks has better effects in acceleration. With the development of high-performance supercomputers, it has become a prevailing trend to migrate and develop deep learning algorithms by employing high-performance supercomputers. Consequently, making full use of the heterogeneous computing resources of high-performance supercomputers to realize parallel algorithms is an effective way to train deep learning algorithms.

The Sunway Bluelight II supercomputer and Sunway many-core processors are representative works of Chinese high-performance computers and processors. The Sunway Bluelight II supercomputer has more computing resources and memory space on a single node than the Sunway Taihulight supercomputer [10]. It can provide more support for training deep learning models. In addition, the Sunway Bluelight II supercomputer provides more efficient bandwidth, which can facilitate the expansion of the model training scale, especially for the data-parallel mode. The data-parallel mode can increase the training data in a single iteration and reduce the number of iterations required to traverse the dataset. It is an effective solution to implement the neural network parallel training based on the Sunway Bluelight II supercomputer.

In this paper, we implement a parallel algorithm of CNN named SW-LeNet, based on the Sunway Bluelight II supercomputer. The main contributions of this paper are as follows:

(1) We propose a new parallel algorithm of LeNet-1 based on the Sunway Bluelight II supercomputer, named SW-LeNet.
(2) A two-level parallelization scheme is proposed, including the thread-level optimization scheme and process-level optimization. In thread-level optimization, the following optimization methods are used, including CPEs parallelism, hybrid scheme and DMA optimization, register optimization and SIMD data parallelism. Data parallelism optimization and parameter packing optimization are used in process-level optimization.

We use the MNIST [3,11] dataset to test SW-LeNet. SW-LeNet can achieve 4.94x speedups in a single core group. SW-LeNet can be scaled up to 2,048 processes, with 133,120 cores, and achieves 84.93% parallel efficiency.

The rest of this paper is as follows. Section 2 shows the related work. Section 3 introduces the Sunway Bluelight II Supercomputer and LeNet-1. Section 4 presents the SW-LeNet and its two-level parallelization scheme. Section 5 presents the evaluation. Finally, Sect. 6 provides the conclusions.

2 Related Work

2.1 Convolutional Neural Network

Deep learning is developing rapidly in autonomous driving [12], image recognition and classification [4–6], natural language processing [13], and other fields

[14]. After LeNet [3] appeared, for a long time, the convolutional neural network was only suitable for identifying small-size images, and the result of processing large-scale data was not good. In 2012, Krizhevsky et al. proposed AlexNet [4] in the ImageNet Large-scale visual Recognition Challenge competition and achieved the best classification effect. Szegedy et al. proposed GoogLeNet [5]. Its parameters are only 1/12 of AlexNet's, and its accuracy is higher than AlexNet's. In 2015, He et al. proposed ResNet [6], which solved the degradation problem in neural networks.

With the development of convolutional neural networks, the number of parameters and the scale of datasets have become larger and larger. Parallel training of convolutional neural networks has become an effective solution. [15] detects malware through a parallel convolutional neural network. [16] predicts traffic conditions in specific areas through a parallel convolutional neural network.

The most basic computation of a neural network is the convolution operation of matrices, [17] improving the speed of matrix computation on multi-core processors. There are many variations in the structure of different neural networks, [18] mapping and optimizing irregular applications onto a multi-core processor architecture. There is also research into parallel training of deep learning on Chinese supercomputers. [7] implements a library for accelerating deep learning applications on Sunway TaihuLight. Compared with cuDNNV5 [19] based on Tesla K40m, swDNN achieves 1.91x–9.75x speedups in more than 100 parameter configurations. [20] implements a parallel framework swCaffe based on Caffe [21] for accelerating deep learning on Sunway Taihulight.

2.2 The Works Based on the Sunway Supercomputers

Sunway supercomputers have obtained outstanding achievements across many fields in the last ten years [22–24]. [22] successfully scaled the fully implicit solver to the entire system of the Sunway taihulight supercomputer and achieved fast and accurate atmospheric simulation at 488 m of horizontal resolution. [23] implements a highly scalable nonlinear seismic simulation tool on the Sunway Taihulight supercomputer. [24] implements a random quantum circuit simulator based on tensors and completes simulated sampling in 304 s with the new Sunway supercomputer. All of these three research projects have won the Gordon Bell Award, which is considered the highest international award in supercomputing.

There are a lot of good works based on the new Sunway supercomputer [10,24–27]. Based on ab initio, [25] achieves Raman spectroscopy simulation of natural biological systems with up to 3,006 atoms. [26] implements a deep learning model named BaGuaLu containing 174 trillion parameters. [27] proves the validity of quantum state representation based on CNN in a new Sunway supercomputer. [10] studies and evaluates HPCG benchmark performance optimization technology for the new Sunway supercomputer.

In conclusion, distributed machine learning has achieved good performance in traditional CPUs, GPUs, and the Sunway Taihulight supercomputer. In this paper, the parallel algorithm of the traditional convolutional neural network

LeNet-1 is implemented and optimized on the Sunway Bluelight II supercomputer, and SW-LeNet is proposed.

3 The Sunway Bluelight II Supercomputer and LeNet-1

3.1 The Architecture of the Sunway Bluelight II Supercomputer

Figure 1 shows the Sunway Bluelight II supercomputer, which is the third-generation supercomputer of the Sunway supercomputer. The Sunway Bluelight II supercomputer includes one cabinet and four computing supernodes. Each supernode has 64 computing plugins, and each computing plugin contains two computing nodes. Each supernode contains 128 nodes. Each node contains two SW26010pro processors, two Sunway message processing chips, and 192 GB of DDR4 memory. The whole system has 512 nodes and 1,024 SW26010pro processors.

Fig. 1. The Sunway Bluelight II Supercomputer.

Figure 2 shows the architecture of SW26010pro. The SW26010pro processor consists of six core groups (CGs), which are connected in a ring architecture via a network on chip (NoC). Each core group comprises a management processing element (MPE) and 64 computing processing elements (CPEs) in an 8×8 mesh format. Four neighboring CPEs share one CPE cluster management component. One SW26010pro contains six MPEs and 384 CPEs.

Each core group has 16 GB DDR4 memory with a bandwidth of 51.2 GB/s. The SW26010pro has 96 GB of memory. The total memory bandwidth is 307.2 GB/s. The CPEs communicate with the corresponding MPE with the Direct Memory Access (DMA). The data exchange between each two CPEs is achieved through the Remote Memory Access (RMA). Each CPE has 256 KB

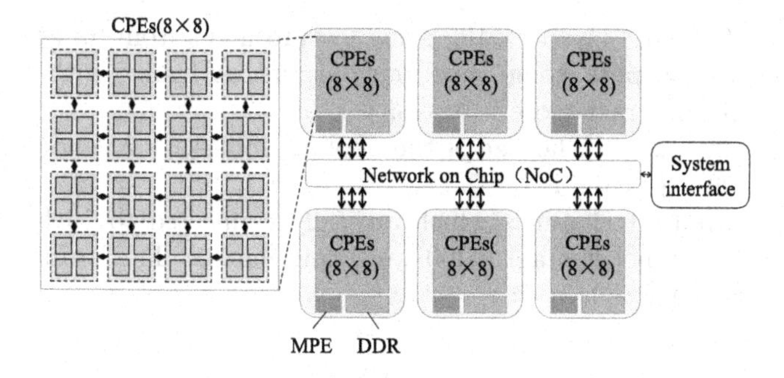

Fig. 2. The architecture of SW26010pro.

of high-speed local data memory (LDM). The MPE and CPEs support 256-bit and 512-bit SIMD vector operations respectively.

3.2 The Architecture of LeNet-1

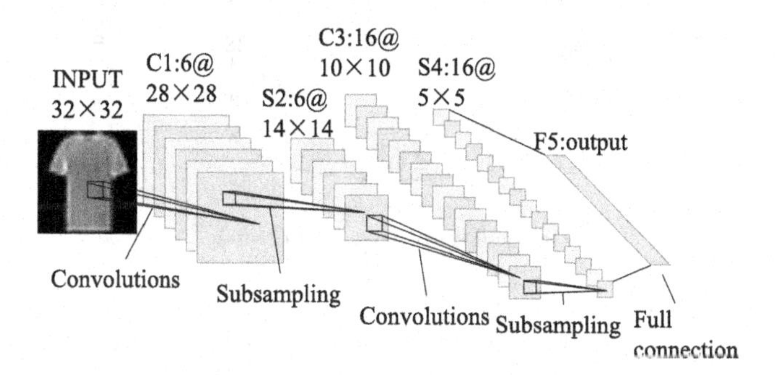

Fig. 3. The architecture of LeNet-1.

Figure 3 shows the architecture of LeNet-1. The model has five layers based on LeNet-1 [3].

The LeNet-1 model includes the C1 convolution layer, the S2 pooling layer, the C3 convolution layer, the S4 pooling layer, and the F5 full connection layer.

C1 uses six convolution kernels, each with a size of 5×5. Each convolution kernel convolves with the original input image (32×32). It outputs six feature maps, and the size of each feature map is $(32 - 5 + 1) \times (32 - 5 + 1) = 28 \times 28$. The activation function is sigmoid.

S2 uses mean-pooling. The size of the pooling unit is 2×2. The output size of 6 feature maps after pooling is 14×14.

C3 uses 16 convolution kernels, each with a size of 5×5. Each convolution kernel accumulates several input images (14×14) from S2 after convolution. It outputs 16 feature maps, and the size of each feature map is

$(14-5+1)\times(14-5+1)=10\times 10$. Some feature maps from S2 are connected to C3 with three outputs, some with four outputs, and some with six outputs. This allows C3 to extract more features.

S4 is the same as S2. The size of the pooling unit is 2×2. The output of 16 feature maps after pooling is 5×5.

F5 has ten output channels which are fully connected to 16 feature maps from S4. Each channel has a 1×400 weighting matrix. The result is a 1×10 classification matrix.

3.3 Serial Execution Mode of LeNet-1

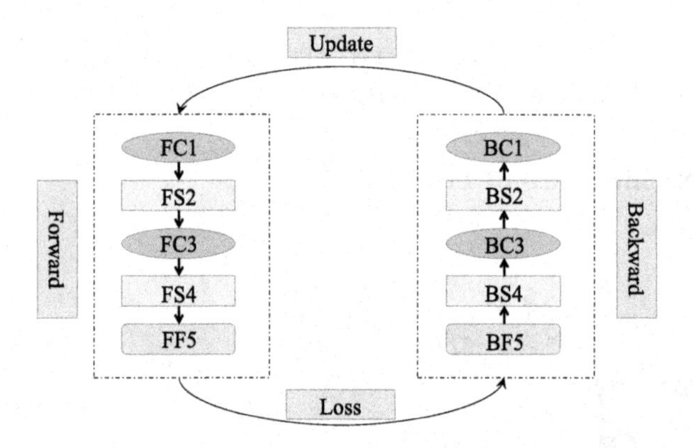

Fig. 4. The process of LeNet-1.

The training process of an image via LeNet-1 includes three stages: forward, backward, and update. Figure 4 shows the process of LeNet-1.

The notations used in LeNet-1 are shown in Table 1.

FC1, FS2, FC3, FS4, and FF5 are performed in forward order. The equation for the most critical convolution operation is

$$Z^{[l]} = W^{[l]} \bigotimes A^{[l-1]} + b^{[l]} \tag{1}$$

where

$$A^{[l]} = f\left(Z^{[l]}\right) \tag{2}$$

BF5, BS4, BC3, BS2, and BC1 are performed in backward order. The equation for the error and gradient in backward is

$$\mathrm{d}Z^{[l]} = ROT180\left(W^{[l+1]}\right) \bigotimes \mathrm{d}Z^{[l+1]} \tag{3}$$

$$\frac{\partial L}{\partial W^{[l]}} = \mathrm{d}Z^{[l]} \bigotimes f'\left(A^{[l-1]}\right) \tag{4}$$

Table 1. The notations used in LeNet-1.

Symbols	Descriptions
α	Learning Rate
f	The activation function
L	Loss
l	The lth layer
$Z^{[l]}$	The output of the lth layer
$W^{[l]}$	The convolution kernel of the lth layer
$A^{[l-1]}$	The input feature map of the lth layer
$b^{[l]}$	The bias of the lth layer
$\otimes$	Convolution product
E_p	The parallel efficiency of only process-level optimization
E_{p+t}	The parallel efficiency of SW-LeNet
T_p	The execution time of only process-level optimization
T_{p+t}	The execution time of SW-LeNet
T_s	The execution time of the serial program
T_{CPEs}	The execution time of only thread-level optimization on a single core group.
N_p	The number of processes

In the update, the convolution kernel parameters of UC1, UC3, and UF5 need to be updated. The parameter update equation is

$$W^{[l]} = W^{[l]} - \frac{\alpha}{batch\_size} \sum \frac{\partial L}{\partial W^{[l]}} \tag{5}$$

$$b^{[l]} = b^{[l]} - \frac{\alpha}{batch\_size} \sum \frac{\partial L}{\partial W^{[l]}} \tag{6}$$

4 SW-LeNet - Parallel Optimization of LeNet-1 on the Sunway Bluelight II Supercomputer

In this section, we implement two-level parallel optimization strategies, including thread-level and process-level optimization. The parallel optimization algorithm with all optimization strategies based on LeNet-1 is named SW-LeNet.

In thread-level optimization, we use three optimization strategies in this paper. One is optimization from CPEs. We move the convolutions of FC1, FC3, BS2, UC1, and UC3 to the CPEs for parallel optimization. For the few output channels of FC1 and the smaller feature maps of FC3, we propose a hybrid scheme and the data transmission optimization via DMA. The last strategy is register and SIMD optimization, which reduces the number of memory access.

For process-level optimization, this paper adopts a data parallelism approach and parameter packing optimization.

4.1 Thread-Level Optimization

CPEs Parallelism. In this paper, FC1, FC3, BS2, UC1, and UC3 are optimized via CPEs. We divide missions equally to each slave core of CPEs, with the remainder to be performed by MPE. Figure 5 shows this scheme, and we call it scheme A.

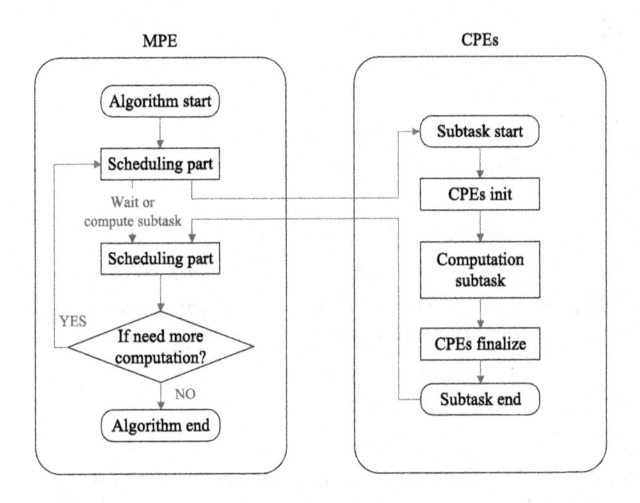

Fig. 5. Sunway many-core hierarchical scheme (scheme A).

Hybrid Scheme and DMA Optimization. For some smaller feature maps and fewer output channels, a generic acceleration library is not always the best choice. In particular, FC1 has only six output channels and FC3 has only a 14×14 feature map. Using the same scheme for all convolutional operations, whether forward, backward or update, does not give the best result.

Next, we describe scheme B. For example, in forward, the input feature map of FC1 requires 28×28 convolution operations. The scheme divides these convolution operations into 49 subtasks by CPEs. Each CPE gets an 8×8 data block and a 5×5 convolution kernel matrix via DMA. Finally, the results are transmitted back to the main memory via DMA. Figure 6 shows this scheme.

Scheme C, which divides tasks into subtasks according to the number of output channels. The more output channels, the more effective this method is. For example, in forward, FC3 requires 10×10 convolution operations. The input feature maps of FC3 are too small. If it is the same as FC1, it will slow down

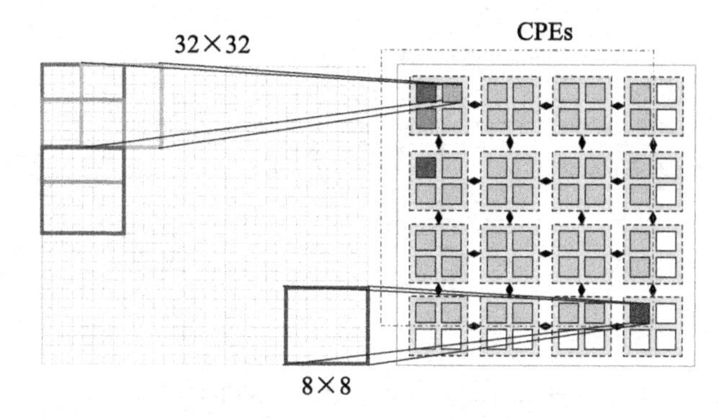

Fig. 6. CPE task division method (scheme B).

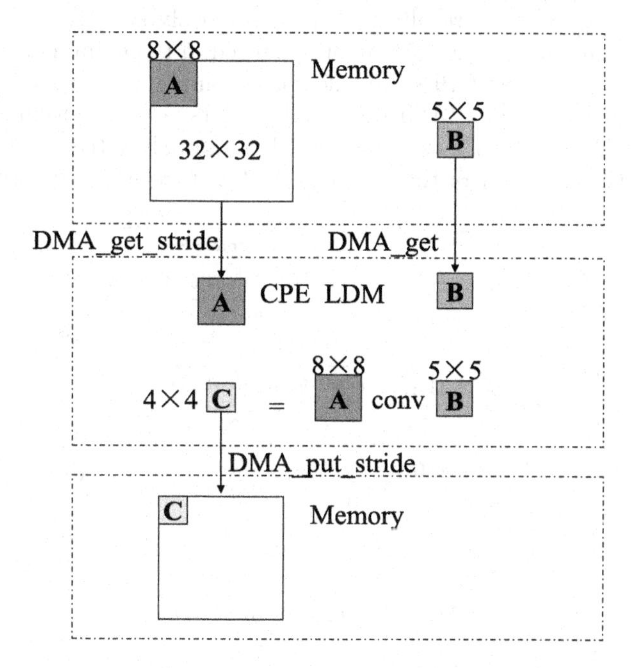

Fig. 7. DMA process of CPE.

the computation due to DMA. Therefore, the parallel strategy of the CPEs of FC3 is to divide these convolution operations into 16 subtasks. Each CPE gets an entire 14×14 input feature map and a 5×5 convolution kernel matrix via DMA. Finally, the 10×10 output feature map is transmitted back to the main memory via DMA.

Schemes B and C use data transmission optimization with DMA. The LDM of the CPE gets the data block of the feature map from memory with DMA_get_stride and gets the data of the convolution kernel with DMA_get. After convolution, the data is transmitted back to memory with DMA_put_stride. Figure 7 shows the detailed process.

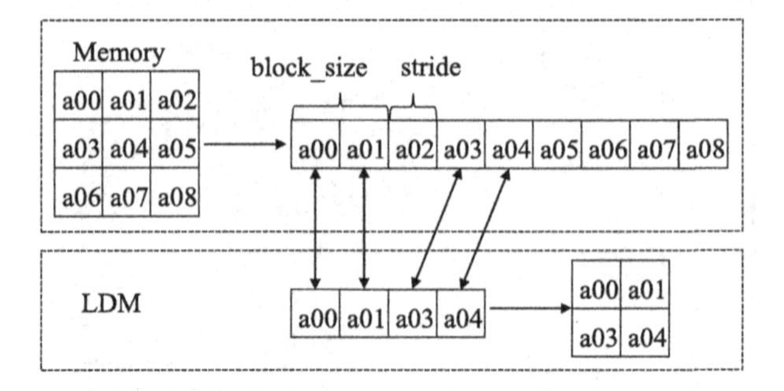

Fig. 8. DMA_GET/PUT_STRIDE.

Scheme B requires stride reading via DMA because the feature map data block and the convolution result are not continuously stored in main memory. Two-dimensional matrices are continuously stored in the main memory. According to the split data, the CPEs find the corresponding addresses and calculate the size of the stride. The LDM continuously gets the discontinuous data via DMA and combines them into a new two-dimensional matrix. In this way, we can improve the efficiency of the convolution. Figure 8 shows the process.

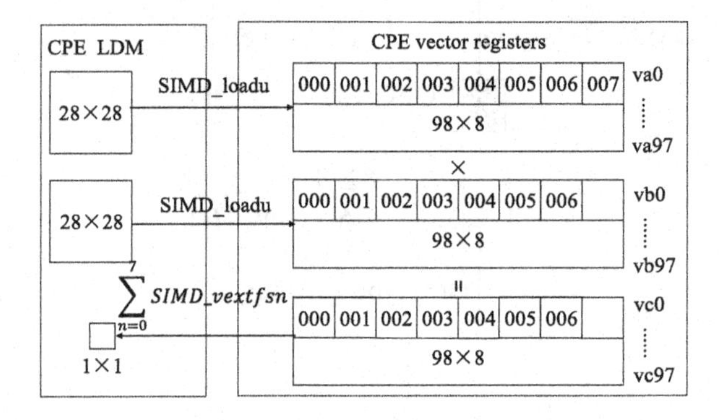

Fig. 9. The process of SIMD.

Register and SIMD Optimization. The MPE of the new generation Sunway processors supports 256-bit SIMD vector operation and each CPE supports 512-bit SIMD vector operation. SIMD can improve the speed of the convolution product. Putting some shared data (such as convolution kernel data) into register variables can improve memory access efficiency.

We implement 512-bit SIMD for FC1 and FC3, UC1 and UC3 in the CPEs. We also use 256-bit SIMD for FF5 in the MPE. For example, two 28×28 matrices are continuously stored in the LDM. The process of SIMD is to load the data from the matrices into 512-bit vector registers, which are VA and VB. VC is the result of multiplying VA and VB. Finally, we get the result by extracting and accumulating the data from the VC. Figure 9 shows the process of SIMD.

4.2 Process-Level Optimization

Algorithm 1 Process-level optimization algorithm on process k

Require: dataset MNIST, sub mini batch size B per process, parameters w, bias b, the number of process n.

1: **for** $t = 0, 1, ..., max\_iter$ **do**
2: CPEs Init()
3: **for** $i = 0, 1, ..., B$ **do**
4: Forward() //Eqs. (1), (2)
5: Backward() //Eqs. (3), (4)
6: **end for**
7: Update() //Eqs. (5), (6)
8: CPEs Finalize()
9: All_Reduce $W_t^k : W_t \leftarrow \frac{1}{n} \sum_{k=0}^{n-1} W_t^k$
10: **end for**

As shown in Fig. 10, we adopt data parallelism on process-level optimization, and the tasks are evenly divided into processes for parallel training and aggregation. It does not change the architecture of LeNet-1 in a single process. After completing a sub-mini-batch, SW-LeNet uses an all-reduce operation to update the entire network parameters of each process. The process-level optimization algorithm is described in Algorithm 1.

In the LeNet-1, the parameters for each network layer are distributed discretely in memory. We use parameter packing optimization to improve the efficiency of all-reduce. This method can aggregate them into a contiguous piece of memory. The operation of parameter packing is shown in Fig. 11. We can reduce

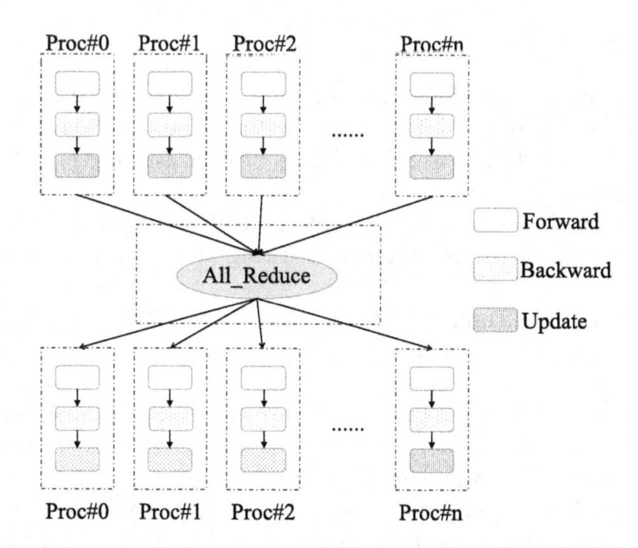

Fig. 10. The process-level optimization of LeNet-1.

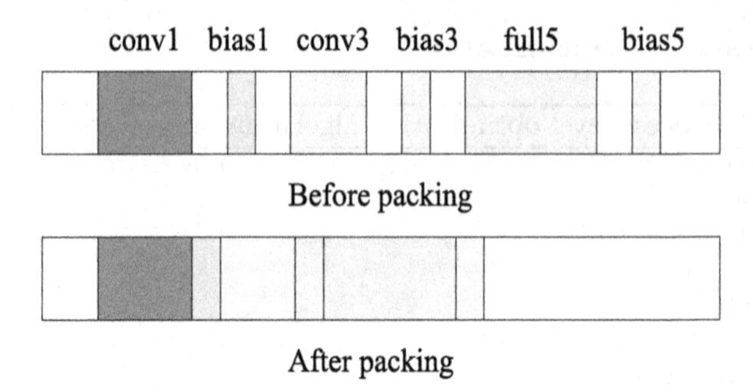

Fig. 11. The operation of parameter packing.

the time of repeated use of all-reduce by packing the parameters. Parameter packing optimization can further improve the speed of process-level optimization and increase the parallel efficiency of SW-LeNet.

5 Evaluation

Table 2. The hardware and software used in the evaluation.

Hardware	Computer	Sunway Bluelight II Supercomputer
	Processor model	SW26010pro
	Number of cores	133,120 (2,048 MPEs and 131,072 CPEs)
	Memory	32,768 GB
Software	System	Sunway Linux 4.4.15
	Compiler	mpicc and sw9gcc

Our experiments are performed on the Sunway Bluelight II Supercomputer. Table 2 shows the hardware and software used in the evaluation. Because of the characteristics of SW26010pro, SW-LeNet is written entirely in C language for better performance.

This paper uses the MNIST [3,8] dataset to test the original LeNet-1 and SW-LeNet. We complete three rounds of testing and take the average value as the final data.

Section 5.1 tests the running time and speedups of thread-level optimization on a single core group. Thread-level optimization includes three optimization schemes.

Section 5.2 tests the running time and speedups of data parallelism optimization and parameter packing optimization, compared with only data parallelism optimization. These are on process-level optimization.

Section 5.3 tests the performance of SW-LeNet and only process-level optimization with a different number of processes and batches. SW-LeNet includes thread-level and process-level optimization. The performance is about running time and speedups. The processes are 1, 2, 4, 8, 16, 32, 64, 128, 256, 512, 1,024, and 2,048 respectively. The numbers of sub-mini-batches are respectively equal to 4, 16, and 64.

Section 5.4 tests the scaling of SW-LeNet and only process-level optimization with a different number of processes and batches. We use one process as the base efficiency to calculate the parallel efficiency with other process numbers.

5.1 The Performance of Thread-Level Optimization

Figure 12 shows the speedups of local with three thread-level optimization schemes, compared with the original program. The speedups are obvious in most layers after using CPEs. FC3 and BS2 are both small input feature maps, so the hybrid scheme and DMA are more effective in optimizing. The results are perfect for FC1, FC3, UC1, and UC3 with register and SIMD optimization. Because the primary operations in these layers are the convolutional product. None of these three thread-level optimizations are optimized for the shorter time layers FS2, FS4, BC1, BC3, BS4, BF5, and UF5. So the speedups of these sections are vacant. In particular, although FF5 has a small percentage of the time, register and SIMD optimization is still effective for this part.

Figure 13 shows the speedups of the forward, backward, update, and total with thread-level optimization schemes. Compared with the original program, the forward, backward, and update achieve 5.88x, 3.16x, and 6.37x speedups, respectively. The total running time of the optimized program achieves 4.94x

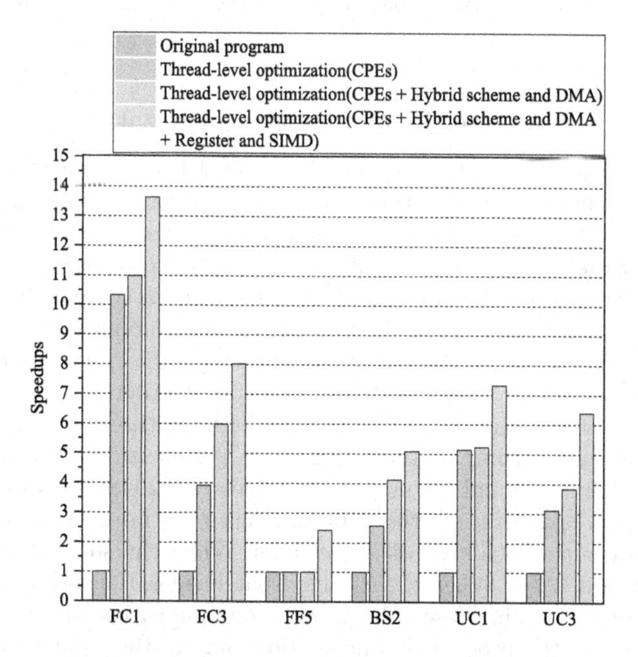

Fig. 12. The speedups of local with thread-level optimization schemes.

speedups. Because only BS2 of backward implements parallel optimization, the speedup of the backward is minimal. Accordingly, the speedup of the update with the highest degree of optimization is the largest.

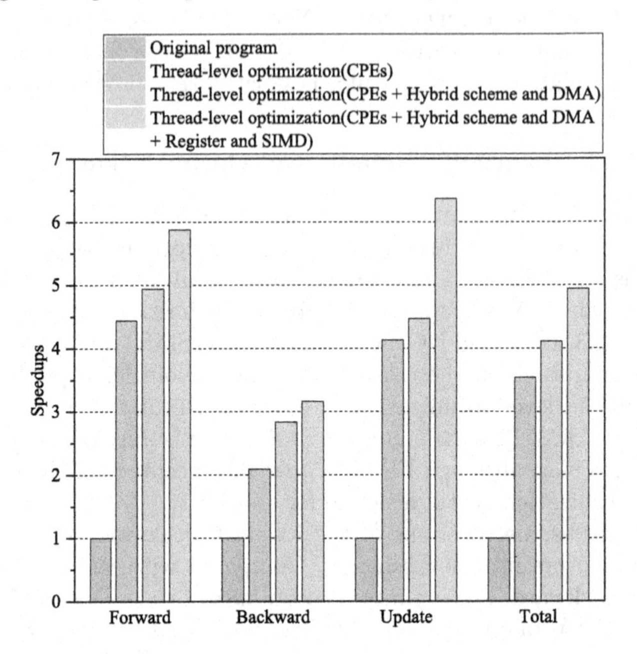

Fig. 13. The speedups of forward, backward, update, and total with thread-level optimization schemes.

5.2 The Performance of Process-Level Optimization

Table 3 shows the running time of only data parallelism optimization compared with data parallelism optimization and parameter packing optimization. The first column of Table 3 indicates the number of processes which are from 1 to 2048. We test the performance for different numbers of processes and batch sizes. There are three groups of sub-mini-batches. They are 4, 16, and 64 respectively. For each batch size, the former column shows the running time of only data-parallel optimization, and the latter column shows the running time of data-parallel optimization and parameter packing optimization.

Figure 14 shows the speedups of data parallelism optimization and parameter packing optimization compared with only data parallelism optimization. When the sub-mini-batch is 4 and the number of processes is 2048, compared with only data parallelism optimization, the data parallelism optimization and parameter packing optimization achieve 1.63x speedups. When the sub-mini-batch is 16, the speedups are 1.20x. When the sub-mini-batch is 64, the speedups are 1.05x. As the sub-mini-batch increases, the process communication time decreases, and the proportion of the process communication time in the total time decreases. The effect of parameter packing optimization slowly disappears.

Table 3. The running time of only data parallelism optimization compared with data parallelism optimization and parameter packing optimization. (second)

Processes	Batch Size = 4		Batch Size = 16		Batch Size = 64	
1	9493.94	**9493.94**	8574.75	**8574.75**	8344.36	**8344.36**
2	5120.77	**4857.66**	4386.15	**4319.77**	4196.38	**4182.60**
4	2699.85	**2446.67**	2225.27	**2166.07**	2106.76	**2092.03**
8	1421.45	**1231.50**	1134.50	**1087.01**	1060.42	**1048.30**
16	752.41	**620.78**	577.85	**545.46**	534.27	**525.59**
32	396.69	**314.26**	296.86	**274.63**	269.29	**263.89**
64	214.96	**159.99**	151.89	**138.09**	135.85	**132.79**
128	116.39	**82.06**	78.09	**69.89**	68.35	**66.67**
256	61.64	**41.84**	40.44	**35.53**	34.78	**33.76**
512	33.54	**22.24**	21.21	**18.25**	17.54	**16.99**
1024	17.84	**11.54**	10.98	**9.22**	8.90	**8.55**
2048	10.37	**6.38**	5.83	**4.87**	4.53	**4.30**

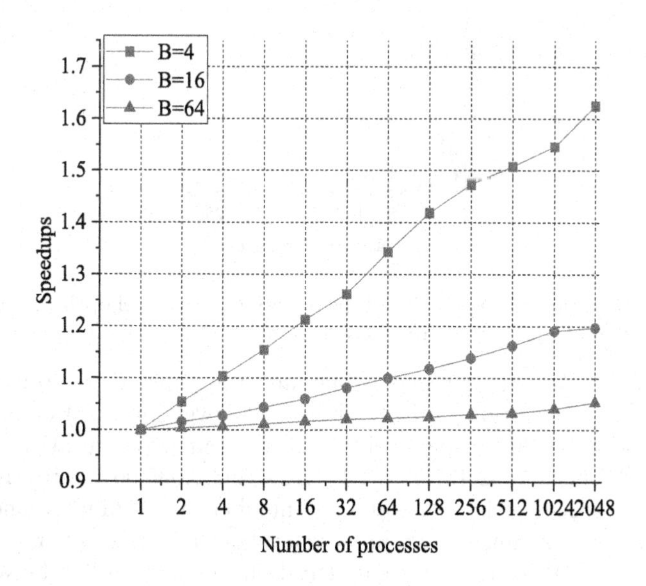

Fig. 14. The speedups of data parallelism optimization and parameter packing optimization vs. only data parallelism optimization.

5.3 The Performance of SW-LeNet

SW-LeNet includes thread-level optimization (a hybrid optimization scheme, data transmission optimization via DMA, memory access optimization with processor register, and SIMD optimization) and process-level optimization (data-parallel optimization and parameter packing optimization).

We evenly divided the images of the dataset into parallel training tasks for each process. The number of processes are 1, 2, 4, 8, 16, 32, 64, 128, 256, 512, 1,024, and 2,048 respectively. The number of sub-mini-batches are respectively equal to 4, 16, and 64.

Figure 15, 16 and 17 shows the execution time of only process-level optimization and SW-LeNet. Figure 15, 16 and 17 also shows the speedups of SW-LeNet, compared with only process-level optimization. They are tested with different numbers of sub-mini-batches and processes. The results are averaged over 3 rounds of training on the MNIST [3,8] dataset. There are 100 epochs in each training.

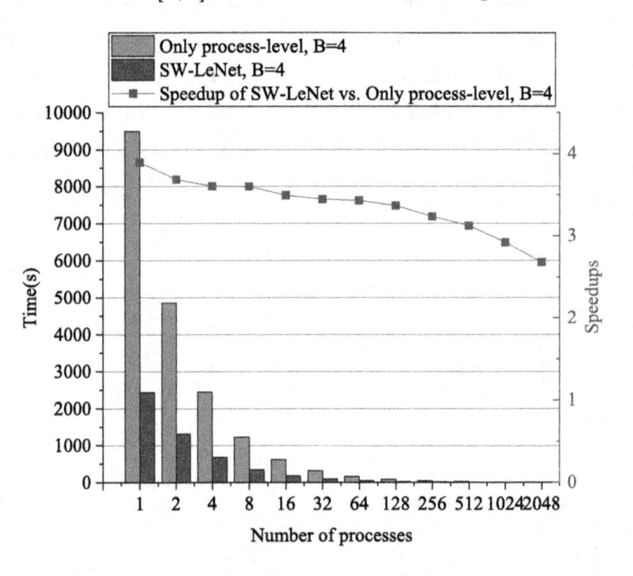

Fig. 15. The execution times and speedups of only process-level optimization and SW-LeNet with different number of processes (sub-mini-batch = 4).

When the number of processes is one, and the number of sub-mini-batch is 4, 16, and 64, the execution time of only process-level optimization is 9,493.94 s, 8,574.75 s, and 8,344.36 s respectively. With the same configuration, the execution time of SW-LeNet is 2,437.45 s, 2,141.12 s, and 2,035.06 s respectively. When the number of processes is 2,048, and the number of sub-mini-batches is 4, 16, and 64, the execution time of only process-level optimization is 6.38 s, 4.87 s, and 4.30 s respectively. With the same configuration, the time of SW-LeNet is 2.38 s, 1.53 s, and 1.17 s respectively.

When the number of processes is one, and the number of sub-mini-batches is 4, 16, and 64, the speedups of SW-LeNet compared with only process-level optimization are 3.90x, 4.00x, and 4.10x respectively. When the number of processes

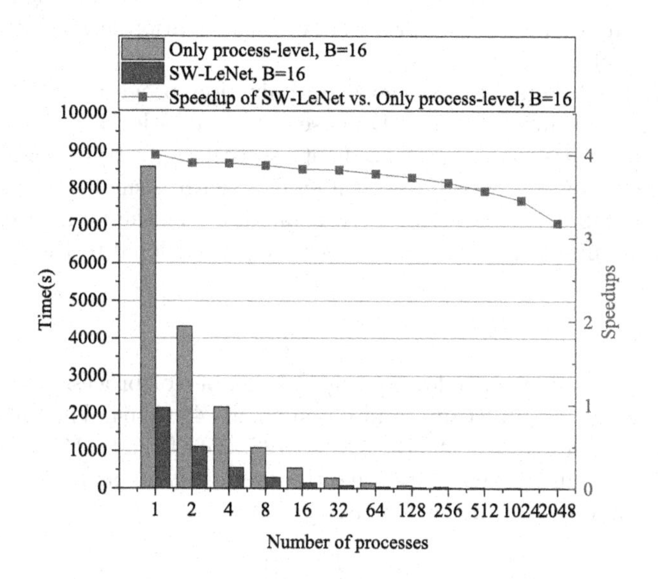

Fig. 16. The execution times and speedups of only process-level optimization and SW-LeNet with different number of processes (sub-mini-batch = 16).

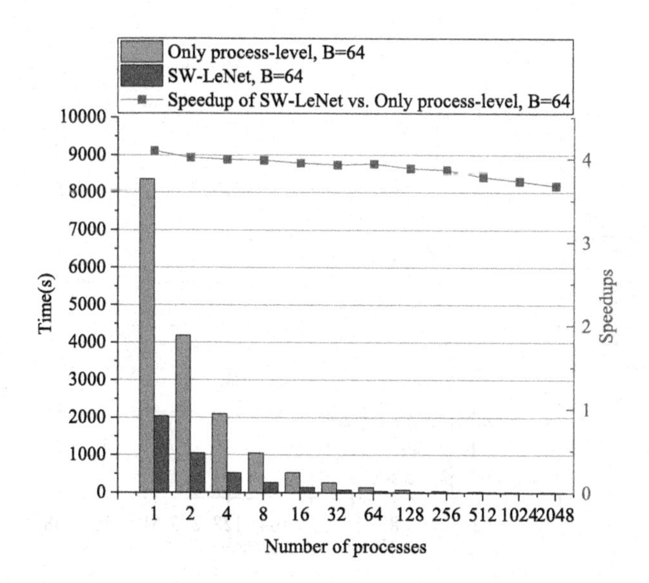

Fig. 17. The execution times and speedups of only process-level optimization and SW-LeNet with different number of processes (sub-mini-batch = 64).

is 2,048, and the number of sub-mini-batches is 4, 16, and 64, the speedups of SW-LeNet compared with only process-level optimization are 2.68x, 3.18x, and 3.67x respectively.

Compared with only process-level optimization program, SW-LeNet does not reach the speedups of 4.94x. Because the process-level optimization adds the calculation of multiple image gradients and the data transmission between processes. However, the speedup is still evident with a smaller number of processes. Furthermore, the speedups show some decay as the number of processes increases. The sub-mini-batch size is inversely proportional to the decay rate.

5.4 Scaling

SW-LeNet is highly efficient for scaling the number of processes. This section shows the scaling test when sub-mini-batch equals 4, 16, 64. The different numbers of processes are 1, 2, 4, 8, 16, 32, 64, 128, 256, 512, 1,024, and 2,048. We complete three rounds of tests and take the average value as the final result.

The equation for the efficiency is as

$$E_p = \frac{T_s}{T_p} \times \frac{1}{N_p} \tag{7}$$

$$E_{p+t} = \frac{T_{CPEs}}{T_{p+t}} \times \frac{1}{N_p} \tag{8}$$

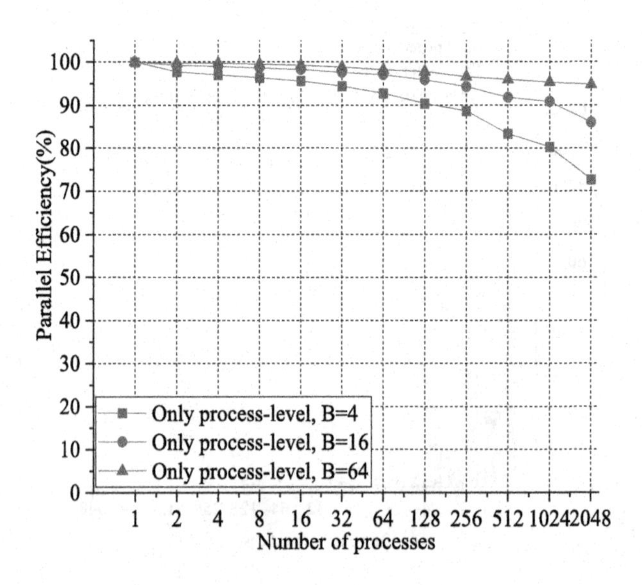

Fig. 18. The scaling of only process-level optimized LeNet-1 without thread-level optimization with different sub-mini-batches and processes.

Figure 18 shows the scaling of only process-level optimized LeNet-1 without thread-level optimization with different sub-mini-batches and processes. When the number of processes is 2,048, and the sub-mini-batch is 4, 16, and 64, the parallel efficiency is 72.66%, 85.97% and 94.75% respectively.

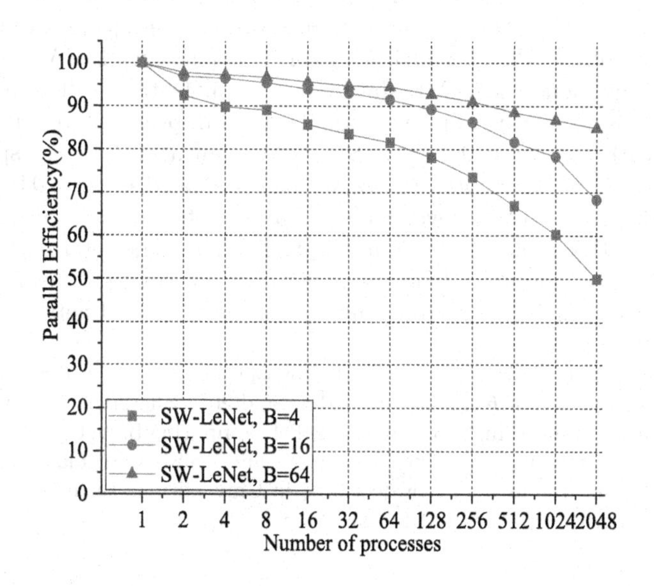

Fig. 19. The scaling of SW-LeNet with different sub-mini-batches and processes.

Figure 19 shows the scaling of SW-LeNet with different sub-mini-batches and processes. When the number of processes is 2,048, and the number of sub-mini-batches is 4, 16, and 64, the parallel efficiency of SW-LeNet is 50.01%, 68.33% and 84.93% respectively.

In data parallelism, after a sub-mini-batch is completed, the all-reduce operation is used to update the network parameters of each process. Therefore, as the size of sub-mini-batches increases, the all-reduce operation time is reduced. Finally, the efficiency of the algorithm is improved.

Compared with only process-level optimization, SW-LeNet significantly improves the time consumption, but it is less efficient than only process-level optimization in parallel. Compared with only process-level optimization, the speedups of SW-LeNet on 2048 processes are smaller than the program on one single core group. As shown in Eq. (7) and Eq. (8), $\frac{T_s}{T_{CPEs}}$ is larger than $\frac{T_p}{T_{p+t}}$, so $\frac{T_s}{T_p}$ is larger than $\frac{T_{CPEs}}{T_{p+t}}$. As a result, the parallel efficiency of SW-LeNet is smaller than only process-level optimization. However, with the thread-level optimization, SW-LeNet achieves better acceleration, especially in terms of the number of small processes, thus saving more time.

6 Conclusion

In this paper, we propose a parallel algorithm of LeNet-1 based on the Sunway Bluelight II supercomputer, named SW-LeNet. Moreover, we propose a two-level parallelization scheme, including thread-level optimization and process-level optimization. In thread-level optimization, the following optimization methods are used, including CPEs parallelism, hybrid scheme and DMA optimization, register optimization, and SIMD data parallelism. Data parallelism optimization and parameter packing optimization are used in process-level optimization. Compared with the original LeNet, SW-LeNet can achieve 4.94x speedups in a single core group. Moreover, SW-LeNet can be scaled up to 2,048 processes, with 133,120 cores, and achieves 84.93% parallel efficiency. This work in this paper explores the implementation and optimization of LeNet-1 on the Sunway Bluelight II supercomputer. Moreover, this work proves that the new Sunway many-core processor can effectively run convolutional neural networks.

Acknowledgement. This work is supported by the National Natural Science Foundation of China under Grant 62002186, 12105150, and 2021 Shandong Youth Innovation Talent Introduction and Education Plan (Parallel Computing Industrial Software Innovation Team Based on Chinese supercomputer), and TaiShan Scholars Program NO. tsqnz20221148, and the unveiling project of Qilu University of Technology (Shandong Academy of Sciences) under Grant 2022JBZ01-01.

References

1. Vaswani, A., et al.: Attention is all you need. In: Guyon, I., et al. (eds.) Advances in Neural Information Processing Systems, vol. 30. Curran Associates, Inc. (2017). https://proceedings.neurips.cc/paper/2017/file/3f5ee243547dee91fbd053c1c4a845aa-Paper.pdf
2. Simonyan, K., Zisserman, A.: Very deep convolutional networks for large-scale image recognition. arXiv preprint arXiv:1409.1556 (2014)
3. Lecun, Y., Bottou, L., Bengio, Y., Haffner, P.: Gradient-based learning applied to document recognition. Proc. IEEE **86**(11), 2278–2324 (1998). https://doi.org/10.1109/5.726791
4. Krizhevsky, A., Sutskever, I., Hinton, G.E.: ImageNet classification with deep convolutional neural networks. In: Advances in Neural Information Processing Systems, vol. 25 (2012)
5. Szegedy, C., et al.: Going deeper with convolutions. In: IEEE Conference on Computer Vision and Pattern Recognition (CVPR), pp. 1–9 (2015)
6. He, K., Zhang, X., Ren, S., Sun, J.: Deep residual learning for image recognition. In: Proceedings of the IEEE Conference on Computer Vision and Pattern Recognition, pp. 770–778 (2016)
7. Fang, J., Fu, H., Zhao, W., Chen, B., Zheng, W., Yang, G.: swDNN: a library for accelerating deep learning applications on sunway TaihuLight. In: 2017 IEEE International Parallel and Distributed Processing Symposium (IPDPS). pp. 615–624. IEEE (2017)
8. Oh, K.S., Jung, K.: GPU implementation of neural networks. Pattern Recogn. **37**(6), 1311–1314 (2004)

9. Ma, K., Han, L., Shang, J., Xie, J., Zhang, H.: Optimized realization of Quantum Fourier Transform for domestic DCU accelerator. J. Phys: Conf. Ser. **2258**, 012065 (2022)
10. Zhu, Q., Luo, H., Yang, C., Ding, M., Yin, W., Yuan, X.: Enabling and scaling the HPCG benchmark on the newest generation Sunway supercomputer with 42 million heterogeneous cores. In: Proceedings of the International Conference for High Performance Computing, Networking, Storage and Analysis, pp. 1–13 (2021)
11. http://yann.lecun.com/exdb/mnist/
12. Kuutti, S., Bowden, R., Jin, Y., Barber, P., Fallah, S.: A survey of deep learning applications to autonomous vehicle control. IEEE Trans. Intell. Transp. Syst. **22**(2), 712–733 (2020)
13. Strubell, E., Ganesh, A., McCallum, A.: Energy and policy considerations for deep learning in NLP. arXiv preprint arXiv:1906.02243 (2019)
14. James, S., Wada, K., Laidlow, T., Davison, A.J.: Coarse-to-fine q-attention: efficient learning for visual robotic manipulation via discretisation. In: Proceedings of the IEEE/CVF Conference on Computer Vision and Pattern Recognition, pp. 13739–13748 (2022)
15. Bakhshinejad, N., Hamzeh, A.: Parallel-CNN network for malware detection. IET Inf. Secur. **14**(2), 210–219 (2020)
16. Rao, G.M., Ramesh, D.: Parallel CNN based big data visualization for traffic monitoring. J. Intell. Fuzzy Syst. **39**(3), 2679–2691 (2020)
17. Kabir, H., Booth, J.D., Raghavan, P.: A multilevel compressed sparse row format for efficient sparse computations on multicore processors. In: 2014 21st International Conference on High Performance Computing (HiPC), pp. 1–10. IEEE (2014)
18. Kirmani, S., Park, J., Raghavan, P.: An embedded sectioning scheme for multiprocessor topology-aware mapping of irregular applications. Int. J. High Perform. Comput. Appl. **31**(1), 91–103 (2017)
19. Chetlur, S., et al.: cuDNN: efficient primitives for deep learning. arXiv preprint arXiv:1410.0759 (2014)
20. Li, L., et al.: swCaffe: a parallel framework for accelerating deep learning applications on Sunway TaihuLight. In: 2018 IEEE International Conference on Cluster Computing (CLUSTER), pp. 413–422. IEEE (2018)
21. Jia, Y., et al.: Caffe: convolutional architecture for fast feature embedding. In: Proceedings of the 22nd ACM International Conference on Multimedia, pp. 675–678 (2014)
22. Yang, C., et al.: 10M-core scalable fully-implicit solver for nonhydrostatic atmospheric dynamics. In: Proceedings of the International Conference for High Performance Computing, Networking, Storage and Analysis, SC 2016, pp. 57–68. IEEE (2016)
23. Fu, H., et al.: 18.9-Pflops nonlinear earthquake simulation on Sunway TaihuLight: enabling depiction of 18-Hz and 8-meter scenarios. In: Proceedings of the International Conference for High Performance Computing, Networking, Storage and Analysis, pp. 1–12 (2017)
24. Liu, Y., et al.: Closing the "quantum supremacy" gap: achieving real-time simulation of a random quantum circuit using a New Sunway Supercomputer. In: Proceedings of the International Conference for High Performance Computing, Networking, Storage and Analysis, pp. 1–12 (2021)
25. Shang, H., et al.: Extreme-scale ab initio Quantum Raman Spectra Simulations on the leadership HPC system in China. In: Proceedings of the International Conference for High Performance Computing, Networking, Storage and Analysis, pp. 1–13 (2021)

26. Ma, Z., et al.: BaGuaLu: targeting brain scale pretrained models with over 37 million cores. In: Proceedings of the 27th ACM SIGPLAN Symposium on Principles and Practice of Parallel Programming, pp. 192–204 (2022)
27. Li, M., et al.: Bridging the gap between deep learning and frustrated quantum spin system for extreme-scale simulations on new generation of Sunway Supercomputer. IEEE Trans. Parallel Distrib. Syst. **33**(11), 2846–2859 (2022)

Multi-label Detection Method for Smart Contract Vulnerabilities Based on Expert Knowledge and Pre-training Technology

Chi Jiang[1], Guojin Sun[2], Jinqing Shen[2], Binglei Yue[1], and Yin Zhang[1,2](✉)

[1] School of Information and Communication Engineering, University of Electronic Science and Technology of China, Chengdu 610000, China
[2] Shenzhen Institute for Advanced Study, UESTC, University of Electronic Science and Technology of China, Shenzhen 518000, China
zhangyin123@uestc.edu.cn

Abstract. Since the establishment of the global decentralized application platform Ethereum in 2015, decentralized applications based on smart contracts have developed rapidly. While smart contracts are widely used in blockchain, they also face more and more security risks, and smart contract vulnerability detection becomes more and more important. Therefore, aiming at the problems that the existing bytecode-based vulnerability multi-label detection methods use a large number of length violence stages, which may lose key vulnerability information and cause misjudgment, resulting in low accuracy of contract vulnerability detection results and lack of multi-label classification, this paper proposes an intelligent contract vulnerability multi-label detection method based on expert knowledge and pre-training technology. This method combines expert knowledge, Bi-LSTM and attention mechanism, and uses smart contract opcode to construct pre-training language model and multi-label classification model. The experimental results show that the accuracy, precision, recall and F1 score of the proposed scheme are improved, and five types of smart contract vulnerabilities can be accurately identified.

Keywords: Blockchain · Smart Contract · Vulnerability Detection · Multi-label Classification · Expert Knowledge

1 Introduction

The concept of smart contracts was proposed by SZABO [1] in 1996, which defined smart contracts as a set of digital commitments, in which the contract participants can implement the agreement. With the widespread popularity of blockchain applications, smart contracts have been found to have multiple vulnerabilities, resulting in multiple incidents of contract vulnerabilities being attacked, resulting in tens of millions of dollars in losses [2–4]. Therefore, in the face of the security problems of smart contracts, it is of great significance

Z. Tari et al. (Eds.): ICA3PP 2023, LNCS 14491, pp. 299–312, 2024.
https://doi.org/10.1007/978-981-97-0808-6_17

to study efficient multi-label detection methods for smart contract vulnerabilities. Furthermore, most of the existing smart contracts are written in advanced programming languages such as Solidity, rather than directly writing bytecode. Unopened bytecodes account for a considerable proportion of smart contracts, which also brings difficulties and challenges to the detection of contract vulnerabilities. Therefore, it is of great significance to protect the legitimate rights and interests of users by detecting whether the contract contains loopholes through the bytecode on the blockchain. At present, bytecode-based vulnerability detection methods have achieved certain results, but there are some problems. First of all, these methods usually use length violent truncation, which may lose key vulnerability information, resulting in misjudgment, making the detection result less accurate and less effective. Secondly, due to the lack of pre-trained language models for smart contract opcodes, these methods are difficult to fully mine the semantic information of smart contracts, thus limiting the accuracy and efficiency of detection. Therefore, these methods have the disadvantages of limited application, high false alarm rate, poor scalability, and long time consumption. In recent years, some researchers have proposed the use of machine learning and deep learning methods for smart contract vulnerability detection to automate vulnerability detection. However, these methods can only use binary detection for smart contracts, and face problems such as few detectable types and poor interpretability of smart contracts containing multiple vulnerabilities [5].

Therefore, aiming at the above problems, this paper proposes a multi-label detection method for smart contract vulnerabilities based on expert mode and pre-training technology. This method combines expert knowledge, Bi-LSTM and attention mechanism, and uses smart contract opcode to construct pre-training language model and multi-label classification model. Specifically, this method first performs data preprocessing, parses the smart contract bytecode binary file into an opcode, and uses the Word2 vec model to preprocess the data, thereby converting the opcode data set into a feature matrix, which can be used as the input of the neural network. At the same time, the core slicing method based on expert knowledge is used to obtain the key code fragments of the smart contract as the input of the pre-training model, thereby reducing the length of the contract opcode sequence. Then, it implements a multi-label classification task through a vulnerability detection module. The task consists of a Bi-LSTM layer, an attention layer, and a fully connected layer. The Bi-LSTM layer is used to obtain the semantic relationship between the previous and the latter opcodes in the smart contract. The attention mechanism layer adds weights to important features to ensure the effectiveness of the smart contract vulnerability detection model, and finally realizes multi-label classification through the fully connected layer. In general, the research contents and contributions of this paper are summarized as follows:

- A core slicing method based on expert knowledge and a pre-trained language model based on Word2vec are proposed. The former improves the accuracy and reliability of the pre-training model, thereby enhancing the effect of smart contract vulnerability detection. The latter can better mine the vulnerability

information in smart contracts and provide strong support for subsequent multi-label classification tasks.
- A multi-label classification model of smart contract opcode is proposed. The existing model performs binary classification detection on contracts, which can only identify one vulnerability, while our model can simultaneously identify multiple vulnerabilities of smart contracts.
- The multi-label detection method of smart contract vulnerability proposed in this paper is verified by a large number of experiments on real data sets. Experimental results show that the proposed method can detect smart contract vulnerabilities efficiently and accurately.

The rest of this article is organized as follows: In Sect. 2, we will explore the latest technologies for smart contract vulnerability detection methods. In Sect. 3, we propose a multi-label detection method for smart contract vulnerabilities based on expert mode and pre-training technology. In Sect. 4, we evaluate the proposed detection method and show the results. Finally, we summarize the full text and discuss some future work.

2 Related Work

In recent years, researchers have proposed some smart contract vulnerability detection methods based on machine learning or deep learning. Momeni P [6] proposed a machine learning-based model to detect security vulnerabilities in smart contracts on the Ethereum platform, and used static code analysis as the underlying technology to train a series of machine learning models for different security vulnerabilities. Eshghie M [7] proposed the Dynamit framework, a monitoring framework for detecting reentrant vulnerabilities in Ethereum smart contracts. The novelty of the proposed framework is that it only relies on transaction metadata and balance data from blockchain systems, and does not require domain knowledge, code detection or special execution environment. In addition, Yu X [8] uses deep-first traversal to convert it into a sequence, uses Bi-LSTM for feature learning, and proposes a smart contract vulnerability detection technology based on deep learning. The bytecode of the smart contract is converted into high-level language code for analysis and research. However, the label is not manually completed, but the software complexity index is used, because the complex software has a high possibility of vulnerabilities.

In order to effectively manage information in a structured way, label technology came into being. However, in the face of the trend of data diversification, a single label cannot fully explain and summarize the data information. Multiple annotation of data information can make the original information adapt to more application scenarios.Alhuzali et al. [9] proposed the SpanEmo model, which uses multi-label sentiment classification as span prediction to help sentiment analysis models learn the association between labels and words in sentences, and also introduces a loss function for modeling multiple coexisting emotions. Yogarajan et al. [10] studied the multi-symptoms of patients, used multi-label variants of medical text classification to enhance the prediction of concurrent medical codes, demonstrated new embeddings in health-related texts, and addressed the unbalanced comparison set variant of the embedding model for multi-label medical text classification. We found that multi-label classification methods have achieved more and more results in the field of natural language processing. Smart contract vulnerability detection is a natural language processing problem. Most of the existing research in this field is based on binary classification methods, which can only detect whether there are loopholes in the contract. Therefore, we use a multi-label classification method to detect smart contracts and multiple types of vulnerabilities, so as to accurately, effectively and quickly detect vulnerabilities in smart contracts and ensure the security of blockchain platforms.

3 Multi-label Detection Method for Smart Contract Vulnerability Based on Expert Knowledge and Pre-training Model

3.1 Smart Contract Vulnerability Analysis and Expert Knowledge

With the wide application of smart contracts, the research on the security of smart contracts is becoming more and more important. The model proposed in this paper detects the following five types of vulnerabilities, namely reentrant, dangerous delegate calls, timestamp dependencies, integer overflow, and unchecked function call return values.

Some traditional smart contract vulnerability detection methods, such as [12–14], use classical static analysis or dynamic execution techniques to identify vulnerabilities. They fundamentally rely on several expert-defined patterns. In addition, manually defined patterns have inherent risks that are error-prone, and some complex patterns need to be covered. Rough use of several rigid modes can lead to high false positive and false negative rates, and cunning attackers may use techniques to easily bypass mode checking. In addition, with the rapid increase in the number of smart contracts, it is not possible for a few experts to screen all contracts to design accurate models. A possible solution might be to have each expert tag some contracts and then collect all tagged contracts from many experts to train a model that automatically predicts whether a contract has a specific type of vulnerability. Therefore, this paper uses expert knowledge to extract the key vulnerability sequence of smart contracts, and uses this to assist vulnerability detection and improve the performance of vulnerability detection.

3.2 Overview of Multi-label Detection Methods for Smart Contract Vulnerabilities Based on Expert Knowledge and Pre-training Model

The overall framework of the smart contract vulnerability detection method based on expert knowledge and pre-training model is shown in Fig. 1 and Fig. 2. The overall framework is divided into two modules, namely the pre-training part

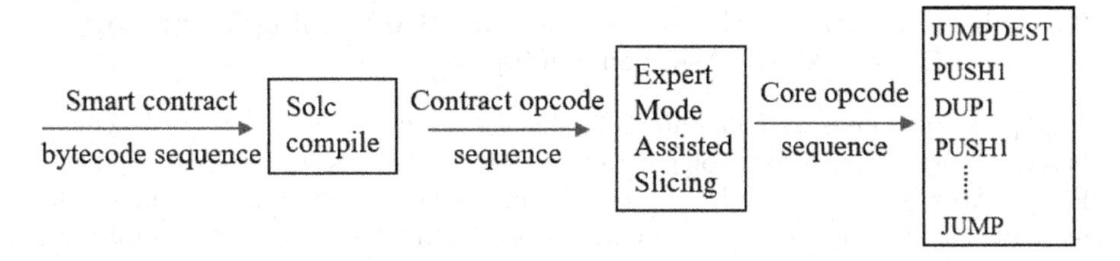

Fig. 1. A framework based on expert mode

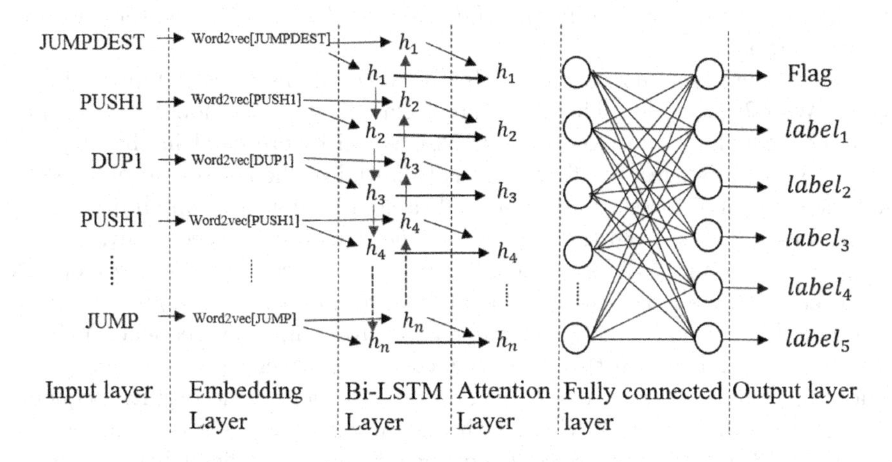

Fig. 2. Smart contract vulnerability detection framework

and the contract detection module. In the pre-training phase, the method first performs data preprocessing to parse the binary file of the smart contract bytecode into the opcode. At the same time, the key code fragments of the smart contract are obtained by the core slicing method based on expert knowledge as the input of the pre-training model, thereby reducing the length of the contract opcode sequence. Then, the obtained core operation sequence is input into the second part, that is, through the contract vulnerability detection module, which is composed of six parts, namely, input layer, word embedding layer, Bi-LSTM layer, attention layer, full connection layer and output layer. The module implements multi-label classification tasks. Specifically, after the input layer passes through the Word2vec model, the opcode is converted into a word vector for vulnerability detection, and the word is represented by a low-dimensional vector, which can avoid the problem of matrix sparsity and capture the semantic information of the word. The opcode data set is converted into a feature matrix, which can be used as the input of the neural network. The Bi-LSTM layer trains the opcode word vector generated by the input layer to capture the sequence features of the opcode. The attention mechanism layer adds weights to important features to ensure the effectiveness of the smart contract vulnerability detection model, and finally realizes multi-label classification through the fully connected layer.

3.3 Single Label Classification Based on Pre-training Technology and Expert Mode Assisted Slicing

This paper proposes a single-label classification vulnerability detection scheme based on pre-training technology and Expert Mode Assisted Slicing (EMAS). Firstly, Word2vec is trained with a large number of compiled and preprocessed unlabeled smart contract opcode data to obtain a pre-training model suitable for smart contract opcode. Secondly, through the ablation experiment, the word vector is input into a variety of deep learning models to classify the smart contract vulnerabilities, so as to verify the effectiveness of the expert pattern assisted slicing method.

This paper uses the current mainstream word distributed representation model Word2vec model. The single-label classification method based on pre-training technology and expert mode assisted slices proposed in this paper is divided into two stages: the first stage is to initialize the key operation code set and the stop condition set ; in the second stage, the sequence containing the key opcodes is extracted from the original opcode sequence X. Specifically, starting from the beginning of the sequence, if the current opcode is in the key opcode set, a recent stopping condition is found forward, and the subsequence from the opcode to the current opcode is extracted and stored in a new sequence Y. Then continue to traverse until the entire sequence X is traversed. If no subsequence containing key opcodes is found in the entire sequence, the original sequence X is returned directly as Y. Finally, the extracted sequence Y is returned. The extraction of smart contract opcode sequences by auxiliary slicing in expert mode can effectively reduce the length of contract opcode sequences, and reduce the average length of extracted contract opcode sequences by more than 20%.

3.4 Multi-label Classification of Core Slices and Bi-LSTM Based on Expert Knowledge

Aiming at the problems of poor detection, incomplete automation, and slow detection speed of traditional smart contract vulnerability detection methods, a method that can accurately and automatically detect whether smart contracts contain multiple vulnerabilities is designed. In this paper, a multi-label classification method based on expert knowledge of core slices and Bi-LSTM plus attention mechanism is proposed. The vulnerability detection solution consists of two parts: data preprocessing and vulnerability detection. Data preprocessing first compiles the smart contract bytecode into opcodes, and then converts the opcodes into word embedding matrices using the pre-trained word embedding model Word2vec to input to the neural network model. The training set is input to the vulnerability detection model for training, and the accuracy of the neural network model with high classification is obtained, and the model is finally verified on the test set. The vulnerability detection model in this paper consists of Bi-LSTM layer, attention layer, fully connected layer and sigmoid function, which is used to realize multi-label vulnerability detection for smart contract classification. The model presented in this paper can detect five types of vulnerabilities, namely integer overflow, unchecked function call return value, reentrant, dangerous delegate call, and timestamp dependency. A smart contract corresponds to six tags, each with a value of 0 or 1. The label at the top of the tag list indicates whether the contract contains vulnerabilities, the remaining tag value is 0, it means that the contract has no specific vulnerabilities, and when the value is 1, it indicates that there are vulnerabilities of the type, and each vulnerability label is independent of each other.

The error loss measurement method used in this experiment is the method used in logistic regression to measure the error between the predicted probability and the real label, specifically, by calculating the sigmoid cross-entropy between the output layer and the label as the measurement standard of error, the specific calculation formula is as follows:

$$\begin{aligned} loss(y,\hat{y}) = -\frac{1}{C}\sum_{i=1}^{m}[y^{(i)} \cdot log(\frac{1}{1+exp(-\hat{y}^{(i)})}) \\ + (1-y^{(i)}) \cdot log(\frac{exp(-\hat{y}^{(i)})}{1+exp(-\hat{y}^{(i)})})]. \end{aligned} \tag{1}$$

where C is the number of label categories, $y^{(i)}$ and $\hat{y}^{(i)}$ both are vectors, which are used to represent the real label and the network output value without any activation function, respectively.

4 Experiments and Analysis of Results

4.1 Dataset

The dataset used in the single-label experiment contains a total of 148,363 pieces of data, which contain 9 properties of the smart contract, namely address, contract code, timestamp, creation value, block number, transaction hash, code, creator, and code [16]. Next, redundant, duplicate, invalid, and empty data is removed through data cleansing operations. Single-label classification divides the vulnerabilities in the smart contract into 2 categories: whether the contract has a vulnerability and a specific type of vulnerability , and then divides the dataset into training sets and verification sets according to 8:2.

In the multi-label classification experiment, the dataset is first verified and deleted some missing values and duplicate contracts, and then five smart contract vulnerabilities with high frequency are selected for experimentation: reentrant, timestamp dependency, dangerous delegate call, integer overflow and unchecked function call return value. Before the experiment, we first perform data balancing processing, in which reentrant, timestamp dependency, dangerous delegate call, integer overflow and unchecked function call return value, and the number of these five vulnerabilities is shown in Fig. 3. After data balancing, 2000 contract data without vulnerabilities are added, and the five vulnerabilities are resampled to 1500 respectively, and the total amount of data is 9500. Among them, a smart contract corresponds to six tags, each with a value of 0 or 1.

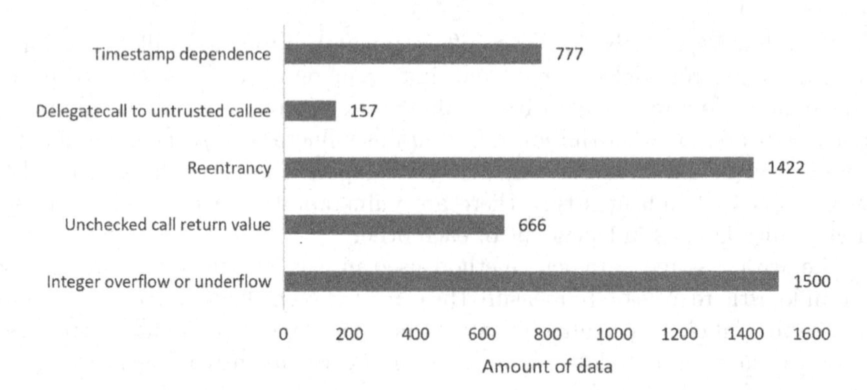

Fig. 3. Smart contract vulnerability data

4.2 Evaluation Indicators

The single-label classification experiment used the following four indicators to evaluate the model, namely accuracy, precision, recall, and F1 score (F1_ score), as shown in Eqs. (11), (12), (13) and (14). The results of true positive (TP), true negative (TN), false positive (FP), and false negative (FN) are the basis for calculating other indicators, where TP represents the number of positive samples predicted as positive samples, TN represents the number of positive samples predicted as negative samples, FP represents the number of negative samples predicted as positive samples, and FN represents the number of positive samples predicted as negative samples. Accuracy represents the proportion of data that is correctly predicted to the total sample, precision indicates the proportion of samples that are correctly predicted to the proportion of samples that are actually correct, and recall represents the proportion of samples that are actually positive to samples that are predicted to be positive. The F1 score considers both precision and recall, and balances precision and recall to take into account both metrics.

$$Accuracy = \frac{TP + TN}{TP + TN + FP + FN}. \tag{2}$$

$$Precision = \frac{TP}{TP + FP}. \tag{3}$$

$$Recall = \frac{TP}{TP + FN}. \tag{4}$$

$$F1\_score = 2 \times \frac{Precision \times Recall}{Precision + Recall}. \tag{5}$$

In multi-label classification, the prediction for an instance is a set of labels, so the prediction can be completely correct, partially correct (with varying levels of correctness), or completely incorrect. Similar to [17], we use the following metrics to evaluate the effectiveness of the proposed method: accuracy, precision, recall, and F1 score (F1_score). For each sample, accuracy is the proportion of the number of correctly predicted labels in the entire number of predicted correct and true labels, recall represents the proportion of the number of correctly predicted labels to the total number of predicted labels, accuracy represents the ratio of the number of predicted correct labels to the total number of actual labels, and the F1 score takes into account both precision and recall two indicators, indicating the harmonic average of precision and recall.

$$Accuracy = \frac{1}{S}\sum_{i=1}^{s}\frac{|y_i \cap \hat{y}_i|}{|y_i \cup \hat{y}_i|}. \tag{6}$$

$$P(y_s, \hat{y}_s) = \frac{|y_s \cap \hat{y}_s|}{|\hat{y}_s|}. \tag{7}$$

$$Precision = \frac{1}{|S|}\sum_{s \in S} P(y_s, \hat{y}_s). \tag{8}$$

$$R(y_s, \hat{y}_s) = \frac{|y_s \cap \hat{y}_s|}{|y_s|}. \tag{9}$$

$$Recall = \frac{1}{|S|}\sum_{s \in S} R(y_s, \hat{y}_s). \tag{10}$$

$$F1\_score = \frac{1}{|S|}\sum_{s \in S}\frac{2 * |y_s \cap \hat{y}_s|}{|y_s| + |\hat{y}_s|}. \tag{11}$$

where y_s is the correct label data for the true value, and $\hat{y}_s$is the value predicted by the classifier to be the correct value. $S = (x_s, Y_s), 1 \leq s \leq |S|$ is a set of training data, and $|S|$ represents the size of the training dataset.

4.3 Compare Experimental Results

We select integer overflow or underlow for detection, and the results are shown in Table 1. It is clear from the table that the vulnerability detection performance of this model is significantly improved compared to the benchmark model. In terms of integer overflow vulnerability detection, compared with the benchmark model, the accuracy of our model is improved by 23%, 0.08%, and 18.92%, and the precision is improved by 33.59%, 1.41% and 20.42%, respectively.

Table 1. Experimental results of the baseline model

	Integer Overflow or Underflow			
Method	Accuracy	Recall	Precision	F1_score
Word2vec+ LSTM	0.5000	0.3030	0.5263	0.3846
CharCNN	0.7292	0.7600	0.7159	0.7373
Word2vec + Bi-LSTM	0.5408	0.8333	0.5258	0.6445
Our Method	0.7300	0.4972	0.8622	0.6307

Table 2. Experimental results under length 800

Method	Accuracy	Precision	Recall	F1_score
Word2vec+LSTM	0.5000	0.6154	0.2286	0.3333
CharCNN	0.7208	0.7162	0.7317	0.7238
Word2vec+Bi-LSTM	0.5492	0.5316	0.8283	0.6476
EMAS+Word2vec+LSTM	0.5868	0.5715	0.7367	0.6400
EMAS+CharCNN	0.7267	0.7105	0.7650	0.7368
EMAS+Word2vec+Bi-LSTM	0.7333	0.6834	0.8683	0.7651

Since only subsequences containing key vulnerability opcodes are extracted in the single-label classification algorithm, instead of extracting all opcodes, the extracted sequence length should be moderate, and too small sequence length may cause some important opcodes to be ignored, so that the accuracy of the model and other performance indicators decrease, on the contrary, the extracted sequence length is too large, which may affect the efficiency of subsequent analysis. Therefore, this paper sets the sequence length to 400,600,800, and then compares some performance indicators such as the accuracy of Word2vec+LSTM, CharCNN and Word2vec+Bi-LSTM before and after using Expert Mode to assist sectioning, as shown in Table 2, 3 and 4.

From the three tables, it can be seen that the method based on pre-training technology and expert mode assisted slicing has achieved good results at different sequence lengths, especially in terms of accuracy. In addition, the Word2vec+Bi-LSTM method has greatly improved the four performance indicators under different sequence lengths under the use of expert mode auxiliary sectioning, for example, when the sequence length is 400, the accuracy, precision, recall rate and F1 score are improved by 26.58%, 17.44%, 36.45% and 25.48%, respectively. This may be due to a relatively high proportion of critical vulnerability opcodes contained in sequences of that length extracted. Although the LSTM model has good performance and can capture the features of long sequences, the overall performance of the LSTM is still poor and cannot highlight the key features in the data. The proposed model uses the Bi-LSTM network to capture more

Table 3. Experimental results under length 600

Method	Accuracy	Precision	Recall	F1_score
Word2vec+LSTM	0.5000	0.5263	0.3030	0.3846
CharCNN	0.7292	0.7159	0.7600	0.7373
Word2vec+Bi-LSTM	0.5408	0.5258	0.8333	0.6445
EMAS+Word2vec+LSTM	0.6094	0.5775	0.8195	0.6756
EMAS+CharCNN	0.7417	0.7139	0.8067	0.7575
EMAS+Word2vec+Bi-LSTM	0.7442	0.6806	0.9200	0.7824

Table 4. Experimental results under length 400

Method	Accuracy	Precision	Recall	F1_score
Word2vec+LSTM	0.5000	0.5517	0.4571	0.5000
CharCNN	0.7108	0.7347	0.6600	0.6953
Word2vec+Bi-LSTM	0.5000	0.5278	0.5588	0.5429
EMAS+Word2vec+LSTM	0.6528	0.5982	0.9387	0.7277
EMAS+CharCNN	0.7150	0.7115	0.7233	0.7174
EMAS+Word2vec+Bi-LSTM	0.7658	0.7022	0.9233	0.7977

comprehensive sequence features, while the auxiliary slice based on expert mode pays more attention to the local feature features of the sequence. Therefore, the model using Expert Mode assisted slices has higher accuracy and better vulnerability detection performance.

Next, we compare the performance of the multi-label classification method using Word2vec + Bi-LSTM + Attention with the multi-label classification method using One-hot + Bi-LSTM + Attention in various vulnerabilities on the test set. The evaluation results of the two methods in different vulnerabilities are shown in Table 5 and Table 6. It can be seen from the two tables that compared with the classification method of One-hot + Bi-LSTM + Attention, the multi-label classification method of Word2vec + Bi-LSTM + Attention has better overall performance and higher accuracy. In terms of delegate call vulnerability detection, the multi-label classification method of Word2vec + Bi-LSTM + Attention outperforms the One-hot+ Bi-LSTM + Attention classification method in all four indicators, especially in terms of recall, which achieves 100%. Therefore, the method proposed in this paper can more accurately detect five types of vulnerabilities, and all evaluation indicators are improved.

Figure 4 and Table 7 is the results of the multi-label classification method using One-hot + Bi-LSTM + Attention using multi-label classification evaluation index. It can be seen from the table and figure that the multi-label classification method of Word2vec + Bi-LSTM + Attention has better performance than the multi-label classification method of One-hot + Bi-LSTM + Attention on the multi-label evaluation index, which verifies the effectiveness of the proposed scheme.

Table 5. Assessment results of different types of vulnerabilities(One-hot+ Bi-LSTM + Attention)

Vulnerabilities	Accuracy	Recall	Precision	F1_score
Whether vulnerabilities are existed	0.5284	0.5213	0.8146	0.6358
Integer overflow or underflow	0.6163	0.4177	0.6301	0.5023
Unchecked call return value	0.8321	–	–	–
Reentrancy	0.8332	0.8399	0.7935	0.8160
Delegatecall to untrusted callee	0.9350	0.9300	0.9394	0.9347
Timestamp dependence	0.8658	0.8514	0.6843	0.7588

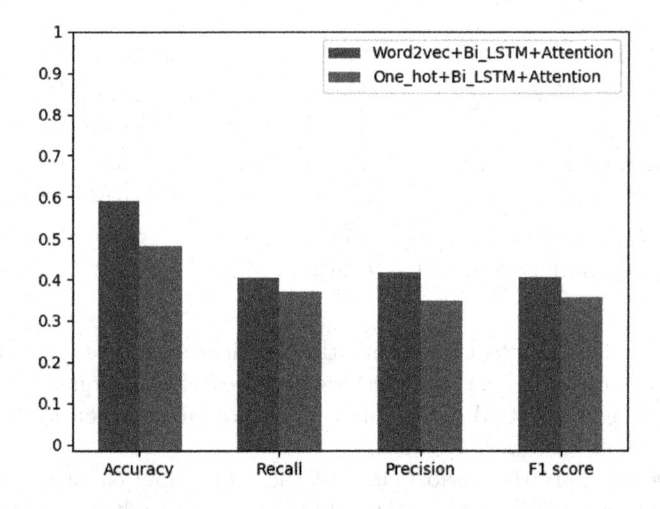

Fig. 4. The result graph under the multi-label classification indicator

Table 6. Assessment results of different types of vulnerabilities(Word2vec + Bi-LSTM + Attention)

Vulnerabilities	Accuracy	Recall	Precision	F1_score
Whether vulnerabilities are existed	0.7832	0.7773	0.9373	0.8499
Integer overflow or underflow	0.7300	0.4972	0.8622	0.6307
Unchecked call return value	0.8516	0.1536	0.8033	0.2579
Reentrancy	0.9079	0.8172	0.9688	0.8866
Delegatecall to untrusted callee	0.9994	1.0000	0.9966	0.9983
Timestamp dependence	0.9452	0.8259	0.9465	0.8821

Table 7. Results under multi-label classification metrics

Method	Accuracy	Recall	Precision	F1_score
Word2vec + Bi-LSTM + Attention	0.6057	0.4193	0.4312	0.4195
One-hot + Bi-LSTM + Attention	0.4973	0.3850	0.3629	0.3697

5 Summary and Future Works

This paper proposes a smart contract vulnerability detection scheme based on expert knowledge and pre-training technology. The expert model is used to assist the slice to extract the subsequence containing the key vulnerabilities of the smart contract, and the sequence features and context information of the opcode are captured by constructing the Bi-LSTM network model. In order to improve the detection accuracy, the attention mechanism is introduced to assign different weights to different features, so as to highlight the key features and realize the detection and recognition of smart contract vulnerabilities. At the same time, considering that the single-label vulnerability detection scheme makes the vulnerability detection performance poor, this paper uses a multi-label classification method to further improve the performance of contract vulnerability detection. The experimental results show that the performance of this model is better than that of machine learning model and traditional detection tools, and can accurately identify five types of vulnerabilities. In the future work, we will build a smart contract vulnerability data set, further optimize the model architecture, learn more smart contract features, and improve the performance of the model to detect contract vulnerabilities.

Acknowledgments. This research is funded by the National Key R&D Program of China (No. 2020YFB1006002).

References

1. Szabo, N.: Smart contracts: building blocks for digital markets. EXTROPY: J. Transhumanist Thought, (16) **18**(2), 28 (1996)
2. Zhuang, Y., Liu, Z., Qian, P., Liu, Q., Wang, X., He, Q.: Smart contract vulnerability detection using graph neural network. In: IJCAI, pp. 3283–3290 (2020)
3. Mehar, M.I., et al.: Understanding a revolutionary and flawed grand experiment in blockchain: the DAO attack. J. Cases Inform. Technol. (JCIT) **21**(1), 19–32 (2019)
4. Fu, M., Wu, L., Hong, Z., Feng, W.: Research on vulnerability mining technique for smart contracts. J. Comput. Appl. **39**(7), 1959 (2019)
5. Hu, Z., Tsai, W.-T., Zhang, L.: Smart-contract vulnerability detection method based on deep learning. In: Smart Computing and Communication: 7th International Conference, SmartCom 2022, New York City, NY, USA, November 18–20, 2022, Proceedings, pp. 450–460. Springer, Cham (2023). https://doi.org/10.1007/978-3-031-28124-2_43
6. Momeni, P., Wang, Y., Samavi, R.: Machine learning model for smart contracts security analysis. In: 2019 17th International Conference on Privacy, Security and Trust (PST), pp. 1–6. IEEE (2019)
7. Eshghie, M., Artho, C., Gurov, D.: Dynamic vulnerability detection on smart contracts using machine learning. In: Evaluation and Assessment in Software Engineering, pp. 305–312 (2021)
8. Yu, X., Zhao, H., Hou, B., Ying, Z., Wu, B.: DeeSCVHunter: a deep learning-based framework for smart contract vulnerability detection. In: 2021 International Joint Conference on Neural Networks (IJCNN), pp. 1–8. IEEE (2021)

9. Alhuzali, H., Ananiadou, S.: SpanEmo: casting multi-label emotion classification as span-prediction. arXiv preprint arXiv:2101.10038 (2021)
10. Yogarajan, V., Montiel, J., Smith, T., Pfahringer, B.: Seeing the whole patient: using multi-label medical text classification techniques to enhance predictions of medical codes. arXiv preprint arXiv:2004.00430 (2020)
11. Irving, G., Holden, J.: How blockchain-timestamped protocols could improve the trustworthiness of medical science. F1000Research **5** (2016)
12. Mikolov, T., Sutskever, I., Chen, K., Corrado, G.S., Dean, J.: Distributed representations of words and phrases and their compositionality. In: Advances in Neural Information Processing Systems, vol. 26 (2013)
13. Jiang, B., Liu, Y., Chan, W.K.: ContractFuzzer: fuzzing smart contracts for vulnerability detection. In: Proceedings of the 33rd ACM/IEEE International Conference on Automated Software Engineering, pp. 259–269 (2018)
14. Tsankov, P., Dan, A., Drachsler-Cohen, D., Gervais, A., Buenzli, F., Vechev, M.: Securify: practical security analysis of smart contracts. In: Proceedings of the 2018 ACM SIGSAC Conference on Computer and Communications Security, pp. 67–82 (2018)
15. Bahdanau, D., Cho, K., Bengio, Y.: Neural machine translation by jointly learning to align and translate. arXiv preprint arXiv:1409.0473 (2014)
16. Huang, J., Zhou, K., Xiong, A., Li, D.: Smart contract vulnerability detection model based on multi-task learning. Sensors **22**(5), 1829 (2022)
17. Sorower, M.S.: A literature survey on algorithms for multi-label learning. Or. State University, Corvallis **18**(1), 25 (2010)

PM-Migration: A Page Placement Mechanism for Real-Time Systems with Hybrid Memory Architecture

Lidang Xu, Gengbin Chen, Dingding Li(✉), and Haoyu Luo

School of Computer Science, South China Normal University, Guangzhou 510631, Guangdong, China
dingly@m.scnu.edu.cn

Abstract. Due to its higher storage density and lower energy consumption compared to DRAM, persistent memory (PM) holds the potential to address the growing memory demands of applications, such as Deep Neural Network (DNN) training. However, PM also suffers from longer latency and lower bandwidth, making it impractical to completely replace DRAM. The hybrid memory architecture, which combines DRAM and PM, is expected to improve this issue. Nevertheless, it also introduces a challenging problem: the state-of-the-art page placement mechanism designed for DRAM-only systems with NUMA ignores the performance disparities between DRAM and PM, resulting in sub-optimal performance. To improve this problem, we propose PM-Migration, a page placement mechanism tailored for real-time systems with hybrid memory architecture. PM-Migration prioritizes placing frequently accessed pages in DRAM and increases the access frequency of write-intensive pages to leverage the read-write asymmetry of Intel Optane DC persistent memory module (DCPMM), a commercially available PM hardware. It also incorporates a transmission handover strategy to select the transfer engine according to the size of the page and then utilizes DMA technology for migrating pages of size 2 MB. Experimental results demonstrate that PM-Migration provides an average throughput improvement of 1.31× to 3.6× compared to existing mechanisms proposed for the hybrid memory architecture.

Keywords: Hybrid memory architecture · Real-time system · Non-uniform memory access · Persistent memory

1 Introduction

As the development of artificial intelligence, applications such as graph processing, DNN training, and key-value store are characterized by rapidly increasing memory footprints, causing insufficient memory space [1]. This can lead to an increase in swap operations, resulting in sub-optimal system performance. Scaling up the capacity of Dynamic Random Access Memory (DRAM) appears to be an effective approach. However, it is limited by cost, power consumption, area constraints and physical space [2].

Z. Tari et al. (Eds.): ICA3PP 2023, LNCS 14491, pp. 313–324, 2024.
https://doi.org/10.1007/978-981-97-0808-6_18

PM is an emerging memory technology. DCPMM is a commercially available product built upon persistent memory (PM), providing byte-addressability, greater storage density compared to DRAM, and reduced power consumption [3] [4]. As a result, DCPMM is capable to provide a larger memory capacity to alleviate the issue of insufficient memory space. Nonetheless, there is still a performance gap between DCPMM and DRAM [5]. For example, the latency of *load* and *store* instructions in DCPMM is 2.2× and 3.4× higher than that of DRAM, respectively [6]. Thus, using a DCPMM-only system would result in performance degradation.

The DRAM-DCPMM hybrid memory architecture, which offers a tiered and byte-addressable main memory that aggregates the memory capacity of both DRAM and DCPMM, is anticipated to effectively mitigate the aforementioned issues. Nevertheless, the state-of-the-art page placement mechanism in Linux is unsuitable for the hybrid memory architecture. To comprehend the reasons behind this problem, it is essential to understand the Non-Uniform Memory Access (NUMA) architecture and the state-of-the-art page placement mechanism.

NUMA is a memory organization scheme specifically designed for multi-processor systems. In a NUMA architecture, each processor is equipped with its own dedicated local memory [7]. Modern server systems employ NUMA architecture for several reasons: 1. *Enhanced memory access performance*: NUMA architecture enables faster access to local memory by each processor, effectively reducing memory access latency. 2. *Scalability support*: NUMA architecture offers excellent scalability capabilities, making it well-suited for systems with a large number of processors. 3. *Flexibility*: NUMA architecture grants each processor independent access to its local memory, allowing the system to adapt to different types of workloads.

The existing page placement mechanism is designed for the DRAM-only system with NUMA and treats all memory nodes as DRAM. Its primary objective is to minimize memory access latency by relocating pages close to the processor executing the process that is accessing the memory, such as moving pages from remote memory nodes to local memory nodes [8]. In a DRAM-only system, it is reasonable to place pages in the local memory nodes as much as possible because the performance of remote memory is inferior to that of local memory. However, in a hybrid memory architecture with NUMA, the remote DRAM still outperforms the local DCPMM [10]. Thereby, adopting the existing page placement mechanism in a hybrid memory architecture would obscure its performance advantages and incurs additional overheads.

Table 1. Asymmetric load/store bandwidth.

Memory Type	Load Bandwidth	Store Bandwidth
DCPMM	30 (GB/s)	8 (GB/s)
DRAM	76 (GB/s)	42 (GB/s)

In recent years, there has been considerable attention from the research community to fully exploit the advantages of the hybrid memory architecture. Despite the proposal of several schemes, different defects still exist in these schemes so that their achievement of this goal is low. For example, Nimble [9] has been designed based on the simulated PM hardware, and its experimental evaluations also rely on simulation approaches, which fail to accurately capture the performance characteristics of real PM hardware. Additionally, when calculating the access frequency of a page, AutoTiering [10] does not differentiate between the number of write and read operations for the page. As a result, write-intensive pages cannot be prioritized for placement in DRAM and fail to effectively utilize the asymmetry in reads and writes of DCPMM (As depicted in Table 1, Gugnani et al. have found the *store* bandwidth of DCPMM is significantly lower than its *load* bandwidth [6]). Furthermore, most proposed mechanisms for the hybrid memory architecture solely rely on *memcpy* (memory copy) to migrate pages between memory nodes, which results in significant performance degradation when the CPU contention is high.

To overcome the limitations mentioned above, we propose PM-Migration, a novel dynamic page placement mechanism for the hybrid memory architecture. In view of the performance disparity between DCPMM and DRAM, PM-Migration prioritizes upper-tier memory for page placement and migrates less frequently accessed pages to lower-tier memory when the upper-tier memory is full. It evaluates the access frequency of a page by analyzing the number of write and read operations, instead of solely counting the number of times the page has been accessed, thus avoiding the issues caused by the read-write performance asymmetry of DCPMM. Moreover, PM-Migration incorporates a handover strategy that allows for switching between CPU and DMA transmission modes, which reduces CPU consumption during page migration and frees up CPU resources for CPU-intensive tasks.

We implemented PM-Migration in the Linux kernel 5.15. We leveraged the AutoNUMA mechanism [11] to drive the page migration process and extended it to support huge pages and transparent huge pages (THPs). The I/OAT technology of Intel [12] has been used to construct the DMA engine for PM-Migration. Experimental results show that under the CPU contention-free environment, the average throughput of PM-Migration is 1.31× that of AutoTiering and 3.12× that of Nimble. Meanwhile, under the high CPU contention environment, its average throughput is 1.94× that of AutoTiering and 3.6× that of Nimble.

The main contributions of this paper are as follows:

- We identify the limitations of the existing DRAM-only system and introduce a DRAM-DCPMM hybrid memory architecture built with NUMA architecture, highlighting the challenges associated with it.
- We propose PM-Migration, a high-efficient dynamic page placement mechanism designed to address the aforementioned challenges in the hybrid memory architecture.

- We construct a real experimental environment to evaluate the performance of PM-Migration and our results demonstrate its superiority over existing solutions.

The rest of this paper is organized as follows. In Sect. 2, we introduce the background and related work of this paper. The design and implementation of PM-Migration are presented in Sect. 3 and Sect. 4. The experimental results are discussed in Sect. 5. Finally, we conclude this paper in Sect. 6.

2 Background and Related Work

2.1 Performance Characteristics of the Hybrid Memory Architecture with NUMA

As demonstrated in Table 2, Kim et al. conducted experiments to investigate the read access latency and bandwidth of local and remote memory in a hybrid memory architecture with NUMA [10]. Their findings indicated that the read latency and bandwidth of other memory nodes are inferior to that of local DRAM, which is a well-established characteristic of traditional NUMA architecture. Furthermore, they also observed that the read latency and bandwidth of remote DRAM can outperform that of local DCPMM. This finding challenges the traditional belief that the local memory always outperforms remote memory and highlights the performance disparity between DCPMM and DRAM.

Table 2. Performance characteristics of the hybrid memory architecture. L: Local, R: Remote.

Metrics	L-DRAM	L-DCPMM	R-DRAM	R-DCPMM
Bandwidth	18.5 (GB/s)	7.1 (GB/s)	15.7 (GB/s)	0.27 (GB/s)
Latency	91 (ns)	184 (ns)	138 (ns)	225 (ns)

2.2 Page Placement Mechanisms for the Hybrid Memory Architecture

In addition to Nimble and AutoTiering, researchers have proposed the following page placement mechanisms for the hybrid memory architecture. Marques et al. introduced HyPlacer [15], which periodically identifies frequently accessed, modified, and infrequently accessed pages and prioritizes their placement in DRAM to maximize overall system performance. Liu et al. proposed a memory management framework called Memos [13], which has been developed for OS in horizontally integrated DRAM and NVM (Non-Volatile Memory). However, they are all designed based on DRAM-simulated persistent memory, which can not accurately capture the hardware characteristics of DCPMM. Huang et al. proposed a novel mechanism, Autonuma [14], which builds upon the original AutoNUMA mechanism to enhance the management of the DRAM-DCPMM hybrid memory architecture, but it only works properly on a system with a single CPU socket and is unsuitable for a system with NUMA architecture. Besides, the above page placement mechanisms adopt *memcpy* to perform page migration, without decoupling CPU and page data transfers.

3 Design

In this section, we will introduce the design and implementation of PM-Migration. We will begin with an overview of PM-Migration, followed by a detailed discussion of the design of page placement and transmission handover strategy.

3.1 Overview

The fundamental procedure of PM-Migration is illustrated in Fig. 1. The *Binding Module* is responsible for binding and unbinding with the application, where binding implies that the application will hand over its memory pages to PM-Migration for management, while unbinding means that it will use the Linux system's default page placement mechanism to manage its memory pages. After the application is bound to PM-Migration, the *Page Marking Module* will periodically scan the memory pages of these applications and mark the pages residing in non-local DRAM as inaccessible. Once a page is marked as inaccessible and

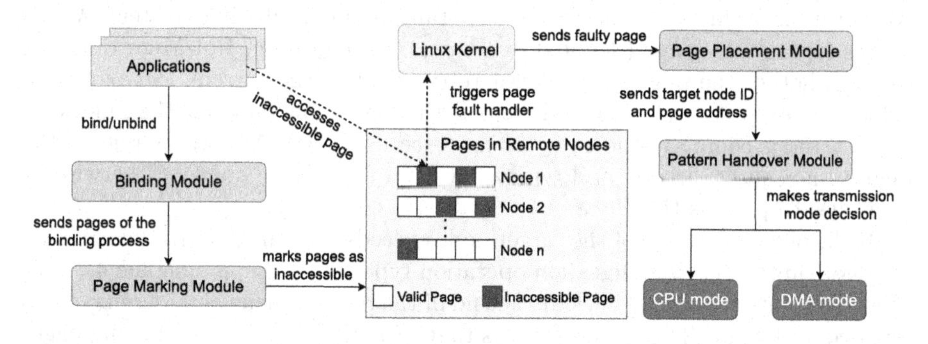

Fig. 1. Overview of PM-Migration. Applications can decide whether to entrust their pages to PM-Migration by binding or unbinding with the *Binding Module.*

then re-accessed by applications, a *NUMA hinting fault* is triggered, and the Linux kernel's page fault handler sends the faulty page (denoted by $Page_f$) to the *Page Placement Module*. The first step of the *Page Placement Module* is to identify the target node for $Page_f$ and assess its suitability for migration. If migration to the target node is feasible, it forwards the address of $Page_f$ and the ID of the target node to the *Pattern Handover Module* for further processing. Otherwise, it continues to search for the next target node or rejects the page migration request directly.

If the *Pattern Handover Module* is activated, it will select an appropriate transmission mode based on several metrics, such as the number of occupied DMA channels, the occupancy rate of the target memory node and the operation type of page migration. If the DMA mode is chosen, the system will employ I/OAT to migrate the page to the target node, whereas if the CPU mode is selected, the system will utilize the native *memcpy* to migrate the page.

3.2 Page Placement Strategy

The execution process of the page placement strategy is as follows: When a memory page (referred to as $Page_f$) triggers a *NUMA hinting fault*, PM-Migration first considers the local DRAM as the target memory node and checks if it has reached its capacity. If not, $Page_f$ will be migrated to the local DRAM. If the local DRAM is at capacity, PM-Migration selects the least accessed page (referred to as $Page_l$) from the local DRAM and compares its access frequency with $Page_f$. If the former is greater than the latter, PM-Migration will swap their positions. Otherwise, PM-Migration then selects the optimal performance node from the remaining memory nodes as the target and repeats the process. When the performance of the current target memory node is equal to or inferior to the node holding $Page_f$, it will remain in place.

3.3 Transmission Handover Strategy

Algorithm 2 outlines the process of the transmission switching strategy. When processing a faulty page (denoted by $Page_f$) for migration, PM-Migration first checks whether the migration request retry count of the current $Page_f$ (initial value is 1) exceeds α (Line 1), where α is the maximum value for the migration request retry count. If the retry count exceeds α, PM-Migration rejects the migration request directly and removes it from the receive queue. Otherwise, it continues to process the $Page_f$.

If the occupancy rate of the target node exceeds β (equivalent to the meaning in Algorithm 1) and the migration operation type is not swapping (Line 4), PM-Migration puts the $Page_f$ back at the end of the receive queue and increments its retry count by one. The main reason is that swapping involves exchanging pages between two memory nodes and does not change the memory occupancy rate of the nodes. However, other page migration operations will inevitably increase the memory occupancy rate of the target node. Therefore, when the occupancy rate of the target node is high, only swapping operations are allowed.

After satisfying the aforementioned conditions, PM-Migration proceeds to check the type of $Page_f$. If it is a 4 KB regular page, it will be migrated directly using the CPU (Lines 8–9). However, if the page type is a huge page or THP, PM-Migration will further evaluate whether the occupied DMA channel is less than γ (Line 11), where γ is used to determine if the DMA channel is fully occupied. If the remaining DMA channels can fulfill the page migration requirement, the *Migrate_ with_ DMA* function will be invoked for page migration(Line

---

Algorithm 1. Transmission Handover Strategy for PM-Migration.

Input: *Count*, page migration request retry count; $Page_f$, faulty page;
$Type_f$, the type of faulty page; $Type_{op}$, the operation type of page migration;
U_d, the number of occupied DMA channels;
U_t, the occupancy rate of target memory node;
Queue, queue for receiving faulty pages.
Output: $Page_f$ is either migrated to target node or page migration request is canceled.
1: **if** $Count > \alpha$ **then**
2: reject_Migration_Request($Page_f$);
3: **end if**
4: **if** $Type_{op} \neq swapping$ && $U_t > \beta$ **then**
5: increase_Attempts_Number($Page_f$, 1);
6: put_Back_To_Queue($Page_f$, *Queue*);
7: **end if**
8: **if** $Type_f = regularPage$ **then**
9: Migrate_with_CPU($Page_f$);
10: **else**
11: **if** $U_d < \gamma$ **then**
12: Migrate_with_DMA($Page_f$);
13: **else**
14: increase_Retry_Count($Page_f$, 1);
15: put_Back_To_Queue($Page_f$, *queue*);
16: **end if**
17: **end if**

---

12). Otherwise, the current $Page_f$ will be re-enqueued at the end of the receive queue, and its retry count will be incremented by one (Lines 14–15).

4 Implementation

This section describes how to support the awareness of huge pages and THPs and discusses how to implement different transmission modes to complete page migration.

4.1 Using Different Transmission Modes to Complete Page Migration

Whenever the Linux operating system migrates a page [8], it calls the *migrate_page()* function which uses *memcpy* to complete the page migration and employs the *try_split_thp()* function to divide a 2 MB THP into multiple 4 KB regular pages. In order to implement the transmission handover strategy, we need to integrate a new function, migrate_page_huge(), into the existing function call path. This function employs DMA to carry out the page migration. We utilize the I/OAT DMA device of Intel to construct the DMA engine, with its driver located in the /driver/dma/ioat directory. For 4KB regular pages, the system continues to call the *migrate_page()* function and uses *memcpy* to complete the page migration.

4.2 Supporting Awareness of Huge Pages and THPs

The native AutoNUMA facility does not accommodate huge pages or THPs, which means it does not label huge pages or THPs located in non-local memory nodes as inaccessible. As a result, we need to make alterations to the native AutoNUMA facility, such as incorporating support for huge pages and THPs in the *task_numa_work()* function and managing *NUMA hinting fault* initiated by huge pages and THPs in the *do_numa_page()* function.

5 Evaluation

5.1 Experimental Setup

We established a real experimental environment to evaluate the performance of PM-Migration. All experiments were conducted on a server equipped with DCPMM, featuring 2 Intel Xeon Gold 5218 processors (2.3 GHz) and 3 SAS hard disk drives (HDD) with a capacity of 600 GB. Each CPU is connected to 32 GB DDR4-DRAM and 128 GB DCPMM. The server runs Ubuntu 20.04 with Linux kernel 5.15.

Table 3. NPB workloads description.

Workload	Read/Write Ratio	Dataset Sizes (GB)
MG	4 Read : 1 Write	28.4 (S), 39.1 (M), 53.9 (L)
FT	1.7 Read : 1 Write	20 (S), 40 (M), 80 (L)
CG	>60 Read : 1 Write	26.5 (S), 74.3 (M), 131 (L)
BT	3.5 Read : 1 Write	18 (S), 39.8 (M), 150 (L)

For our experiments, we selected BT, FT, MG, and CG applications from NPB [16] (Table 3 provides descriptions of these applications) and configured their dataset sizes to exceed the capacity of a single DRAM memory node (32 GB). The reason for doing so is that most page placement mechanisms prioritize storing application data in local DRAM and only utilize other memory nodes when local DRAM capacity is insufficient. Therefore, when the dataset size is smaller than the capacity of a single DRAM memory node, the performance differences between various page placement mechanisms are not significant.

5.2 Throughput Under CPU Contention-Free and High CPU Contentions

We conducted experiments in two types of environments: CPU contention-free and high CPU contention, to measure the throughput of applications using different page placement mechanisms. The former refers to a scenario where the average CPU utilization is relatively low, while the latter, which we simulated using stress-ng [17], involves a large number of applications competing for CPU resources, resulting in high average CPU utilization. As a baseline, we used the DRAM-only page placement mechanism (referred to as Linux-default) in Linux, and compared PM-Migration with Nimble and AutoTiering against this baseline.

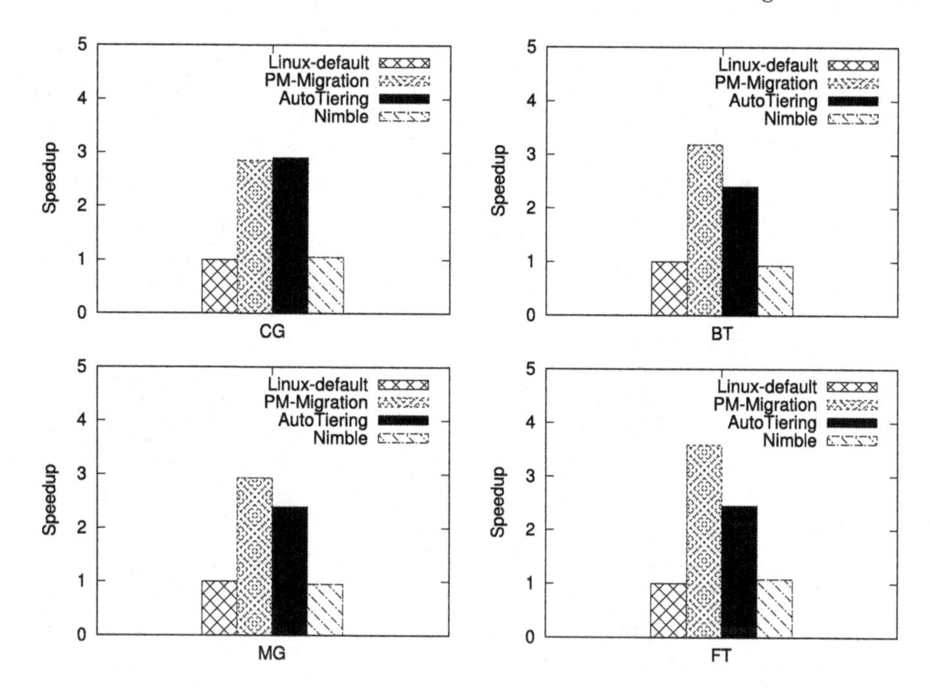

Fig. 2. Throughput speedup compared to Linux-default under the CPU contention-free environment.

CPU Contention-Free Environment. Figure 2 depicts the throughput speedup of Nimble, AutoTiering, and PM-Migration compared to Linux-default under the CPU contention-free environment. In order to exert sufficient pressure on memory capacity, we set the dataset size of all test cases to "Large" (denoted as "L" in Table 3). In most test cases, Nimble performs worse than other page placement mechanisms, particularly in the MG and BT test cases where its performance is relatively worse than Linux-default. This can be attributed to the fact that Nimble was designed based on DRAM-simulated persistent memory, which at that time was only considered as the memory that was slower than DRAM and had persistence. As a result, its design may adapt poorly to the hardware characteristics of real persistent memory, such as the DCPMM, leading to poor performance on actual hardware.

In all test cases, AutoTiering outperformed Linux-default and Nimble due to its design that can adapt well to the characteristics of hybrid memory architectures and some hardware features of DCPMM. However, except for CG, the performance of PM-Migration is superior to that of AutoTiering. As the write ratio in the test case increased, the performance difference between the two became more pronounced. For instance, in the MG, BT, and FT test cases, the throughput of PM-Migration is 1.22×, 1.32×, and 1.46× higher than AutoTiering, respectively. This can be attributed to the page access frequency calculation strategy of AutoTiering, which does not take into consideration the read-write

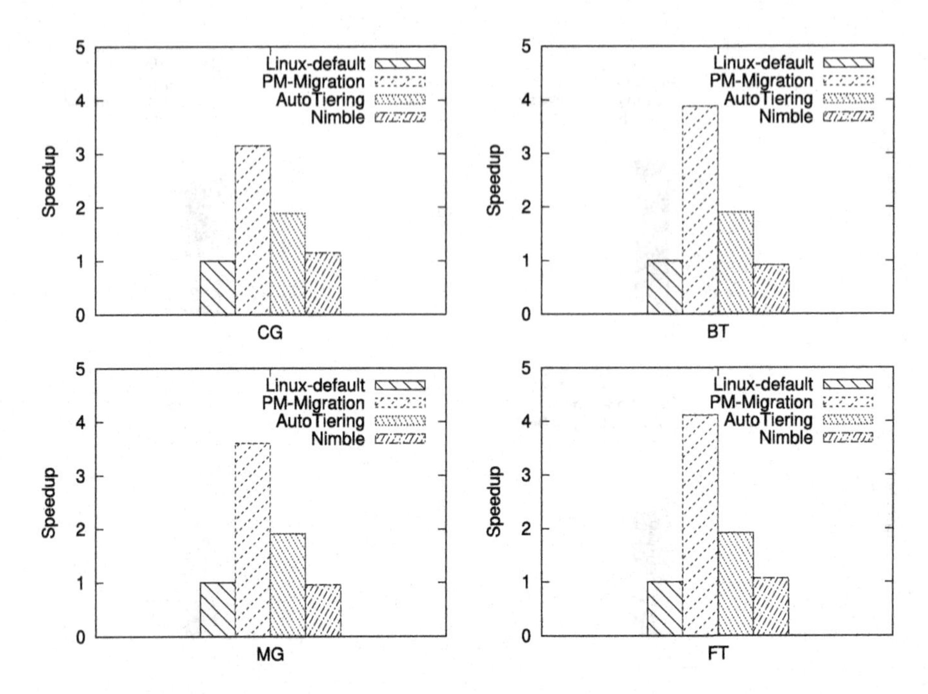

Fig. 3. Throughput speedup compared to Linux-default under the high CPU contention environment.

asymmetry of DCPMM. As a consequence, write-intensive memory pages cannot be prioritized to be stored in DRAM when migrating pages between memory nodes, resulting in inevitable performance degradation.

High CPU Contention Environment. Figure 3 displays the throughput speedup of Nimble, AutoTiering, and PM-Migration in comparison to Linux-default under the high CPU contention environment. In contrast to the CPU contention-free environment, AutoTiering experiences an average decrease in throughput speedup by 20%. This is primarily due to AutoTiering's reliance on CPU-based *memcpy* techniques for page migration, which results in reduced efficiency in an environment with high CPU contention where multiple applications compete for CPU resources. Therefore, AutoTiering is unable to fully leverage its potential advantages compared to Linux-default. On the other hand, PM-Migration demonstrates stable performance in such an environment, with an average increase in throughput speedup of 18% compared to the CPU contention-free environment. This can be attributed to the transmission handover strategy. It employs different transmission modes for transferring pages with varying sizes, such as using DMA for 2 MB huge pages and THP, effectively decoupling CPU from page migration operations. As a result, the average throughput of PM-Migration is 1.93× that of AutoTiering under the high CPU contention environment.

6 Conclusion

The DRAM-only system suffers from limited memory capacity, while the hybrid memory architecture that combines the capacities of both DRAM and DCPMM can effectively mitigate this issue. However, adopting the state-of-the-art page placement mechanism in a hybrid memory architecture introduces additional overhead. As a result, We propose PM-Migration, a novel dynamic page placement mechanism for the hybrid memory architecture. PM-Migration is designed with a dynamic page placement algorithm suitable for hybrid memory architectures, a novel approach for calculating page access frequency, and utilizes the transmission handover strategy to fully exploit the potential advantages of the hybrid memory architecture. Extensive experiments have been conducted to validate the advantages of PM-Migration. We plan to expand PM-Migration to seamlessly support hybrid memory architectures with more than two CPU slots, and offload some of the work to user space to reduce the overhead caused by kernel traps in the future.

Acknowledgment. This work was funded by the Key-Area Research and Development Program of Guangdong Province under grant number 2020B0101650001, by the National Natural Science Foundation of China under grant number 61972164.

References

1. Li, D., Liao, X., Jin, H., Zhou, B., Zhang, Q.: A new disk I/O model of virtualized cloud environment. IEEE Trans. Parallel Distrib. Syst. **24**(6), 1129–1138 (2013)
2. Luo, H., et al.: CLR-DRAM: a low-cost dram architecture enabling dynamic capacity-latency trade-off. In: 2020 ACM/IEEE 47th Annual International Symposium on Computer Architecture (ISCA), pp. 666–679 (2020)
3. Li, D., Zhang, N., Dong, M., Chen, H., Ota, K., Tang, Y.: PM-AIO: an effective asynchronous I/O system for persistent memory. IEEE Trans. Emerg. Top. Comput. **10**(3), 1558–1574 (2022)
4. Beeler, B.: Intel Optane DC persistent memory module (PMM). https://www.storagereview.com/news/intel-optane-dc-persistent-memory-module-pmm
5. Weiland, M., et al.: An early evaluation of intel's optane DC persistent memory module and its impact on high-performance scientific applications. In: Proceedings of the International Conference for High Performance Computing, Networking, Storage and Analysis, ser. SC 2019. New York, NY, USA: Association for Computing Machinery (2019). https://doi.org/10.1145/3295500.3356159
6. Gugnani, S., Kashyap, A., Lu, X.: Understanding the idiosyncrasies of real persistent memory. In: Proceedings of the VLDB Endow, vol. 14, no. 4, pp. 626–639 (2021). https://doi.org/10.14778/3436905.3436921
7. Awati, R.: Non-uniform memory access (NUMA). https://www.techtarget.com/whatis/definition/NUMA-non-uniform-memory-access
8. Linux Memory Management Documentation —— Page migration. https://www.kernel.org/doc/html/v5.4/vm/page_migration.html

9. Yan, Z., Lustig, D., Nellans, D., Bhattacharjee, A.: Nimble page management for tiered memory systems. In: Proceedings of the Twenty-Fourth International Conference on Architectural Support for Programming Languages and Operating Systems. p. 331–345. ASPLOS 2019, Association for Computing Machinery, New York, NY, USA (2019). https://doi.org/10.1145/3297858.3304024
10. Kim, J., Choe, W., Ahn, J.: Exploring the design space of page management for multi-tiered memory systems. In: USENIX Annual Technical Conference (2021)
11. AutoNUMA: the other approach to NUMA scheduling. https://lwn.net/Articles/835402/
12. Vaidyanathan, K., Chai, L., Huang, W., Panda, D.K.: Efficient asynchronous memory copy operations on multi-core systems and I/OAT. In: 2007 IEEE International Conference on Cluster Computing, pp. 159–168 (2007). https://doi.org/10.1109/CLUSTR.2007.4629228
13. Liu, L., Yang, S., Peng, L., Li, X.: Hierarchical hybrid memory management in OS for tiered memory systems. IEEE Trans. Parallel Distrib. Syst. **30**(10), 2223–2236 (2019)
14. Ying, H.: AutoNUMA: optimize memory placement for memory tiering system. https://lwn.net/Articles/835402/
15. Marques, M., Kuzmin, I., Barreto, J., Monteiro, J., Rodrigues, R.: Dynamic page placement on real persistent memory systems (2021). https://arxiv.org/abs/2112.12685
16. NAS parallel benchmarks. https://www.nas.nasa.gov/software/npb.html
17. Stress-ng (stress next generation). https://github.com/ColinIanKing/stress-ng

Explaining Federated Learning Through Concepts in Image Classification

Jiaxin Shen[1], Xiaoyi Tao[1,4(✉)], Liangzhi Li[2,3], Zhiyang Li[1], and Bowen Wang[3]

[1] Dalian Maritime University, Dalian, China
[2] Meetyou AI Lab (MAIL), Suzhou, China
[3] Osaka University, Osaka, Japan
[4] Key Laboratory of Symbolic Computation and Knowledge Engineering of Ministry of Education, Jilin University, Changchun, China
xytao@dlmu.edu.cn

Abstract. Federated learning is a machine learning framework that solves the problem of data silos under secure data protection measures and is gradually becoming a machine learning paradigm for future AI development. In recent years, federated learning has evolved in research areas such as security, model aggregation, and incentive mechanisms. However, the direction of interpretability of the model in the federated learning framework has not been explored. To bridge this gap, this paper proposes an interpretable model of Federated Concept Learning (FCL). FCL is trained on the client side using Bottleneck Concept Learner (BotCL) to generate human-understandable concepts. Each client uploads the co-occurrence scores of the concepts and classes obtained from the training to the server, and in order to mitigate the influence of possible malicious clients on the model, the server aggregates the obtained co-occurrence scores after optimization. The aggregated scores are then sent to the clients to update the model, which performs the classification task only by the presence or absence of the concepts. Experimental results show that our model has equivalent performance to other federated learning methods and successfully mitigates the impact of malicious client degree on the model performance, as well as provides an interpretation of the model classification results.

Keywords: Federated Learning · Explainable AI · Concept Learning · Security · Classification

1 Introduction

With the rapid development of artificial intelligence, the availability of high-quality data has become an important factor limiting its further development. In this context, there is a growing demand for data sharing and integration, but the process of data sharing has met with the problem of data silos, where data

Code is available at https://github.com/jiaxin-shen/FCL..

Z. Tari et al. (Eds.): ICA3PP 2023, LNCS 14491, pp. 325–340, 2024.
https://doi.org/10.1007/978-981-97-0808-6_19

is not allowed to be shared directly between different users [1,2]. Due to security issues, competition, and approval processes, data exists in "silos" in the industry and even within companies. The existence of data silos prevents companies or organizations from maximizing the value of the data that they own, which often leads to incorrect decisions and backward development.

In addition, in traditional machine learning approaches, the training data needs to be centralized in a particular machine or a single data center, and additional machines have to be added to accommodate the gradually increasing amount of data. In the process of data centralization, there is a risk of data leakage.

To solve these problems, federated learning has been proposed. Federated learning is a machine learning framework that can effectively help multiple organizations perform data usage and machine learning modeling while meeting the requirements of user privacy protection, data security, and government regulations. As a distributed machine learning paradigm, federated learning can effectively solve the problems of data silos and data leakage. While deep learning using federated settings can solve data silos and data leakage problems, there are explainability problems and security problems in federated learning. This makes the model subject to many limitations in terms of application.

The purpose of explainable models is to create a set of machine learning techniques that produce explainability while maintaining a high level of learning performance [3]. A DNN model [4], for example, is like a black box that is given input to get a decision result, but we cannot know exactly the basis of the decision behind it and whether the decision it makes is reliable [5]. The lack of explainability will potentially pose a serious threat to many DNN-based applications in real-world tasks, especially security-sensitive tasks. Only explainable machine learning models are likely to be more widely adopted and avoid discriminatory predictions and malicious attacks on decision systems. In many domains, model explainability is necessary. For example, in the medical field, experts need explainability from models in order to support diagnosis. AI algorithmic models also need to be explainable in self-driving cars, as well as in critical areas such as transportation, security, and finance [3]. In 2021, a trust AI project showed that XAI can be used for interdisciplinary application problems, including psychology, statistics and computer science, and may provide explanations to increase users' trust [6].

Consequently, the introduction of explainability in federated learning is necessary. First, participants in federated learning may come from different backgrounds and have different levels of expertise, and participants are better able to trust the model when the features and fundamentals of the model output are explained. Second, federated learning is often applied to certain sensitive domains, and the models used in these domains should be explainable, transparent, and understandable in order to be credible, reliable, and consistent. Finally, being explainable in a federated learning framework can help researchers gain insight into the inner workings of the model, allowing it to be improved and refined.

While federated learning inherently has privacy-preserving properties, it plays an important role in industry domains involving sensitive data. However, there are still model security issues. Security issues are mainly caused by vulnerabilities in FL systems targeted by curious or malicious attackers, which can lead to significant performance degradation or even model failure [7]. Federated learning is vulnerable to data poisoning attacks [8,9] and model poisoning attacks [10,11]. A malicious client can attack the global model by intentionally altering its local data or its gradients [12]. For example, a Byzantine attack is an untargeted poisoning attack that uploads a malicious gradient to the server, which causes global model failure [13–15].

In this paper, we design a model that can explain model classification results in a federated learning framework. We enable the model to classify images using concept activation without concept supervision by using a bottleneck concept learner(BotCL [16]). In federated learning, the client uses local data and BotCL to obtain co-occurrence scores for different concepts and classes, and the co-occurrence scores only represent the importance of the concept to the class. We simulate the attack behavior of the malicious client by varying the co-occurrence scores. The server side optimizes the scores uploaded by the clients and aggregates them to get the updated scores, while the clients apply the updated scores obtained from the server side to update the local model. By showing the different concepts trained in the image, we can observe what the model has learned and enhance the interpretability of the model.

Contributions. To summarize, the main contributions of this paper are fourfold:

- We found that current federated learning methods lack explainability. Therefore, we propose an interpretable model in a federated learning framework. Our approach allows for the explanation of classification results.
- We investigate existing federated learning methods and find that there may be malicious clients among the clients, and the malicious clients degrade the model performance by attacking. We reduce the impact of possible malicious clients on the model by improving the scores uploaded by the clients to the server.
- We evaluate FCL against existing federated learning methods and the centralized method BotCL on widely adopted datasets (MNIST, CUB200, CIFAR-10). Numerous experiments demonstrate the existence of explainability and robustness of our approach.
- We compare this with existing explainability methods. By looking at the generated heat map we can clearly see the generation of concepts.

2 Related Works

2.1 Federated Learning

Federated Learning (FL) is a machine learning framework that meets the requirements of user privacy protection, data security, and government regulations, that enables multiple organizations to use data and model machine learning. The core idea of Federated Learning is to achieve a balance between data privacy protection and data sharing computation by performing distributed model training across multiple data sources with local data, and constructing global models without the need to exchange local data by aggregating model parameters or intermediate results only on the server side [17–19]. Federated Learning is divided into three categories: Horizontal Federated Learning, Vertical Federated Learning, and Federated Transfer Learning [20]. In horizontal federated learning, data sets share the same feature space, but the samples are different. In vertical federated learning, data sets share the same sample ID space, but the feature space is different. In federated transfer learning, the sample space and feature space of the data set are different. Federated learning is commonly used in healthcare, finance, the Internet of Things (IoT), and other fields that contain sensitive information. For example, FL is used to classify electroencephalography (EEG) signals collected from various devices [21], and in the field of IoT, researchers have proposed an intrusion detection system based on IoT anomaly detection. Such a system can work without user intervention and is capable of detecting new attacks [22].

2.2 Explainable AI

As machine learning models play an increasingly important role in many daily scenarios, the interpretability of models becomes a key factor in determining whether users can trust them. Explainable artificial intelligence (XAI) is when researchers interpret the behavior of black-box models in ways that people can understand. The goal of XAI is to make AI algorithms more transparent and reliable. XAI's methods can be divided into 3 categories: the first category is data-based interpretable methods, which are mainly based on data analysis and visualization techniques to highlight the input features that have a strong impact on the DNN output, visualize the deep model, and visually demonstrate the key basis for obtaining the model results. For example, the visualization method of creating a class activation map (CAM) using the global average pool (GAP) in CNN [23]. A method called gradient-weighted class-activation Mapping (Grad-CAM) visualizes input regions that are important for prediction [24], from these values, we can learn where exactly the machine learning model focuses when making predictions. The second category is model-based interpretable methods, which are also called interpretable model methods, mainly by constructing interpretable models that make the model itself interpretable and output the results along with the reasons for getting the results. The concept of prototype is proposed in [16,25,26], which classifies images according to the human way of thinking. In designing the model, the way of human thinking is simulated by decomposing the image to get different prototype parts and then by combining this information to get the correct classification. MTUNet [27] interprets models using visual representations of backbone models and patterns generated from self-attentive interpretable modules. The third category is the result-based interpretable approach, the idea of this approach is to treat the existing model as a black box and infer the reasons for the corresponding results based on the given batch of inputs and the corresponding outputs, combined with the observed behavior of the model. As an example, the idea of LIME algorithm can be understood as follows: try to fit the original complex and poorly understood deep model with a simpler model that is easier to explain, and if it can produce a model with similar results as the original complex model, then the representational state of this simple model can be used to explain the original model [28].

2.3 Bottleneck Concept Learner

The explanation of the results can be used to look for concepts that are comprehensible to humans, and these concepts can be shared between the different target classes in the task. The concepts are mental categories that help us categorize objects, events, or ideas, and are based on the understanding that each object, event, or idea has a common set of related characteristics. Concept learning is a strategy that requires learners to compare groups or categories that contain concept-related features with groups or categories that do not contain concept-related features. Concept bottlenecks are used to detect a set of concepts and perform downstream tasks based on their presence or absence [29]. Recently researchers have incorporated conceptual bottlenecks into neural networks. This structure uses concept activation, representing images only by training the presence or absence of concepts learned by the target task, without explicit supervision of these concepts [16,29]. Botcl uses a slot-attention [30] based mechanism to discover the region where each concept is located.

Our study uses concepts to explain the classification results of the federated learning model. The client uploads the scores of different concepts in each category obtained through training, and the server aggregates the scores and then sends the aggregated scores to the client for updating. This process only shows how important the concept is to the category, and does not reveal the data of each client, enhancing privacy protection.

3 Federated Concept Learning

The training scheme of federated concept learning consists of multiple clients and a server. An overview of the FCL training process is depicted in Fig. 1. The client contains the local data set and a bottleneck concept learner. Define N clients $\{C_i\}_{i=1}^{N}$, each client holds its own data $\{D_i\}_{i=1}^{N}$. Set the number of classes as w and the number of concepts as k, the client C_i uses local data D_i and bottleneck concept learner training to get concepts, and uploads the co-occurrence scores $Z^i \in \mathbb{R}^{w \times k}$ to the server. Each value of the Z^i is the co-occurrence score of concepts and categories, denoted by Z_{wk}. The server aggregates the scores by operating M and FedAvg() to obtain the updated scores $Z' \in \mathbb{R}^{w \times k}$, then sends the updated scores Z' to the client to update the local model, and repeats the above steps until the ideal model is obtained. The client uses the co-occurrence

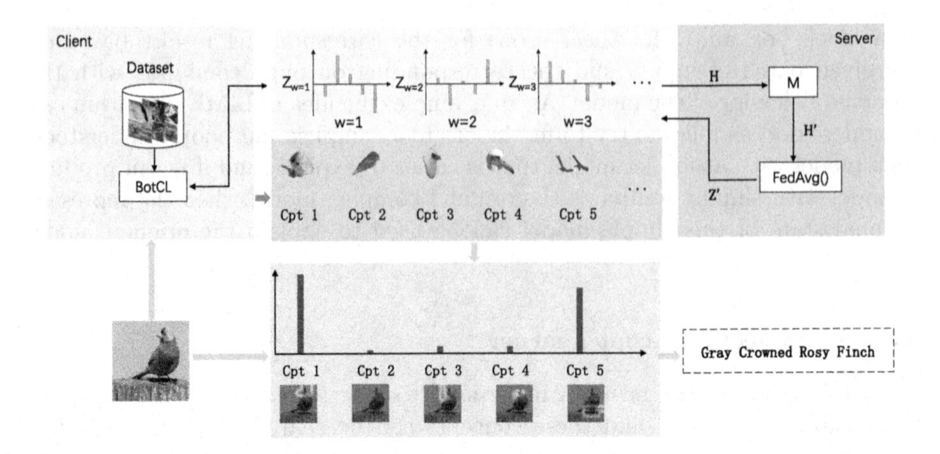

Fig. 1. An overview of FCL

score of each concept and class to perform the classification task. Taking an image of a Gray Crowned Rosy Finch in Fig. 1 as an example, the image is fed to the client and FCL finds concepts that can be interpreted as gray crowned and feet.

3.1 Federated Learning

Federated learning mitigated privacy concerns by allowing users to collaboratively train sharing models while keeping personal data on their devices. There are two main entities in the federated learning system: the clients and the server. The clients are the data owners, and the server is the model owner. A basic federated learning framework consists of a central server and multiple clients. Individual data owners C_i work together to train a model M_{fed}. The training process consists of the following steps:

- The server sends the initialized co-occurrence scores Z^{co} to the clients.
- The clients update the local model with the parameters it receives from the server. Each client C_i then trains the model using local data D_i and sends the calculated score information Z^i to the server.
- The server side aggregates the score information $Z^1, ...Z^n$ uploaded by the client and then updates the co-occurrence score $Z^{'}$. Then the co-occurrence score is delivered to the clients.

Figure 2 shows the pipeline of federated learning mainly through the communication between client 1 and the server. In reality, there are malicious clients between clients. Malicious clients can alter the co-occurrence of concepts and classes, causing the model to train into the wrong correspondence. For example, for images of class w_1, concepts k_1 and k_3 get high scores, indicating that k_1 and k_3 play a crucial role in classification, while malicious clients will reduce the co-occurrence scores of k_1 and k_3, making the image obtain the wrong class label.

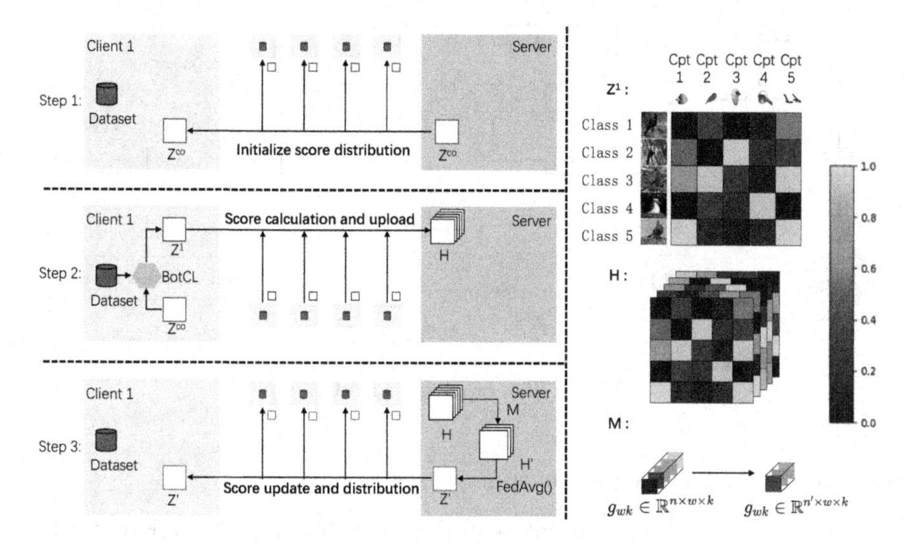

Fig. 2. The Federated Learning pipeline. Z^1 represents the co-occurrence scores of different classes and different concepts uploaded to the server by client 1. H represents the matrix of co-occurrence scores from different clients aggregated at the server side. M represents the operation of removing the corresponding percentage of the highest and lowest scores for a certain co-occurring score.

For privacy purposes, we assume that no more than 20% of malicious clients exist. On the server side, the server gets the score matrix $H \in \mathbb{R}^{n \times w \times k}$ uploaded by n clients, and for the co-incidence score of each class and concept, we delete the highest and lowest scores proportioned to ensure that we reduce the impact of malicious clients.

The improved score matrix is $H' \in \mathbb{R}^{n' \times w \times k}$. In H', the account of two-dimensional matrices is

$$n' = n - 2 \times \lceil 20\% n \rceil . \tag{1}$$

Each value in the matrix is a co-occurrence of a concept and a category. For example, the score g_{48} represents the score for concept 8 under category 4, with higher scores indicating that the concept is more important to this category. The FedAvg() [17] algorithm is used to aggregate the score information. The aggregate score is

$$Z' = \sum_{w} (\sum_{k} \sum_{n'} \frac{q_n}{q} g_{wk}), \tag{2}$$

where q represents the amount of all data owned by all clients, and q_n represents the total amount of data owned by all clients.

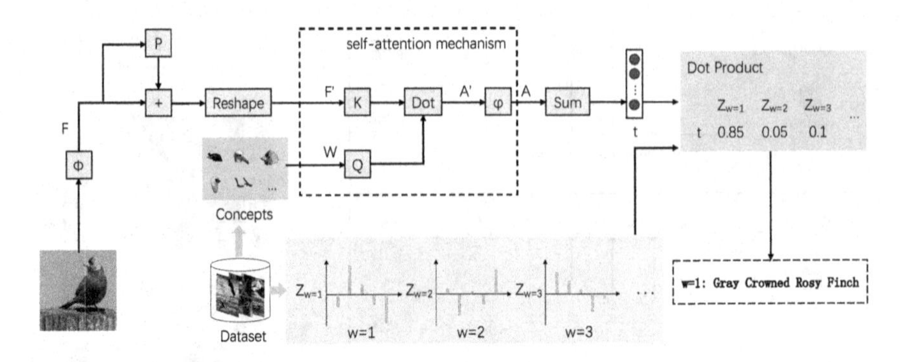

Fig. 3. The pipeline of BotCL

3.2 Bottleneck Concept Learner

On each client, we do the concept acquisition, as shown in Fig. 3. For the data $Di = \{(x_{ij}, y_{ij})|j = 1, 2, ...\}$ owned by each client C_i, where x_j is an image and y_j is the target class label associated with x_j, BotCL uses y_j's as weak supervision to learn a set of k concepts. For the new image x, we use the backbone convolutional neural network Φ to extract the feature map $F = \Phi(x)$. And then input F to the concept learner. In order to keep the space information, we add position embedded P in the feature map F to get F'. Set the concept prototype matrix W, in which each column vector is the concept prototype to be learned. Based on the dot-product similarity between the concept prototype W and F', we use the self-attentive mechanism to give attention to the spatial dimension of F'. Take the nonlinear transformation of W and F', with Q and K representing nonlinear variations, and then dot the results of the nonlinear transformation

$$A' = Q(W)^\top K(F'). \tag{3}$$

The attention A indicates where the k concepts are present in the image. A is given using a normalization function ϕ as

$$A = \phi(A'). \tag{4}$$

The normalization ϕ determines the spatial distribution of each concept, which may depend on the target domain. For concepts in an image, there are two cases:

- The concepts do not overlap in space. For example, for an image of a hand-written digit recognition dataset, the concept is the shape of a stroke. For the non-overlapping case, we design ϕ as

$$\phi(A') = \sigma(A') \odot softmax_{row}(A'), \tag{5}$$

where σ is the sigmoid function and $\odot$ is the Hadamard product. Softmax is applied to concepts (i.e. each row vector) such that different concepts cannot be detected at the same spatial location.

- The concepts may overlap in the same spatial position. For example, in natural images, concepts may overlap due to the presence of color, texture, and shape. In this case, we define the function ϕ to be

$$\phi(A') = \sigma(A'). \tag{6}$$

The attention generates a concept activation t by reducing the spatial dimension as

$$t = \tanh(A1_l). \tag{7}$$

Concept activation t indicates the existence of each concept. We want t to be a list of binary values to indicate the presence or absence of each concept, but, due to the training based on gradient descent, a continuous value is used. We use quantization loss to make the value close to 0 or 1. Let $\mathcal{B}$ be a small batch of data, and quantization loss is

$$l_{qua} = \frac{1}{|\mathcal{B}|k} \sum_{x \in \mathcal{B}} \| \, |2t - 1_k| - 1_k \|^2, \tag{8}$$

where $||$ gives the element-wise absolute value, $\|\|$ gives the Euclidean norm. The concept activation in t is used as the input to the classifier f to calculate the score s. We use a single fully connected layer without the bias term as our classifier. Let $Z \in \mathbb{R}^{c \times k}$ be a learnable matrix. Score s is computed by

$$s = softmax(Zt). \tag{9}$$

Training this simple classifier can be interpreted as finding the co-occurrence of each concept and the class. That is, the element z_{wk} of Z, which corresponds to class w and concept k, will a positive value when w and k co-occur a lot while it will be a negative value otherwise. We use cross-entropy for training, given by

$$l_{cls} = \frac{1}{|\mathcal{B}|} \sum_{(x,y) \in \mathcal{B}} \sum_{w} \hat{y_w} \log s_w, \tag{10}$$

where $\hat{y_w}$ and s_w are the element of one-hot representation $\hat{y}$ of y corresponding to w and the element of s corresponding to w.

4 Results

4.1 Experiments Settings

Datasets. We conducted experiments on three classification datasets, including a handwritten digit recognition task, namely MNSIT [31], a small dataset for recognizing universal objects, namely CIFAR-10 [32], and a natural image recognition task, namely CUB200 [33].

Comparison Methods. For MNSIT, we compare our approach with two federated learning methods and one centralized learning method: the two federated learning methods are the ordinary federated learning method and the federated learning method using the Flower framework [34], and the centralized method is BotCL. For CUB200, We compare our approach with the centralized approach BotCL on 5 and 8 clients respectively.

Implementation Details. We implement all the methods in Pytorch1.13 and all experiments are conducted on NVIDIA RTX A6000 GPU. On the three datasets, we set local epoch = 20, while for the global communication round, we use round = 50 on MNSIT and round = 30 on CUB200 and CIFAR-10. On 5 clients, we discuss the impact of removing the corresponding proportion of the highest and lowest scores on the classification accuracy. For MNSIT, we used the number of concepts k=20. For CUB200, we chose different numbers of concepts depending on the number of classes. On the CIFAR-10 dataset, we set the number of concepts k = 30.

4.2 Classification Performance

We observe accuracy on the MNSIT dataset to ensure the availability of our method in cases where concepts do not overlap spatially. We observe the test accuracy achieved on different federated learning methods and compare it with the centralized method BotCL. It can be observed that our experimental method obtains better classification results compared to other federated learning methods, indicating that our method is a competitive federated learning model. Compared with the centralized method BotCL, the accuracy of our method is slightly lower because the dataset is not stored centrally but locally in the client, and the data owned by different clients are diverse. In the federated learning environment, the performance is degraded by the dataset. So the performance of our model is acceptable in terms of classification performance. As shown in Table 1,

Table 1. Accuracy analysis on MNSIT dataset

Method	Global (round)	Local (epoch)	Accuracy	gain
FL	50	20	94.93	–
Flower	50	20	97.55	+2.62
FCL	50	20	95.18	+0.25
BotCL	–	60	97.82	+2.89

On the CUB200 dataset, we fix or alter the number of concepts over different number of classes to observe the availability of our method in the cases where the concepts may overlap spatially. We compare with the centralized method BotCL to observe their test accuracy. By observing Fig. 4 we can conclude that as the number of classes increases, the accuracy decreases for either fixing or altering the number of classes. However, by observing fixed or altered number of concepts, we can see that a higher number of concepts gives a higher accuracy for the same number of classes. Compared with the centralized method BotCL, our method and BotCL obtain about the same accuracy.

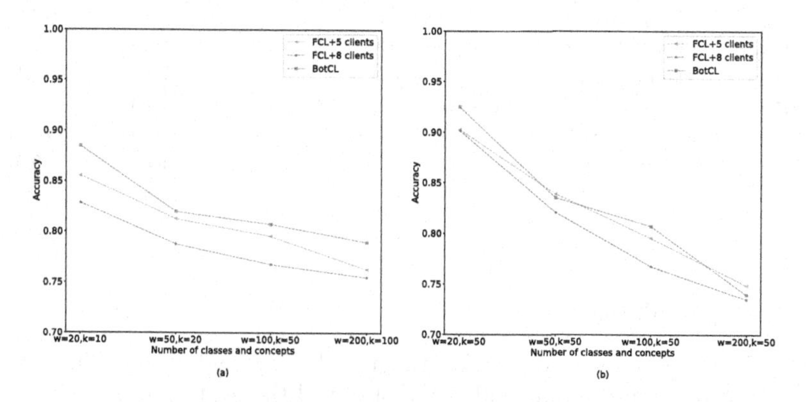

Fig. 4. Accuracy on the CUB200 dataset. (a) Altering the number of concepts as the number of categories increases. (b) Fix the number of concepts.

4.3 Robustness

We observe the test accuracy when malicious clients have existed, and the test accuracy after removing the corresponding percentage of the highest and lowest scores at the server side, respectively, to ensure that our model is robust. We assume that malicious clients do not exceed 20% and that the purpose of malicious clients is to reduce the accuracy of model classification. We model the behavior of malicious clients by varying the co-occurrence scores, and changing the scores affects the importance of concepts to classes, thus changing the correspondence between images and classes for the purpose. We observe the robustness of the model on the MNSIT dataset, CFIAR-10 datasets and the CUB200 dataset with 5 clients, respectively. On the MNSIT dataset, we set the number of concepts k = 20. Similarly, on the cifar-10 dataset, we set the number of concepts k = 30, and on the CUB200 dataset we use the number of classes w = 20 and the number of concepts k = 50. Through Table 2 we can observe that there is a large impact of malicious clients on the model, but the accuracy of the improved model reaches the normal level. On the CIFAR-10 dataset, we first used the original method of modeling malicious clients, and the accuracy remained almost unchanged, but when we enhanced the co-occurrence scores of the errors, which increased the impact of the malicious clients, we can see that the accuracy decreased. This shows that our improved approach is effective in mitigating the impact of malicious clients.

Table 2. Robustness analysis of FCL on different datasets.

Datasets	Method	Global (round)	Local (epoch)	Accuracy	Gain
MNIST	Presence of malicious client FCL	50	20	79.64	–
	The modified FCL	50	20	92.93	+13.29
CUB200	Presence of malicious client FCL	30	20	72.94	–
	The modified FCL	30	20	90.06	+17.12
CIFAR-10	Presence of malicious client FCL	30	20	94.61	–
	Enhanced malicious clients FCL	30	20	67.64	−26.97
	The modified FCL	30	20	95.18	+0.57

4.4 Interpretability

We visualize each concept to see what the model learns and to show the existence of the interpretability of our model. We observed the heat map of each concept with its top 5 activated samples in the dataset image superimposed on it. As shown in Fig. 5, on the MNSIT dataset, the images are typically black and white, and only the shapes formed by the strokes are significant. In this case, concepts are unlikely to overlap each other spatially. We can see that each concept has a consistent region of attention in different samples (even in different classes of samples). For example, concept 4 is located in the lower middle of the image and its attention is on the vertical strokes, which is understandable. But there are concepts that humans cannot fully understand, for example, concept 1 may complete a semicircle of strokes in the lower part of the image, and we can assume that they can help the model to understand and distinguish the numbers.

As shown in Fig. 6, on the CUB200 dataset, concepts may overlap at the same spatial location because natural images have color, texture, shape, etc. We can see that different concepts focus on different body parts of the bird and each concept is consistent. For example, concept 1 shows the model learning to yellow the head of the bird, and concept 3 focuses on the feet of the bird, which proves that the model can learn valid concepts from natural images. Explanation is provided for the model.

By looking at the concepts learned by the model, we can see that the importance of different concepts to the class is different. For example, on the MNSIT dataset, concept 4 is important for the number 1, so the co-occurrence score of

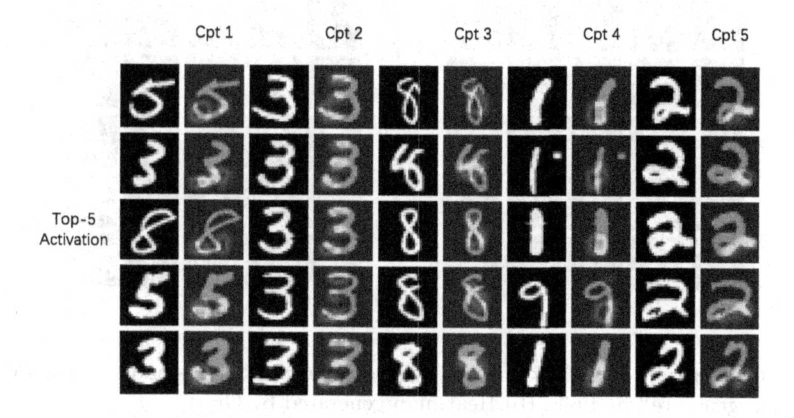

Fig. 5. Concept demonstration with top-5 activation in MNSIT.

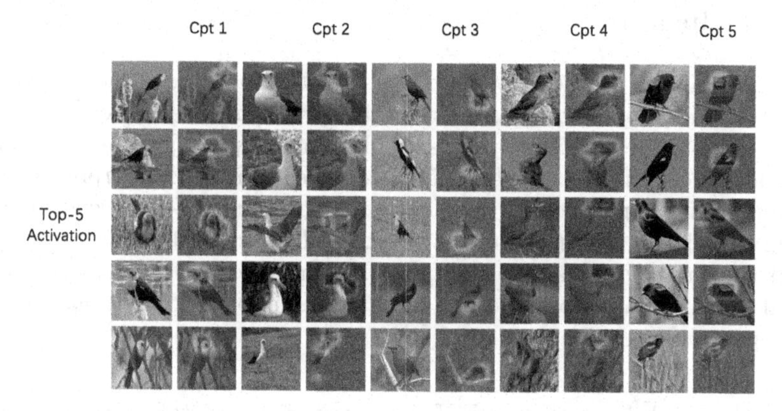

Fig. 6. Concept demonstration with top-5 activation in CUB200.

concept 4 and the class for the number 1 is a high score. For number 3, both concept 1 and concept 2 have high influence, so if both co-occurrence scores are high, we can decide that the category is number 3. Similarly, on the CUB200 dataset, if both concept 1 and concept 3 get high scores, then the category will be decided as a yellow-headed blackbird.

We also compare our method FCL with the existing explainable method Grad-CAM. On the MNIST dataset, with the heatmap generated in Fig. 7, we can observe that our method mainly focuses on the stroke part, with the aim of helping the model to understand and differentiate between numbers. Whereas Grad-CAM focuses mostly on the blank part next to the strokes. We can clearly observe the concept generation by looking at the heat map generated by FCL.

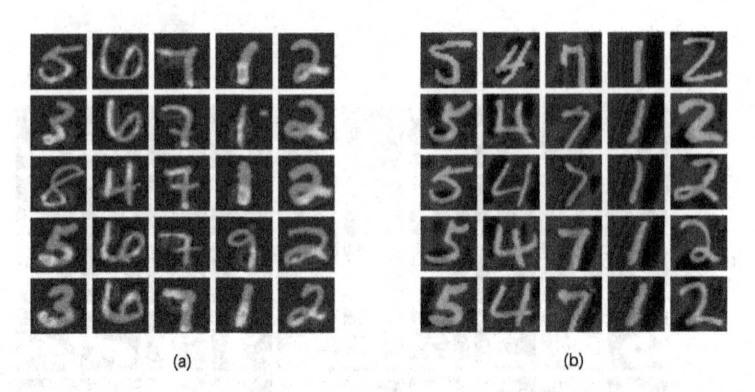

Fig. 7. Comparison of FCL with the existing interpretable method Grad-CAM. (a) Heat map generated by FCL. (b) Heat map generated by Grad-CAM.

5 Conclusion

Federated learning is gradually showing its great advantages in the field of artificial intelligence, however, the interpretability problem for federated learning has not been well addressed. To solve this problem, in this paper, we propose an interpretable model FCL in the context of federated learning. FCL interprets concepts by training them on the client side, and we remove some co-occurrence scores on the server side considering the existence of malicious clients. Our experimental results show that FCL has the ability to explain model classification results in a federal learning framework and can mitigate the impact of malicious clients on model performance.

Limitations. A limitation of FCL is that it does not take into account the Non-IID case, where we assume that each client has data for all classes. But the reality is that most clients only have data of some categories. We are trying to solve this problem.

Acknowledgment. This work is supported by the High-Performance Computing Center of Dalian Maritime University, and the Fundamental Research Funds for the Central Universities, JLU.

References

1. Kim, J., Ha, H., Chun, B.-G., Yoon, S., Cha, S.K.: Collaborative analytics for data silos. In: 2016 IEEE 32nd International Conference on Data Engineering (ICDE), pp. 743–754, IEEE (2016)
2. Speiser, S., Harth, A.: Taking the lids off data silos. In: Proceedings of the 6th International Conference on Semantic Systems, pp. 1–4 (2010)
3. Arrieta, A.B., et al.: Explainable artificial intelligence (XAI): concepts, taxonomies, opportunities and challenges toward responsible AI. Inform. Fusion **58**, 82–115 (2020)

4. LeCun, Y., Bengio, Y., Hinton, G.: Deep learning. Nature **521**(7553), 436–444 (2015)
5. Fleisher, W.: Understanding, idealization, and explainable AI. Episteme **19**(4), 534–560 (2022)
6. Gunning, D., Aha, D.: DARPA's explainable artificial intelligence (XAI) program. AI Mag. **40**(2), 44–58 (2019)
7. Long, G., Shen, T., Tan, Y., Gerrard, L., Clarke, A., Jiang, J.: Federated learning for privacy-preserving open innovation future on digital health. In: Chen, F., Zhou, J. (eds.) Humanity Driven AI, pp. 113–133. Springer, Cham (2022). https://doi.org/10.1007/978-3-030-72188-6_6
8. Wang, H., et al.: Attack of the tails: yes, you really can backdoor federated learning. Adv. Neural. Inf. Process. Syst. **33**, 16070–16084 (2020)
9. Xie, C., Huang, K., Chen, P.-Y., Li, B.: DBA: distributed backdoor attacks against federated learning. In: International Conference on Learning Representations (2020)
10. Bagdasaryan, E., Veit, A., Hua, Y., Estrin, D., Shmatikov, V.: How to backdoor federated learning. In: International Conference on Artificial Intelligence and Statistics, pp. 2938–2948. PMLR (2020)
11. Sun, Z., Kairouz, P., Suresh, A.T., McMahan, H.B.: Can you really backdoor federated learning?. arXiv preprint arXiv:1911.07963 (2019)
12. Lyu, L., et al.: Privacy and robustness in federated learning: Attacks and defenses. IEEE Trans. Neural Netw. Learn. Syst. (2022)
13. Blanchard, P., El Mhamdi, E.M., Guerraoui, R., Stainer, J.: Machine learning with adversaries: Byzantine tolerant gradient descent. In: Advances in Neural Information Processing Systems, vol. 30 (2017)
14. Yin, D., Chen, Y., Kannan, R., Bartlett, P.: Byzantine-robust distributed learning: Towards optimal statistical rates. In: International Conference on Machine Learning, pp. 5650–5659. PMLR (2018)
15. Lamport, L., Shostak, R., Pease, M.: The byzantine generals problem. In: Concurrency: the works of leslie lamport, pp. 203–226 (2019)
16. Wang, B., Li, L., Nakashima, Y., Nagahara, H.: Learning bottleneck concepts in image classification. In: Proceedings of the IEEE/CVF Conference on Computer Vision and Pattern Recognition (CVPR), pp. 10962–10971, June (2023)
17. McMahan, B., Moore, E., Ramage, D., Hampson, S., y Arcas, B.A.: Communication-efficient learning of deep networks from decentralized data. In: Artificial intelligence and statistics, pp. 1273–1282, PMLR (2017)
18. Li, T., Sahu, A.K., Zaheer, M., Sanjabi, M., Talwalkar, A., Smith, V.: Federated optimization in heterogeneous networks. Proc. Mach. Learn. Syst. **2**, 429–450 (2020)
19. Jiang, Y., Konečný, J., Rush, K., Kannan, S.: Improving federated learning personalization via model agnostic meta learning. arXiv preprint arXiv:1909.12488 (2019)
20. Wang, G.: Interpret federated learning with shapley values. arXiv preprint arXiv:1905.04519 (2019)
21. Gao, D., Ju, C., Wei, X., Liu, Y., Chen, T., Yang, Q.: HHHFL: hierarchical heterogeneous horizontal federated learning for electroencephalography. arXiv preprint arXiv:1909.05784 (2019)
22. Nguyen, T.D., et al.: A federated self-learning anomaly detection system for IoT. In: 2019 IEEE 39th International Conference on Distributed Computing Systems (ICDCS), pp. 756–767. IEEE (2019)

23. Zhou, B., Khosla, A., Lapedriza, A., Oliva, A., Torralba, A.: Learning deep features for discriminative localization. In: Proceedings of the IEEE Conference on Computer Vision and Pattern Recognition, pp. 2921–2929 (2016)
24. Selvaraju, R.R., Cogswell, M., Das, A., Vedantam, R., Parikh, D., Batra, D.: Grad-cam: Visual explanations from deep networks via gradient-based localization. In: Proceedings of the IEEE International Conference on Computer Vision, pp. 618–626 (2017)
25. Chen, C., Li, O., Tao, D., Barnett, A., Rudin, C., Su, J.K.: This looks like that: deep learning for interpretable image recognition. In: Advances in Neural Information Processing Systems, vol. 32 (2019)
26. Alvarez Melis D., Jaakkola, T.: Towards robust interpretability with self-explaining neural networks. In: Advances in Neural Information Processing Systems, vol. 31 (2018)
27. Wang, B., Li, L., Verma, M., Nakashima, Y., Kawasaki, R., Nagahara, H.: Match them up: visually explainable few-shot image classification. Appl. Intell., 1–22 (2023). https://doi.org/10.1007/s10489-022-04072-4
28. Ribeiro, M.T., Singh, S., Guestrin, C.: "why should i trust you?" explaining the predictions of any classifier. In: Proceedings of the 22nd ACM SIGKDD International Conference on Knowledge Discovery and Data Mining, pp. 1135–1144 (2016)
29. Koh, P.W., et al.: Concept bottleneck models. In: International Conference on Machine Learning, pp. 5338–5348. PMLR (2020)
30. Li, L., Wang, B., Verma, M., Nakashima, Y., Kawasaki, R., Nagahara, H.: SCOUTER: slot attention-based classifier for explainable image recognition. In: IEEE International Conference on Computer Vision (ICCV) (2021)
31. Deng, L.: The MNIST database of handwritten digit images for machine learning research [best of the web]. IEEE Signal Process. Mag. **29**(6), 141–142 (2012)
32. Krizhevsky, A., Hinton, G., et al.: Learning multiple layers of features from tiny images (2009)
33. Welinder, P., et al.: Caltech-UCSD birds 200 (2010)
34. Beutel, D.J., et al.: Flower: a friendly federated learning research framework (2022)

FEAML: A Mobile Traffic Classification System with Feature Expansion and Autonomous Machine Learning

Qing Yang[1], Xiangyu Kong[1], Yilei Xiao[1], Yue Lin[2], Rui Wen[1], and Heng Qi[1](✉)

[1] School of Computer Science and Technology, Dalian University of Technology, Dalian, China
hengqi@dlut.edu.cn
[2] Hisense Group Holdings Company, Qingdao, China

Abstract. Network traffic classification is a crucial component in network protocol and application identification, playing a pivotal role in various network and security-related activities. However, conventional traffic classification techniques are not suitable for mobile app traffic. This is primarily because mobile traffic exhibits a considerable disparity from Internet traffic, primarily in terms of the inconsistent traffic characteristics generated by the same app and imbalanced traffic samples, among others. To overcome these challenges, this paper presents a new mobile traffic classification system called FEAML(Feature Expansion and Autonomous Machine Learning), which leverages feature expansion and autonomous machine learning. The proposed system employs the SMOTE tool to address the imbalance problem and a hybrid architecture based on the self-attentive mechanism's CNN and Stacked LSTM layers to achieve feature expansion, thereby improving the generalization and classification accuracy of models. Through a series of data-driven experiments conducted on three public mobile traffic datasets, the proposed system demonstrates a 13% improvement in accuracy compared to state-of-the-art classification solutions.

Keywords: Mobile Traffic Classification · Auto-ML · Feature Expansion

1 Introduction

As a typical application of AI-driven networking, network traffic classification with machine learning draws more and more attentions from academia and industry [12], especially mobile traffic classification (TC) [3,22,25,26,31]. Mobile TC refers to identifying types of network traffic, as well as the Apps that generate this traffic.

Analyzing mobile traffic is crucial in today's network landscape due to the vast number of mobile application installations and downloads. Neglecting to scrutinize this traffic can create challenges for mobile network operators and

Z. Tari et al. (Eds.): ICA3PP 2023, LNCS 14491, pp. 341–360, 2024.
https://doi.org/10.1007/978-981-97-0808-6_20

service providers in terms of service differentiation, security policy enforcement, and traffic engineering. Efficient resource allocation and support for quality of service become difficult for software service providers. Mobile traffic analysis also plays a vital role in network management, traffic engineering, cloud resource allocation, quality-of-service, and anomaly monitoring.

However, existing machine learning and deep learning models for traffic classification may not be suitable for mobile traffic. Mobile traffic classification techniques face three key challenges:

(1) Mobile traffic datasets, derived from devices deployed in natural settings, typically exhibit imbalanced data classes consisting of a majority and minority class. These differences are due to the fact that human habits of use are for the most part similar and uniform, some of which should be widely used, while others are the opposite. The presence of imbalanced data classes has a notable influence on the accuracy of traffic classification.
(2) Mobile applications are frequently updated to incorporate novel features, thereby generating disparate traffic patterns within the same application class. This variability in traffic is indicative of a lack of consistency between the traffic patterns generated by applications of the same class. Conversely, certain applications with similar functionalities generate comparable traffic patterns irrespective of their assigned class. Furthermore, the traffic patterns generated by diverse application classes exhibit considerable similarities.
(3) In the context of mobile traffic classification, a dearth of information within a small subset of mobile flows, indicated by null values or zeros, may compromise the integrity of the existing model, consequently resulting in reduced classification accuracy. In order to mitigate such a scenario, it is imperative to enhance the model's generalization capacity and fortify its resilience towards abnormal data points. Furthermore, the efficacy of feature extraction may be hampered by the inherent encryption protocols employed in mobile traffic, which obscures the contents of packets, thereby resulting in a restricted feature space that may significantly impact the classification performance.

We effectively improve the classification accuracy while providing a generalizable, flexible, and robust approach. In a nutshell, We can sum up our contribution in the following areas:

- We propose a novel mobile traffic classification system called FEAML, which leverages three key strategies to improve classification reliability. Firstly, we employ distance-based oversampling to address the problem of class imbalance and excessive differences between data categories, resulting in improved model accuracy.
- We design a hybrid model architecture that combines Convolutional Neural Networks (CNN) and Stacked Long Short-Term Memory (LSTM) layers, featuring a self-attention mechanism that extends the original packet features and enriches the information extracted from the mobile flow. This methodology improves both the generalization ability and the robustness of the model,

leading to enhanced classification precision. Finally, we apply Auto-ML to mobile flow classification to achieve high-performance model selection based on optimization.
- To evaluate the effectiveness of our methodology, we conducted experiments on three separate mobile traffic datasets. The results demonstrate that our proposed system outperforms existing classification approaches, achieving a 13% increase in mobile application classification accuracy.

The rest of the paper is organized as follows. In Sect. 2 we discuss some related work to mobile TC or TC classification or identification; Sect. 3 outlines the design of FEAML and how the different components can work in concert to accomplish the task of mobile application classification; In Sect. 4 we present our main work on the FE algorithm in detail and give the working procedure; In Sect. 5 we address the performance of the FE algorithm and evaluate the performance of the whole FEAML; In Sect. 6 we conclude the current work.

2 Related Work

This section highlights the limits of current literature by briefly describing the ML-based and DL-based traffic classification studies linked to our approach.

2.1 Machine Learning Based Mobile TC

Mobile application traffic classification has become a subject of increasing interest in recent literature. To this end, Taylor et al. [25] proposed AppScanner, a system capable of identifying mobile application fingerprints from encrypted traffic and classifying applications through the use of ML classifiers, achieving higher accuracy. By setting the prediction probability threshold to a modest 0.5, the authors were able to increase precision, recall, and accuracy to 95.1%, 92.4%, and 95.0% respectively. Notably, classifiers were able to make judgments on almost half (45.5%) of the unlabelled flows. To further investigate the impact of different devices and fingerprint versions on the accuracy of application fingerprinting, [26] extended this work, addressing the shortcomings of previous experiments.

Sengupta et al. [22] proposed an innovative approach for mobile application traffic classification. They extracted bit-sequence-based features from TLS encrypted traffic, combined them with conventional packet-level features, and employed the Random Forest (RF) algorithm for classification. The introduced bit-sequence-based features outperformed traditional packet-level features, achieving F1-scores of 69% and accuracy of 71%. Combining these features improved classification accuracy further, yielding an impressive F1-score of 95% and high accuracy.

Wang et al. [31] designed a lightweight architecture that integrates network traffic analysis and C4.5 for Android malware classification. Their approach achieved high classification accuracy, ranging from 99% to 25% on different datasets.

Aceto et al. [3] proposed a fusion multi-classification approach to enhance the classification performance of mobile and encrypted traffic. They intelligently combined state-of-the-art classifiers and evaluated four combined systems based on classifier output, learning philosophy, and training requirements. Experimental results showed performance improvements of up to +9.5% in recall score compared to the best state-of-the-art classifier.

However, these approaches have limitations, such as neglecting imbalanced data and relying heavily on manual feature engineering for machine learning. Manual feature engineering often requires domain-specific expertise and time-consuming techniques like grid search for model selection and parameter tuning.

2.2 State-of-the-Art Deep Learning-Based TC

Deep learning (DL) has been widely explored in mobile traffic classification, showing high-performance potential [6,9]. Convolutional neural networks (CNNs) have become popular in traffic classification due to their ability to automatically learn data features and reduce network complexity [7,23,32,36]. However, CNNs have limitations in capturing temporal correlation information, which can affect their classification performance.

To overcome this limitation, hybrid architectures combining CNNs and recurrent neural networks (RNNs) have emerged. Several studies have successfully combined CNNs and RNNs to capture spatial and temporal correlations and achieve improved performance [5,13,19,21]. For example, App-Net by Wang et al. utilized a parallel hybrid neural network based on RNN and CNN to extract features from the original TLS stream [33]. Aceto et al. proposed a multimodal DL framework incorporating CNN, Gated Recurrent Unit (GRU), and LSTM to surpass the limitations of single-modal DL-based methods [4].

Other approaches include the combination of CNN and stacked autoencoder (SAE) by Wang et al., which achieves effective classification but is computationally expensive [29]. Xie et al. introduced a self-attentive mechanism combined with CNN to consider the interdependence of bytes in packets for more efficient application classification, but it raises privacy concerns [35].

Furthermore, DL-based traffic classification models like FS-Net by Liu et al., based on stacked bidirectional GRUs (bi-GRU) with a multilayer coding structure, outperform state-of-the-art models [18]. Graph Neural Networks (GNNs) have also been applied for traffic classification [24].

However, these techniques can be computationally expensive and time-consuming, limiting their practicality for real-world deployment, especially when processing massive datasets.

3 System Architecture

With the aim of overcoming the limitations of existing approaches, we propose a novel mobile traffic classification (TC) approach based on feature expansion and automatic machine learning. Our proposed approach seeks to (i) minimize model instability in the ever-changing traffic environment while maximizing model generalization; (ii) significantly enhance the richness of traffic information in the current mobile communication landscape, where encryption protocols are evolving, and encrypted traffic is increasing, thereby improving the ability to classify mobile applications; and (iii) reduce time consumption and costs by utilizing automatic model selection and tuning techniques, enabling the efficient performance of different dataset scaling tasks and coping with arbitrary classification tasks. Our proposed approach comprises three critical components in the general structure of the FEAML system illustrated in Fig. 1:

- **Data Preprocessing**: Our approach aims to balance the categories of collected features in mobile traffic datasets by synthesizing a small number of category samples. Subsequently, the data is scaled and normalized to fit into a certain digital space, facilitating efficient feature extraction and processing.
- **Feature Extension**: The categories of the data stream generate new unique abstract features through supervised learning of the combined model. These new features are then combined with the original feature data to extend the sample features and gain extra information for each sample. The extended feature set is subsequently fed into the classification model.
- **Classification**: The stage following feature expansion (FE) processes the data stream and performs classification. This involves utilizing optimization techniques along with model combination and parameter adjustment to identify the optimal model for a given labeled sample.

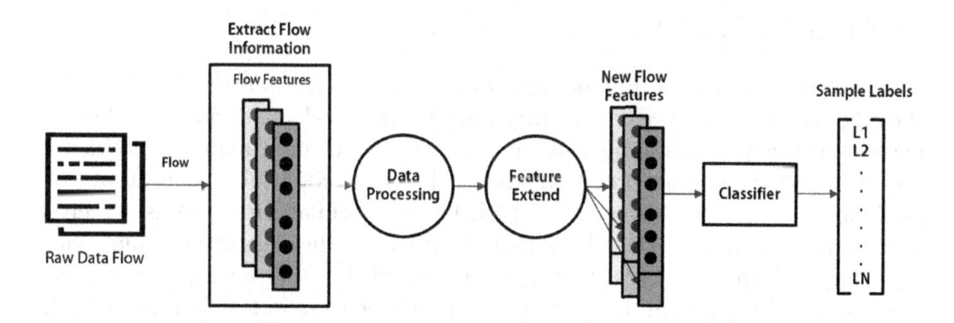

Fig. 1. The FEAML system has a clear workflow for mobile traffic classification. It starts by extracting features from the network stream. Next, the FE algorithm analyzes the extracted features to generate new feature information. The new features are then combined with the original features and provided to the classifier for decision-making. This systematic approach enhances the accuracy and effectiveness of mobile traffic classification.

3.1 Data Pre-processing

The accuracy of the mobile TC classification model is dependent on the quality of the dataset used. To ensure reliable classification results, our FEAML system focuses on producing high-quality data from the bi-directional stream (i.e., source and destination data pairs) dataset processed by the original packet. To achieve this, we perform a logical pre-processing of the initial data. Our pre-processing steps consist of two serial processing pipelines, as follows:

- **Resolving Data Category Imbalance**: To mitigate this issue, we adopt distance_SMOTE [8], a sampling technique that synthesizes a small number of class samples. The technique initially isolates a few category samples, then applies random weighted distance to locate n samples closest to the sample location, whose values are subsequently averaged. The discrepancy between the sample points and the average is multiplied by a random value of either 0 or 1, resulting in fresh synthetic samples. By creating a more balanced dataset, our proposed approach enables the classifier to better adapt to the complex traffic patterns of real-world mobile applications.
- **Data Normalization**: In order to facilitate subsequent calculations, the bidirectional flow dataset is transformed to ensure that all feature data is in a consistent value range. Our proposed FEAML system employs standard distribution normalization.

Although the unit of data pre-processing may add substantially to the computational cost and cause deployment delays, it can enhance the classifier's performance by increasing the number of samples from fewer categories, thereby positively impacting the classifier.

3.2 Feature Extension

The feature extension technique we present is a key contribution that enhances the efficacy and resilience of arbitrary classification tasks in dynamic and evolving settings, while simultaneously enriching the valid information contained in the acquired data. The proposed approach is rooted in the concept of sub-modular learning, wherein each set of DL-based hybrid architectures is perceived as a distinct module that can be classified into a specific class, generating a class label for the input samples. Our approach employs small batches for training and operates online in an iterative manner, culminating in the final layer of the feature expansion step that produces the expanded feature values of the samples in N dimensions (where N is arbitrarily set), thereby achieving the aim of enhancing the dataset's information. The fused extended feature values and original data are subsequently directed to the classification engine to observe the outcomes.

The purpose of our feature extensions is as follows:

- **Scalability to handle large-scale mobile traffic datasets**: The ubiquity of mobile devices and applications has led to an explosion of mobile traffic data, posing significant challenges to mobile traffic classification.
- **Improving the reliability of extracted feature information**: Due to the increasing use of encryption protocols, particularly TLS, as a means of

preserving user privacy, we refrain from using packet payload as feature information. Thus, it is crucial to ensure the dependability of the limited feature information extracted from the data.

- **Enhancing the stability of classifiers in dynamic environments**: By improving the adaptability of classifiers to dynamically changing traffic patterns, we can bolster their robustness and stability.
- **Generating extended feature values**: This is the cornerstone of our technique and the key to achieving its efficacy and resilience.

The aforementioned properties represent the precise goals of our current endeavor to establish a modular deep learning (DL)-based feature extension framework. The specifics of our feature expansion methodology will be elucidated in Sect. 4.

3.3 Classification

For the purpose of classification, the label corresponding to the current flow is determined by our classifier based on the common feature values extracted during the feature expansion stage. The efficacy of our proposed technique for mobile traffic classification is influenced by two primary factors:

(a) The merged dataset, combining the original data stream with the feature-expanded output, is processed by the FEAML classifier. This classifier has the capability to provide labels for all samples and exhibits robustness against dynamic label changes in the traffic. Moreover, the incorporation of individualistic learning of supplementary feature information from each category enhances dissimilarity between traffic types, resulting in a more adaptable and generalizable classifier.

(b) FEAML's classification approach sets it apart from traditional ML and DL methods by utilizing the Auto-ML tool, AutoGluon-Tabular, renowned for its efficiency and accuracy in analyzing tabulated data. Our approach aims to extend beyond mobile traffic classification tasks by automating the process of feature development, model selection, and parameter tuning with AutoGluon-Tabular. The tool's exceptional performance in tabulated data classification is demonstrated by its superiority over other methods under similar data conditions.

Formally, we consider its specific steps as follows:

Step (i): Upon completion of the automated feature extension process, the resultant dataset is fed into AutoGluon-Tabular to finalize the classification training.

Step (ii): The automated model utilizes a variety of machine learning models, stacks, and their ensembles, such as LightGBM [14], KNeighborsUnif, NeuralNetFastAI, CatBoost [10], XGBoost, and a dozen other models, to train the classification on the input dataset and evaluate the training scores.

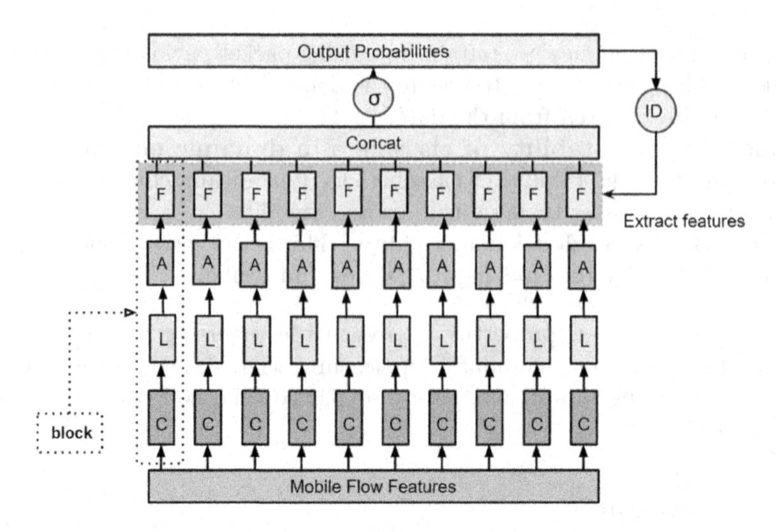

Fig. 2. Overall architecture for feature extend.

Step (iii): Following the selection of the best performing model, it proceeds to fine-tune the model using a grid search approach to automatically adjust the selected model's parameters for optimal performance.

Step (iv): Upon completion of training using the training set, the optimal model is obtained and subsequently tested using the test set. With no imposed constraints on memory or time, we split the input dataset into 70% training and 30% test sets to enhance the TC task for mobile applications. The proposed FEAML technique improves the accuracy of various mobile application classification datasets.

In conclusion, the FEAML model has exhibited a certain degree of enhancement for the mobile TC task in our proposed architecture. This assertion is substantiated by the experimental data presented in Sect. 4.

4 Feature Extension

In modern mobile environments, the task of classifying mobile TCs is incredibly challenging. This is due to the fact that legitimate information is often compressed, as well as the fact that mobile traffic (i) frequently contains a large, high-dimensional volume of data, and (ii) there is no clear pattern among distinct class members. In such cases, machine learning-based solutions are a viable option. However, they are not automated, and the updating process is time-consuming. Furthermore, they are becoming outdated with advancements. To address these concerns, FEAML employs an information enrichment process using a hybrid DL-based architecture. The proposed design consists of several blocks, each of which is associated with a certain category that, upon learning the block, can produce a set of extended N-dimensional features. Figure 2 illustrates our hybrid architecture, which comprises the following components:

(1) CNN: The convolutional neural network (CNN) is deployed as the first layer in FEAML's hybrid architecture to extract features from the initial bidirectional flow. Previous studies [17,23,32,34] have shown the efficacy of CNN in analyzing network traffic data. Therefore, 1D-CNN is used as the initial layer in the FEAML's input processing pipeline.
(2) LSTM: The second layer in our proposed hybrid network architecture is composed of Long Short-Term Memory (LSTM) units. This approach addresses the limitation of CNN in capturing the relevant information about the arrival time of each packet in the flow.
(3) Self-Attention: To address the challenge of capturing dependencies in N-dimensional features when combined with long serialized data, our FEAML system employs an attention technique. By incorporating attention, the model enhances its ability to extract relevant information from both the N-dimensional features and the original elements, resulting in improved classification accuracy.

Despite its ability to reduce dependence on external input and capture internal data characteristics of a system, Self-Attention [28] is particularly well-suited for mobile traffic classification. Its Q=K=V input formula signifies the mapping function's significance of Q to a sequence of key-value pairs (Key, Value). The resulting output provides definitive answers:

$$Attention(Q, K, V) = Softmax(\frac{QK^T}{\sqrt{d_k}})V \tag{1}$$

Algorithm 1 is presented to illustrate the operation of our proposal and provide a more detailed explanation of our method. Additionally, as depicted in Fig. 3, we outline the concrete procedure of the method, which is outlined below::

Algorithm 1 Proposal Feature Extend

Require:
  $Instance \quad X = \{x_1^M, x_2^M, x_3^M ..., x_m^M\}$
  $Classifier \quad C = \{c_1, c_2, c_3 ..., c_n\}$
Ensure:
  function FEATURE-EXTEND(X,C)
    for each $x_i, c_i \in \{X, C\}$ **do**
      $x_i^k \leftarrow Mixedstructure\left(x_i^M, c_i\right)$
      $x_i^k \leftarrow Selfattention\left(x_i^k\right)$
    end for
    $allFeature \leftarrow Concat(\sum_{i=1}^{n} x_i^M)$
    $LabelID \leftarrow argmax\,(Probability)$
    for $i = 1; i < n; i++$ **do**
      $BlockID \leftarrow GetNumberOfBlock\,(i, LabelID)$
      $ExtendFeature \leftarrow getFeature(BlockID)$
    end for
    $x_i^{M+N} \leftarrow United\left(x_i^M, ExtendFeature\right)$
    return x_i^{M+N}
  end function

Step(i): First of all, we start by defining some assumptions. Let us consider the sampled ensemble $X = \{x_1, x_2...x_m\}$. And the total number of categories in this set $C = \{c_1, c_2...c_n\}$, whereby n is the total number of categories as well as the total number of feature expansion blocks. For each of the m samples in X,there are raw features of M dimensions,i.e.,X_m^M.

Step(ii): After that, for any $x_i^M \in X$, it is input to the model as the Mobile Flow Features layer in the figure.Where x_i corresponds to the category c_i. The M-dimensional features are added to the first layer of the block, i.e., 1D-CNN, where the number of blocks is set to n, representing a total of n 1D-CNNs in the first layer of the model.

Step(iii): The output of the 1D-CNNs is then sent into the LSTM and Self-attention layers, with the output value of the Self-attention layer being the feature value of c_i with weights added.

Step(iv): Following that, x_i obtains the expanded, N-dimensional (can be randomly set) features belonging to category c_i through the fully connected layer. The output is passed through the Concat layer and the softmax layer function

$$\delta(z)_j = \frac{e^{z_j}}{\sum_{k=1}^{K} e^{z_k}} for j = 1...K \tag{2}$$

to obtain a set of probability sets for x_i^M, representing the probability of x_i^M belonging to the n courses.

Step(v): We can counteract the search and find the block numbered i based on the current probability value. The Extracted Features are extracted from this block, and the original features are combined with the Extracted Features. Additionally, the sample's $(N + M)$-dimensional characteristics are determined, i.e., $x_i^{N+M} \in X_i^{N+M}$. The proposed system enables the quick learning of inherent pattern information of flows, facilitating the identification of similar flows through the feature extension strategy.

5 Implementation and Evaluation

In this section, we present the experimental implementation details and evaluate the performance of the proposed FE algorithm on various classification tasks. We further assess the efficacy of the entire FEAML architecture using three publicly available mobile datasets. Our validation of FEAML's superior accuracy is supported by a comparison with other benchmark techniques, specifically on one of the larger datasets.

5.1 Implementation

The FEAML system was implemented using Python 3.7 [27], along with the TensorFlow [1] and Scikit-Learn [20] libraries. The FE technique employs two layers of 1D-CNNs with a filter size of 32*1 and 64*1 and stride values of 2 and 3, respectively. Furthermore, two layers of LSTMs with 128 units each are utilized.

The fully connected layer's output in each block is set to a feature expansion length of 10. During the training phase, the Adam optimizer [15] is used. The AutoGluon-Tabular is utilized for classification, employing automatic classification learning, and its parameters were not over-engineered. The proposed method is demonstrated using three publicly available mobile datasets, and its performance is compared with other benchmark methods, demonstrating its superior accuracy.

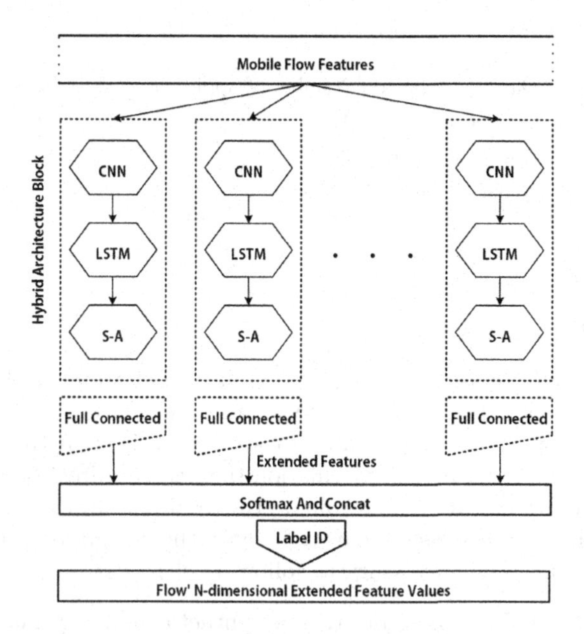

Fig. 3. The whole procedure of FE. Specifically, it commences with the extraction of the flow's comprehensive features, which are further processed using the hybrid architecture block. Subsequently, a fully connected layer is utilized to facilitate a dimensionality transformation of the data. Finally, the Softmax technique, which employs the labels generated by the block, is leveraged to facilitate the identification of the sample class's extended features.

5.2 Dataset Description

We have evaluated the efficacy of FEAML using three distinct mobile traffic datasets, as presented in Table 1. It is worth noting that these datasets were collected by human users through the utilization of genuine applications, rather than robot-generated mobile traffic, thus ensuring the authenticity and reliability of the data.

- **Mobilegt**: Based on the traffic generated by the monitored smartphones collected by Mobilegt [30], which is implemented by deploying mgtClient on the smartphones of 10 volunteers and mgtServer on a remote server.

- **MIRAGE-2019**: Based on the reproducible MIRAGE architecture captured mobile traffic dataset [2].
- **UTMobileNetTraffic2021 (UTM)**: UTM [11] is a freely released network traffic classification dataset. The dataset is made up of mobile data gathered by both the automated platform and human users from a variety of 16 popular apps.

Table 1. Detailed Information On Three Datasets

Dataset	Data Granularity	Number of applications	Number of Samples	Year
Mobilegt [30]	transport layer	12	118020	2016
MIRAGE-2019 [2]	transport layer	38	57730	2019
UTM [11]	transport layer	14	5498	2019

5.3 Data Preprocessing

The pre-processing phase plays a crucial role in determining the accuracy of the FEAML system. In this phase, the feature information extracted from the bi-directional flow of the originating packets is utilized. Although individual packets provide more temporal information than the bi-directional flow, we also incorporate a uni-flow dataset to evaluate the feasibility of the FEAML system. As a result, data pre-processing aims to balance the categories of the dataset for benchmarking purposes, involving the following elements:

- **Distance_SMOTE**: As a priority, a synthetic minority class sample was utilized to balance the dataset. Table 2. shows a comparison of the dataset before and after preprocessing as an example.
- **Standardization Of Normal Distribution**: The preprocessed data undergo partial normalization to ensure that the feature data conforms to a specific numerical range, fulfilling computational prerequisites and diminishing computational time consumption.

5.4 Evaluation Metrics

In order to evaluate the performance of the FEAML system, we employed four different evaluation metrics based on the number of True Positives (TP), False Positives (FP), False Negatives (FN), and True Negatives (TN). Based on these metrics, four different assessment measures were utilized to analyze the accuracy of our classification results. We used Accuracy, Precision, Recall, F_1-*Score*, and give the method of calculating F1 as an example:

$$\mathbf{F_1\text{-}Score} = 2 \cdot \frac{Precision \cdot Recall}{Precision + Recall} \tag{3}$$

Table 2. Comparison of Mobilegt-based preprocessing and no preprocessing results

	Before Data Process		After Data Process	
Category	Count	Percentage (%)	Count	Percentage (%)
Facebook	1400	1.10	25512	8.30
Weibo	25407	21.50	25512	8.30
YahooMail	3485	2.90	25512	8.30
JdShop	4008	3.30	25512	8.30
QQMail	1432	1.20	25512	8.30
QQ	17104	14.40	25512	8.30
TencentVideo	1593	1.30	25512	8.30
Youku	5825	4.90	25512	8.30
Browser	25512	21.60	25512	8.30
WeChat	13631	11.50	25512	8.30
MgTV	14046	11.90	25512	8.30
VipShop	4577	3.80	25512	8.30
Total	118020	100	306144	100.00

5.5 Feature Extension Results

We initially evaluated the performance of our FE algorithm on three distinct mobile traffic datasets, comparing it to the properties of three widely-used, public machine learning models: the K-nearest neighbor algorithm (KNN), random forest (RF), and decision tree (DT). To ensure a thorough analysis, default parameters were utilized for both the ML and FE algorithms.

The results of our evaluation, as presented in Fig. 4, revealed that the RF-Extend, DT-Extend, and KNN-Extend classification models - wherein the input datasets were processed by our FE model to extend the features of each sample - exhibited improved accuracy. For instance, while the Mobilegt dataset produced an accuracy rate of 89% on RF, the Mobilegt-Extend dataset achieved an accuracy improvement of approximately 5% following the FE algorithm. Although the effect was less pronounced, a similar trend was observed for UTM, indicating that all three machine learning models benefited from the FE phase. In conclusion, our FE algorithm proves advantageous in extending the primary traffic information, thus enabling the capture of valuable insights to enhance the features and boost the performance of classification models.

5.6 Mobile Application TC Results

In the following step, we proceed to assess the overall performance of our FEAML model in the context of mobile traffic classification.

To scrutinize the efficacy of our proposed FEAML model, we conduct multi-classification on the three datasets mentioned above, classifying the traffic types

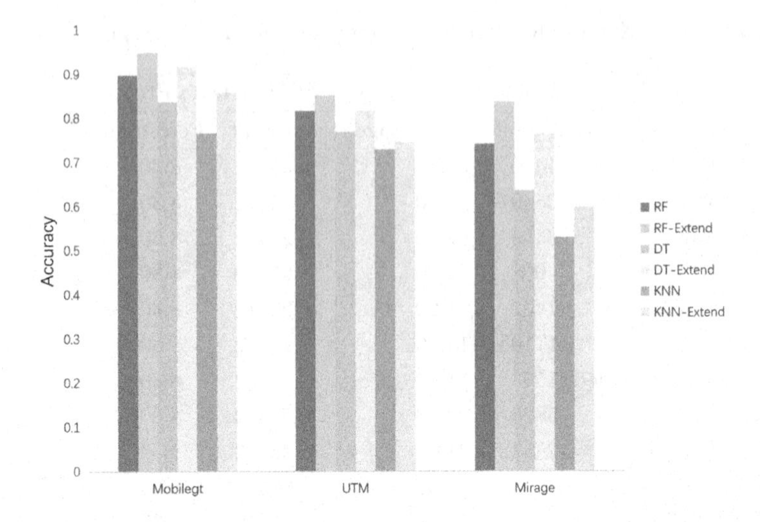

Fig. 4. The performance of the FE algorithm is evaluated on three different datasets based on RF, KNN, DT machine learning techniques.

that emanate from various applications. Furthermore, to evaluate the strengths and weaknesses of FEAML vis-à-vis other application-based traffic classification methodologies, we juxtapose it with a few state-of-the-art algorithms. Specifically, we compare FEAML with FS-NET [18], which deploys a bi-directional Gate Recurrent Unit (bi-GRU) within a multi-layer encoder-decoder structure to tackle the encrypted traffic classification problem; SAM [35], which exploits a novel self-attentive-based multi-layer encoder-decoder architecture, comprising a CNN and a full encoder that is composed of CNN and fully connected layers for traffic classification; 1D-CNN [16], which endeavors to automatically learn the nonlinear relationship between raw input and expected output; and HYBRID [19], which employs a joint model of CNN and LSTM for classification based on header features extracted from packets without any encrypted data.

In order to evaluate the performance of our FEAML model for mobile traffic classification, we utilized the confusion matrix for the Mobilegt, as shown in Figs. 5. The results demonstrate that FEAML has the ability to effectively categorize various mobile applications, even when faced with similar mobile application traffic.

Moreover, we computed the accuracy, precision, recall, and F_1-*Score* of FEAML on the aforementioned datasets and presented them in Table 3. According to the table, our proposed FEAML model delivers impressive performance in large-scale mobile traffic classification tasks, with all four metrics surpassing 96% for the Mobilegt dataset. Furthermore, our model exhibits excellent outcomes on the UTM dataset, which is relatively small in size. Notably, FEAML demonstrates strong capability in handling mobile traffic with a large number of application categories, such as the MIRAGE-2019 dataset, which comprises 38 different types of mobile applications, despite the metrics being slightly lower

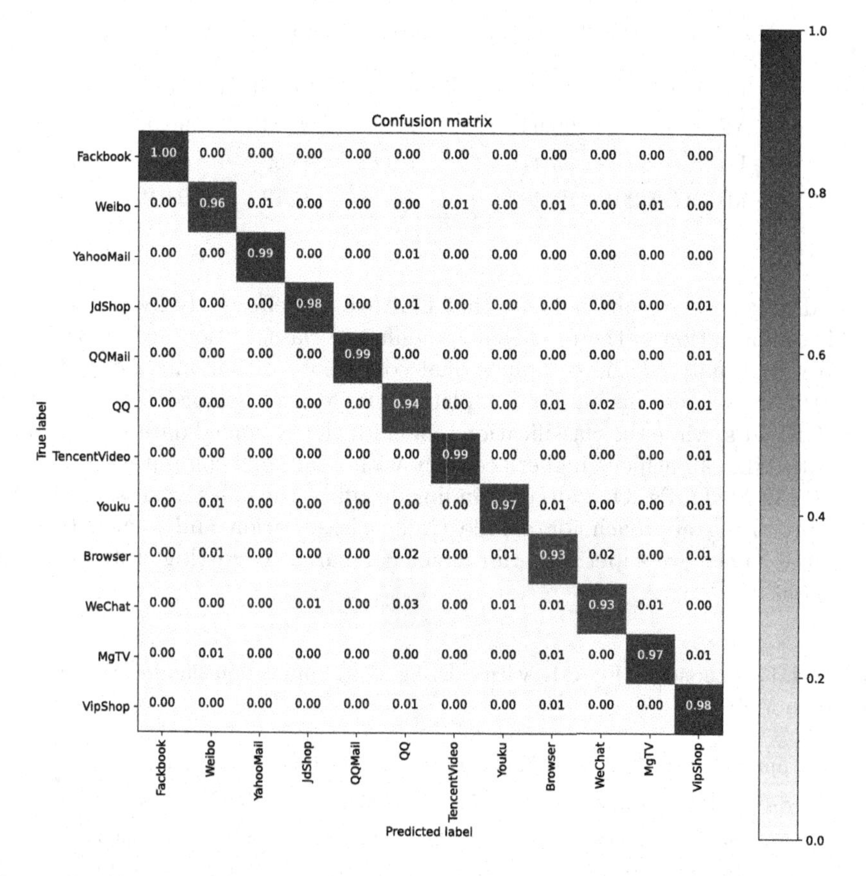

Fig. 5. Confusion matrix obtained by using FEAML for mobile application multiclassification on the Mobilegt dataset. Approximately perfect results were obtained.

when compared to the datasets with fewer categories (12 and 14 categories, respectively). Nevertheless, the achieved metrics remain within an acceptable range.

In order to evaluate the effectiveness of our proposed FEAML model, we compared it with other state-of-the-art mobile traffic classification (TC) benchmarking methods on the Mobilegt dataset. The compared methods are all based on multi-application classification approaches. The results are presented in Table 4, where it can be observed that FEAML outperforms all other methods in terms of all four metrics, including accuracy, which is 13% higher than SAM that employs self-attentiveness and CNN, and F1-SCORE, which is about 15% higher. Our unique data preprocessing and feature extraction steps, enabled by the FE algorithm, are the key factors for this remarkable improvement. We balanced the application classes by oversampling and extended the data by using a two-step classification process, which includes generating enriched features in the first stage, and automatically classifying the apps based on the traffic source in the

Table 3. FEAML evaluation results on three datasets

Dataset	Accuracy (%)	Pre (%)	Recall (%)	F1 (%)
Mobilegt	96.93	96.93	96.93	96.92
UTM	95.41	95.42	95.39	95.40
MIRAGE-2019	94.18	94.20	94.19	94.19

second stage. The results indicate that FEAML can effectively learn and utilize hidden information in the input data to enhance classification accuracy.

We also analyzed the computational complexity of our method and found that the model responsible for completing the feature expansion part requires 9.32 GFLOPs, while the classification model for the expanded dataset, generated by Auto-ML, can achieve higher accuracy with a minimal computational cost of only 0.132 MFLOPs. Overall, our findings confirm the effectiveness of FEAML as a promising approach for mobile traffic classification and demonstrate its potential to achieve superior performance compared to existing state-of-the-art techniques.

Table 4. Comparison of FEAML with existing multi-application classification methods based on Mobilegt

Approach	Accuracy (%)	Pre (%)	Recall (%)	F1 (%)	FLOPs (M)
FS-Net [18]	78.92	75.91	73.99	73.97	138.76
SAM [35]	83.98	83.58	79.59	81.12	8626.63
1D-CNN [32]	62.41	65.09	51.31	54.90	0.42
HYBRID [33]	41.22	48.94	23.38	23.51	0.63
FEAML	96.93	96.931	96.934	96.92	0.135

5.7 Importance of Features

To gain deeper insights into the impact of the extended information produced by the FE algorithm on the subsequent classification process, we conducted an analysis of the feature importance on the classifier using the Mobilegt dataset. Specifically, we identified fifteen feature values that are relatively important for the classifier among the numerous features extracted by our FEAML, as shown in Fig. 6. Our analysis suggests that the extended features generated by the FE algorithm have a significant impact on the performance of the classification process. Importantly, the effect of the extended features appears to be positive, indicating that the additional information provided by our FE algorithm is valuable in improving the accuracy of the classification task.

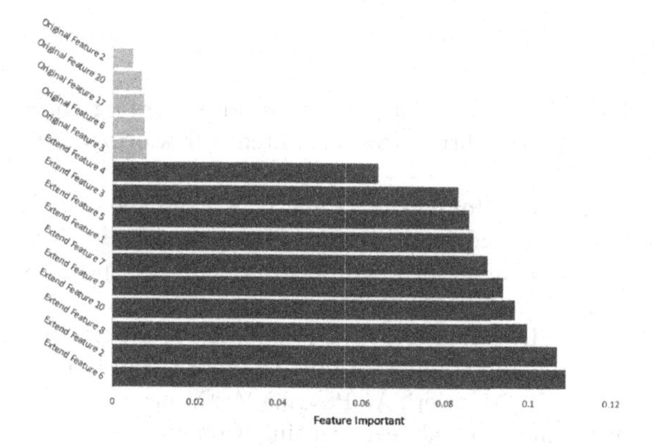

Fig. 6. The 15 most important feature values in the classification process of Mobilegt dataset.

6 Conclusions

In conclusion, this paper proposes a novel approach called FEAML for multi-application classification based on mobile application traffic. Our method extends the feature space of traffic flow data by leveraging the deep learning-based hybrid architecture of the FE algorithm, thereby improving the ability to differentiate between similar network traffic. Furthermore, we address the issue of class imbalance in the dataset by oversampling a few classes of samples. Additionally, an Auto-ML-based classifier is employed to complete the autonomous classification task to reduce manual intervention and time consumption.

Experimental results on three standard datasets show that our approach outperforms existing methods, achieving high accuracy, precision, recall, and F1-Score metrics. FEAML can handle a wide range of traffic analysis challenges while improving robustness and generalization. We also show the importance of the extended features generated by our FE algorithm in improving the performance of the subsequent classification process.

Overall, FEAML provides a valuable contribution to the field of multi-application classification based on mobile application traffic, and we hope that it will inspire further research in this area.

Acknowledgments. This work was supported in part by the National Key Research and Development Program of China under Grant 2019YFB2102404, in part by the NSFC under Grant 62072069, in part by Hisense Group Holdings Company.

References

1. Abadi, M., et al.: Tensorflow: large-scale machine learning on heterogeneous distributed systems (2015). http://download.tensorflow.org/paper/whitepaper2015.pdf
2. Aceto, G., Ciuonzo, D., Montieri, A., Persico, V., Pescapé, A.: Mirage: mobile-app traffic capture and ground-truth creation. In: 2019 4th International Conference on Computing, Communications and Security (ICCCS), pp. 1–8 (2019). https://doi.org/10.1109/CCCS.2019.8888137
3. Aceto, G., Ciuonzo, D., Montieri, A., Pescapé, A.: Multi-classification approaches for classifying mobile app traffic. J. Netw. Comput. Appl. **103**, 131–145 (2018)
4. Aceto, G., Ciuonzo, D., Montieri, A., Pescapè, A.: Mimetic: mobile encrypted traffic classification using multimodal deep learning. Comput. Netw. **165**, 106944 (2019). https://doi.org/10.1016/j.comnet.2019.106944
5. Akbari, I., et al.: A look behind the curtain: traffic classification in an increasingly encrypted web. In: Proceedings of the ACM on Measurement and Analysis of Computing Systems, vol. 5, pp. 1–26 (2021). https://doi.org/10.1145/3447382
6. Al-Naami, K., Chandra, S., Mustafa, A., Khan, L., Lin, Z., Hamlen, K., Thuraisingham, B.: Adaptive encrypted traffic fingerprinting with bi-directional dependence. In: Proceedings of the 32nd Annual Conference on Computer Security Applications, ACSAC 2016, pp. 177–188. Association for Computing Machinery (2016). https://doi.org/10.1145/2991079.2991123
7. Beliard, C., Finamore, A., Rossi, D.: Opening the deep pandora box: explainable traffic classification. In: IEEE INFOCOM 2020 - IEEE Conference on Computer Communications Workshops (INFOCOM WKSHPS), pp. 1292–1293 (2020). https://doi.org/10.1109/INFOCOMWKSHPS50562.2020.9162704
8. De La Calleja, J., Fuentes, O.: A distance-based over-sampling method for learning from imbalanced data sets. In: FLAIRS Conference, pp. 634–635 (2007)
9. Diallo, A.F., Patras, P.: Adaptive clustering-based malicious traffic classification at the network edge. In: IEEE INFOCOM 2021 - IEEE Conference on Computer Communications, pp. 1–10 (2021). https://doi.org/10.1109/INFOCOM42981.2021.9488690
10. Dorogush, A.V., Ershov, V., Gulin, A.: Catboost: gradient boosting with categorical features support. CoRR arxiv:1810.11363 (2018)
11. Heng, Y., Chandrasekhar, V., Andrews, J.G.: Utmobilenettraffic 2021: a labeled public network traffic dataset. IEEE Network. Lett. **3**(3), 156–160 (2021). https://doi.org/10.1109/LNET.2021.3098455
12. Horchulhack, P., Viegas, E.K., Santin, A.O.: Toward feasible machine learning model updates in network-based intrusion detection. Comput. Netw. **202**, 108618 (2022). https://doi.org/10.1016/j.comnet.2021.108618
13. Huo, Y., Ge, H., Jiao, L., Gao, B., Yang, Y.: Encrypted traffic identification method based on multi-scale spatiotemporal feature fusion model with attention mechanism. In: Proceedings of the 11th International Conference on Computer Engineering and Networks, pp. 857–866. Springer, Singapore (2022). DOI: https://doi.org/10.1007/978-981-16-6554-7_92
14. Ke, G., et al.: Lightgbm: a highly efficient gradient boosting decision tree. In: Guyon, I., Luxburg, U.V., Bengio, S., Wallach, H., Fergus, R., Vishwanathan, S., Garnett, R. (eds.) Advances in Neural Information Processing Systems, vol. 30. Curran Associates, Inc. (2017)

15. Kingma, D.P., Ba, J.: Adam: a method for stochastic optimization. In: ICLR (Poster) (2015)
16. Lecun, Y., Bottou, L., Bengio, Y., Haffner, P.: Gradient-based learning applied to document recognition. Proc. IEEE **86**(11), 2278–2324 (1998). https://doi.org/10.1109/5.726791
17. LeCun, Y., Bengio, Y., Hinton, G.: Deep learning. Nature **521**(7553), 436–444 (2015)
18. Liu, C., He, L., Xiong, G., Cao, Z., Li, Z.: FS-NET: a flow sequence network for encrypted traffic classification. In: IEEE INFOCOM 2019 - IEEE Conference on Computer Communications, pp. 1171–1179 (2019). https://doi.org/10.1109/INFOCOM.2019.8737507
19. Lopez-Martin, M., Carro, B., Sanchez-Esguevillas, A., Lloret, J.: Network traffic classifier with convolutional and recurrent neural networks for internet of things. IEEE Access **5**, 18042–18050 (2017). https://doi.org/10.1109/ACCESS.2017.2747560
20. Pedregosa, F., et al.: Scikit-learn: machine learning in python. J. Mach. Learn. Res. **12**, 2825–2830 (2011)
21. Rezaei, S., Kroencke, B., Liu, X.: Large-scale mobile app identification using deep learning. IEEE Access **8**, 348–362 (2020). https://doi.org/10.1109/ACCESS.2019.2962018
22. Sengupta, S., Ganguly, N., De, P., Chakraborty, S.: Exploiting diversity in android tls implementations for mobile app traffic classification. In: The World Wide Web Conference, WWW 2019, pp. 1657–1668. Association for Computing Machinery, New York (2019). https://doi.org/10.1145/3308558.3313738
23. Shahraki, A., Abbasi, M., Taherkordi, A., Kaosar, M.: Internet traffic classification using an ensemble of deep convolutional neural networks, pp. 38–43. Association for Computing Machinery (2021)
24. Shen, M., Zhang, J., Zhu, L., Xu, K., Du, X.: Accurate decentralized application identification via encrypted traffic analysis using graph neural networks. IEEE Trans. Inf. Forensics Secur. **16**, 2367–2380 (2021). https://doi.org/10.1109/TIFS.2021.3050608
25. Taylor, V.F., Spolaor, R., Conti, M., Martinovic, I.: Appscanner: automatic fingerprinting of smartphone apps from encrypted network traffic. In: 2016 IEEE European Symposium on Security and Privacy (EuroS&P), pp. 439–454 (2016). https://doi.org/10.1109/EuroSP.2016.40
26. Taylor, V.F., Spolaor, R., Conti, M., Martinovic, I.: Robust smartphone app identification via encrypted network traffic analysis. IEEE Trans. Inf. Forensics Secur. **13**(1), 63–78 (2018). https://doi.org/10.1109/TIFS.2017.2737970
27. Van Rossum, G., Drake, F.L.: Python 3 Reference Manual. CreateSpace, Scotts Valley (2009)
28. Vaswani, A., et al.: Attention is all you need. In: Advances in Neural Information Processing Systems, vol. 30. Curran Associates, Inc. (2017)
29. Wang, M., Zheng, K., Luo, D., Yang, Y., Wang, X.: An encrypted traffic classification framework based on convolutional neural networks and stacked autoencoders. In: 2020 IEEE 6th International Conference on Computer and Communications (ICCC), pp. 634–641 (2020). https://doi.org/10.1109/ICCC51575.2020.9344978
30. Wang, R., Liu, Z., Cai, Y., Tang, D., Yang, J., Yang, Z.: Benchmark data for mobile app traffic research. In: Proceedings of the 15th EAI International Conference on Mobile and Ubiquitous Systems: Computing, Networking and Services, MobiQuitous 2018, pp. 402–411. Association for Computing Machinery, New York (2018). https://doi.org/10.1145/3286978.3287000

31. Wang, S., Chen, Z., Yan, Q., Yang, B., Peng, L., Jia, Z.: A mobile malware detection method using behavior features in network traffic. J. Netw. Comput. Appl. **133**, 15–25 (2019)
32. Wang, W., Zhu, M., Wang, J., Zeng, X., Yang, Z.: End-to-end encrypted traffic classification with one-dimensional convolution neural networks. In: 2017 IEEE International Conference on Intelligence and Security Informatics (ISI), pp. 43–48 (2017). https://doi.org/10.1109/ISI.2017.8004872
33. Wang, X., Chen, S., Su, J.: Automatic mobile app identification from encrypted traffic with hybrid neural networks. IEEE Access **8**, 182065–182077 (2020). https://doi.org/10.1109/ACCESS.2020.3029190
34. Wang, X., Chen, S., Su, J.: Real network traffic collection and deep learning for mobile app identification. Wirel. Commun. Mobile Comput. **2020** (2020)
35. Xie, G., et al.: Sam: self-attention based deep learning method for online traffic classification. In: Proceedings of the Workshop on Network Meets AI & ML, pp. 14–20 (2020)
36. Zhang, J., Li, F., Ye, F., Wu, H.: Autonomous unknown-application filtering and labeling for dl-based traffic classifier update. In: IEEE INFOCOM 2020 - IEEE Conference on Computer Communications, pp. 397–405 (2020). https://doi.org/10.1109/INFOCOM41043.2020.9155292

Task Offloading and Resource Allocation for Edge-Cloud Collaborative Computing

Yaxing Wang[1], Jia Hao[1], Gang Xu[1](✉), Baoqi Huang[1], and Feng Zhang[2]

[1] Inner Mongolia A.R. Key Laboratory of Wireless Networking and Mobile Computing, Hohhot, China
32109234@mail.imu.edu.cn, {csxugang,cshbq}@imu.edu.cn

[2] School of Computer and Information Technology, Shanxi University, Taiyuan, China
zhangfeng@sxu.edu.cn

Abstract. Current research mainly focuses on two-tier cloud networks with edge clouds and users. The computational power relocation and resource allocation problem is relatively under-researched in three-tier networks consisting of multiple mobile users (MUs), multiple edge servers, and a centralized cloud. This paper constructs a system energy minimization problem based on a three-tier network by optimizing the computational resource allocation and offloading decision. Given the complexity of the problem, the optimization problem is decomposed into a resource allocation problem and a offloading decision problem, the Kuhn-Munkres Offloading Algorithm (KOA) is proposed to minimize the total system energy consumption by optimizing the offloading decisions. The simulation results suggest that the method performs well.

Keywords: Mobile edge computing · Offloading decision optimization · Edge-Cloud collaboration

1 Introduction

Applications like smart access control, intelligent vehicle networks, and virtual reality based on facial recognition have increasingly become a reality with the advancement of microelectronics and communication technologies. [1]. However, these applications often require large amounts of data computation while being very sensitive to response speed. Although the new generation of mobile devices has more computational power, they still suffer from high latency and energy consumption problems. To address these issues, researchers have proposed Mobile Edge Computing (MEC) [2]. The core idea of mobile edge computing is to set up a server with some computing power near the mobile device. By sending some or all of the data that would otherwise be sent to a cloud server to the edge server for processing, Mobile Edge Computing can shorten user response

Supported by organization x.

Z. Tari et al. (Eds.): ICA3PP 2023, LNCS 14491, pp. 361–372, 2024.
https://doi.org/10.1007/978-981-97-0808-6_21

time, reduce network load, and reduce the amount of data transmission and network resources used by the core network. Therefore, mobile edge computing is widely used in various application scenarios such as smart cities, smart healthcare, and smart homes, and provides high-quality services and data security for these applications. [3].

Current research on energy consumption and optimization of offloading decisions for cloud computing systems has yielded some results, and in the literature [4] the authors consider a generic multi-user mobile edge computing system with multiple MEC servers and propose an efficient two-stage algorithm, including a one-dimensional search (ODS) and alternate optimization (AO), that jointly optimizes offloading decisions for all user tasks and the computational and allocation of communication resources. In the literature [5], the authors optimize joint radio resource allocation and edge offloading decisions in a multi-cell orthogonal frequency division multiple access cellular networks by investigating the use of variable relaxation and minimization methods to find locally optimal solutions to minimize the energy consumption of the system. To reduce the delay and energy cost of MEC systems, the researchers in [6] address the computationally intensive and time-sensitive mobile edge computing computational offloading problem for tasks, and a joint partial offloading and task prioritization computational offloading algorithm was proposed.

However, most of these studies only consider a two-tier network with users and edge servers. Therefore, this paper addresses the problem of minimizing the system energy consumption based on the optimization of computational resources, offloading ratios, and offloading decisions in a three-tier network scenario consisting of multiple mobile users (MUs), multiple edge servers, and a centralized cloud, considering latency constraints. Since the problem is difficult to solve, the optimization problem is decomposed into a resource allocation problem and a offloading decision problem, which is solved using a convex optimization approach, and a KOA algorithm is proposed. The main work of this paper is as follows:

(1) Construct an energy minimization system in a three-tier network consisting of multiple mobile users (MUs), multilateral edge servers, and centralized clouds, considering delay constraints, and decompose it into two subproblems: the resource allocation problem and the offloading decision problem.
(2) Since the original resource allocation problem is a non-convex function, this paper transforms it into a convex function and solves it using a convex optimization method. For the problem of optimizing the off, the KOA algorithm is proposed to find the best offloading decision that minimizes the total energy consumption of the system.

2 Related Work

In recent years, research on cloud computing systems has been intensified. In the literature [7], the authors proposed an energy-efficient dynamic offloading and resource scheduling decision to reduce energy consumption and shorten application completion time in cloud computing scenarios. In the literature [8], the authors study the multi-user computing offloading problem for mobile cloud computing in a multi-channel wireless competitive environment and propose a fully distributed computing offloading (FDCO) algorithm based on machine learning techniques to reduce the execution cost. To improve the efficiency of edge clouds with limited communication and computational capacity, the literature [9] investigates the collaboration between cloud and edge computing and proposes a collaborative computing system that combines MEC and MCC(Mobile Cloud Computing) techniques. In the literature [10] the overall system overhead of the MEC system is minimized by jointly optimizing the computing power offloading and communication channel allocation decisions within the MEC user system.

Based on the existing work, this paper proposes the KOA algorithm for the under-studied three-tier network scenario with multiple users, multiple edges, and a central cloud to find the optimal offloading decision and reduce the overall energy consumption and time complexity of the system by reasonably allocating computational resources and offloading ratios in the three-tier network.

3 System Models

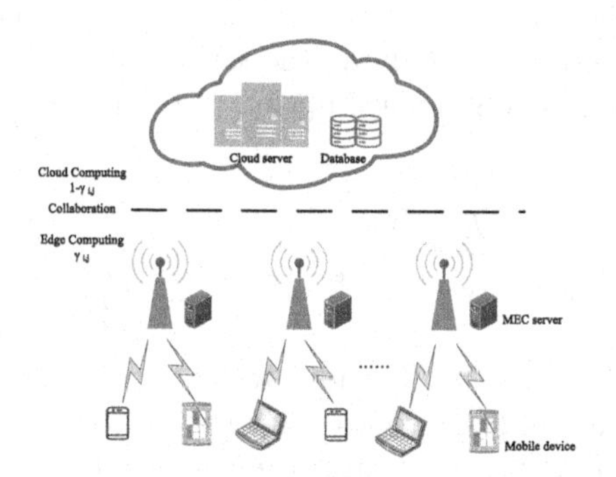

Fig. 1. Edge-Cloud collaboration system.

As shown in Fig. 1, this paper establishes an edge-cloud collaborative mobile edge computing model, which consists of three parts, namely "cloud computing center, edge server layer and mobile user layer". We use $\boldsymbol{N} = \{1, 2, \ldots, n, \ldots, N\}$ to denote mobile users.$\boldsymbol{M} = \{1, 2, \ldots, m, \ldots, M\}$ to represent edge nodes. Each mobile user is connected to the edge server and the cloud computing center through a wireless channel, and the communication between the edge server and the cloud computing center is wired.

Randomly generate task $Z_i = \{L_i, C_i, T_{i,max}\}$ will be generated for each mobile user. Where L_i denotes the input data size, C_i denotes the CPU cycles required to complete the task calculation, and $T_{i,max}$ denotes the maximum allowable transmission delay.

3.1 Local Computing Model

The time consumption and energy consumption of local computation can be expressed as:

$$T_i^{local} = \frac{C_i L_i}{f_{local}} \tag{1}$$

$$E_i^{local} = k_l f_{local} C_i L_i \tag{2}$$

where f_{local} denotes the local computing power of the mobile user and k_l is the effective switching capacity [11].

3.2 Edge-Cloud Collaborative Computing

Edge-cloud collaborative computing is a combination of MEC and MCC. Assumptions: (1) Each computing application can be split arbitrarily. (2) The edge cloud and the central cloud work in parallel, and the central cloud allocates its complete computational resources to all MDs for parallel computation. Therefore each MEC needs to determine whether the offloaded tasks should be executed by itself or by the edge server and the central cloud in concert. In the case where the central cloud is required to perform collaboratively, the edge server needs to determine the percentage $\gamma_{i,j}$ of applications to be executed on the central cloud server after receiving the task.

Therefore, the total energy consumption of the edge node j performing the task i:

$$E_{i,j} = E_{i,j}^{tran,d} + E_{i,j}^{comp} \tag{3}$$

$$E_{i,j}^{comp} = E_{i,j}^{comp,e} + E_{i,j}^{tran,e} + E_{i,j}^{comp,c} \tag{4}$$

Among them$E_{i,j}^{tran,d} = p_i T_{i,j}^{tran,d}$indicates the transmission energy consumption of the user offloading tasks to the edge server,
$E_{i,j}^{comp,e} = (1 - \gamma_{i,j}) k_j C_i L_i f_{i,j}$indicates the computational energy consumption of the task at the edge server, $E_{i,j}^{tran,e} = p_j T_{i,j}^{tran,e}$
represents the energy consumption for transporting tasks from the edge server to the central cloud,p_j for the transmission power from the edge server to the central cloud,$E_{i,j}^{comp,c} = k_j \gamma_{i,j} C_i L_i f_c$represents the computational energy consumption of a task in the central cloud,$f_{i,j}$is the computing resource allocated by edge server j to user i ; f_c is the computing capacity of the cloud center; and $f_j \ll f_c$.

The total delay of edge node j to execute task i is

$$T_{i,j} = T_{i,j}^{tran,d} + \max\{T_{i,j}^{comp,e}, T_{i,j}^{tran,e} + T_{i,j}^{comp,c}\} \tag{5}$$

3.3 Communication Model

Since it is a wireless channel between the mobile user and the edge server node, the channel communication rate is obtained from Shannon's formula as:

$$R_{i,j} = B \log_2 \left(1 + \frac{p_i g_{i,j}}{\sum_{n=1}^{N} \sum_{m=1,m \neq j}^{M} p_n g_{n,m} x_{n,m} + \sigma} \right) \tag{6}$$

where B is the channel bandwidth, p_i is the transmitted power, $g_{i,j}$ is the channel gain, σ is the Gaussian white noise power, and $x_{i,j} = \{0,1\}$ denotes the result of the computation offloading decision for task i. When $x_{i,j} = 0$, it means that the task is not offloaded to this server; when $x_{i,j} = 1$, it means that the task Z_i is offloaded to this server j to perform edge computation.

The transmission delay for offloading task Z_i to edge node j is:

$$T_{i,j}^{tran,d} = \frac{L_i}{R_{i,j}} \tag{7}$$

The communication rate between the edge server and the cloud computing center is wired, and there is no need to consider the interference between their channels, thus the communication rate between the edge server and the cloud computing center is denoted by W_C .

Then the transmission delay required to offloading the part of task Z_i to the cloud computing center is:

$$T_{i,j}^{tran,e} = \gamma_{i,j} \frac{L_i}{W_c} \tag{8}$$

where $\gamma_{i,j}$ is the percentage of data offloaded to the cloud center, $\gamma_{i,j} \in [1,0]$.

4 Problem Formulation

In order to minimize system energy consumption, the problem of minimizing energy consumption is achieved by optimizing resource allocation and mobile edge offloading decisions. When $x_{i,0} = 1$, $E_{i,0} = E_i^{Local}$.

Our goal is to optimize the user's offloading decision $x_{i,j}$, the computational resource allocation $f_{i,j}$ at the edge nodes, and to adjust the variables of the task division ratio $\gamma_{i,j}$ at the edge nodes to achieve the minimization of the overall system energy consumption under the condition that the tasks meet the delay requirements. The final optimization problem is expressed as:

$$\begin{aligned} P &: \min \sum_{i=1}^{N} \sum_{j=0}^{M} x_{i,j} E_{i,j} \\ C1 &: T_{i,j} \leq T_{i,max} \\ C2 &: \sum_{i}^{M} f_{i,j} \leq F_j \end{aligned} \tag{9}$$

$$C3: \sum_{j=0}^{M} x_{i,j} \leq 1$$

$$C4: x_{i,j} \in \{0,1\}, \gamma_{i,j} \in [0,1]$$

Since the proposed optimization problem is a mixed integer nonlinear programming (MINLP) problem that is difficult to solve directly, an efficient algorithm is given by decomposing the problem into different subproblems.

4.1 Resource Allocation

Under the condition that the offloading decision $x_{i,j}$ is determined, the objective function becomes convex, but the constraint C1 is still nonconvex, so the problem remains nonconvex.

In the case where the central cloud is needed for collaborative execution, the edge server needs to determine the proportion $\gamma_{i,j}$ of the task to be executed on the central cloud server after receiving the task, and thus the minimum latency can be achieved by considering the design of the optimal partition ratio $\gamma_{i,j}$. Since the edge cloud and the central cloud process the task in parallel, the total time consumption of the edge server to execute the task Z_i can be expressed as:

$$T_{i,j} = T_{i,j}^{tran,d} + \max\{T_{i,j}^{comp,e}, T_{i,j}^{tran,e} + T_{i,j}^{comp,c}\} \tag{10}$$

where $T_{i,j}^{tran,d} = \dfrac{L_i}{R_{i,j}}$ denotes the transmission delay for offloading task Z_i to edge node j, $T_{i,j}^{comp,e} = \dfrac{(1-\gamma_{i,j})C_i L_i}{f_{i,j}}$ Indicates the computational latency of task offload to the edge server, $T_{i,j}^{tran,e} = \gamma_{i,j}\dfrac{L_i}{W_c}$ represents the transfer latency required to offload a portion of task Z_i to a cloud computing center, $T_{i,j}^{comp,c} = \gamma_{i,j}\dfrac{C_i L_i}{f_c}$represents the computational latency required to offload a portion of task Z_i to a cloud computing center .When $T_{i,j}^{comp,e} = T_{i,j}^{tran,e} + T_{i,j}^{comp,c}$ the minimum value of this formula is obtained, so the optimal partition ratio can be expressed as:

$$\gamma_{i,j} = \frac{f_{i,j}(f_c + C_i W_c)}{f_{i,j}(f_c + C_i W_c) + f_c C_i W_c} \tag{11}$$

The partition ratio can be eliminated by substituting $\gamma_{i,j}$ into C1, and the constraint C1 can be written as:

$$C1: \frac{L_i}{R_{i,j}} + \frac{C_i L_i}{f_{i,j}} - \frac{C_i L_i (f_c + C_i W_c)}{f_{i,j}(f_c + C_i W_c) + f_c C_i W_c} \leq T_{i,\max} \tag{12}$$

It can be verified that the equation is convex and the optimization problem is thus transformed into:

$$P1: \min \sum_{i=1}^{N} \sum_{j=0}^{M} x_{i,j} E_{i,j} \tag{13}$$

$$C1: \frac{L_i}{R_{i,j}} + \frac{C_i L_i}{f_{i,j}} - \frac{C_i L_i (f_c + C_i W_c)}{f_{i,j}(f_c + C_i W_c) + f_c C_i W_c} \leq T_{i,\max} \tag{14}$$

$$C2: \sum_{i}^{M} f_{i,j} \leq F_j \tag{15}$$

The problem is transformed into a convex optimization problem under the condition that $x_{i,j}$ is determined and the optimal solution can be found.

4.2 Optimal Offloading Decision

Optimization of the offloading decision based on the constraints of a given resource allocation. Finding the optimal offloading decision by enumerating all possible offloading decisions is very computationally expensive. Therefore, the KOA algorithm is proposed to find the suboptimal offloading decision to reduce the total energy consumption of the system.

Under the condition of known resource allocation, the optimization problem can be transformed into:

$$P2: \min \sum_{i=1}^{N} \sum_{j=0}^{M} x_{i,j} E_{i,j} \tag{16}$$

$$C3: \sum_{j=0}^{M} x_{i,j} \leq 1 \tag{17}$$

$$C4: x_{i,j} \in \{0,1\}, \gamma_{i,j} \in [0,1] \tag{18}$$

A common approach to obtain the minimization of the energy consumption of the whole system by optimizing the value of the offloading decision is to calculate the corresponding system energy consumption by enumerating all possible task offloading decisions and then selecting the offloading decision that minimizes the system energy consumption. That is, the Enumeration Offloading Algorithm(EOA) proposed in this paper.

However, such an approach suffers from high computational complexity when the number of users M and edge servers N is large, for which the KOA matching algorithm is proposed to find the suboptimal offloading decision x. The offloading decision x can be transformed into a user-edge server matching problem with known edge server computational resource allocation, and we model this matching problem as a bipartite graph matching problem where the task and the edge server correspond to two parts of the graph, respectively. The matching problem is then solved using the Kuhn - Munkres algorithm, a classical maximum-weight matching algorithm that finds a set of optimal matches in the bipartite graph such that the sum of the matched edge weights is maximized. However, in the task offloading problem, we need to find a set of minimum-weight matches such that the sum of the edge weights of the matches is minimized. Therefore, the cost matrix is first processed to transform the minimum-right matching problem into a maximum-right matching problem. Then the energy consumption of the edge server is used as the cost of performing tasks from that server and the Kuhn-Munkres algorithm is used to solve the minimum-right matching. This minimum-weight matching scheme tells us which edge server each task should be assigned to in order to minimize the total energy consumption of the whole system.

EOA algorithm: Find all possible task offloading decisions, The system energy consumption is then calculated separately, i.e., step 1 in Algorithm 1, Find the task offloading decision that minimizes the total system energy consumption in all cases i.e. Step 2.

Algorithm 1: EOA

Input: Number of mobile users containing tasks M, number of edge servers N
Output: Minimum total system energy consumption E, task offload decision x
  for the Total number of all offloading decisions **do**
    generates the offloading decision x
    Calculate the total energy consumption E of the corresponding x according to Equation (6)
  end for
  Minimum total system energy consumption E, task offloading decision x

---

KOA algorithm: in this algorithm, not all task relocation decisions need to be found, so the complexity of the algorithm is less than that of the EOA algorithm. The elements in the cost matrix of the algorithm are the energy consumption of the task while it is executed on each server. Based on the cost matrix, a bipartite graph is created in which the nodes of the mobile user and the nodes of the edge server correspond to the rows and columns, respectively, and the edge weights between the nodes are the values of the corresponding items in the cost matrix. The adjacency matrix of the bipartite graph is then constructed, i.e., step 1, and for each node of a mismatched mobile user, it is marked as the maximum weight of all edge server nodes adjacent to it. A while loop is then used to continuously search for extended paths until all mobile user nodes are matched, i.e., steps 2–12. The predecessor nodes of all edge server nodes are searched to obtain the task offloading decision, and the minimum energy consumption of the system is determined by computing the sum of the weights of all matched edges, i.e., steps 13–14.

5 Simulation Results

Simulation results are given in this section to evaluate the effectiveness of the proposed algorithm. Multiple edge clouds and multiple MUs are assumed to exist and the bandwidth is equally distributed. The wireless channel gain is a Rayleigh random variable with unit variance.

Figure 1 shows the relationship between the total energy consumption of the system and the bandwidth, for since the tasks are computed locally without transmission energy consumption, therefore, the change in bandwidth has

Algorithm 2 :KOA

Input: Number of mobile users with tasks M, number of edge servers N
Output: Minimum total system energy consumption E, task offloading decision x

```
Initialize cost matrixCost, nodelabels, slack variables, visited labels, the predecessor.
Construct the adjacency matrix of the bipartite graph according to Cost
for All mobile users with tasks M do
  Mark labels(i) as the maximum weight of all edge server nodes adjacent to it
  while true do
    Initialize min_slack
    for All notes do
      Initialize min_slack
      if visited(j)=0 then
        Calculate slack(j)
        if slack(j) changes then
          Calculate the new predecessor(j) and slack(j)
        end if
        if slack(j) ¡ min_slack then
          min_slack ← slack(j)
          Calculate the new min_slack and its corresponding edge server node
        end if
      end if
    end for
    if Find the edge server node corresponding to min_slack then
      for all nodes do
        if visited(j) =1 then
          Compute the new labels(j)
        else
          Compute the new slack(j)
        end if
      end for
      Set visited ← 1 for the edge server node corresponding to min_slack
    else
      Exit the loop
    end if
  end while
end for
for All edge server nodes do
  Find the precursor node predecessor
  Calculate the sum of the weights of all matching edges
end for
```

no effect on the total energy consumption of the system, and secondly it can be seen from the figure that the total energy consumption decreases gradually with the increase in bandwidth, although the energy consumption of our proposed KOA algorithm is slightly higher than that of the EOA algorithm, the gap decreases gradually with the gradual increase in bandwidth. Figure 2 shows the relationship between the number of users and the total system energy consump-

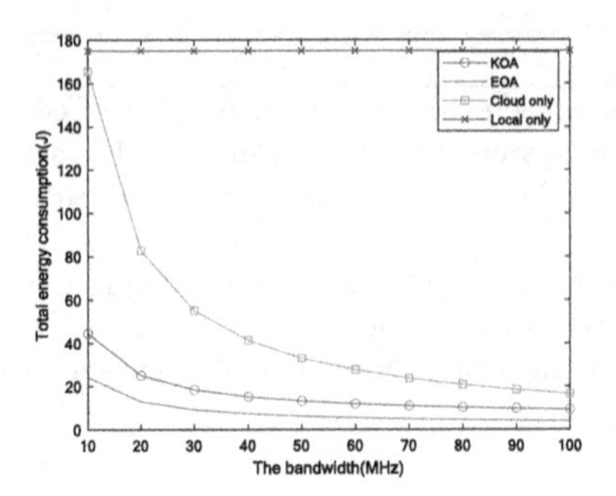

Fig. 2. Bandwidth and energy consumption.

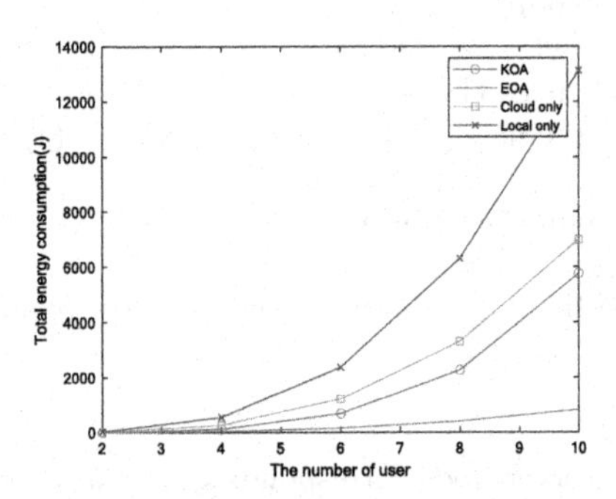

Fig. 3. Number of users and energy consumption.

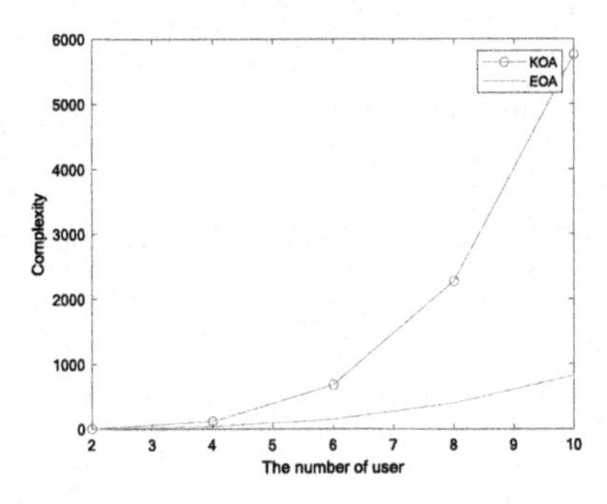

Fig. 4. Number of users and algorithm complexity.

tion, as we expected, the total system energy consumption increases with the number of users, and EOA and KOA algorithms always outperform the other two algorithms.Figure 3 compares the complexity of the two algorithms, with the growth of the number of users EOA algorithm complexity is much greater than the KOA algorithm, From this, we can obtain that the complexity of the KOA algorithm is much lower than that of the EOA algorithm if a certain amount of energy consumption is used as a cost (Fig. 4).

6 Conclusion

In this paper, we study a system energy consumption minimization problem in a three-tier network scenario consisting of multiple mobile users (MUs), multiple edge clouds, and a central cloud, while considering the condition of delay constraint. Since the problem is difficult to solve directly, we decompose it into two subproblems: the resource allocation problem and the offloading decision problem. For the resource allocation problem, since the original problem is nonconvex, we transform it into a convex function and solve it using a convex optimization method. For the offloading decision problem, we propose the KOA algorithm to find the optimal offloading decision. Finally, the simulation results show the effectiveness of the proposed algorithm.

Acknowledgments. This work was supported by the National Natural Science Foundation of China under Grants 62061036 and 62077032; The Self-Open Project of Engineering Research Center of Ecological Big Data, Ministry of Education; Natural Science Foundation of Shanxi Province under Grant 201901D111035.

References

1. Soyata, T., Muraleedharan, R., Funai, C., Kwon, M., Heinzelman, W.: Cloud-vision: real-time face recognition using a mobile-cloudlet-cloud acceleration architecture. In: 2012 IEEE Symposium on Computers and Communications (ISCC), pp. 000059–000066 (2012)
2. Hu, Y.C., Patel, M., Sabella, D., Sprecher, N., Young, V.: Mobile edge computing-a key technology towards 5G. ETSI White Paper, 11 (2015)
3. Liu, Y., Peng, M., Shou, G., Chen, Y., Chen, S.: Toward edge intelligence: multiaccess edge computing for 5G and internet of things. IEEE Internet Things J. **7**(8), 6722–6747 (2020)
4. Zhang, K., Gui, X., Ren, D.: Joint optimization on computation offloading and resource allocation in mobile edge computing. In: 2019 IEEE Wireless Communications and Networking Conference (WCNC), pp. 1–6 (2019)
5. Khalili, A., Zarandi, S., Rasti, M.: Joint resource allocation and offloading decision in mobile edge computing. IEEE Commun. Lett. **23**(4), 684–687 (2019)
6. Pan, M., Li, Z.: Multi-user computation offloading algorithm for mobile edge computing. In: 2021 2nd International Conference on Electronics, Communications and Information Technology (CECIT), pp. 771–776 (2021)
7. Guo, S., Liu, J., Yang, Y., Xiao, B., Li, Z.: Energy-efficient dynamic computation offloading and cooperative task scheduling in mobile cloud computing. IEEE Trans. Mob. Comput. **18**(2), 319–333 (2019)
8. Dong, H., Zhang, H., Li, Z., et al.: Computation offloading for service workflows in mobile edge computing environment. Comput. Eng. Appl. **55**, 42–49 (2018)
9. Ren, J., Yu, G., He, Y., Li, G.Y.: Collaborative cloud and edge computing for latency minimization. IEEE Trans. Veh. Technol. **68**(5), 5031–5044 (2019)
10. Du, C., Chen, Y., Li, Z., Rudolph, G.: Joint optimization of offloading and communication resources in mobile edge computing. In: 2019 IEEE Symposium Series on Computational Intelligence (SSCI), pp. 2729–2734 (2019)

11. Lyu, X., Tian, H., Sengul, C., Zhang, P.: Multiuser joint task offloading and resource optimization in proximate clouds. IEEE Trans. Veh. Technol. **66**(4), 3435–3447 (2017)
12. Gao, Z., Hao, W., Han, Z., Yang, S.: Q-learning-based task offloading and resources optimization for a collaborative computing system. IEEE Access **8**, 149011–149024 (2020)
13. Cao, H., Cai, J.: Distributed multiuser computation offloading for cloudlet-based mobile cloud computing: a game-theoretic machine learning approach. IEEE Trans. Veh. Technol. **67**(1), 752–764 (2018)
14. Ly, M.H. Dinh, T.Q., Kha, H.H.: Joint optimization of execution latency and energy consumption for mobile edge computing with data compression and task allocation. In: 2019 International Symposium on Electrical and Electronics Engineering (ISEE), pp. 113–118 (2019)
15. Tan, T., Zhao, M., Zhu, Y., Zeng, Z.: Joint offloading and resource allocation of uav-assisted mobile edge computing with delay constraints. In: 2021 IEEE 41st International Conference on Distributed Computing Systems Workshops (ICDCSW), pp. 21–26 (2021)
16. Lan, Y., Wang, X., Wang, D., Liu, Z., Zhang, Y.: Task caching, offloading, and resource allocation in D2D-aided fog computing networks. IEEE Access **7**, 104876–104891 (2019)

SW-TRRM: Parallel Optimization Research of the Random Ray Method Based on Sunway Bluelight II Supercomputer

Zenghui Ren, Tao Liu(✉), Zhaoyuan Liu, Ying Guo, Jingshan Pan, Dawei Zhao, Xiaoming Wu, and Meihong Yang

Shandong Computer Science Center (National Supercomputer Center in Jinan), Qilu University of Technology (Shandong Academy of Sciences), Jinan, China
liutao@sdas.org

Abstract. The Random Ray Method (TRRM) is a new approach to solving partial differential equations (PDEs) based on the method of characteristics (MOC). It employs stochastic rather than deterministic discretization of characteristic tracks and can be used for the numerical simulation of nuclear reactors. In this paper, we propose SW-TRRM, a parallel optimization program for TRRM based on the Sunway Bluelight II Supercomputer for the first time. We present a two-level parallelization scheme that consists of thread-level and process-level optimization. At the thread-level, we introduce three schemes for speeding up within a single core group, including direct parallelization, parallelization by energy groups, and loop structure optimization. At the process-level, we implement task parallelization among multiple processes using domain replication. Moreover, we devise an algorithm to optimize the MPI collective communication across super-nodes. Experimental results show that SW-TRRM achieves a 17.40× speedup within a single core group compared to the original TRRM program. When scaled up to 2,048 processes and 133,120 cores, SW-TRRM maintains good strong and weak scalability.

Keywords: High performance computing · Parallel optimization · Sunway supercomputer · The random ray method

1 Introduction

The method of characteristics (MOC) solves partial differential equations (PDEs) by defining characteristic lines. Along these lines, PDEs are reduced to ordinary differential equations (ODEs). PDEs can be estimated numerically by iterating on the initial conditions for the lines. Typical MOC applications require the simulation of many deterministic characteristic lines in space. In [1], a new stochastic method of track selection was introduced. It uses stochastic lines to simulate the tracks and iterates until termination criteria are met. The new approach is called The Random Ray Method (TRRM). It can continuously improve computational

Z. Tari et al. (Eds.): ICA3PP 2023, LNCS 14491, pp. 373–393, 2024.
https://doi.org/10.1007/978-981-97-0808-6_22

accuracy over successive iterations and reduce memory requirements. Since it does not need to store the data of deterministic integration and initial conditions.

The Sunway Bluelight II Supercomputer and the Sunway many-core processor are outstanding achievements of China in high-performance computing. TRRM utilizes its distinctive algorithmic features to reduce memory usage. In contrast to the Sunway taihulight supercomputer, the Sunway bluelight II supercomputer has more powerful computing capabilities and larger memory space on each node [2]. The Sunway bluelight II supercomputer can fully harness the potential of TRRM, and enhance the precision and efficiency of nuclear reactor numerical simulation.

In this paper, we design a parallel program of the random ray method based on the Sunway bluelight II supercomputer, which is called SW-TRRM. Our parallelization scheme is universal and well suited to the hardware of the Sunway supercomputers. The main contributions of this paper are as follows.

(1) We propose SW-TRRM, a parallel optimization program for TRRM based on the Sunway bluelight II supercomputer for the first time.
(2) We present a two-level parallelization scheme that consists of thread-level and process-level optimization. At the thread-level, we introduce three schemes for speeding up within a single core group, including direct parallelization, parallelization by energy groups, and loop structure optimization. At the process-level, we implement task parallelization among multiple processes using domain replication. Moreover, we devise an algorithm to optimize MPI collective communication across super-nodes of the Sunway bluelight II supercomputer.

We evaluate SW-TRRM using 2D C5G7 [3] benchmark. SW-TRRM achieves a speedup of 17.40× within a single core group. When scaled up to 2,048 processes and 133,120 cores, SW-TRRM maintains good strong and weak scalability.

The rest of this paper is organized as follows. Section 2 introduces the random ray method and Sunway bluelight II supercomputer. Section 3 presents the thread-level parallel optimization of SW-TRRM. Section 4 presents the process-level parallel optimization of SW-TRRM. Section 5 shows the experimental results and analysis. Section 6 introduce the related work. Finally, Sect. 7 concludes this paper.

2 The Random Ray Method and Sunway Bluelight II Supercomputer

2.1 The Random Ray Method

Reactor numerical simulation aims to calculate real-time data of the reactor and predict its operational behavior. These data help engineers to design and modify the structure of reactors. Two of the most important phenomena are the eigenvalue or criticality and its spatial power distribution. These phenomena are computed by Boltzmann Neutron Transport Equation. In the case of a steady state reaction Boltzmann Neutron Transport Equation is as in Eq. (1):

$$\begin{aligned}\vec{\Omega}\cdot\vec{\nabla}\psi(\vec{r},\vec{\Omega},E)+\Sigma_t(\vec{r},E)\psi(\vec{r},\vec{\Omega},E) = \\ +\int_0^{\infty}dE'\int_{4\pi}d\Omega'\,\Sigma_s(\vec{r},\vec{\Omega}'\rightarrow\vec{\Omega},E'\rightarrow E)\psi(\vec{r},\vec{\Omega}',E') \\ +\frac{\chi(\vec{r},E)}{4\pi K_{eff}}\int_0^{\infty}dE'\nu\,\Sigma_f(\vec{r},E')\int_{4\pi}d\Omega'\psi(\vec{r},\vec{\Omega}',E')\end{aligned} \tag{1}$$

ψ is the angular neutron flux.
$\vec{\Omega}$ is the angular direction unit vector that represents the direction of travel for the neutron.
$\vec{r}$ is the spatial position vector.
E is the neutron energy (or speed) in continuous space.
Σ_t is the total neutron cross section.
Σ_s is the scattering neutron cross section.
Σ_f is the fission neutron cross section.
χ is the energy spectrum.
ν is the average number of neutrons born per fission.
K_{eff} is the eigenvalue of the equation.

If the right hand of Eq. (1) is condensed into a single term, its form is shown in Eq. (2).

$$\overbrace{\vec{\Omega}\cdot\vec{\nabla}\psi(\vec{r},\vec{\Omega},E)}^{\text{streaming term}}+\overbrace{\Sigma_t(\vec{r},E)\psi(\vec{r},\vec{\Omega},E)}^{\text{absorption term}} = \overbrace{Q(\vec{r},\vec{\Omega},E)}^{\text{total neutron source term}} \tag{2}$$

MOC is a method for solving the Boltzmann neutron transport Eq. [4]. MOC works by solving Eq. (2) along a single characteristic line. For Eq. (2), when the ray passes through a constant cross-section region of the reactor geometry with a specific length and angle, it can be written as Eq. (3):

$$\vec{\Omega}\cdot\vec{\nabla}\psi(\vec{r_0},s\vec{\Omega},\vec{\Omega},E)+\Sigma_t(\vec{r_0},s\vec{\Omega},E)\psi(\vec{r_0},s\vec{\Omega},\vec{\Omega},E) = Q(\vec{r_0},s\vec{\Omega},\vec{\Omega},E) \tag{3}$$

We can assume the dependence of s on $\vec{r_0}$ and $\vec{\Omega}$ such that $\vec{r_0}+s\vec{\Omega}$ simplifies to s. When the differential operator is also applied to the angular flux ψ, the characteristic form of the Boltzmann neutron transport equation is obtained given in Eq. (4):

$$\frac{d}{ds}\psi(s,\vec{\Omega},E)+\Sigma_t(s,E)\psi(s,\vec{\Omega},E) = Q(s,\vec{\Omega},E) \tag{4}$$

An analytical solution to this characteristic equation can be achieved with the use of an integrating factor:

$$e^{-\int_0^s ds'\,\Sigma^T(s',E)}$$

to arrive at the final form of the characteristic equation shown in Eq. (5):

$$\psi(s,\vec{\Omega},E) = \psi(\vec{r_0},\vec{\Omega},E)e^{-\int_0^s ds' \Sigma_t(s',E)} + \int_0^s ds'' Q(s'',\vec{\Omega},E)e^{-\int_{s''}^s ds' \Sigma_t(s',E)} \tag{5}$$

MOC uses the multi-group approximation to discretize the continuous energy spectrum of neutrons traveling through the reactor into a fixed set of energy groups G. Each group $g \in G$ has its own specific cross-section parameters. Another assumption is that a large and complex problem can be broken up into small constant cross-section regions, and these regions have group-dependent isotropic sources (fission + scattering), Q_g. Under these assumptions, the multi-group MOC form of the neutron transport equation is shown in Eq. (6).

$$\psi_g(s,\vec{\Omega}) = \psi_g(\vec{r_0},\vec{\Omega})e^{-\int_0^s ds' \Sigma_{tg}(s')} + \int_0^s ds'' Q_g(s'',\vec{\Omega})e^{-\int_{s''}^s ds' \Sigma_{tg}(s')} \tag{6}$$

As shown in Eq. (7), the normal approximation of a spatially constant isotropic fission and scattering source q_0 leads to simple exponential attenuation along an individual characteristic of length s.

$$\psi_g(s) = \psi_g(0)e^{-\Sigma_{t,g}\, s} + \frac{q_0}{\Sigma_{t,g}}(1 - e^{-\Sigma_{t,g}\, s}) \tag{7}$$

The above is a series of changes made by [1] to Eq. (2), resulting in the basic equation suitable for single-track nuclear reactor simulation.

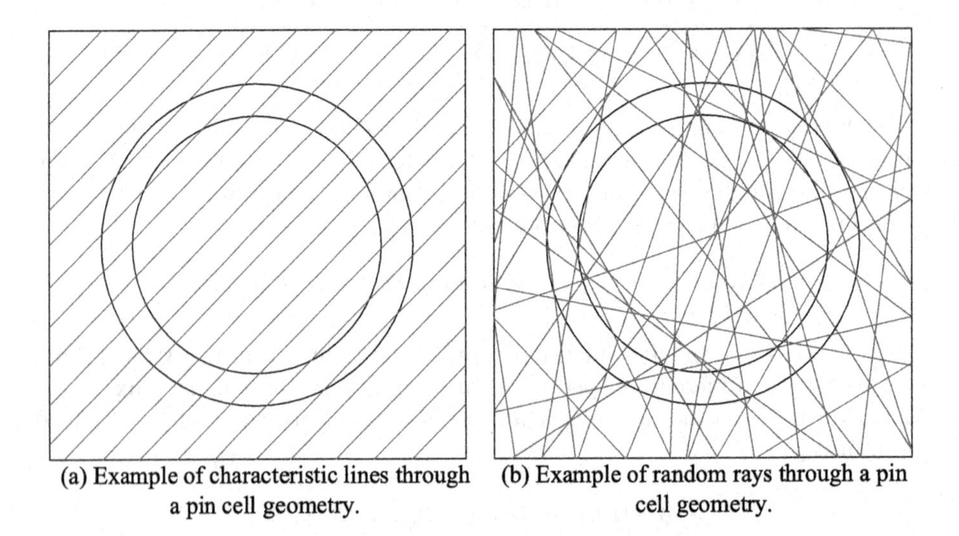

(a) Example of characteristic lines through a pin cell geometry. (b) Example of random rays through a pin cell geometry.

Fig. 1. Examples of characteristic lines and random rays.

TRRM is a new algorithm based on MOC. In MOC, the characteristic lines are tracked along deterministic starting points and directions, as shown in Fig. 1(a). TRRM uses a different strategy from MOC. It tracks rays in space randomly, as shown in Fig. 1(b). TRRM follows randomly sampled rays and solves the characteristic form of the neutron transport equation (Eq. (7)) along the way until certain termination conditions are satisfied.

$$Q_{r,e} = \frac{1}{\sum_{t,r,e}}\left[S_{r,e} + \frac{1}{K}F_{r,e}\right] \tag{8}$$

Equation (8) calculates the source at each iteration. Where $Q_{r,e}$ is the total source in source region cell r and neutron energy group e, $\sum_{t,r,e}$ is the total cross section, $S_{r,e}$ is the scattering source, k is the eigenvalue, and $F_{r,e}$ is the fission source.

To be able to set any number and length of rays in the program, after completing the transport sweep, the scalar flux is normalized to the total distance traveled by all rays during the transport sweep. The scattering and fission sources are added to the scalar flux, as shown in Eq. (9):

$$\phi_{r,e} = \frac{\phi_{r,e}}{\sum_{t,r,e} V_r} + Q_{r,e} \tag{9}$$

Where $\phi_{r,e}$ is the new scalar flux of different regions r and different energy groups e.

Each ray in TRRM is independent and can be calculated by different processors separately. Ray tracing operations are performed on the geometric regions of the problem, which are very similar to ray tracing in Monte Carlo methods [5]. When ray tracing is completed and the nearest adjacent surface is found, flux attenuation calculations are performed. The feature of this algorithm is that there are no azimuthal, track, and polar weighting computations in traditional MOC algorithms.

2.2 The Sunway Bluelight II Supercomputer

The Sunway bluelight II supercomputer is the new generation of Sunway supercomputers. It includes one cabinet and four computing super-nodes. Each super-node has 64 computing plugins, and each computing plugin contains two computing nodes. Each super-node contains 128 nodes. Each node contains two SW26010pro processors, two Sunway message processing chips, and 192 GB of DDR4 memory. The whole system has 512 nodes and 1,024 SW26010pro processors. The peak performance of the machine can reach 3.13 PFLOPS.

Figure 2 shows the architecture of the SW26010pro processor. The SW26010pro processor has six core groups (CGs). Each processor has a total of 96 GB of memory. The main memory can be evenly distributed to one, two, three, or six processors. Each core group contains a manage processing element (MPE) and 64 computing processing elements (CPEs). An SW26010pro processor has six MPEs and 384 CPEs. Each core has its own instruction cache and

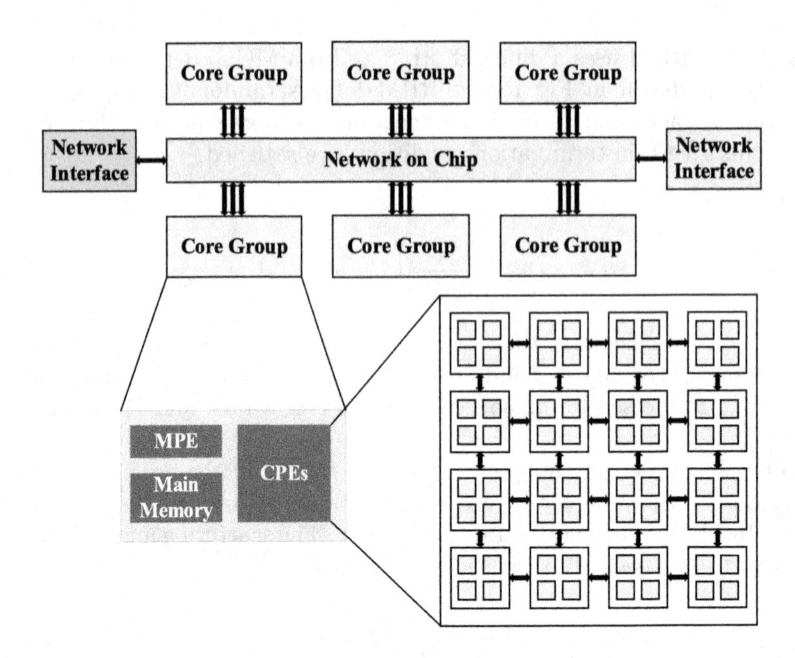

Fig. 2. The architecture of SW26010pro.

local data memory (LDM) or local data cache. The data transfer mode between main memory and LDM uses direct memory access (DMA). The data transfer mode between the main memory and local data cache uses load/store instructions. The data transfer mode between the LDM of cores uses remote memory access (RMA). Both MPE and CPEs use the SW64 instruction set [6].

3 Thread-Level Parallel Optimization of SW-TRRM

3.1 Direct Parallelization

Direct parallelization is the most basic thread-level parallel optimization scheme. According to our tests, 90% of the total time spent in Transport Sweep. Therefore, we put this hotspot into the CPEs for parallelization. A Sunway many-core parallelization model of TRRM is shown in Fig. 3. Each core group of the SW26010pro processor includes one MPE and 64 CPEs. To take full advantage of the core group's computing resources, we assign the computing tasks to the CPEs as much as possible, and the MPE is mainly responsible for preprocessing and control. The LDM of CPE can be logically divided into three regions, local LDM, shared LDM, and cache. When the program running, the CPE loads the data from the main memory into the CPE cache via the load/store instructions to read and write data.

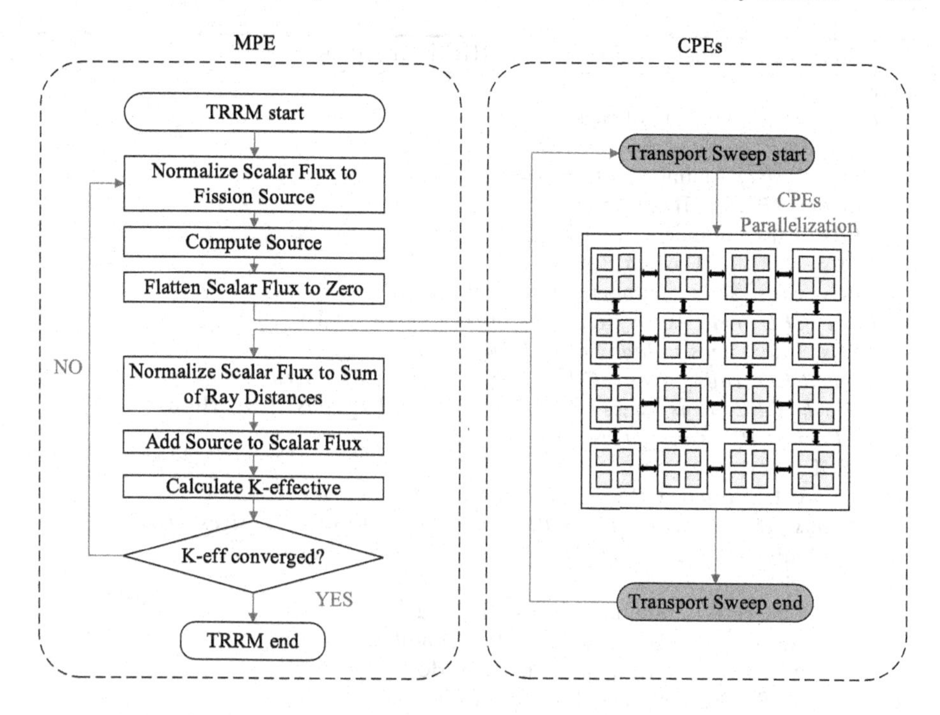

Fig. 3. Sunway many-core parallelization model of TRRM.

Algorithm 1. Attenuate Segment via Direct Parallelization

Input:

main_memory_scalar_flux : the data of scalar_flux in main memory;

1: **for all** Energy Groups $g \in G$ **do**

2: $\quad \Delta\psi_g = (\psi_g - Q_{r,g})(1 - e^{-\sum_{t,r,g} s})$

3: $\quad \phi_{r,g} = \phi_{r,g} + \Delta\psi_g$ ▹ Update *main_memory_scalar_flux*[*r*]

4: $\quad \psi_g = \psi_g - \Delta\psi_g$

5: **end for**

Algorithm 1 is the operation of attenuate segment via direct parallelization. The parallelization of the transport sweep on the CPEs is shown in Algorithm 2. We distribute the random rays in the program to different CPEs for simulation and try to load balance the system by ensuring that each CPE simulates the same number of rays.

Algorithm 2. Transport Sweep for TRRM on CPEs

Input:
$Rays\_sizes$: numbers of rays;
$Rays\_ID$: the ID of rays in each iteration;
$Process\_sizes$: numbers of processes;
$Process\_ID$: the ID of process;
$CPEs\_sizes$: numbers of CPEs;
$CPEs\_ID$: the ID of CPE;
$CPEs\_rays\_sizes$: numbers of rays for CPEs;
$Each\_CPE\_rays\_sizes$: numbers of rays for each CPE;
1: $CPEs\_rays\_sizes = Rays\_sizes/Process\_sizes$
2: $Each\_CPE\_rays\_sizes = CPEs\_rays\_sizes/CPEs\_sizes$
3: **if** $Each\_CPE\_rays\_sizes * CPEs\_sizes! = CPEs\_rays\_sizes$ **then**
4: $Each\_CPE\_rays\_sizes++$
5: **end if**
6: **for** $i = 0, 1, ..., Each\_CPE\_rays\_sizes$ **do**
7: $Rays\_ID = Process\_ID * CPEs\_rays\_sizes + i * CPEs\_sizes + CPEs\_ID$
8: **if** $Rays\_ID < (Process\_ID + 1) * CPEs\_rays\_sizes$ **then**
9: Distance Traveled $D = 0$
10: Generate Randomized Rays_ID $(\hat{r}, \hat{\Omega})$
11: Apply Rays_ID Boundary Flux Condition
12: **while** $D <$ TerminationDistance **do**
13: Set Nearest Neighbor distance $s = \infty$
14: **for all** Cell Neighbors **do**
15: Rays_ID Trace to Find Distance s_n to Neighbor Surface
16: **if** $s_n < s$ **then**
17: $s = s_n$
18: **end if**
19: **end for**
20: Attenuate Segment s (Algorithm 1)
21: $D = D + s$
22: Move Rays_ID Forward or Reflect
23: **end while**
24: **end if**
25: **end for**

3.2 Parallelization by Energy Groups

When we use direct parallelization, every read and write of data from CPEs requires direct access to the main memory. Using local LDM can increase the speed for accessing the main memory of CPEs. But the LDM is only 256 KB. We cannot create a complete data copy from the main memory in each CPE's LDM. In this section, because of the independence of different energy groups in TRRM, we parallelize the computation by energy groups within the core group.

As shown in line 1 of Algorithm 3, we create an array for only one energy group in the local LDM of the CPE. We can write data directly in the local LDM without accessing the main memory. After completing the data operation,

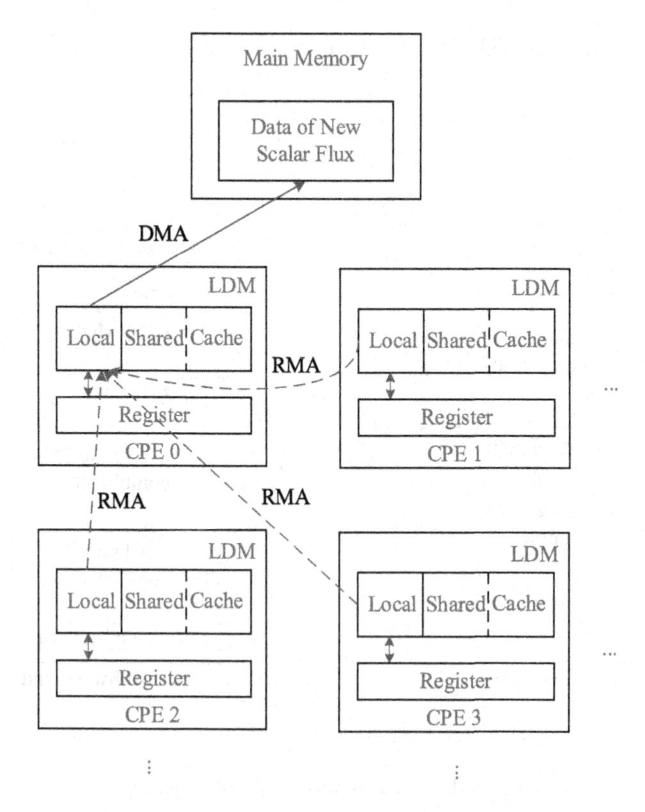

Fig. 4. Allreduce via RMA within CPEs.

we aggregate the data in the local LDM of all CPEs via RMA. Finally, we write back the data to the main memory via DMA. It is shown in Fig. 4.

Algorithm 3. Parallel for Attenuate Segment by Energy Groups

Input:

Cell_sizes : numbers of cells;

main_memory_scalar_flux : the data of scalar_flux in main memory;

1: __thread_local *local_scalar_flux*[*Cell_sizes*]

2: **for all** Energy Groups $g \in G$ **do**

3: $\Delta\psi_g = (\psi_g - Q_{r,g})(1 - e^{-\Sigma_{t,r,g}\, s})$

4: $\phi_{r,g} = \phi_{r,g} + \Delta\psi_g$ ▷ Update *local_scalar_flux*[*r*]

5: *athread_redurt*(*local_scalar_flux*) ▷ Allreduce the data in CPEs via RMA

6: *athread_DMA_put*(*main_memory_scalar_flux*, *local_scalar_flux*)

7: $\psi_g = \psi_g - \Delta\psi_g$

8: **end for**

3.3 Loop Structure Optimization

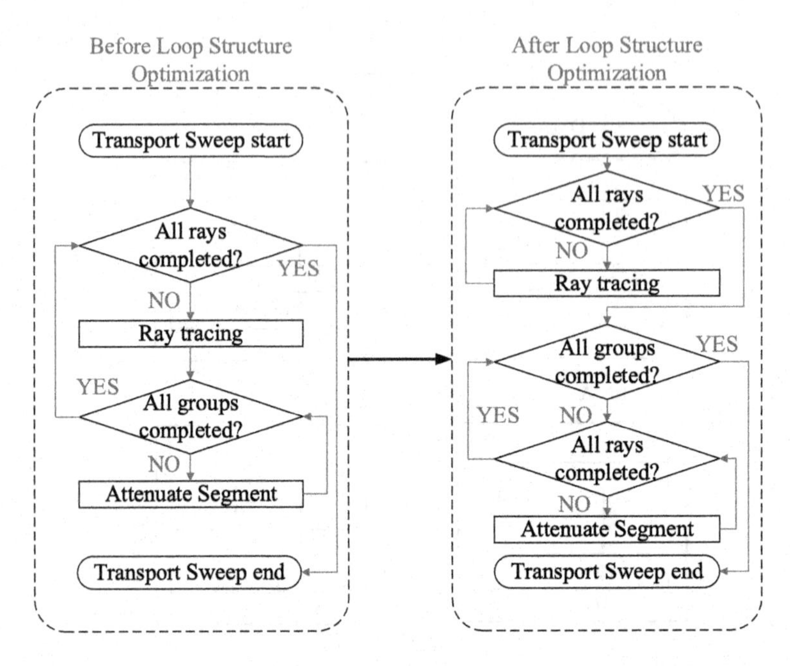

Fig. 5. Comparison of process diagrams before and after the loop structure optimization.

There are many data fetches in the CPE. They need to directly access the main memory. These data accesses are highly discrete, and it is impossible to increase the access speed by loading a whole block of contiguous data into the local LDM.

Figure 5 shows the comparison of process diagrams before and after the loop structure optimization. To improve the speed of discrete data access, we optimize the loop structure of Algorithm 2. This is shown in Algorithm 4. After running the ray-tracing simulation for each ray, we record the ray information about the intersection in the geometric region. This information is used in Algorithm 4.

4 Process-Level Parallel Optimization of SW-TRRM

This section introduces the process-level parallelization scheme of TRRM and proposes an algorithm to optimize MPI collective communication across supernodes based on the Sunway bluelight II supercomputer.

Algorithm 4. Optimizing the Loop Structure of Transport Sweep for TRRM on CPEs

Input:
Refer to Algorithm 2 and Algorithm 3 for predefined operations
$Rays\_intersections$: number of intersections for rays;
1: $CPEs\_rays\_sizes = Rays\_sizes / Process\_sizes$
2: $Each\_CPE\_rays\_sizes = CPEs\_rays\_sizes / CPEs\_sizes$
3: **if** $Each\_CPE\_rays\_sizes * CPEs\_sizes! = CPEs\_rays\_sizes$ **then**
4: $Each\_CPE\_rays\_sizes++$
5: **end if**
6: Ray Tracing (Algorithm 2 (Remove Attenuate Segment))
7: **for all** Energy Groups $g \in G$ **do**
8: **for** $i = 0, 1, \ldots, Each\_CPE\_rays\_sizes$ **do**
9: $Rays\_ID = Process\_ID * CPEs\_rays\_sizes + i * CPEs\_sizes + CPEs\_ID$
10: **if** $Rays\_ID < (Process\_ID + 1) * CPEs\_rays\_sizes$ **then**
11: **for** $i = 0, 1, \ldots, Rays\_intersections$ **do**
12: $\Delta\psi_g = (\psi_g - Q_{r,g})(1 - e^{-\sum_{t,r,g} s})$
13: $\phi_{r,g} = \phi_{r,g} + \Delta\psi_g$ ▹ Update $local\_scalar\_flux[r]$
14: $\psi_g = \psi_g - \Delta\psi_g$
15: **end for**
16: **end if**
17: **end for**
18: $athread\_redurt(local\_scalar\_flux)$ ▹ Allreduce the data in CPEs via RMA
19: $athread\_DMA\_put(main\_memory\_scalar\_flux, local\_scalar\_flux)$
20: **end for**

4.1 Domain Replication

In process-level parallel optimization, we propose domain replication. Because the main memory between different processes is distributed and cannot be directly shared, we need to complete domain replication in each process. As shown in Fig. 6, we copy the data in the main memory to each process. Each process's main memory has the same data.

The corresponding parallel algorithm is shown in Algorithm 5. We evenly distribute all the rays to different processes. After processes execute transport sweep, we use all-reduce on each process to aggregate the data.

4.2 Optimizing the Communication Across Super-Nodes

Algorithm 5 requires all-reduce after each iteration. Much communication is an important factor that affects parallel efficiency. The Sunway bluelight II supercomputer has 1,024 CPUs, and each super-node has 256 CPUs. Each CPU has six core groups or 6 processes. A super-node has 1,536 processes. When the number of processes becomes large, the processes are distributed to different super-nodes inevitably. The MPI communication between different super-nodes has higher latency than the MPI communication within one super-node. There-

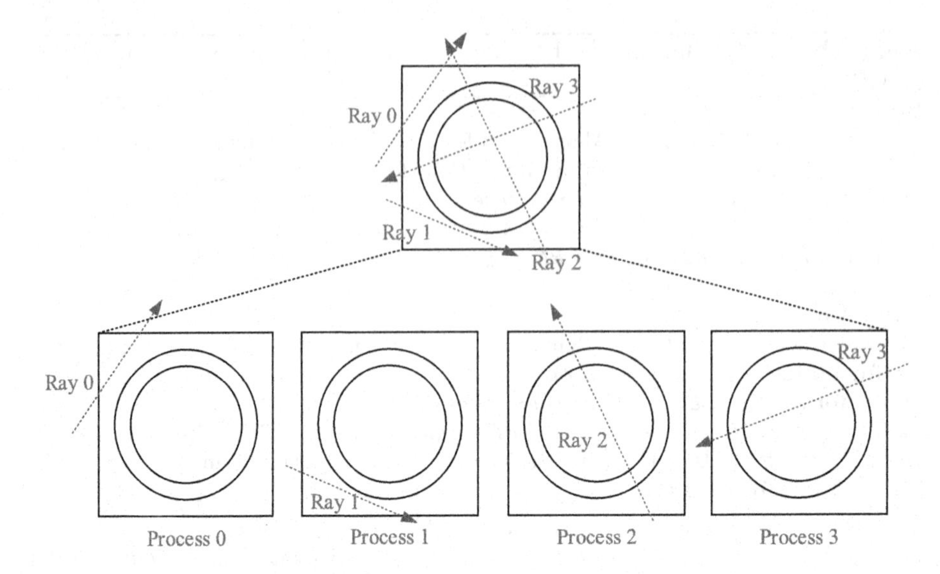

Fig. 6. Process-level parallel scheme of TRRM via domain replication.

Algorithm 5. Process-Level Parallelization of the Iteration for TRRM

Input:
  Rays_sizes : numbers of rays;
  Rays_ID : the ID of rays in each iteration;
  Process_sizes : numbers of processes;
  Process_ID : the ID of process;
  Each_process_rays_sizes : numbers of rays for each process;
1: Initialize Scalar Fluxes to 1.0
2: **while** K-effective and Scalar Flux Unconverged **do**
3:   Normalize Scalar Flux to Fission Source
4:   Compute Source (Equation (8))
5:   Flatten Scalar Flux to Zero
6:   Sample Rays in Local MPI Subdomain
7:   $Each\_process\_rays\_sizes = Rays\_sizes/Process\_sizes$
8:   **if** $Each\_process\_rays\_sizes * Process\_sizes! = Rays\_sizes$ **then**
9:     $Each\_process\_rays\_sizes++$
10:   **end if**
11:   **for** $i = 0, 1, ..., Each\_process\_rays\_sizes$ **do**
12:     $Rays\_ID = i * Process\_sizes + Process\_ID$
13:     Transport Sweep *Rays_ID*
14:   **end for**
15:   MPI Allreduce the Data of Scalar Flux in Main Memory
16:   Normalize Scalar Flux to Sum of Ray Distances
17:   Add Source to Scalar Flux (Equation (9))
18:   Calculate K-effective
19: **end while**

fore, we develop a new algorithm to reduce the latency of MPI based on the Sunway bluelight II supercomputer.

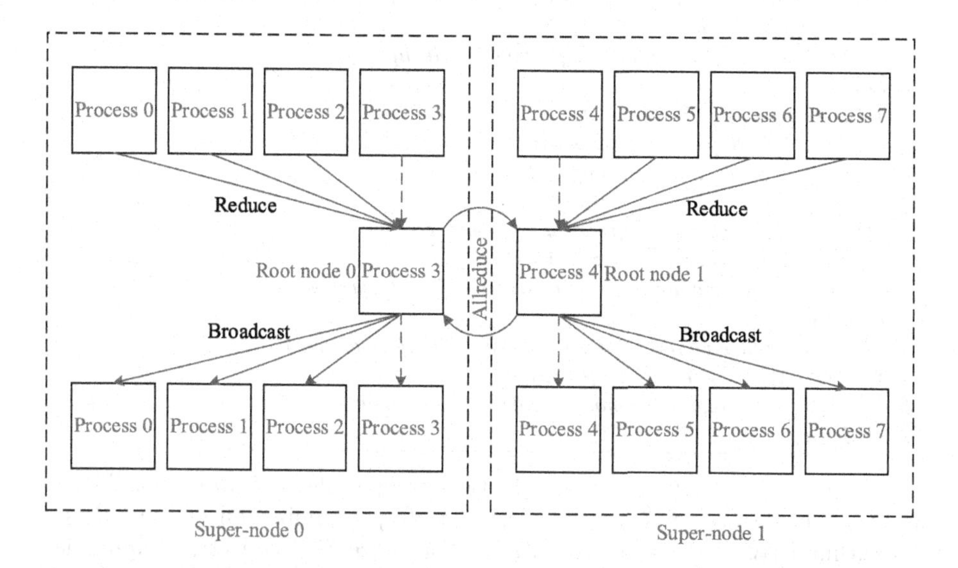

Fig. 7. The optimization of MPI collective communication across super-nodes.

We propose an algorithm to optimize the MPI collective communication across super-nodes. It reduces the latency of MPI aggregation operations in multiple processes. As shown in Figure 7, we need to optimize the MPI communication across two super-nodes. Firstly, we select a root node in each super-node and aggregate the data of the processes in the super-node to the root node. Then, we use all-reduce on the root nodes. Finally, we broadcast the information of the root nodes to the processes within each super-node. As shown in Algorithm 6, the processes are divided into several groups within different super-nodes. All the root nodes form a group within each super node, and we create communicators for these groups.

In Algorithm 6:

Process_sizes is the number of processes within each super-node;
Supernode_sizes is the number of super-nodes;
Group_i is the ID of the MPI group;
Group_root is the ID of the group which is the root node in each MPI group;
Comm_i is the ID of the MPI communicator;
Comm_root is the ID of the communicator which is the root node in each MPI communicator;
Rank_root[supernode_sizes] is the array of root process IDs for all super-node;
Rank_supernode_i[process_sizes] is the array of processes IDs within each super-nodes.

Algorithm 6. Optimization for MPI Communication Across Super-nodes

Input:

$main\_memory\_scalar\_flux$: the data of scalar_flux in main memory;

1: **for** $i = 0, 1, ..., Supernode\_sizes$ **do**
2: $MPI\_Group\_incl(Rank\_supernode\_i, Group\_i)$
3: **end for**
4: $MPI\_Group\_incl(Rank\_root, Group\_root)$
5: **for** $i = 0, 1, ..., Supernode\_sizes$ **do**
6: $MPI\_Comm\_create(Group\_i, \&Comm\_i)$
7: **end for**
8: $MPI\_Comm\_create(Group\_root, \&Comm\_root)$
9: **for** $i = 0, 1, ..., Supernode\_sizes$ **do**
10: $MPI\_Reduce(main\_memory\_scalar\_flux, Comm\_i)$
11: **end for**
12: $MPI\_Allreduce(main\_memory\_scalar\_flux, Comm\_root)$
13: **for** $i = 0, 1, ..., Supernode\_sizes$ **do**
14: $MPI\_Bcast(main\_memory\_scalar\_flux, Comm\_i)$
15: **end for**

In summary, we have completed the two-level parallel optimization of TRRM, as shown in Algorithm 7. We implement a deep optimization of the CPEs in Algorithm 4. We propose an algorithm to optimize MPI collective communication across super-nodes in Algorithm 6.

Algorithm 7. Two-Level Parallelization of the Iteration for TRRM

1: Initialize Scalar Fluxes to 1.0
2: **while** K-effective and Scalar Flux Unconverged **do**
3: Normalize Scalar Flux to Fission Source
4: Compute Source (Equation (8))
5: Flatten Scalar Flux to Zero
6: Sample Rays in Local MPI Subdomain
7: Transport Sweep (Algorithm 4)
8: MPI Allreduce the Data of Scalar Flux in Main Memory (Algorithm 6)
9: Normalize Scalar Flux to Sum of Ray Distances
10: Add Source to Scalar Flux (Equation (9))
11: Calculate K-effective
12: **end while**

5 Evaluation

5.1 Configuration

We evaluated the two-level parallel optimization program named SW-TRRM based on the Sunway bluelight II supercomputer. Table 1 shows the configuration used for the evaluation. We used 2D C5G7 [3] to test the performance of SW-TRRM at various parallelizations. We set the number of random rays to 131,072. Ensure that each CPE can execute at least one random ray. We set the ray travel distance to small, medium, and large scales, respectively 1,340 cm, 2,680 cm, and 4,180 cm. All the experimental results use the average results from three rounds.

Table 1. The hardware and software used in the evaluation.

Hardware	Computer	Sunway Bluelight II Supercomputer
	Processor model	SW26010pro
	Number of nodes	171
	Number of cores	133,120 (2,048 MPEs and 131,072 CPEs)
	Memory	32,768 GB
Software	System	Sunway Linux 4.4.15
	Compiler	mpicc and sw9gcc

5.2 Performance with Various Parallelizations

In the Thread-Level Optimization. The initialization parameters of the program are 131,072 rays and one iteration.

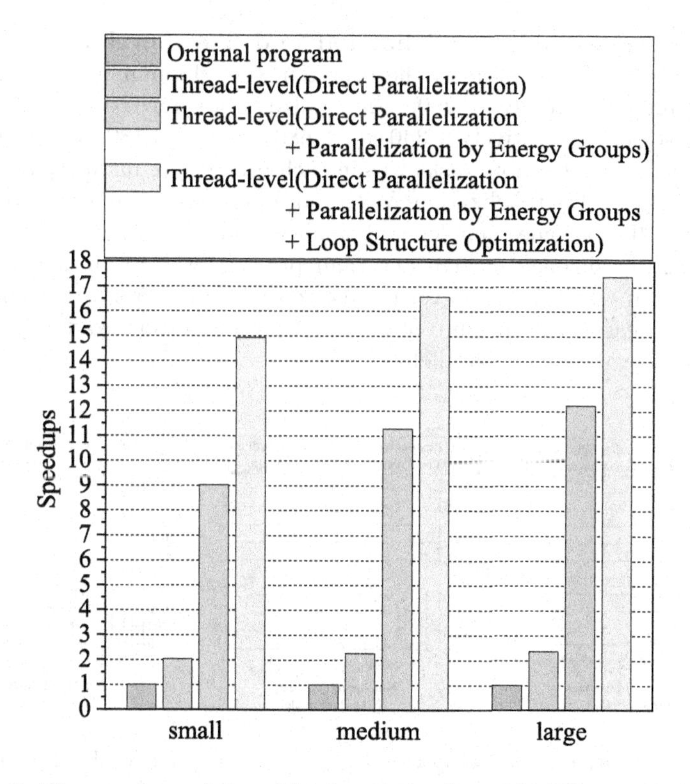

Fig. 8. The speedups of thread-level optimization with different schemes.

Figure 8 shows the speedups of thread-level optimization with different schemes. The first scheme is direct parallelization. We obtained 2.03–2.36× speedups with different scales In this scheme. The second scheme is parallelization by energy group, and we obtained 9.01–12.22× speedups at different scales in direct parallelization and parallelization by energy group. The third scheme is loop structure optimization. In direct parallelization, parallelization by energy group, and loop structure optimization, we obtained 14.93–17.40× speedups at different scales.

In the scheme of direct parallelization, CPEs access the main memory directly via load/store instruction. CPEs are less efficient in this way of accessing main memory. In the scheme of parallelization by energy group, according to the characteristic of independence between energy groups in TRRM, we parallelize the rays by energy group. This significantly reduces the memory space required for data in a single loop. We copy the data in the local LDM of each CPE, increasing the speed of CPEs for accessing the data. In the scheme of the loop structure optimization, by optimizing the loop structure of Algorithm 2, the access block is increased by accessing the data under the same energy group in one loop. This can improve the data reuse rate in CPEs.

In the Process-Level Parallelization and Two-Level Parallelization. The initialization parameters of the program were the number of rays 131,072 and the number of iterations 200. The ray travel distances are small, medium, and large scales, respectively 1,340 cm, 2,680 cm, and 4,180 cm. Compared to process-level parallelization, tasks within each process are further parallelized in CPEs via two-level parallelization. With more processes increasing, the number of rays in CPEs becomes smaller and smaller. The percentage of hotspot time in the program shrinks much faster than process-level optimization after the acceleration of CPEs in two-level parallelization. Therefore, as the number of processes increases, the speedup of two-level parallelization gradually decreases compared to process-level parallelization.

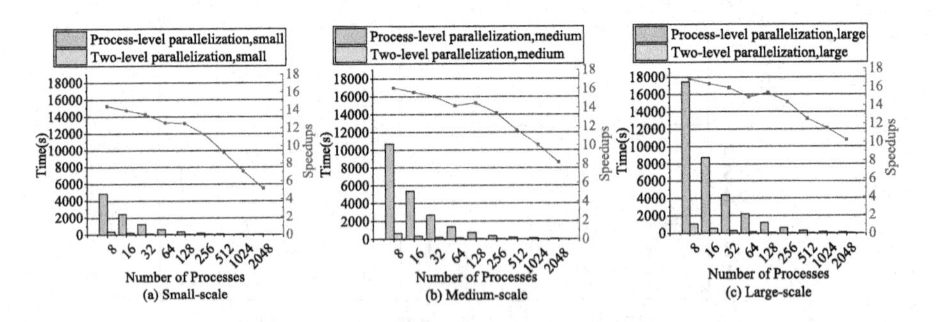

Fig. 9. The execution time and speedups of process-level and two-level parallelization with different scales.

The execution time and speedups for process-level parallelization and two-level parallelization at different scales are shown in Fig. 9. When the number of processes is 2,048, compared to process-level parallelization, the two-level parallelization maintains 5.2×, 8.18×, and 10.21× speedups at different scales. It shows that a two-level parallelization program can make full use of the hardware resources of the Sunway bluelight II supercomputer. It can handle more computational tasks. As the number of computing tasks increases, the two-level parallelization program can show better acceleration in multiple processes.

As described in Sect. 4.2, when the number of processes becomes large, the processes are distributed to different super-nodes. We tested the latency of the program under 1,024 processes and 2,048 processes at different scales. The initialization parameters of the program were the number of rays 131,072 and the number of iterations 200. Compared with the original MPI, when the number of processes is 1,024 and 2,048, we obtain 1.07×–1.42× speedups via the algorithm to optimize MPI collective communication across super-nodes at different scales.

5.3 Scalability

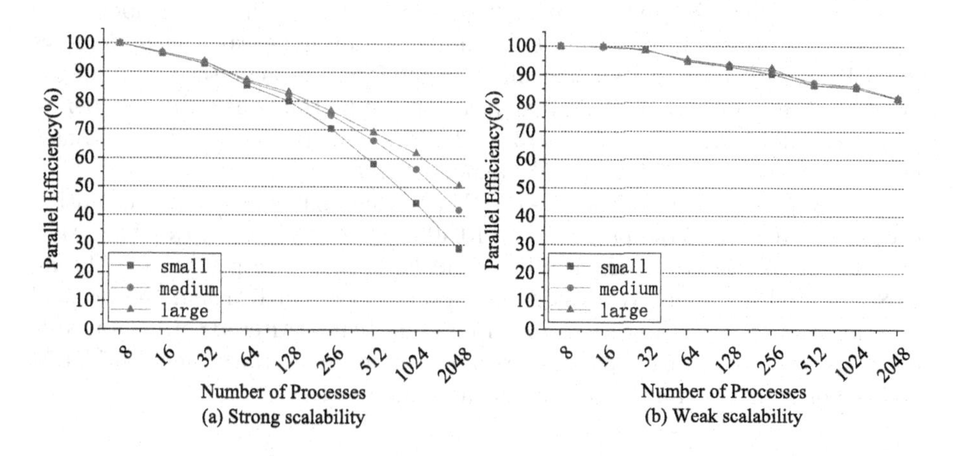

Fig. 10. Strong and weak scalability of SW-TRRM.

For strong scalability, we set three groups of ray travel distances. They are 1,340 cm, 2,680 cm, and 4,180 cm as small, medium, and large scales respectively. And the execution time of the program with 8 processes (8 MPEs + 512 CPEs) is used as the benchmark for evaluation. The strong scalability of SW-TRRM is shown in Figure 10(a). When the number of processes increases to 1,024 (1,024 MPEs + 65,536 CPEs) and above, the parallel efficiency of SW-TRRM drops rapidly. On the one hand, the percentage hotspot time of function in the two-level parallelization decreases as the number of processes increases. The MPI communication latency is constant. This makes the percentage time of MPI communication increase. On the other hand, when using more nodes, it is inevitable to cross super-nodes, which also increases the MPI communication latency. After using the algorithm to optimize MPI collective communication across super-nodes, we improved the parallel efficiency of the program by 3%~5%

when the number of processes is 1,024 (1,024 MPEs + 65,536 CPEs) and 2,048 (2,048 MPEs + 131,072 CPEs). In this case, even on a small scale, the parallel efficiency of the program has reached 28.61%. In the medium scale, the parallel efficiency is 42.1%. On a large scale, the parallel efficiency is 50.57%.

For weak scalability, we set the number of rays at 2,048 processes (2,048 MPEs + 131,072 CPEs) to 131,072. We proportionally reduce the number of processes in the program according to the number of random rays. When the number of processes is 8 (8 MPEs + 512 CPEs), the number of random rays in the program is also 512. This setting ensures that under multiple groups of processes, each CPE simulates one random ray. We use the execution time of the program with 8 processes (8 MPEs + 512 CPEs) as the benchmark for evaluation. The weak scalability of SW-TRRM is shown in Figure 10(b). When the number of processes increases, the weak scalability of the program decays to a certain extent. We change the number of rays in the program for different numbers of processes. This changes the number of computing tasks in the program. However, it does not affect the amount of MPI communication. With more nodes, the number of processes involved in MPI communication increases, which increases the MPI communication time consumed. Therefore, the weak scalability of the program does not strictly follow a linear speedup. But when the number of processes is 2,048 (2,048 MPEs + 131,072 CPEs), the program still maintains a parallel efficiency of 81.59%~81.71% under different scales.

We compared the results of SW-TRRM and the original program. The initialization parameters of the program are 10,404 cells and 131,072 rays. The travel distance of the rays is small, medium, and large scales, respectively 1,340 cm, 2,680 cm, and 4,180 cm. We tested the results of SW-TRRM and the original program after different iterations under different scales. After experimental testing, when the initialization parameters of the program are the same, SW-TRRM and the original program have exactly the same results.

6 Related Work

The random ray method is a new approach based on MOC. It was first proposed by John R. Tramm in [1] and can be used in many computational physics problems that use MOC. For example, neutron transport [7,8], and gamma ray transport [9]. The MIT team has developed a high-fidelity full-reactor numerical simulation software based on MOC, OpenMOC [7]. It is written in C/C++ and CUDA and provides a Python interface. It achieves good acceleration effects on GPUs compared to single-threaded CPU versions. The Institute of Nuclear and New Energy Technology (INET) has developed a new Three-Dimensional method of characteristics neutron transport code, ARCHER. It aims to obtain a high-fidelity transport solution of a pebble-bed High-Temperature gas-cooled Reactor (HTR) with explicit pebble-bed geometry [8].

John R. Tramm developed an advanced random ray code (ARRC) [10] that uses the random ray method and has very high memory utilization efficiency based on the domain decomposition method of task partitioning. It can simulate three-dimensional nuclear reactors and demonstrate good scalability on multiple processes. [11] is another program based on the random ray method that operates on the GPU architecture. Called immortal rays, it achieves better acceleration effects on both a single GPU and the Summit supercomputer for simulation problems ranging from small two-dimensional to large three-dimensional.

Sunway supercomputers have achieved outstanding achievements in many fields in the past decade [12–14]. [12] successfully scaled the fully implicit solver to the entire system of the Sunway taihulight supercomputer and achieved fast and accurate atmospheric simulation at 488 m of horizontal resolution. [13] implements a highly scalable nonlinear seismic simulation tool on the Sunway taihulight supercomputer. [14] implements a random quantum circuit simulator based on tensors and completes simulated sampling in 304 s with the new Sunway supercomputer. The above three research projects have won the Gordon Bell Prize.

There are a lot of good works based on the new Sunway supercomputer [14–20]. [15] completes a dynamic Monte Carlo simulation of up to 54 trillion atoms on a new Sunway supercomputer. [16] simulates a magnetized annular plasma with 111.3 trillion particles and 25.7 billion grids using the symplectic dynamic PIC method. Based on ab initio,[17] achieves Raman spectroscopy simulation of natural biological systems with up to 3,006 atoms. [18] implements a deep learning model named BaGuaLu containing 174 trillion parameters. [19] proves the validity of quantum state representation based on CNN in a new Sunway supercomputer. [20] implements a quantum simulator SW_Qsim based on a tensor network, which can simulate up to 400 qubits and extend to 28.75 million cores.

In summary, there are a lot of projects based on the Sunway supercomputers and much parallelization research on TRRM. However, TRRM has never been migrated and optimized on the Sunway supercomputer. Therefore, it is significant to implement the parallel optimization research of TRRM on the Sunway supercomputer. The SW TRRM proposed in this paper can satisfy the speed and accuracy for the numerical simulation of reactors based on the Sunway supercomputers. Moreover, this work can enrich the application ecology of the Sunway supercomputers.

7 Conclusion

In this paper, we implemented a parallel optimization program named SW-TRRM based on the Sunway bluelight II supercomputer. For the parallel optimization of TRRM, we proposed a two-level parallel scheme, including thread-level and process-level optimizations. At the thread-level, we introduce three schemes for speeding up within a single core group, including direct parallelization, parallelization by energy groups, and loop structure optimization. At the process-level optimization, we implemented task parallelization among multiple processes based on domain replication. In addition, we also proposed an algorithm to optimize the MPI collective communication across super-nodes.

Experimental results show that, compared with the original TRRM program, SW-TRRM can achieve 17.40× speedups within a single core group. When scaled up to 2,048 processes and 133,120 cores, SW-TRRM maintains good strong and weak scalability.

Acknowledgements. This work is supported by the National Natural Science Foundation of China under Grant 62002186, 12105150, and 2021 Shandong Youth Innovation Talent Introduction and Education Plan (Parallel Computing Industrial Software Innovation Team Based on Chinese supercomputer), and TaiShan Scholars Program NO. tsqnz20221148, and the unveiling project of Qilu University of Technology (Shandong Academy of Sciences) under Grant 2022JBZ01-01.

References

1. Tramm, J.R., Smith, K.S., Forget, B., Siegel, A.R.: The Random Ray Method for neutral particle transport. J. Comput. Phys. **342**, 229–252 (2017)
2. Zhu, Q., Luo, H., Yang, C., Ding, M., Yin, W., Yuan, X.: Enabling and scaling the HPCG benchmark on the newest generation Sunway supercomputer with 42 million heterogeneous cores. In: Proceedings of the International Conference for High Performance Computing, Networking, Storage and Analysis, pp. 1–13 (2021)
3. Lewis, E., Smith, M., Tsoulfanidis, N., Palmiotti, G., Taiwo, T., Blomquist, R.: Benchmark specification for Deterministic 2-D/3-D MOX fuel assembly transport calculations without spatial homogenization (C5G7 MOX). NEA/NSC **280**, 2001 (2001)
4. Askew, J.: A characteristics formulation of the neutron transport equation in complicated geometries. Technical report, United Kingdom Atomic Energy Authority (1972)
5. Romano, P.K., Forget, B.: The OpenMC monte carlo particle transport code. Ann. Nucl. Energy **51**, 274–281 (2013)
6. Hu, W., et al.: 2.5 million-atom ab initio electronic-structure simulation of complex metallic heterostructures with DGDFT. In: 2022 SC22: International Conference for High Performance Computing, Networking, Storage and Analysis (SC), pp. 48–60. IEEE Computer Society (2022)
7. Boyd, W., Shaner, S., Li, L., Forget, B., Smith, K.: The OpenMOC method of characteristics neutral particle transport code. Ann. Nucl. Energy **68**, 43–52 (2014)
8. Zhu, K., Kong, B., Hou, J., Zhang, H., Guo, J., Li, F.: ARCHER-a new Three-Dimensional method of characteristics neutron transport code for Pebble-bed HTR with coarse mesh finite difference acceleration. Ann. Nucl. Energy **177**, 109303 (2022)
9. Chao, N., Yang, H., Liu, Y.K., Xia, H., Ayodeji, A.: A local method of characteristics for dose assessment. Radiat. Phys. Chem. **173**, 108869 (2020)
10. Tramm, J.R., Smith, K.S., Forget, B., Siegel, A.R.: ARRC: a random ray neutron transport code for nuclear reactor simulation. Ann. Nucl. Energy **112**, 693–714 (2018)
11. Tramm, J.R., Siegel, A.R.: Immortal rays: rethinking random ray neutron transport on GPU architectures. Parallel Comput. **108**, 102832 (2021)
12. Yang, C., et al.: 10M-core scalable fully-implicit solver for nonhydrostatic atmospheric dynamics. In: SC 2016: Proceedings of the International Conference for High Performance Computing, Networking, Storage and Analysis, pp. 57–68. IEEE (2016)

13. Fu, H., et al.: 18.9-Pflops nonlinear earthquake simulation on sunway TaihuLight: enabling depiction of 18-Hz and 8-meter scenarios. In: Proceedings of the International Conference for High Performance Computing, Networking, Storage and Analysis, pp. 1–12 (2017)
14. Liu, Y., et al.: Closing the "Quantum Supremacy" gap: achieving real-time simulation of a random quantum circuit using a new sunway supercomputer. In: Proceedings of the International Conference for High Performance Computing, Networking, Storage and Analysis, pp. 1–12 (2021)
15. Shang, H., et al.: TensorKMC: kinetic monte carlo simulation of 50 trillion atoms driven by deep learning on a new generation of sunway supercomputer. In: Proceedings of the International Conference for High Performance Computing, Networking, Storage and Analysis, pp. 1–14 (2021)
16. Xiao, J., et al.: Symplectic structure-preserving particle-in-cell whole-volume simulation of tokamak plasmas to 111.3 trillion particles and 25.7 billion grids. In: Proceedings of the International Conference for High Performance Computing, Networking, Storage and Analysis, pp. 1–13 (2021)
17. Shang, H., et al.: Extreme-scale ab initio quantum Raman spectra simulations on the leadership HPC system in China. In: Proceedings of the International Conference for High Performance Computing, Networking, Storage and Analysis, pp. 1–13 (2021)
18. Ma, Z., et al.: BaGuaLu: targeting brain scale pretrained models with over 37 million cores. In: Proceedings of the 27th ACM SIGPLAN Symposium on Principles and Practice of Parallel Programming, pp. 192–204 (2022)
19. Li, M., et al.: Bridging the gap between deep learning and frustrated quantum spin system for extreme-scale simulations on new generation of sunway supercomputer. IEEE Trans. Parallel Distrib. Syst. **33**(11), 2846–2859 (2022)
20. Li, F., et al.: SW_Qsim: a minimize-memory quantum simulator with high-performance on a new sunway supercomputer. In: Proceedings of the International Conference for High Performance Computing, Networking, Storage and Analysis, pp. 1–13 (2021)

An Empirical Study of Memory Pool Based Allocation and Reuse in CUDA Graph

Ruyi Qian, Mengjuan Gao, Qinwen Shi, and Yuanchao Xu(✉)

College of Information Engineering, Capital Normal University, Beijing, China
{qianruyi,gaomengjuan,shiqinwen,xuyuanchao}@cnu.edu.cn

Abstract. As the size of deep neural network models continues to increase, it places higher demands for memory capacity and allocation efficiency. NVIDIA GPUs are widely used in deep learning systems. CUDA has proposed many new techniques in recent years to make memory management more efficient. However, it is not easy to achieve *correct* and *efficient* memory reuse. To the best of our knowledge, we have not yet seen any literature that comprehensively, clearly, and unambiguously analyzes and discusses the key points about memory reuse in CUDA graph. This paper attempts to provide a systematic analysis of memory reuse in CUDA graph by performing an empirical study of related issues. We clarified a lot of unclear details and observed some key points in programming. We believe that our work would help programmers better unlock the potential of memory reuse, while avoiding inadvertent mistakes.

Keywords: Deep neural network · CUDA graph · Memory reuse · Memory allocation · Memory pool

1 Introduction

Deep neural network models have been successfully deployed in a variety of application domains. In recent years, as the size of models continues to increase, both for training and inference, higher demands have been placed on the capacity of device memory [1]. In addition, memory allocation/deallocation may be very frequent during model execution, thus becoming the main bottleneck affecting execution efficiency of neural network models.

NVIDIA GPUs are widely used in the field of deep learning. CUDA, a well-known heterogeneous programming model that goes along with it, has proposed many new techniques to make memory management more efficient, while laying a solid foundation for improved memory usage. These techniques effectively overcome the deep-rooted shortcomings of existing memory management mechanisms, including inevitable data copy, expensive synchronization costs, and low memory usage.

Among these technologies, one extremely important innovation is the CUDA graph, which is designed to allow work to be defined as a graph rather than single operations. CUDA graph has been used in the well-known benchmark MLPerf,

Z. Tari et al. (Eds.): ICA3PP 2023, LNCS 14491, pp. 394–406, 2024.
https://doi.org/10.1007/978-981-97-0808-6_23

showing significant performance speedup [2–4]. However, using CUDA graphs correctly and efficiently is not easy, especially how to reuse memory.

Memory optimization has always been a complex and challenging topic [1,5–12]. From the perspective of alleviating the insufficient memory capacity of devices and the huge demand for memory capacity in models, many methods have been proposed, including memory swapping [1,8], recomputation [7,9], model compression [6], memory reuse [5,11–14], etc. Therefore, memory reuse has become an ideal method for improving memory usage without any overhead.

The following example is to show the advantage of memory reuse. As illustrated in Fig. 1, the CUDA graph consists of memory allocation nodes, kernel nodes, and memory free nodes. If there is no reuse at all, total memory consumption of the whole graph is *A1+A2+A3*. If there is partial or full reuse, the total memory consumption is less than *A1+A2+A3*, which depends on the execution order closely. It indicates that memory reuse can significantly reduce the memory consumption, making it possible to cope with large models with small memory.

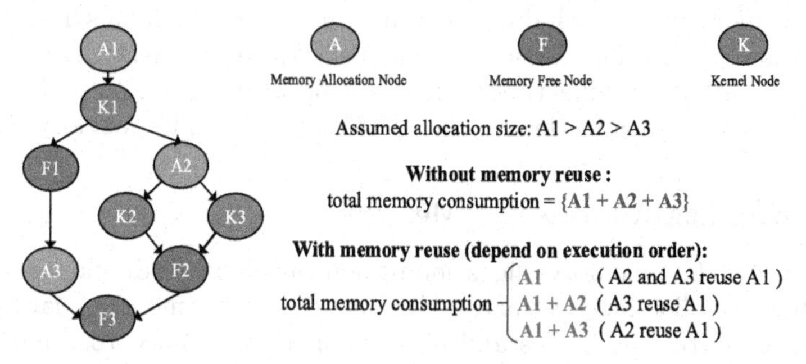

Fig. 1. CUDA graph without and with memory reuse. Reuse depends on allocation size and execution order of nodes.

While the benefits of memory reuse in CUDA graph are encouraging, it is not easy to achieve *correct* and *efficient* memory reuse. Although some concepts and interfaces can be found in programming manuals [15–18], they only elaborate on how to use these API interfaces themselves. To our knowledge, we have not yet seen a systematic and comprehensive literature discussing some key points affecting memory reuse. Some key questions could not be answered clearly and unambiguously. For instance, *When the unused memory should be freed? Should release threshold be set and how much? How to determine allocation size? When a graph is instantiated into multiple executable graphs, how memory is reused? What is the impact of reuse algorithms on memory reuse?*

To figure out some details in memory reuse, we have performed a comprehensive empirical study of issues related to memory reuse in CUDA graph. We focus on how to reuse memory correctly and efficiently to achieve lower allocation latency and higher memory usage. The experiments were conducted on two NVIDIA GPU servers (Ubuntu 20.04): V100 GPU with CUDA 11.7 and A100 GPU with CUDA 11.8. The source code is publicly available at https://github.com/CNU-XU/memoryreuse.

Our contribution lies in several observations:

- `cudaFreeAsync` should be called in time to free allocated memory into the pool once the lifetime of corresponding tensor is over, so that subsequent allocation within the stream can reuse memory.
- The trim operation affects both the available system memory size and address remapping overhead, so it should be used judiciously throughout the execution of the graph to avoid overuse or no use at all.
- If synchronization is required to ensure response timeliness, release threshold should be set a proper value.
- To execute concurrently, a graph should be instantiated for multiple times into multiple executable graphs, then launched into different streams.

2 Background

For many years, there have been several long-standing obstacles to achieving more efficient memory usage, including inevitable memory copy, high synchronization overhead, and low memory usage. To overcome these problems, CUDA has successively proposed a series of features as follows [15], which create potential opportunities for memory optimization and lay a solid foundation for memory reuse.

2.1 Decoupling Address and Memory

Before CUDA 10.2, memory allocation options available to developers has been limited to the malloc-like abstractions like `cuMalloc` and `cuMallocManaged`. The allocation of virtual addresses and physical space is coupled together, making execution efficiency low in many cases. For instance, in order to increase an allocation's size, developers had to explicitly allocate a larger buffer, copy data from the initial allocation, free it and then continue to keep track of the newer allocation's address. Obviously, if memory is reallocated frequently, the overhead will be significant.

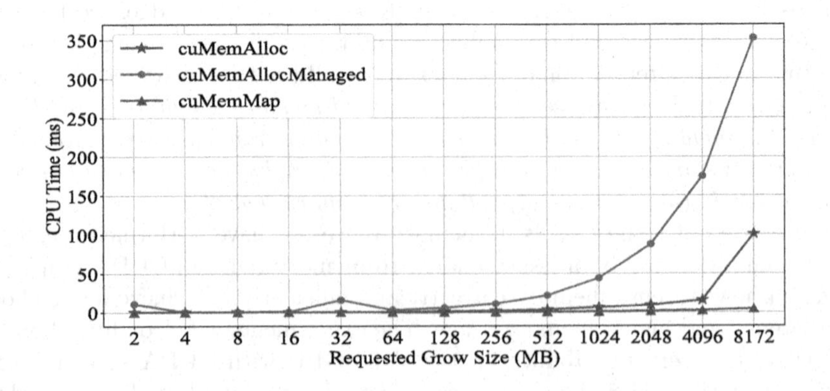

Fig. 2. `cuMemMap` shows the optimal performance.

Virtual memory management (VMM) decouples address and memory by separating the reservation of virtual addresses (`cuMemAddressReserve`) from the allocation of physical memory (`cuMemCreate`) into independent APIs as illustrated in [17]. Applications can map (`cuMemMap`) and unmap (`cuMemUnmap`) memory from virtual address ranges as needed. Therefore, when larger memory is required, the `cuMemMap` function avoids memory copies and hence is obviously more efficient, as shown in Fig. 2. Most importantly, decouple of address and memory lays a solid foundation for high allocation efficiency and memory reuse.

2.2 Implicit Device Synchronization

Almost every CUDA program uses `cudaMalloc` and `cudaFree` to allocate and release GPU memory. However, we may not be aware that calling `cudaFree` will induce an implicit and unintended device synchronization across all the GPUs [17]. It implies that any in-flight work on the device is completed and the CPU thread calling the function is blocked until all this work is completed.

This can be demonstrated through a simple example. Assuming that there are N independent threads launching work on separate streams, and there is also a thread allocating/freeing its own memory. Intuitively, they should not interfere with each other, and there is no gap between the kernels. However, due to device synchronization on `cudaFree`, it leads to a significant gap between kernels, thus sacrificing a lot of throughput, as illustrated in Fig. 3(a). By contrast, if VMM APIs are employed to replace `cudaMalloc` and `cudaFree`, the gaps between kernels are greatly reduced, as illustrated in Fig. 3(b). It shows that VMM APIs can eliminate implicit synchronization.

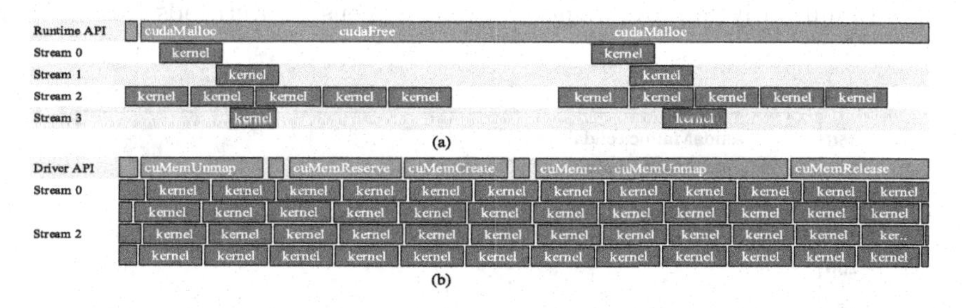

Fig. 3. (a) Timeline for multi-threaded case: one thread launches `cudaMalloc` and `cudaFree` in a loop, while others submit kernels. There are obvious gaps between kernels. `cudaFree` returns only when the previous pending work completes, and (b) Use CUDA VMM APIs. Gaps between kernels are greatly reduced.

2.3 Pool Based Asynchronous Malloc

Memory reuse is based on the memory pool, which has been introduced to the CUDA programming model since CUDA 11.2. It is managed by the driver rather than the OS. There are two kinds of memory pools in CUDA [15]: default pool and explicit pool. Each device has a default pool being created *implicitly* as the current pool, which cannot be shared by processes [18]. CUDA driver will request memory from the pool by calling `cudaMallocAsync` and will return the memory back to the pool rather than to the OS by calling `cudaFreeAsync` [16]. When calling `cudaMallocAsync` again, the requested memory can reuse part or all of the memory freed by the previous one. Therefore, the driver can help keep memory footprint at a low level along with extremely low allocation latency.

3 Key Issues in Memory Reuse

3.1 Impact of Allocation Size

For CUDA graph, each graph has its own memory pool, and the size of the memory pool is not requested all at once, but changes as memory is allocated and freed during the execution of the model. As will be seen later in the analysis, the size of each allocation does not depend only on the size required by the tensor itself, but also affects memory usage and allocation efficiency. If the allocation is too large, it will cause waste of memory space. On the contrary, it will increase the number of system calls.

When `cudaMallocAsync` is called for the first time in a graph, the driver has to issue a request to the OS for system memory because there is no available memory in the pool at this time. After that, the subsequent allocations will be done quickly once the pool can satisfy the request. Otherwise, CUDA driver must request extra memory from the OS again. Figure 4 intuitively shows the difference in performance between asynchronous allocation and synchronous allocation. Obviously, with memory pool, the efficiency of asynchronous allocation is significantly improved since it avoids expensive overheads from system calls.

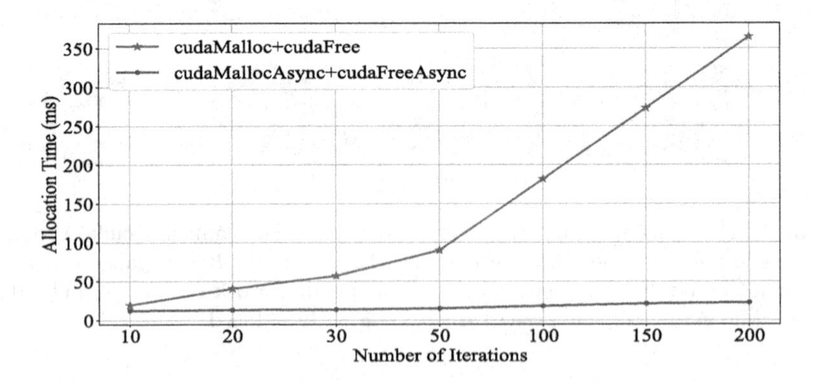

Fig. 4. The difference in performance between synchronous allocation and asynchronous allocation.

In addition, allocation size also has a significant impact on memory reuse in the CUDA graph. If the execution order is `allocA->freeA`, then `allocB` and `allocC` simultaneously depend on `freeA`. As the results in Table 1 shows, if the total size of `allocB` and `allocC` is not greater than `allocA`, both of them can reuse part of virtual address and physical memory of allocA (Case 3 and Case 5). If `allocB` requests a greater size than `allocA`, `allocB` cannot reuse the memory freed by `allocA` (Case 1 and Case 2). As a result, it has to request memory from the OS, leading to a greatly increasing memory footprint. If the requested size of both `allocB` and `allocC` is less than `allocA`, and their total memory size is greater than `allocA`, only one of them can reuse memory (Case 4). *Therefore, the previous memory allocation node in a graph should request a slightly greater size.* Moreover, it should be freed immediately once its lifetime is over so that subsequent nodes can reuse the memory.

Table 1. `allocB` and `allocC` simultaneously depend on `freeA` for memory reuse

	Case1	Case2	Case3	Case4	Case5
allocA size (MB)	1024	1024	1024	1024	1216
allocB size (MB)	1120	1120	512	608	608
allocC size (MB)	928	1120	512	608	608
Memory footprint (MB)	2144	3264	1024	1632	1216
reuse of allocB	N	N	Y	Y	Y
reuse of allocC	Y	N	Y	N	Y

It should be emphasized that CUDA driver may reuse memory within a graph by assigning the same virtual address range to different allocations as long as their lifetimes do not overlap. As virtual addresses may be reused, it should be paid more attention for programmers that pointers to different allocations with disjoint lifetimes are not guaranteed to be unique.

3.2 Impact of Synchronization and Release Threshold

Synchronous operations are common in heterogeneous multicore systems. However, once a synchronous operation, such as event, stream, and device [16], is encountered, unused memory accumulated in the pool will be returned to the OS by default. For example, each face recognition needs to impose a synchronization operation before reading the inference result to ensure the correctness of the inference result. Unfortunately, after each synchronization, memory pool will release all the unused memory, leading to subsequent memory reuse failures. To address this issue, it is essential to set the release threshold of memory pool. If the release threshold is not set, the driver will release all unused memory to the OS during the next synchronization. In contrast, if release threshold is set and current pool size is greater than release threshold, the driver will attempt to release some unused memory back to the OS till the remaining pool size reduces to *Size*, where $Size = \lfloor threshold/granularity \rfloor * granularity$.

We evaluated the impact of release threshold on memory reuse. The results are shown in Fig. 5. If threshold is not set (*zero*), the memory cannot be reused. If the threshold is set to 32 MB, then after synchronization, when allocation size is greater than 32 MB, the reserved memory in the pool will be released to 32 MB. As a result, part or all of this cached memory can be reused. Furthermore, if the threshold is set to the UINT64_MAX, unused memory in the pool will no longer be released after each synchronization. However, it is not appropriate to keep unused memory in the memory pool for a long time. Therefore, it is of great importance to set non-zero release threshold for achieving memory reuse. The size of release threshold depends on whether and how a device is shared [18]. Simply put, ideally, *the threshold should be equal to the peak size of the application's memory footprint.*

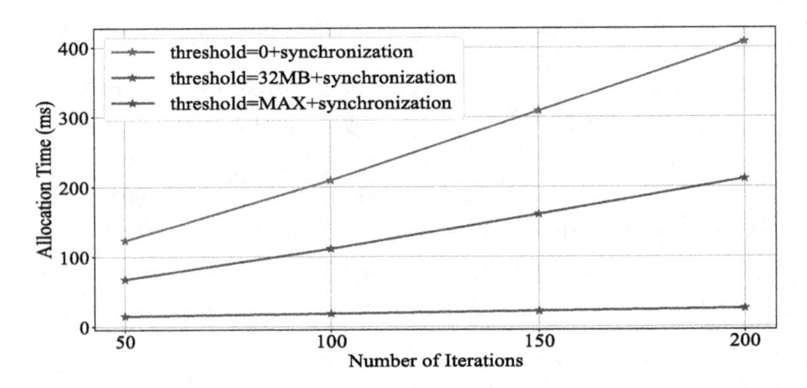

Fig. 5. Allocation time in different cases. Set the threshold to 0 (red), 32MB (green), and UNIT_MAX (blue). (Color figure online)

3.3 Impact of Memory Free and Trim

Memory Reuse within a Stream. Before requesting new memory from the OS, the driver tries to use unused memory in memory pool. To demonstrate this, we construct two different sequences of operations. One consists of asynchronous memory allocation, empty kernel invokes, asynchronous memory free, while the other only performs continuous memory allocation without using `cudaFreeAsync`. Suppose that each allocation size is 64 MB and both of the sequences execute 50, 100, 150, and 200 runs. As shown in Table 2, *if the memory is allocated continuously without free, the latency is obvious higher than that with free.* Therefore, from the perspective of allocation efficiency, it is critical to free allocated memory to the pool in time by calling `cudaFreeAsync` in order that subsequent allocation within the stream can reuse memory with extremely low latency.

Memory Reuse within a Graph. The order of graph memory nodes in a graph has a great impact on the execution time and memory footprint of the

Table 2. Allocation time of calling and not calling `cudaFreeAsync`.

numbers of runs	50	100	150	200
MallocAsync with FreeAsync (ms)	14.78	17.99	20.87	22.26
MallocAsync without FreeAsync (ms)	20.93	30.45	40.61	50.84

whole graph. For example, if the execution order is `allocA->allocB->freeA`, `allocB` cannot reuse the memory of `allocA` since `allocA` is still not freed. As a result, `allocB` has to issue expensive system calls, leading to larger memory footprint, as shown in Table 3. Therefore, programmers should have a clear understanding of the graph memory allocation nodes dependencies and release it promptly at the end of the allocation node's lifetime.

The free operation simply frees the memory into the memory pool. As the graph is executed, the required memory drops from the peak to the trough. At this point, if some of the free memory is properly released to the OS, it can increase the system memory and facilitate the execution of other models.

Memory can be implicitly released to the OS after synchronization by setting the threshold (Sect. 3.2). It should be noted that trim operation for CUDA graph is different from that for stream-ordered memory allocation.

Destroying a graph with memory nodes will not immediately return physical memory to the OS, even if their allocations are freed. Moreover, after `cudaFreeAsync` and `cudaFree` have been called in the graph, the virtual address will be released, but the actual physical memory footprint will not be reduced. In other words, the physical memory cannot be used by other graphs. To explicitly release memory back to the OS, `cudaDeviceGraphMemTrim` can be used. When a memory trim operation is invoked, the physical memory being actively in use that is reserved by graph memory nodes will not be affected. Allocations that have not been freed and graphs that are scheduled or running are considered to be actively using physical memory. Therefore, calling `cudaFreeAsync` is required before calling `cudaDeviceGraphMemTrim`.

Table 3. The order of nodes impacts on the execution time and memory footprint

Execution Order	Size (MB)	Time (ms)	Memory Footprint (MB)
allocA->freeA->allocB	1024	2.854912	1024
	4096	11.276288	4096
	6144	17.18784	6144
allocA->allocB->freeA	1024	5.609472	2048
	4096	23.000065	8192
	6144	35.168255	12288

For stream-ordered memory allocation, it operates on a different pool from CUDA graph. It is done by calling `cudaMemPoolTrimTo` to explicitly release the unused memory in the pool back to the OS. The reserved memory size will be reduced to *minBytesToKeep*.

How to use these APIs is easy, however, it is challenging to determine when to explicitly release unused memory. It is not good to release memory too early or too late. *Usually, the memory space required by a neural network model gradually shrinks in the later stages of model execution, and it is necessary to release the memory pool back to the operating system by appropriate trim operations to increase the size of available system memory.* It is also important to note that trim operations can also have an impact on address remapping (Sect. 3.5), so try to avoid frequent trim operations.

3.4 Graph Memory Reuse in CUDA

CUDA 11.4 integrated the memory pool into graph. Different from stream-ordered memory allocator, graph memory management truly decouples virtual addresses and physical memory. Work submission using CUDA graph is separated into three distinct stages: definition, instantiation, and execution. Physical memory is actually allocated in the execution stage. Therefore, memory reuse in CUDA graph can be subdivided into two types: virtual address reuse and physical address reuse. We discuss graph memory reuse separately below from within and between streams.

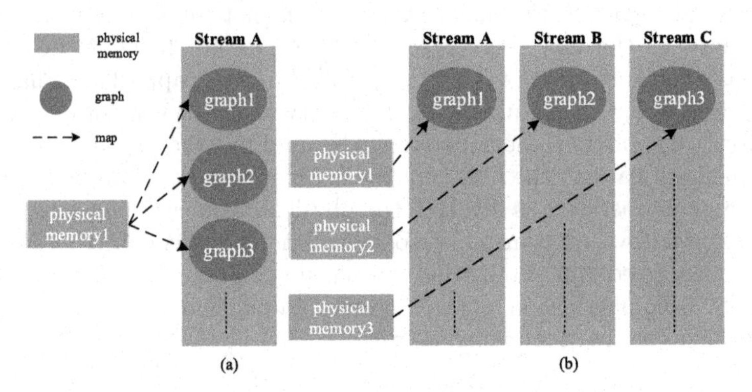

Fig. 6. (a) Launching multiple graphs into the same stream sequentially, CUDA can use the same physical memory to satisfy all the allocations, and (b) Launching multiple graphs in different streams, each graph occupies separate physical memory.

Graph Memory Reuse within a Stream. The key precondition of memory reuse between graphs is that their lifetimes do not overlap. As shown in Fig. 6(a), launch multiple graphs into the same stream sequentially. Each graph frees all allocated memory before the graph ends its execution. Since the graphs in the same stream are executed serially and their lifetimes do not overlap, CUDA can reuse physical memory via virtual aliasing. In other words, all the graphs within a stream use the same physical memory to satisfy all allocation requests. However, as shown in Fig. 6(b), if multiple graphs are launched into different streams, they need different physical memory.

Graph Memory Reuse between Streams. An executable graph will be executed in serial order even if being launched into different streams. The experimental results show that the execution time of an executable graph into N different streams is N times longer than that of executing a graph once. Visual Profiler demonstrates that an executable graph is indeed executed serially in different streams. Therefore, *it can be concluded that only one instance of a graph may exist at any point in time. A launch of an executable graph will be ordered after previous launches of the same executable graph. To execute concurrently, a graph should be instantiated into multiple executable graphs and then launched into different streams.*

It should be noted that there is a restriction to graphs which contain memory allocation and/or memory free nodes, i.e. they cannot be instantiated into multiple executable graphs. If multiple executable graphs can be instantiated, the virtual address range of each executable graph will be the same. Then if multiple executable graphs of a created graph are launched in parallel into multiple streams, multiple physical memory will be mapped to the same virtual address range, resulting in memory leaks. Hence, the graph containing memory allocation and/or memory free nodes cannot instantiate multiple executable graphs. Graphs that do not contain memory allocation and/or memory free nodes can be instantiated into multiple executable graphs, which can be parallelized.

3.5 Impact of Address Remapping

As the experiments of memory reuse in CUDA graph have demonstrated, when multiple graphs are launched into the same stream, CUDA driver attempts to allocate the same physical memory to them. Theoretically, the execution of the graph requires physical address mapping of memory. Each launch requires address mapping. Obviously, duplicate mapping (i.e. remapping) is unnecessary. *The cost of remapping can be eliminated if physical mapping of the graph is retained between two launches.*

Listing 1.1. Iteratively launch one graph/different graphs in a stream

```
1   CUgraphExec g_exec1, g_exec2;
2   cuGraphUpload(g_exec1, s1);
3   cuGraphUpload(g_exec2, s1);
4
5   #if 1
6   for (int i = 0; i < 2000; i++) {
7           cuGraphLaunch(g_exec1, s1);
8   }
9   #else
10  for (int i = 0; i < 1000; i++) {
11          cuGraphLaunch(g_exec1, s1);
12          cuGraphLaunch(g_exec2, s1);
13  }
14  #endif
15  cuStreamSynchronize(s1);
```

Physical memory cannot be allocated or mapped during graph instantiation because it is unknown in which stream the graph will be executed. Mapping is done during graph launch. Calling `cudaGraphUpload` can separate out the cost of allocation from the launch by performing all mappings for that graph immediately and associating the graph with the upload stream. We evaluated the overhead of `cudaGraphUpload` on the A100 GPU server and it was around 4 ms. If the Upload operation is called before launch, the overhead of launch is around 10 ms; if the `cudaGraphUpload` operation is not called before launch, the overhead of launch is around 14 ms. It can be seen that the overhead of launch can be reduced by calling `cudaGraphUpload` in advance. Therefore, moving the physical mapping out of the critical path of graph execution can significantly speed up the launch.

We designed a use case to iteratively launch the same graph in one stream, as well as iteratively launch a different graph in one stream. The code is shown in Listing 1.1, and the execution time of the latter increases by about 100 us, indicating that the average overhead per remapping is about 0.1 us.

Remapping must be performed in the order of execution, but after any previous execution of the graph completed. Otherwise, memory that is still in use could be unmapped. Due to order constraints and system calls, mapping overhead is expensive. For example, launching a graph in a loop, as shown in Fig. 7, if calling a trim operation after execution every time, the performance is dramatically degraded due to remapping. *Hence, from the perspective of reducing remapping overhead, frequent trim operations should be avoided as much as possible.*

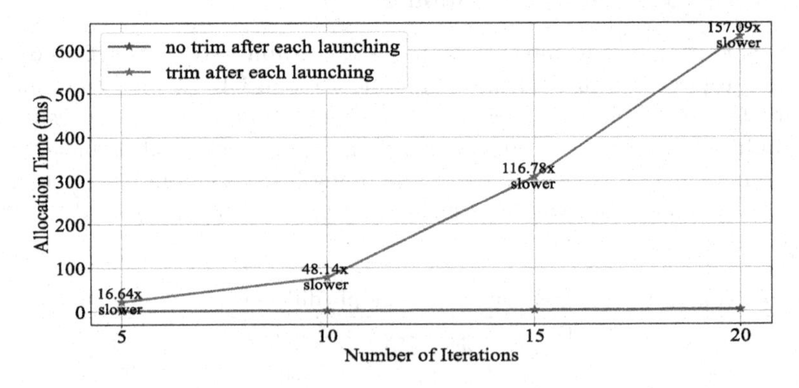

Fig. 7. When trim operation is performed after each graph execution, an remapping will be performed, which greatly increases the cost.

4 Conclusion

Both allocation efficiency and memory usage have a significant impact on the execution of neural network models. To address some ambiguities in memory pool-based allocation and reuse, this paper conducts an extensive empirical study and in-depth analysis, resulting in several observations and a set of programming recommendations. The work in this paper helps programmers to implement memory reuse correctly and efficiently, thus further unlocking the potential of memory reuse for GPU with CUDA.

Acknowledgements. This work is supported by Beijing Natural Science Foundation under grant 4212017 and National Natural Science Foundation of China under grant 62077002.

References

1. Gimelshein, N., Clemons, J., Zulfiqar, A., Keckler, S.W., Rhu, M.: VDNN: virtualized deep neural networks for scalable, memory-efficient neural network design. In: 2016 49th Annual IEEE/ACM International Symposium on Microarchitecture (MICRO), pp. 1–13. IEEE (2016)
2. NVIDIATech. Mlperf v1.0 training benchmarks: Insights into a record-setting nvidia performance (2021). https://developer.nvidia.com/blog/mlperf-v1-0-training-benchmarks-insights-into-a-record-setting-performance/
3. Michael, C., Sukru, B.E., Edward, Y., Vinh, N.: Accelerating pytorch with cuda graphs (2021). https://pytorch.org/blog/accelerating-pytorch-with-cuda-graphs/
4. Li, C., Yao, J.: Cuda graph in tensorflow (2021). https://www.nvidia.com/en-us/on-demand/session/gtcspring21-s31312/
5. Lee, J., Pisarchyk, Y.: Efficient memory management for deep neural net inference (2020). arXiv preprint arXiv:2001.03288
6. O'Connor, M., Chatterjee, N., Pool, J., Kwon, Y., Keckler, S.W., Rhu, M.: Compressing DMA engine: leveraging activation sparsity for training deep neural networks. In: 2018 IEEE International Symposium on High Performance Computer Architecture (HPCA), pp. 78–91. IEEE (2018)
7. Ye, J., et al.: Superneurons: dynamic GPU memory management for training deep neural networks. In: Proceedings of the 23rd ACM SIGPLAN Symposium on Principles and Practice of Parallel Programming, pp. 41–53 (2018)
8. Jin, G., Li, J., Huang, C.C.: Swapadvisor: pushing deep learning beyond the GPU memory limit via smart swapping. In: Proceedings of the Twenty-Fifth International Conference on Architectural Support for Programming Languages and Operating Systems, pp. 1341–1355 (2020)
9. Shi, X., et al.: Capuchin: tensor-based GPU memory management for deep learning. In: Proceedings of the Twenty-Fifth International Conference on Architectural Support for Programming Languages and Operating Systems, pp. 891–905 (2020)
10. Li, M., et al.: Mxnet: a flexible and efficient machine learning library for heterogeneous distributed systems (2015). arXiv preprint arXiv:1512.01274
11. Liu, B., et al.: Layer-centric memory reuse and data migration for extreme-scale deep learning on many-core architectures. ACM Trans. Arch. Code Optim. (TACO) **15**(3), 1–26 (2018)
12. Xu, M., et al.: Melon: breaking the memory wall for resource-efficient on-device machine learning. In: Proceedings of the 20th Annual International Conference on Mobile Systems, Applications and Services (2022)
13. Imamichi, T., Imai, H., Raymond, R., Sekiyama, T.: Profile-guided memory optimization for deep neural networks (2018). arXiv preprint arXiv:1804.10001

14. Yeung, S.H., Shu, Y., He, B., Wang, W., Zhang, J.: Efficient memory management for GPU-based deep learning systems. arXiv preprint arXiv:1903.06631 (2019)
15. NVIDIATech. Cuda c programming guide (2022). https://docs.nvidia.com/cuda/archive/11.7.0/cuda-c-programming-guide/index.html
16. Kini, V., Hemstad, J.: Using the nvidia cuda stream-ordered memory allocator part 1 (2021). https://developer.nvidia.com/blog/using-cuda-stream-ordered-memory-allocator-part-1/
17. Perry, C., Sakharnykh, N.: Introducing low-level GPU virtual memory management (2020). https://developer.nvidia.com/blog/introducing-low-level-gpu-virtual-memory-management/
18. Kini, V., Hemstad, J.: Using the nvidia cuda stream-ordered memory allocator part 2 (2021). https://developer.nvidia.com/blog/using-cuda-stream-ordered-memory-allocator-part-2/

Log Anomaly Detection Based on Semantic Features and Topic Features

Peipeng Wang(✉), Xiuguo Zhang, and Zhiying Cao

School of Information Science and Technology, Dalian Maritime University, Dalian 116026, China
{wpp7,zhangxg,czysophy}@dlmu.edu.cn

Abstract. System logs serve as crucial data sources for monitoring system performance and enhancing service quality. Many existing log-based anomaly detection methods primarily focus on detecting anomalies through parsed log templates. However, they often overlook valuable information such as the components and anomaly levels present in the original logs. These details provide crucial context about the logs' origin and importance, serving as effective auxiliary information for the anomaly detection task. And recent studies only focus on log semantics, with a single log feature and the performance of the model is easily affected by template changes. Therefore, we propose a log anomaly detection method based on semantic feature and topic feature, LogST. LogST combines component and level information on the basis of log templates to construct a multi-information fused log sequence. The BERT model is used to extract log sentence vectors, and the SVD algorithm is used to reduce dimensionality to obtain efficient semantic features. Meanwhile, the LDA topic model is used to extract the topic features of log sequences and fully explore the distribution of different templates in normal and abnormal sequences. In addition, in order to highlight the key features and make full use of the anomaly log knowledge, we design a weighted residual-connected TCN model, which consists of a multilayer convolutional architecture for log sequences and uses attention mechanism to weight and aggregate hidden features of different convolutional layers. The experimental results on public datasets indicate that LogST has better performance than existing methods.

Keywords: Semantic Features · BERT · SVD · Topic Features · TCN

1 Introduction

The log information in a software system truly reflects the operating status of the system, which is of great significance for system management, monitoring, and fault diagnosis [1]. However, large systems generate tens of millions of log messages every day, with a small number of abnormal logs hidden in a large number of normal logs, making manual inspection extremely inefficient. Therefore,

Z. Tari et al. (Eds.): ICA3PP 2023, LNCS 14491, pp. 407–427, 2024.
https://doi.org/10.1007/978-981-97-0808-6_24

in recent years, log anomaly detection based on artificial intelligence technology has become an important means to ensure system reliability and stability. Since logs are typically semi-structured text, they contain partially noisy information based on structured log events. It is necessary to effectively analyze the log messages and train a good anomaly detection model. According to the existing research related to logs, how to make full use of raw log messages and fully extract log sequence features so as to improve anomaly detection effectiveness is still a problem to be studied [2].

According to different anomaly detection models, existing log-based anomaly detection methods can be roughly divided into two categories: machine learning-based methods and deep learning-based methods. Traditional machine learning methods utilize manually extracted log features and use machine learning models such as SVM [3], PCA [4], decision trees [5], and IM [6] to implement anomaly detection tasks. However, such methods still require manual feature extraction and are limited by the algorithm, and the generalization ability of the model is weak. In contrast, deep learning-based methods extract features in a fully automated manner without manual intervention. For example, DeepLog [7] and LogEvent2vec [8] used the log event index as input and use different models for anomaly detection. But such index-based deep learning methods ignore the log semantic information, and the model is less robust when the log is updated. In this regard, existing mainstream methods model the parsed log template as a natural language sequence and train the anomaly detection model by extracting the semantic features of the log template. For example, LogRobust [9] used FastText to model the semantics of log templates, LightLog [10] used Word2vec and PCA to extract the semantic features of log templates. These methods improve the robustness of the model to a certain extent, but they are only limited to the semantic features, and the detection accuracy will be limited by the effectiveness of the language model. Moreover, when a large update of the template occurs or the existence of noisy data causes changes in the log semantics, a single semantic feature often cannot fully express the differences between normal and abnormal logs. Therefore, it is necessary to further analyze the potential features of log sequences using other models and combine them with semantic features to improve the effectiveness of anomaly detection.

In addition, existing studies often take the parsed log templates as input and do not consider other information in the original log messages, such as components and levels, which, as important components of the system logs, can add additional valuable information to the log templates thus providing a powerful aid to the anomaly detection task. Specifically, key information such as the source and location of each log template is recorded in the component, and certain anomalies in the system are not necessarily fully reflected in the log template sequence, but directly in the component workflow. The log level records the importance of the log, usually divided into "DEBUG", "INFO", "WARRING", etc. By combining the log levels, we can analyze the impact of each template on the log sequence in more depth. Therefore, it is necessary to combine component and level information based on log template sequences to improve the effectiveness of anomaly detection.

Taking into account the above factors, this paper proposes LogST, a novel log anomaly detection method based on semantic features and topic features. First, in order to fully utilize various valuable information in the original log messages, LogST combines the component and level on the basis of the log template, thereby determining the source of each template and the importance of the template, and constructing a multi-information fused log sequence. Second, in order to obtain more reliable and complete log semantic, we use BERT [11] to generate template sentence vectors, and uses Singular Value Decomposition (SVD) [12] to reduce vector dimensions. Meanwhile, we use the LDA [13] topic model to extract the topic features of log sequences, which can better describe and distinguish normal and abnormal log sequences by mining the implicit topic information in log sequences, avoiding the problem of over-reliance on log semantic leading to limited detection accuracy and further improving the robustness of the model. Final, this paper designs a weighted residual-connected TCN model for anomaly detection, based on the ordinary TCN [14], which reuses the anomaly log knowledge extracted from different convolutional kernels by means of residual connection, and drawing on the idea of attention mechanism, a weighted method is designed for residual connection, highlighting key hidden layer features.

This paper evaluated the proposed methods on two different datasets, including HDFS and BGL. The experimental results showed that LogST outperformed other baseline methods. We also conducted ablation experiments to verify the impact of topic features on anomaly detection performance, and analyzed the effectiveness of the improved TCN model. The main contributions of this paper can be summarized as follows:

(1) We combine component and level information on the basis of log templates for the first time, constructing a multi-information fused log sequence and enhancing the expression ability of log templates.
(2) We use the BERT model to extract the template sentence vectors, the SVD algorithm is used to reduce the vector dimension and obtain efficient semantic features. We innovatively combine semantic features with topic features, enriching the diversity of features and effectively improving the robustness and detection performance of the model.
(3) We design a weighted residual-connected TCN model for anomaly detection, which not only achieves the reuse of anomaly knowledge, but also highlights the key hidden layer features

2 Related Work

In this section, according to the process of log anomaly detection, we first introduce several log parsing methods, then we present some typical methods and their respective limitations in log anomaly detection related research, and corresponding solutions are proposed to achieve better performance of anomaly detection.

2.1 Log Parsing

The log messages recorded by the system are usually semi-structured, consisting of constant text and variable parts. The variable part contains a large amount of noise data that is not helpful for anomaly detection tasks. Therefore, it is necessary to remove the noise information before subsequent feature extraction. The purpose of log parsing is to convert semi-structured log messages into structured log templates. Many existing works have conducted in-depth research on log parsing technology. IPLoM [15] adopt an iterative partitioning strategy to divide log messages into groups based on message length, token location, and mapping relationships. Spell [16] used the longest common subsequence algorithm (LCS) to parse the logs in a streaming fashion. Shi et al. [17] proposed a log parsing method based on N-gram and Frequent Pattern Mining (FPM) methods to reduce the model complexity. SwissLog [18] designed a dictionary-based log parsing method which did not require any parameter tuning process. There are also LKE [19] based on hierarchical clustering algorithm, MoLFI [20] based on evolutionary algorithm, etc. According to the evaluation results of Zhu [21] et al. Drain [22] applied a fixed-depth tree structure to represent log messages, which could generate less noisy data and had higher detection accuracy compared to other parsers, so we use Drain algorithm to parse raw log messages.

2.2 Anomaly Detection

In the field of log anomaly detection, it can be divided into anomaly detection with supervised learning and anomaly detection with unsupervised learning, depending on whether labeled data is required or not. Due to the time and resources required to obtain a large amount of labeled data in real scenarios, and the difficulty in obtaining valuable labeled data when new systems are deployed, unsupervised anomaly detection methods have important application value. Zhang [23] et al. proposed an unsupervised log sequence anomaly detection network based on local information extraction and global spare Transformer model by analyzing the characteristics of adjacent logs and long distance logs in log sequences, which effectively improved the anomaly detection accuracy, but its efficiency decreases when new log sequences are encountered due to its lack of learning of historical anomaly knowledge. To exploit supervised information in unsupervised methods, Lin et al. [24] proposed PLELog to learn historical anomaly knowledge by probabilistic label estimation and attention-based Bi-GRU model to learn log context information, which achieved an accuracy close to supervised learning. But the method used clustering and dimensionality reduction to assign pseudo-labels, which may lose some accuracy. In addition, Han et al. [25] used a transfer learning method to make log data from different systems with similar distributions using adversarial domain adaptation techniques, and mapped log sequences into a hypersphere using an LSTM model to detect anomaly logs that deviate from the sphere boundary, achieving unsupervised anomaly detection across systems. But this method required manual setting of decision boundaries, which made it difficult to obtain high anomaly detection accuracy.

In contrast, although the supervised learning method requires a large number of labeled data to learn the anomaly detection model, it has achieved high performance by relying on the strong fitting ability of the deep learning model and the effective analysis of the log sequence. LogNADS [26] obtained log semantics by extracting topic words, and used LSTM to obtain log context features, achieving high detection accuracy while effectively reducing time costs. To improve the robustness of anomaly detection, LightLog [10] utilized Word2vec and PCA to extract semantic features of log templates, and used an improved TCN model for anomaly detection, which proved that TCN can achieve high detection performance with fewer parameters. Inspired by this paper, we also chose TCN as the anomaly detection model. LogTransfer [27] extracted template word vectors using Glove, obtains semantic features by weighted averaging, and acquired sequence features using LSTM. However, they only analyzed the semantic features of log sequences, which change when log templates undergo large updates, and fusing multi-angle features can adapt to such changes more effectively. Therefore, Wang et al. [28] further combined the semantic features with the statistical features of template words in normal and abnormal log sequences. LogAnomaly [29] effectively improved anomaly detection performance by combining template counting vectors and sequential features on the basis of semantic features. In order to analyze the log sequence in depth, based on the semantic features, we use the topic model to obtain the topic features of the log sequence and achieve multi-feature extraction.

In addition, existing anomaly detection methods (unsupervised anomaly detection and supervised anomaly detection) mostly use log templates as the basic unit, ignoring other information in the logs. Based on this, DeepSyslog [30] further considered the metadata in the log on the basis of the log template, thereby obtaining more complete log context information. HitAnomaly [31] considered parameter value information and used Transformer as the parameter value encoder. DeepLog [7] also introduced parameter value information of logs based on the index of log templates, and used LSTM for anomaly detection. However, both metadata and parameter values are variable parts of the original log message, without a fixed format. Analyzing variable information can improve anomaly detection performance to a certain extent, but it is easy to introduce noise and increase algorithm complexity. In contrast, information such as log components and levels are limited in each log dataset and represent actual meaning in the system and can be an important source of information to enrich log templates.

3 Method

In order to make full use of the information in log messages other than templates, enrich the knowledge of log templates, and at the same time perform multi-angle feature extraction of multi-information fused log templates to improve the effect of anomaly detection, we propose LogST, a log anomaly detection method based on semantic features and topic features. Its overall structure is shown in Fig. 1. First, we take the original unstructured log dataset as input and parsed by Drain algorithm to obtain structured log template. Based on the log template, the component and level of the template are fused to obtain a multi-information fused log sequence. Second, LogST uses the BERT model to extract the templates sentence vectors, and it is noteworthy that due to the high dimensionality of the vectors extracted by BERT, the computation is too large, and we further adopt the SVD for dimensionality reduction to obtain compact semantic features. At the same time, the LDA is used for topic features extraction, and combined them with semantic features, which enriches the diversity of log features. Final, LogST feeds the fused log features into the improved TCN model, which achieves anomaly knowledge reuse by weighted residual concatenation, and focuses on key hidden feature that is more important for anomaly detection results.

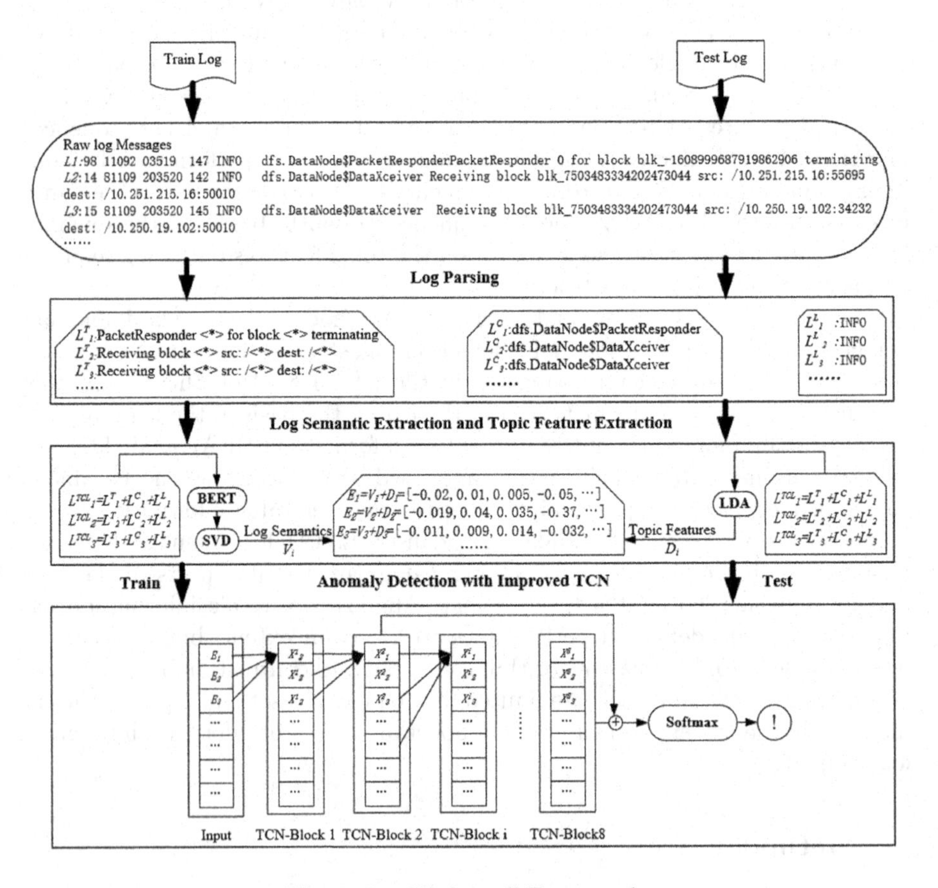

Fig. 1. LogST Overall Framework

3.1 Multi-information Fused Log Sequences Construction

The original log message is unstructured data that contains many specific information, such as Date, Pid, Component, Level, etc. We define a log message as L, from which a sequence of original log messages $S = \{L_1, L_2, ..., L_n\}$ can be obtained, where n represents the number of log templates in the sequence. Correspondingly, as mentioned in Fig. 1, the parsed log template is L^T. Each log template is a log event that represents a specific operation that occurred in the system or application, where <*> represents the variable value in the original log. After parsing the log, the log template sequence is $S^T = \{L_1^T, L_2^T, ..., L_n^T\}$, the component sequence is $S^C = \{L_1^C, L_2^C, ..., L_n^C\}$, the level sequence is $S^L = \{L_1^L, L_2^L, ..., L_n^L\}$.

Large software systems usually consist of multiple components, which represent separate units of the system with specific services, and each component is responsible for a different function. Existing methods detect anomalies by analyzing the sequential relationship of log templates, however, in concurrent large-scale systems, the same log template may come from different services or processes, i.e., belonging to different components. The analysis of component workflows can directly determine certain system failures; conversely, ignoring component information may not effectively analyze log sequences, which in turn leads to reduced anomaly detection effectiveness. In addition, the log level records the importance of each log event, not only can indicate whether the program has potential problems, and even directly as the basis of a log sequence is abnormal, for example, a log event level of "ERROR", it indicates that the program error, and the log sequence where the log event is located is also an anomaly sequence. Therefore, we further obtain the log components L^C and log levels L^L corresponding to each template on the basis of log templates L^T, and constructs a multi-information fused log sequence $S^{TCL} = \{L_1^{TCL}, L_2^{TCL}, ..., L_n^{TCL}\}$ by stitching, as shown in Eq. (1)

$$L_i^{TCL} = concat(L_i^T, L_i^C, L_i^L) \tag{1}$$

Where, $L_i^T = \{t_1, t_2, ..., t_t\}$ is the log template for $i-th$ log. $L_i^C = \{c_1, c_2, ..., c_c\}$ represents the components of the log template, and $L_i^L = \{level\}$ represents the levels of the log template, and t, c represent the number of tokens in the log template and log component, respectively.

3.2 Log Semantic Extraction

The index-based method is easy to lose the semantic information inside the log template, and cannot solve the problem of log update or noise, resulting in poor robustness of the model. Therefore, most existing methods are based on semantic features, which use different language models such as Word2vec [32], Glove [27], etc. to obtain the word vectors of log templates and aggregate or concatenate them into semantic features. According to the characteristics of the log dataset, the process of converting log templates L^T into semantic features V^T needs to meet the following requirements

First, the log template will change with system updates, but the key semantics will remain unchanged. The extraction method of semantic features require adaptability to this change. Second, due to the different hardware and software environments of different systems, different log templates often use the same vocabulary in different contexts, which also leads to the phenomenon of polysemy. In addition, this paper combines component and level on the basis of log templates, which further requires word embedding methods to be able to generate more reliable semantic features based on the log context environment.

To meet the above requirements, LogST uses the BERT, a pre-training language model, to extract sentence vector of log templates, components, and levels. It is capable of dynamically generating vector representations by using a bi-directional Transformer structure [33], thus solving the word polysemy problem. In this paper, we take the a log sequence with multi-information fused $L_i^{TCL} = \{L_i^T, L_i^C, L_i^L\}$ as input, and fully extract the features V_i^1 of log template, component, and level, as shown in Eq. (2).

$$V_i^1 = BERT(L_i^{TCL}) \tag{2}$$

where,$V_i^1 = \{V_i^{T^1}, V_i^{C^1}, V_i^{L^1}\}$, $V_i^{T^1}$,$V_i^{C^1}$ and $V_i^{L^1}$ represent the log template, component, and level sentence vector, respectively.

Because the log sentence vector generated by BERT has a high dimension and the model operation is complex, we perform SVD decomposition on sentence vector. It is a matrix factorization, which approximately represents the original feature matrix by calculating the number of singular values and singular vectors, and reduces dimensions by ranking importance and discarding unimportant features, as shown in Eq. (3).

$$V_i^{1T} V_i^1 = (V\Sigma^T U^T)(U\Sigma V^T) \tag{3}$$

Where, V_i^1 is $i-th$ log sentence vectors, $(U\Sigma V^T)$ is SVD decomposition of V_i^1, U is the left vector matrix, Σ is the singular value diagonal matrix, V the right vector matrix. The log sentence vectors V_i^1 extracted by BERT is decomposed by SVD, and the vector corresponding to the first r maximum singular values is taken to obtain the log semantic $V_i = \{V_i^T, V_i^C, V_i^L\}$. After SVD decomposition, the original high-dimensional complex log sentence vectors are transformed into low-dimensional efficient log semantic features, which effectively reduce the computational burden of the model during training and detection while retaining semantic integrity.

3.3 Log Topic Feature Extraction

Existing deep learning-based log anomaly detection methods often use different language models to extract semantic features and then use different neural networks, such as LSTM, for anomaly detection, which improve the robustness of the models to some extent, but they only consider the semantic of log templates and do not mine log features from a macro perspective (e.g., document topics).

In the NLP field, Peinelt et al. [34] combined the Topic model to enrich the semantic features of the text, improve the efficiency of text matching, and prove that the Topic model can provide additional auxiliary features for BERT. In software systems, normal logs and abnormal logs often have obvious differences in topics, and topic features are less disturbed by single log template updates and noise factors. Therefore, LogST further considers the topic information of logs based on the extracted semantic features, and achieves multi-feature extraction.

Among many topic models, the LDA (Latent Dirichlet Allocation) topic model is a typical unsupervised Bayesian probabilistic model [13], which can effectively identify the topic information in documents by delving into the relationship between documents, topics and vocabulary. We first preprocess the multi-information fused log template $L^{TCL} = \{L_1^{TCL}, L_2^{TCL}, ..., L_M^{TCL}\}$, where, M is the number of types of multi-information fused templates. After text tokenization, deletion of stop words, lower case conversion and other operations, the Bi-gram model is used to model the word vector. Then, the obtained word vector is input into the LDA topic model for training, which deeply analyzing the relationship between the multi-information fused log template - topic - vocabulary. The related symbols and meanings are shown in Table 1. The specific process is as follows:

Table 1. Symbols of the LDA topic model.

Symbols	Meanings
K	The number of topics shared by the multi-information fused log template
M	The number of the multi-information fused log template
N_i	Total number of words in the $i-th$ multi-information fused log template
θ_i	Topic distribution of the first multi-message fusion log template
α	Generate dirichlet prior parameters for θ_i
φ_k	Vocabulary distribution of the $k-th$ topic
β	Generate dirichlet prior parameters for φ_k
$Z_{i,j}$	The $j-th$ topic of the $i-th$ multi-information fused log template
$w_{i,j}$	The $j-th$ word of the $i-th$ multi-information fused log template

1) Sample each topic k to generate a distribution of words θ_i for that topic.
2) For $i-th$ multi-information fused log template, sampling generates its topic distribution θ_i.
3) For $j-th$ word in $i-th$ multi-information fused log template, the $j-th$ topic $Z_{i,j}$ of $i-th$ document is first sampled from θ_i, and subsequently the word $w_{i,j}$ is generated by sampling from that topic.

After the above process, the LDA topic model generates a multi-information fused log template - topic matrix, i.e., topic feature D, as shown in Eq. (4).

$$D_i = LDA([L_i^{TCL}]) \tag{4}$$

where, $L_i^{TCL} = \{L_i^T, L_i^C, L_i^L\}$, D_i for the $i-th$ multi-information fused log template topic features. Based on the semantic features, LogST not only avoids the problem of limited detection accuracy due to over-reliance on log semantic, but also improves the robustness of the model by considering the thematic features of logs.

3.4 Anomaly Detection with Improved TCN

The multi-information fused log sequences are converted into log semantic V and topic features D after log semantic extraction and topic feature extraction, and the log sequence features $E = concat(V, D)$ are obtained by feature stitching, so each sequence is represented as $E = [E_1, E_2, ..., E_n]$. Taking log sequence features as input, LogST designs an improved TCN network for anomaly detection.

Recurrent neural networks (RNN) have become the preferred model for many researchers to process text data, but traditional RNN models suffer from the problem of gradient disappearance and gradient explosion, while convolutional networks can effectively avoid this defect and allowing parallel computation, therefore, researchers propose a temporal convolutional network TCN for sequence modeling. The ordinary TCN is a one-dimensional fully convolutional network that saves computation time and memory consumption through parallel computing. Based on the CNN model, it is extended from three perspectives for sequence structure, making it superior to sequence models such as LSTM and more suitable for log sequence anomaly detection tasks. First, causal convolution is used for operations in TCN, which ensures that only the input data before current timestamp t is used to predict the output of the timestamp t, a strictly time-constrained model. Second, to get rid of the problem that the modeling length of time is limited by the size of the convolutional kernel, TCN adopts a dilated convolution operation, which increases the perceptual field to let each convolution output a larger range of information without doing pooling to lose information. The dilation convolution is calculated as shown in Eq. (5). Final, since TCN introduces causal convolution and dilation convolution to increase the perceptual field, which inevitably increases the network depth, the original TCN uses residual connection to enable the network to pass information in a cross layer fashion, thus effectively training the deep network.

$$F(E_t) = (E * F(d))_{E_t} = \sum_{i=0}^{k-1} f(i) E_{t-d*i} \tag{5}$$

where d denotes the dilation rate, k denotes the size of the convolution kernel, $E_t \in E$ is the input log sequence feature. When $d = 1$, the dilation convolution degenerates to normal convolution.

Although the TCN network can save computation time by parallel computing, it is still computationally intensive, and the network is deep and the number of parameters increases exponentially with the depth of the network, which makes it prone to overfitting when using small data sets for training. In the log sequences, abnormal logs often account for a small portion, for example, only 3% of the approximately 580,000 logs in the HDFS dataset are abnormal. This imbalance will lead to the difficulty for the ordinary TCN model to fully explore the hidden features of the abnormal logs, which is prone to the phenomenon that normal log features cover the abnormal features, thus causing inappropriate offset of the classification boundary and reducing the accuracy of the detection results. Therefore, we design weighted residual connections in the TCN for network optimization. Specifically, First, four convolution kernels $d = 1, 2, 4, 8$ with different dilation rate are used in LogST, and the number of convolution kernel channels is set to $filters = 3$ capture the shortterm dependencies and long term dependencies of log sequences, thus obtaining multi-scale log sequence features, while this layer-by-layer incremental dilation rate balances the model's attention to log sequences of different lengths. Subsequent, LogST uses a convolutional layer with a number of channels $filters = 1$ to abstract the extracted log sequence features after different hidden layers. This not only improves the expression ability of sequence features, but also reduces the number of parameters and complexity of the model. Final, this paper considers that the convolutional kernels with different expansion factors have different size perceptual fields, and the log context information they capture is very different, which leads to different TCN blocks output log features have different effects on anomaly detection results, and different log features contain different anomaly knowledge. Therefore, LogST uses a weighted approach to highlight key features, specifically using an attention mechanism to assign different weight values to the output features of TCN modules containing different dilation rates d, as shown in Eq. (6). And combined with the residual connection for anomaly knowledge reuse, the structure of which is shown in Fig. 2.

$$\alpha_i = \tanh(W_i^a \cdot X_i) \tag{6}$$

where, α_i is the weight value of output feature of the $i - th$ TCN blocks, and $i - 2, 4, 6$ represent TCN modules with $filters = 1$ channels respectively, $\alpha_8 = 1$. X_i is the output feature of the $i - th$ TCN blocks, W_i^a is the weight matrix that can be trained, $tanh(\cdot)$ is the activation function. Through this attention weighted approach, importance can be automatically learned, with a focus on the key hidden layer features of the log.

Then, LogST multiplies the hidden layer features output by TCN modules with different dilation rate by their weight factors to obtain the final hidden state, as shown in Eq. (7).

$$S = \sum_{i=2,4,6,8} \alpha_i \cdot X_i \tag{7}$$

Since S contains log context information extracted from convolutional kernels with different dilation rate, we also achieve reuse of abnormal logs knowledge.

Final, input the final hidden state S into the Softmax layer to output the classification results, as shown in Eq. (8)

$$pred = \text{softmax}\left(W \cdot \left(\sum_{i=0}^{n} \alpha_t \cdot S_t\right)\right) \tag{8}$$

where, W is the weight of softmax layers, n is the length of log sequence. During the training phase, we use the prediction outputs $pred$ and the ground-truth y_i provided by datasets to calculate the cross-entropy as the loss function, as shown in Eq. (9).

$$L = -\frac{1}{M}\sum_{i=1}^{M}[y_i \log pred_i + (1-y_i)\log(1-pred_i)] \tag{9}$$

where, $y_i \in \{0,1\}$ is the label of $i-th$ sample, M is the number of samples.

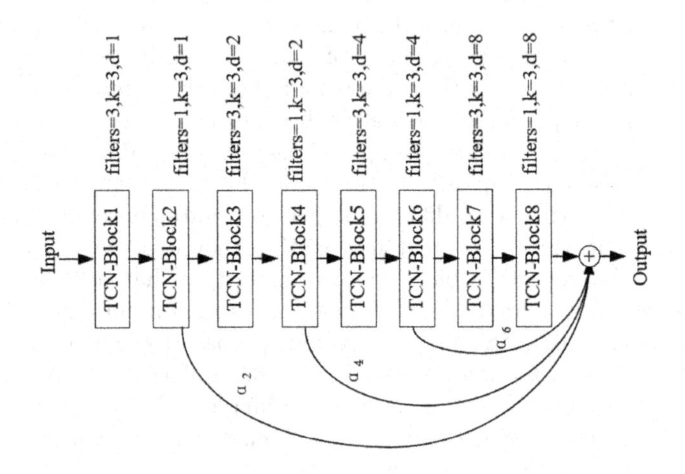

Fig. 2. Weighted residual connection

4 Experiment

In this section, we first describe the experimental dataset, baseline methods, evaluation metrics, and experimental environment. Then, the performance of LogST and other baseline models on different datasets was compared, and the effectiveness of each module in LogST was analyzed.

4.1 Experimental Design

During the experimental process, we evaluate the LogST method by discussing the following research content: (1) Comparing several existing log anomaly detection methods and analyzing the effectiveness of LogST; (2) Deleting LDA based topic feature extraction to verify the effectiveness of topic features; (3) Verifying the effectiveness of the improved TCN model in this paper.

Dataset. We evaluate the LogST method in two publicly available log datasets, namely supercomputer log BGL and distributed system log HDFS [35]. These two datasets are widely used in log anomaly detection research, and their detailed information is described in Table 2.

Baseline Models. This paper compares LogST with several anomaly detection methods, including an unsupervised methods and supervised methods. The brief descriptions of these methods are as follows.

Table 2. Dataset Details.

Datasets	Messages	Window Type	Templates	Component	Level	Anomalies
HDFS	11,175,629	Session	48	9	2	16,838
BGL	4,747,963	Fix	377	14	10	348,460

Invariant Mining (IM) [6]: The model takes the log template index as input, it mines the linear invariance from the template count vector, then the sequences that breaks the invariant is an abnormal.

SVM [3]: This model takes the count vector and label of the log template as input, trains a hyperplane to separate high-dimensional space, judges the position relationship between the log sequence and the hyperplane. If the log sequence is above the hyperplane, it is abnormal, otherwise it is marked as normal.

DeepLog [7]: This model takes the index of a normal log sequence as input, uses LSTM to learn the pattern of a normal log sequence, and predicts the next log template to determine whether it deviates from the normal pattern.

LogRobust [9]: The Fast Text algorithm is used to model log word vectors, TF-IDF is used to aggregate word vectors into log semantic, and an attention-based Bi-LSTM model is used to detect abnormal log sequences.

Evaluation Metrics. Log anomaly detection as a typical classification problem, this paper adopts accuracy, recall and F1-score as evaluation metrics to measure the anomaly detection effectiveness of LogST.

Precision: the proportion of the real anomaly log sequence in the anomaly log sequence detected by the model, as shown in Eq. (10).

$$Precision = \frac{TP}{TP + FP} \tag{10}$$

where, TP is the number of abnormal log sequences accurately detected by the model, FP represents the number of normal logs mis-classified as abnormal.

Recall: the proportion of all real abnormal log sequences that are accurately detected by the model as abnormal log sequences, as shown in Eq. (11).

$$Recall = \frac{TP}{TP + FN} \tag{11}$$

where, FN represents the number of abnormal logs mis-classified as normal.

F1-score: the summed average of precision and recall, as shown in Eq. (12).

$$F1 - score = \frac{2 * Precision * Recall}{Precision + Recall} \tag{12}$$

Experimental Environment. This experiment was conducted on a NVIDIA RTX 3090 24G GPU server, using a Python 3.8 environment, with a deep learning framework of Tensorflow 2.5.0 and CUDA 11.2. We set the dilation rates of different sizes in TCN to 1, 2, 4, 8, with convolution kernel size of 3, and the number of convolution kernels for each dilation rates of 3 and 1. During the experiment, 80% of the dataset was selected as the training set, and the remaining 20% as the test data. In HDFS datasets, the log sequence is obtained based on the "block_id" session window and the corresponding label is used to determine whether the log sequence is abnormal. When extracting log semantic, we leverage an off-the-shelf service bert-as-service to obtain the sentence vector of the log template. Subsequent, when using the SVD algorithm for dimensionality reduction, the vector dimension is reduced from 768 to 270, i.e. $r = 270$; When extracting topic features, set the number of topics to 30, i.e. $K = 30$. In the BGL dataset, we use a sliding window to obtain a log sequence with a window size of 20 and a step size of 20. If there are abnormal logs in the sequence, then the sequence is considered abnormal. When extracting log semantic, the bert-as-service is also used to obtain the sentence vectors of the log template, and the SVD algorithm is used to reduce the dimensionality. The vector dimension is reduced from 768 to 200, i.e. $r = 200$. When extracting topic features, set the number of topics to 100, i.e. $K = 100$. LogST uses Adam [36] as the optimizer and uses a random gradient descent algorithm for training. The batch size is 64 and the epoch count is set to 60.

4.2 Analysis of Anomaly Detection Results

We compare the anomaly detection performance of LogST and other baseline models in the HDFS and BGL datasets. From Table 3, it can be seen that LogST outperforms other baseline models in F1-score. In the baseline model, by comparing two machine learning algorithms, the SVM algorithm with supervised learning is better than the IM algorithm with unsupervised learning, which shows that the model can effectively improve the effect of anomaly detection by learning historical anomaly knowledge. Although SVM has achieved high precision in both datasets, its Recall is low. Because SVM cannot effectively learn the feature of anomaly logs, resulting in the constructed classification hyperplane being too large, and part of the anomaly log being judged as normal. Compared with the two deep learning algorithms, DeepLog only has a slightly higher Recall value than LogRobust in the HDFS dataset, and other metrics are lower than LogRobust, and its prediction effect is better than the two machine learning methods. The analysis reason is that DeepLog can effectively learn the normal log sequence pattern by using the LSTM model, and consider abnormal parameter value, but it ignores the semantic information inside the log, and cannot process log updates or noise factors. LogRobust solves the problem of updating log templates to some extent through word embedding, and can handle noise interference. It not only improves the robustness of the model, but also achieves higher anomaly detection performance compared to other baseline models by supervised learning of historical anomaly knowledge.

Due to IM, SVM, and DeepLog all using the index of the log template as input, it is easy to overlook some anomalies or misjudgments, resulting in a lower F1-score. Compared to this, LogST achieves 0.981 and 0.984 in the F1-score of two datasets, respectively. In the HDFS dataset, compared to the machine learning model IM, it improved by 9.3%, 5%, and 7.2% on three metrics, respectively, which indicates that by considering the semantic features of log templates and combining them with supervised deep learning methods, more anomalies can be detected. In BGL datasets with a larger number of log templates, LogST improves by 4.7% and 2.5% on Recall and F1-score, respectively, compared to the optimal baseline model LogRobust. This is because on the one hand LogST not only uses the BERT model combined with the SVD algorithm to obtain log semantic features, but also extracts log topic features using LDA, whereas LogRobust only uses the Fast-Text algorithm to extract semantic features, lacking in mining log template topics, and the detection effect is easily affected by semantic changes. On the other hand, LogRobust uses attention-based Bi-LSTM model to extract contextual information of log sequences and detect anomalies, but it lacks the extraction of local features of log sequences, compared to the TCN model of LogST, which mines local features through convolutional networks, and extracts long sequential relationships through dilated convolution. In addition, the weighted residual-connected TCN model makes more effective use of log anomaly knowledge and assigns higher weights to key features, thus achieving better detection results.

Table 3. Anomaly detection results in different datasets.

Datasets	HDFS			BGL		
Metrics	Precision	Recall	F1-score	Precision	Recall	F1-score
IM	0.884	0.936	0.909	0.827	0.948	0.883
SVM	0.978	0.911	0.943	0.982	0.885	0.930
DeepLog	0.958	0.933	0.945	0.924	0.959	0.941
LogRobust	0.981	0.958	0.969	0.975	0.942	0.959
LogST	0.977	0.986	0.981	0.980	0.989	0.984

4.3 Analysis of Topic Features

LogST uses the LDA topic model to extract the topic features of logs, and enriches log sequence features by combining semantic features. To explore the effectiveness of the topic features, we conduct ablation experiments by eliminating the topic features. The model only uses the BERT model combined with the SVD algorithm to obtain semantic features, which are used as input to improved TCN model. The model is trained using the same steps and parameters as LogST for anomaly detection. We name the reduced No-Topic, and the experimental results are shown in Fig. 3.

After removing the topic features from the HDFS dataset, the dimensionality of the semantic features after reducing the SVD is still 270, as shown (a) in Fig. 3 both Recall and F1-score are significantly decreased, and Pre is basically unchanged, which indicates that the model can detect more anomalies and reduce the occurrence of misjudgment in normal logs when combined with topic features. In the BGL dataset with a larger number of templates, this paper sets the dimensionality of semantic features after SVD dimensionality reduction to 200, as shown (a) in Fig. 3, and it can be observed that No-Topic is lower than LogST in all metrics, which further illustrates that for complex log structures, over-reliance on semantic features generated by the language model tends to miss certain anomalies, that the model performance degrades when large updates occur in the log template. In LogST, the LDA topic model is able to further analyze the differences between different log templates from a macro perspective by analyzing the relationships between template words, templates, and template topics in logs. In addition, in the HDFS dataset based on the Block_id session window to obtain log sequences, the logs contain fewer component categories and levels, while in the BGL dataset, the sliding window is used to obtain log sequences, the logs contain more component and levels, so the topic feature in the BGL dataset can more fully capture the differences between normal and abnormal logs, and the effect is more obvious. It can be seen that, based on the log semantic, extracting topic features of logs from a macro perspective and achieving multi-feature fusion are more helpful for log anomaly detection tasks.

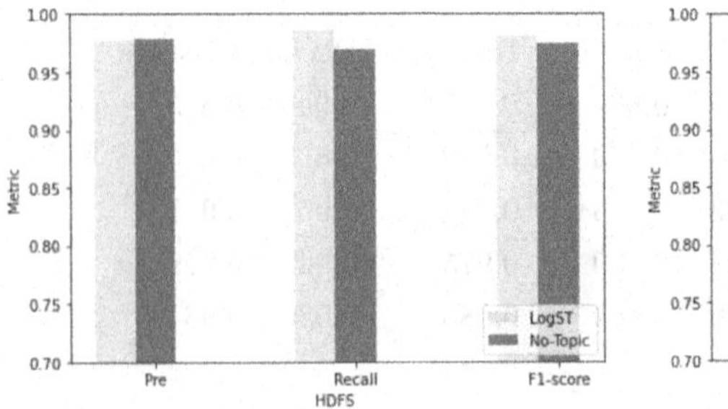

(a) Metrics comparison on HDFS dataset

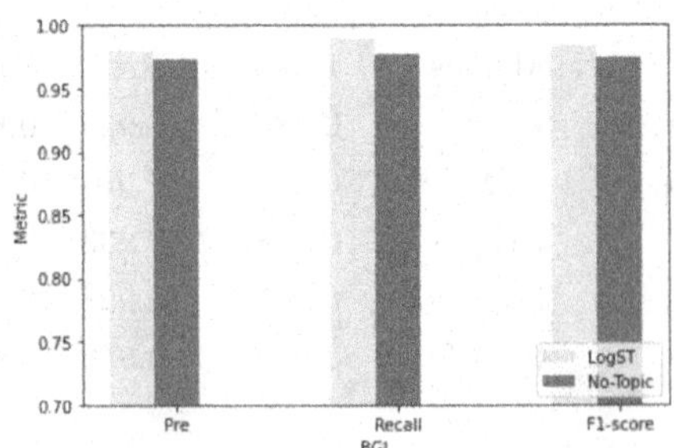

(b) Metrics comparison on BGL dataset

Fig. 3. Comparison results of ablation experiments

4.4 Analysis of Improved TCN

LogST uses convolution with different dilation rate for log feature extraction and weighted residual connection to improve the effect of anomaly detection. In this paper, we remove the weighted residual connections of the improved TCN in LogST, and the model is changed to a TCN model using four different expansion factors, each of which contains two TCN blocks, $filters = 3$ and $filters = 1$. Residual connections are performed in the first and last TCN block, and the variant model was named No-Dense. The results were compared in two different datasets, as shown in Table 4. After removing the weighted residual connection, the number of No-Dense parameters significantly decreased, approximately one-third of that of LogST. However, the anomaly detection effect of No-Dense in both datasets decreases significantly, which indicates that the TCN in LogST obtains higher anomaly detection effect by weighting residual connections although the number of parameters is increased to some extent, and this practice is reasonable and valuable. Further analysis of the experimental effect, in which Recall in the HDFS dataset decreases by 5.6%, indicates that No-Dense does not effectively utilize the anomaly knowledge of log sequences, resulting in insufficient analysis of some of the anomaly features, and fails to identify more anomalous logs compared to LogST. In the BGL dataset with more types of log templates, components and levels, the metrics differ significantly and the F1-score decreases by 3.8%, which further demonstrates that the performance of TCN is significantly improved by the weighted residual connection proposed in this paper. For log sequences of complex data sets, the weighting can highlight the key hidden layer features that play a greater role in anomaly detection results, and assign lower weights to unimportant features, reducing the interference of noise factors, thus improving the anomaly detection effect.

Table 4. Comparison of parameters and experimental effects

Datasets	HDFS				BGL		
Metrics	Params	Precision	Recall	F1-score	Precision	Recall	F1-score
No-Dense	263	0.961	0.930	0.945	0.949	0.945	0.946
LogST	867	0.977	0.986	0.981	0.980	0.989	0.984

4.5 Discussion

The overall process of LogST is divided into three parts: log parsing, feature extraction, and anomaly detection. In practical applications, the format and types of logs generated by the system are complex and diverse, through the Drain algorithm, the complex semi-structured logs structured parsing, can quickly grasp the full picture of the logs, to avoid a lot of time to view the repetitive information; Subsequent, the parsed and pre-processed data will be used to extract topic features based on LDA, and semantic features extracted based on BERT. The combination of the two can mine the distinguishing features of normal and abnormal logs in terms of macro and detail, without the need for operation and maintenance personnel to manually analyze the differences between normal and abnormal logs, and capturing more comprehensive features than other feature extraction algorithms. During system operation, new logs are continuously generated, and topic and semantic features can be obtained by mapping the processed logs with the pre-generated topic and semantic feature library; Final, the improved TCN is used for anomaly detection based on the fusion of log sequence features, which does not need to explicitly and manually configure the log anomaly rules, and is able to distinguish the anomalous logs automatically by analyzing the log sequence features. In addition, since most neural network models such as RNN require expensive computational power to achieve highprecision anomaly detection, the heavy computation is difficult to be applied in practical deployment, while the TCN model used in this paper has a simple structure and a small number of parameters, which belongs to the lightweight neural network, and it is more suitable for deploying in the edge devices.

5 Conclusion

This paper proposes a log anomaly detection method based on semantic features and topic features, LogST. First, to enrich log sequence information, LogST constructs a multi-information fused log sequence by splicing the log components and levels on the basis of log templates. Second, we use the BERT model to extract log sentence vectors and uses the SVD algorithm for feature dimensionality reduction to obtain more efficient log semantic. At the same time, the LDA topic model is used to extract the topic features of logs, further analyze the differences between normal and abnormal log templates, and enrich the log sequence features. Final, we improve the TCN model by using a weighted residual connection, which assign higher weights to key features and achieve reuse of anomaly knowledge. A large number of experimental results indicate that LogST has significantly improved various metrics in different datasets. The analysis of the topic features shows that the topic features of logs can help the model effectively detect abnormal logs in various complex scenarios. The analysis of the effectiveness of improved TCN model shows that the LogST model is more lightweight, and with weighted residual connections, the model is able to highlight key features and utilize anomaly knowledge more effectively, which improves the robustness and accuracy of the model.

LogST can highlight key features and more effectively utilize anomaly knowledge through weighted residual connections,

Acknowledgements. This work is supported by Liaoning Province Applied Basic Research Program Project (Grant No. 2023JH2/101300195).

References

1. He, S., He, P.J., Chen, Z.B., Yang, T.Y., Su, Y.X., Lyu, M.R.: A survey on automated log analysis for reliability engineering. ACM Comput. Surv. **54**(6), 1–37 (2022)
2. Le, V., Zhang, H.Y.: Log-based anomaly detection with deep learning: how far are we? In: 44th IEEE/ACM 44th International Conference on Software Engineering (ICSE), pp. 1356-1367(2022)
3. He, P.J., Zhu, J.M., He, S.L., Li, J., Lyu, M. R.: Towards automated log parsing for large-scale log data analysis. IEEE Trans. Depend. Secure Comput. **15**(6), 931–944(2018)
4. Xu, W., Huang, L., Fox, A., Patterson, D.A., Jordan, M.I.: Detecting large-scale system problems by mining console logs. In: Proceedings of the 27th International Conference on Machine Learning (ICML-10), pp. 37–46 (2010)
5. Chen, M.Y., Zheng, A.X., Lloyd, J., Jordan, M.I., Brewer, E.A.: Failure diagnosis using decision trees. In: 1st International Conference on Autonomic Computing, pp. 36–43 (2004)
6. Lou, J.-G., Fu, Q., Yang, S.Q., Xu,Y., and Li , J.: Mining invariants from console logs for system problem detection. In: USENIX Annual Technical Conference (ATC), pp. 1–14 (2010)
7. Du, M., Li, F., Zheng, G., Srikumar, V.: DeepLog: anomaly detection and diagnosis from system logs through deep learning. In: Proceedings of the 2017 ACM SIGSAC Conference on Computer and Communications Security, pp. 1285–1298 (2017)
8. Wang, J. et al.: LogEvent2vec: LogEvent-To-vector based anomaly detection for large-scale logs in internet of things. Sensors **20**(9) (2020)
9. Zhang, X., et al.: Robust log-based anomaly detection on unstable log data. In: Proceedings of 27th ACM Joint Meeting Eur. Softw. Eng. Conf. Symp. Foundations Softw. Eng., pp. 807–817 (2019)
10. Wang, Z.M., Tian, J.Y., Fang, H., Chen, L.M., Qin, J.: LightLog: a lightweight temporal convolutional network for log anomaly detection on the edge. Comput. Networks **203** (2022)
11. Devlin, J., Chang, M.W., Lee, K., Toutanova., K.: BERT: pre-training of deep bidirectional transformers for language understanding. In: NAACL-HLT, Minneapolis, pp. 4171–4186 (2019)
12. Akritidis, L., Bozanis, P.: How dimensionality reduction affects sentiment analysis NLP tasks: an experimental study. In: 18th AIAI, pp. 301–312 (2022)
13. David, M., Andrew, Y., Michael, I.: Latent Dirichlet allocation. J. Mach. Learn. Res. **3**(4–5), 993–1022 (2003)
14. Bai, S.J., Kolter, J. Z., Koltun, V.: An empirical evaluation of generic convolutional and recurrent networks for sequence modeling (2018)
15. Makanju, A., Zincir-Heywood, A., Milios, E.: A lightweight algorithm for message type extraction in system application logs. IEEE Trans. Know. Data Eng. **24**(11), 1921–1936 (2012)
16. Du, M., Li, F.: Spell: streaming parsing of system event logs. In: IEEE 16th International Conference on Data Mining (ICDM), pp. 859–864 (2016)

17. Shi, Y., et al.: An improved KNN-based efficient log anomaly detection method with automatically labeled samples. ACM Trans. Knowl. Discov. Data **15**(3), 1–22 (2021)
18. Li, X.Y., Chen, P.F., Jing, L.X., He, Z.L., Yu, G.B.: SwissLog: robust and unified deep learning based log anomaly detection for diverse faults. In: 31st IEEE International Symposium on Software Reliability Engineering (ISSRW), pp. 92–103 (2020)
19. Fu, Q., Lou, J.G., Wang, Y., Li, J.: Execution anomaly detection in distributed systems through unstructured log analysis. In: The Ninth IEEE International Conference on Data Mining (ICDM), pp. 149–158 (2009)
20. Messaoudi, S., Panichella, A., Bianculli, D., Biand, L.C., Sasnauakas R.: A search-based approach for accurate identification of log message formats. In: Proceedings of the 26th Conference on Program Comprehension (ICPC), pp. 167–16710 (2018)
21. Zhu, J.M. et al.: Tools and benchmarks for automated log parsing. In: Proceedings of the 41st International Conference on Software Engineering: Software Engineering in Practice, pp. 121–130 (2019)
22. He, P., Zhu, J.M, Zheng, Z.B., Lyu, M. R.: Drain: an online log parsing approach with fixed depth tree. In: IEEE International Conference on Web Services (ICWS), pp. 33–40 (2017)
23. Zhang, C.K., Wang, X.Y., Zhang, H.Y., Zhang, H.Y., Han, P.Y.: Log sequence anomaly detection based on local information extraction and globally sparse transformer model. IEEE Trans. Netw. Serv. Man. **18**(4), 4119–4133 (2021)
24. Lin, Y., et al.: Semi-supervised log-based anomaly detection via probabilistic label estimation. In: 43rd IEEE/ACM International Conference on Software Engineering: Companion Proceedings (ICSE), pp. 230-231 (2021)
25. Han, X., Yuan, S.: Unsupervised cross-system log anomaly detection via domain adaptation. In: The 30th ACM International Conference on Information and Knowledge Management (CIKM), pp. 3068–3072 (2021)
26. Liu, X., et al.: LogNADS: network anomaly detection scheme based on log semantics representation. Future Gener. Comp. Sy. **124**, 390–405 (2021)
27. Chen, R. et al.: LogTransfer: cross-system log anomaly detection for software systems with transfer learning. In: 31st IEEE International Symposium on Software Reliability Engineering (ISSRE), pp. 37–47(2020)
28. Wang, Q.Z., Zhang, X.G., Wang, X.J., Cao, Z.Y.: Log sequence anomaly detection method based on contrastive adversarial training and dual feature extraction. Entropy **24**(1) (2022)
29. Meng, W.B., et al.: LogAnomaly: unsupervised detection of sequential and quantitative anomalies in unstructured logs. In: Proceedings of the Twenty Eighth International Joint Conference on Artificial Intelligence (IJCAI), pp. 4739–4745 (2019)
30. Zhou, J.W., Qian, Y.J., Zou, Q.T., Liu, P., Xiang, J.W.: DeepSyslog: deep anomaly detection on syslog using sentence embedding and metadata. IEEE Trans. Inf. Foren. Sec. **17**, 3051–3061 (2022)
31. Huang, S.H., et al.: HitAnomaly: hierarchical transformers for anomaly detection in system log. IEEE Trans. Netw. Serv. Man. **17**(4), 2064–2076 (2020)
32. Wang, J., Zhao, C.Q., He, S.M., Gu, Y., Alfarraj, O., Abugabah, A.: LogUAD: log unsupervised anomaly detection based on Word2Vec. Comput. Syst. Sci. Eng. **41**(3), 1207–1222 (2022)
33. Vaswani, A., et al.: Attention is all you need. In: Annual Conference on Neural Information Processing Systems, pp. 5998–6008 (2017)

34. Peinelt, N., Dong, N., Maria, L.: tBERT: topic models and BERT joining forces for semantic similarity detection. In: Proceedings of the 58th Annual Meeting of the Association for Computational Linguistics(ACL), pp. 7047–7055 (2020)
35. He, S.L., Zhu, J.M., He, P.J., Lyu, M.R.: Loghub: a large collection of system log datasets towards automated log analytics (2020)
36. Diederik, P., Jimmy, B.: Adam: a method for stochastic optimization. In: 3rd International Conference on Learning Representations (ICLR) (2015)

Popularity Cuckoo Filter: Always Keeping Popular Items in Mind

Xuetan Cheng[1], Lailong Luo[1(✉)], Wei Zou[3], Xiangrui Yang[2], and Deke Guo[1]

[1] School of Systems, National University of Defense Technology, Changsha, China
{chengxuetan18,luolailong09,dekeguo}@nudt.edu.cn
[2] School of Computer, National University of Defense Technology, Changsha, China
[3] School of Computer Science and Engineering, Central South University, Changsha, China

Abstract. A Bloom Filter is a basic and randomized means of storing information that can accurately determine membership status queries with no false negatives and a small probability of false positives. As its improvement, a Cuckoo Filter is a kind of new data structure which can support adding, removing items dynamically and achieving higher performance than a Bloom Filter. But current Cuckoo filters usually handle items assuming they have the same possibility to be queried, and treat them without difference, which is unable to satisfy the demand for querying that most popular items in dataset, such as in web caching.

We propose a new data structure called the popularity cuckoo filter that can make false positive smaller and prioritize storing members with higher popularity. Popularity cuckoo filters use different numbers of hash functions for items with different popularities, so they have better space efficiencies. Our experimental results show that the popularity cuckoo filter can distinguish items with large or small popularities well and suit datasets with irregular query patterns and non-uniform membership likelihood.

Keywords: Cuckoo hashing · popularity · Bloom filters

1 Introduction

A Bloom filter [1] is a Hash-based randomized data structure for representing a dataset in order to support membership queries. It is widely used in many database, caches, routers, and storage systems to decide if a given item is in a dataset. The Bloom filter has no false negative but a low false positive probability. Because of its memory efficiency, the Bloom filter has been studied extensively, and many variants were steadily proposed [2–8]. Bloom filters have been used to search files in networks [5], ensure if a item is in a dataset [1, 3], route false data [9] and so on.

An important limitation of standard Bloom filters is that items added cannot be removed unless rebuilding the entire filter. Some variants extend the standard Bloom filters to try to support deletion [4, 10, 11], but they either introduce an undesirable false negative probability like autoscaling Bloom filters [12] or nonnegligible space and performance overhead [4, 10, 11]. A Cuckoo filter [13] is a compact variant of a cuckoo

Z. Tari et al. (Eds.): ICA3PP 2023, LNCS 14491, pp. 428–445, 2024.
https://doi.org/10.1007/978-981-97-0808-6_25

hash table [14] that stores only fingerprints instead of key-value pairs. It inherits the ideas of the Bloom filter but uses another storage strategy. A Cuckoo filter supports editing items dynamically, provides higher performance than traditional Bloom filters, uses less space than Bloom filters in many applications and is easier to implement than variants like the quotient filter and so on. To determine if an item x is part of a set, a search is done in the hash table of the cuckoo filter for x's fingerprint. If the same fingerprint is found, then x is considered a member of the set and a true value will be returned. The size of the fingerprint in a cuckoo filter is established based on the false positive rate that is aimed for during the construction process, and the size of the smallest possible fingerprint must increase logarithmically with the number of entries in the table in cuckoo filter for keeping the false positive rate unchanged.

The Cuckoo Filter assumes that all the items in the dataset will be viewed and treated identically, and stores them with same two buckets. This is the best strategy without their query frequency distribution or likelihood of being a member in the dataset. Under this assumption, the Cuckoo filter can be proved to provide better space efficiency and lookup performance [13]. However, it's common for some items to be queried more frequently than others in practice, with popular items being queried much more often than the less popular ones. When the probability of a query being made or an item being included in the dataset is not equal for all items, using a standard cuckoo filter configuration may not result in the best performance. Possible situations include but are not limited to mutual squeeze between items with most space utilized, failure to include highly popular items. If we know the query popularity or other similar attributes of items, we can further improve the cuckoo filter to support storing and querying items better and utilize filter space more efficiently [28–31]. In fact, data like the number of times a query is made has been utilized to enhance the effectiveness of caching structures [15]. Related applications include pattern matching in a distributed mobile environment [16], optimizing skewed traffic flow access on web cache [15, 17], *COUNT SKETCH* [18] used to estimate the most frequent items in a data stream using very limited storage space, Locality-Sensitive Bloom Filter [19] and so on.

Specifically, in caching, a typical demand is to represent cache content of each Internet proxy in a compact form. [20] A low-overhead method is to only records the portion of a proxy's cache content that will be of interest to other proxies, like items that have more possibilities to be queried. A proper data structure utilized in wireless communications to hasten the process of querying cache and updating the cache mechanism is very important, which can offer cache content summaries to validate cache membership, preventing negative lookups and reducing computational workload. In extensive wireless networks, every node holds the public key of certain other nodes within the network. But since the storage capacity of each node is limited, it can only save a few public keys. To overcome this issue, developing a brief method to represent the public keys which enables storing and retrieving more keys effectively and quickly is necessary.

Based on this target, this paper presents the implementation of the popularity Cuckoo filter. This model is designed to handle items with varying frequencies and can be seen as a generalization and further improvement of the traditional Cuckoo filter. It grades each item based on its popularity and assigns appropriate k hash functions accordingly to optimize performance, where k depends on the upper limit of the number of hash functions using in the filter and the popularity values of items, and the mapping function algorithms between popularity values and number of the hash functions can freely edited by developers. Therefore, each item in the filter has its own different false positive probability, and the average false positive probability becomes a weighted sum over the query possibilities of the whole items in the filter. It also uses a popularity backoff algorithm to handle collisions between an old and a new item and which item will be kick out to reload in.

In cases where all items in the dataset have equal query frequencies, the popularity Cuckoo filter becomes indistinguishable from the standard Cuckoo filter, but has the worse space efficiency than it. In experiments we have briefly implemented a popularity cuckoo filter and test it, in which the significant screening effects of popularities between successfully inserted and failed items can be observed. In addition, the filter improves several defects in standard cuckoo filters, such as better support of repeated insertion, reduced probability of collisions among items, and higher space utilization.

2 An Overview of Cuckoo Filter

Let us briefly introduce the standard Cuckoo filter [13] first. A Cuckoo filter is a variant version of a cuckoo hash table [14]. A simple cuckoo hash table often consists of several arrays, in which every item has two buckets determined by hash functions $h_1(x)$ and $h_2(x)$ for potential locations. While adding items, if either of the two buckets for an item is empty, it can accept the item by the insert algorithm and be signed as nonempty. During the lookup procedure, both buckets will be examined to ensure that the target item is present. If neither of buckets are empty, the item first kicks out one of the items in and then put itself in, the kicked item will calculate the hash functions to find another alternate location, as showed in Fig. 1. Sometimes this procedure may repeat many times until an empty location is found lastly, or reach the max times of repetition. In case there isn't an available bucket, the hash table is deemed as too congested to add data. Though Cuckoo hashing may perform a series of displacements, but the average time it takes to insert can be still $O(1)$.

Most recently, several cuckoo filter variants have been proposed to further improve its performance. As an example, the parallelized Cuckoo filter [21] focus on cold data, achieves parallelized insertion, accurate representation, and constant-time query/deletion, naturally parallelizes item insertions in parallelized Cuckoo filter. Another structure, Chucky [22], substitutes several Bloom filters with a single Cuckoo filter which associates every data input with a supplementary address representing its position within the LSM-tree, succinctly encode the auxiliary addresses so that the fingerprints can stay large, and finally achieves both a modest access cost and a low false positive rate at the same time. Morton filters [23] present a new method for saving a sparsely distributed filter compactly in memory, using a compressed block format. This technique also utilizes succinct embedded metadata to minimize unnecessary memory usage and biases insertions more heavily to use a single hash function. It supports self-resizing, a feature of quotient filters, but be much faster than rank-and-select quotient filters and keeps a lossless space efficiency. TinyTable [24] is a hash-table-based scheme. It supports membership queries, multiplicity queries and removals. Unlike Counting

Bloom filters which incur a significant memory space overhead, TinyTable is more space efficient than those derived from Bloom filters and other schemes.

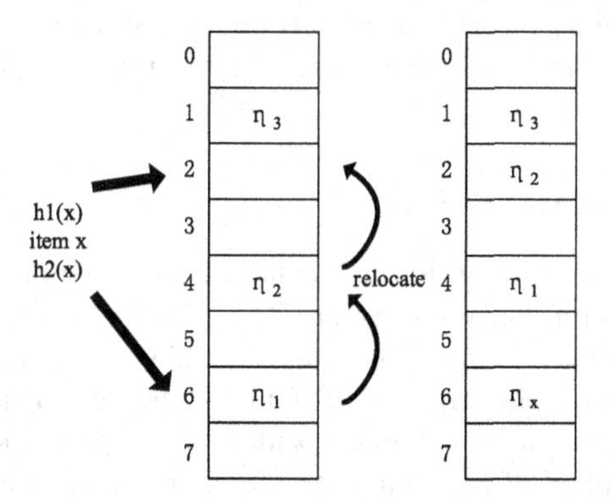

Fig. 1. The framework of Cuckoo filter: before and after inserting an item *x*

Cuckoo hashing can improve space utilization by re-evaluating the initial item placement decisions while inserting new items. The maximum possible load has been analyzed [25] while using *k* hash functions and buckets of size *b*. In many practical applications of cuckoo hashing, most implementations use buckets that can store multiple items in order to make cuckoo hashing more practical, as suggested in [26]. By setting the parameters of the cuckoo hash table appropriately, there is a high likelihood that the table space can be filled up to 95%. Despite this, standard cuckoo hash tables still require much more space than cuckoo filters.

In Cuckoo filters, all items are firstly calculated by algorithms to fingerprints before operating, and each item has indexes of the two candidate buckets as follow:

$$\mathrm{h}_1(x) = \mathrm{hash}(x), \tag{1}$$

$$\mathrm{h}_2(\mathrm{x}) = \mathrm{h}_1(\mathrm{x}) \oplus \mathrm{hash}\left(\mathrm{x}'\mathrm{s\ fingerprint}\right) \tag{2}$$

It's partial-key cuckoo hashing, a variant of standard cuckoo hashing and was first introduced in [27], earlier than standard cuckoo filters. The xor operation ensures an important property: either of the index of buckets can be calculated by another index and fingerprint, in which the original item is no longer necessary. The fingerprint is usually hashed before xor to help distribute the items uniformly in the table, because the size of fingerprints may affect the distance while reloading the kicked items. By hashing the fingerprints, it becomes possible to move these items to totally different buckets of the hash table, which helps to decrease the occurrence of hash collisions and increase the efficiency of the table.

A standard cuckoo filter usually has *m* buckets, in which there are *b* entries, and the length of fingerprint in bits is *f*. For all n items, the false-positive probability is the same of $1 - \left(1 - \frac{1}{2^f}\right)^{2b} \approx \frac{2b}{2^f}$, which is related to the size of the bucket *b* and fingerprint *f*.

Let the upbound of false positive possibility lower than ε, there is $f \geq \lceil log_2(2b/\varepsilon) \rceil$. Furthermore, it can be inferred that the upper bound of space cost $C \leq \lceil log_2(2b/\varepsilon) \rceil/\alpha$.

In order to achieve better performance of the Cuckoo filter, we aim to decrease the false positive probability or enhance its performance at the same false positive probability. Intuitively, we hope to enhance the effectiveness by changing filters' storage structure and algorithm logic.

3 Popularity Cuckoo Filter

The fundamental component of the cuckoo hash tables used in the popularity cuckoo filters for this study is referred to as an entry. These hash tables are arranged in an array, and are composed of buckets that can contain several entries. But what is stored in our entries is no longer a single fingerprint, but a tuple composed of a fingerprint, a popularity value and an ordinal number. The Fig. 2 shows how tuples work while inserting an item into a non-full-filled PCF. In a multitude of applications, various techniques have been developed to estimate or gather query frequencies, which is the main focus of this filter and can be used to optimize the optimal configuration of the filter. Same as the standard Cuckoo filter, the Popularity Cuckoo filter uses hash functions to distribute items. The xor is no longer useful, what is instead is more functions and special popularity algorithm. For each item trying to insert, the popularity cuckoo filter asks it to provide a popularity value (less than the max value set while initializing the filter) and distributes corresponding number of hash functions for it. So, the bigger the popularity is, the more bucket index the item will have. In this case the probability of false positives in the filter is calculated based on the weighted sum of false positive probabilities of individual items. The weight assigned to each item is directly related to how frequently it is queried, namely, the popularity of each item. Besides that, in algorithm complexity, there is not much difference between Popularity Cuckoo filters and standard Cuckoo filters. Popularity Cuckoo filters may execute more displacements than standard Cuckoo filters because of averagely more buckets per item, the amortized insertion time can be still $O(1)$.

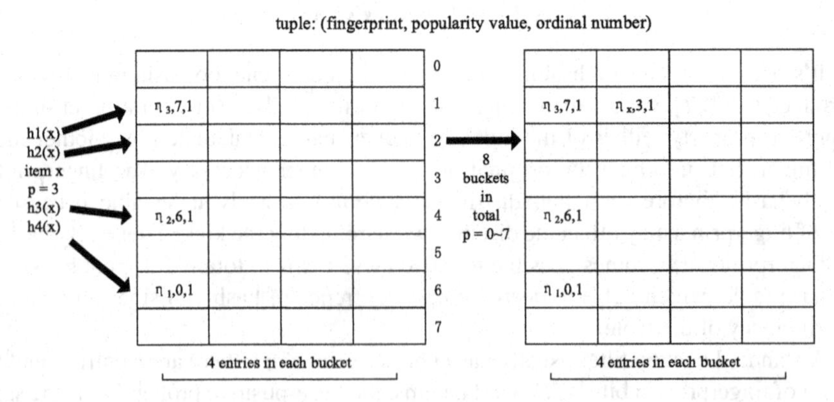

Fig. 2. Insert an item into a non-full-filled PCF

3.1 Insert

In standard cuckoo hashing tables and filters, inserting new items to an existing hash table requires some means of accessing the original existing items in order to determine where to relocate them if needed to make room for the new ones. But cuckoo filters only store fingerprints which cannot restore the original items, so they use a technique called partial-key cuckoo hashing to solve this problem. In popularity cuckoo filters, the numbers of items are stochastic, in which partial-key cuckoo hashing cannot work. We use a tuple (fingerprint, popularity value, ordinal number) to replace a single fingerprint. The popularity value means how many hash functions it can use at most, and the ordinal number means which hash functions is used in calculating the current index of buckets. New design makes an important property invalid: one index of buckets can be calculated by another index and items' fingerprint, which means the process of insertion solely utilizes the fingerprints present within the table.

Using fingerprint tuple is slightly cumbersome but necessary, in which there are two key messages inside: how many functions this item can use and which function the item is using. The index of buckets is calculated by different hash functions and item's fingerprint, so the original item is still unnecessary in re-calculating the index. In hash collisions, the reloading item will be determined by an algorithm named popularity backoff: the filter takes out these two items' popularity values in tuples and calculates them with editable algorithms, such as normalized probability, then obtain a random number to kick out either an item. The Fig. 3 shows a typical inserting into a full-filled PCF and its two possible results. Algorithm 1 shows how it works while inserting.

The popularity backoff algorithm is responsible for mapping popularity values to the number of hash functions, which is similar to histogram transformation. The backoff algorithm receives popularity values, calculates and outputs a suitable number according to a certain function or other rules. Developers can freely define their own rules to handle the mapping. A simple enough method is to make number of functions with bit size of popularity values, which is used in our code. Also, the average time-complexity of each insert is O(1).

Algorithm 1: Insert(x)

---

f = fingerprint(x);
$i_1 \sim i_x$ = hashlib(f); (hashlib is a library used to provide plural hash functions)
if bucket[$i_1 \sim i_x$] has an empty entry **then**
  add f's tuple to that bucket;
  return Done;
no empty and must relocate existing items;
randomly select a bucket[j] from bucket[$i_1 \sim i_x$];
randomly select an entry e from bucket[j];
the fingerprint in e and f randomly collide by their popularity values;
if f wins **then**
  swap f's tuple and the tuple of fingerprint stored in entry e;
  try reinsert f;
  return Done;
f fails in collision;
return Failure;

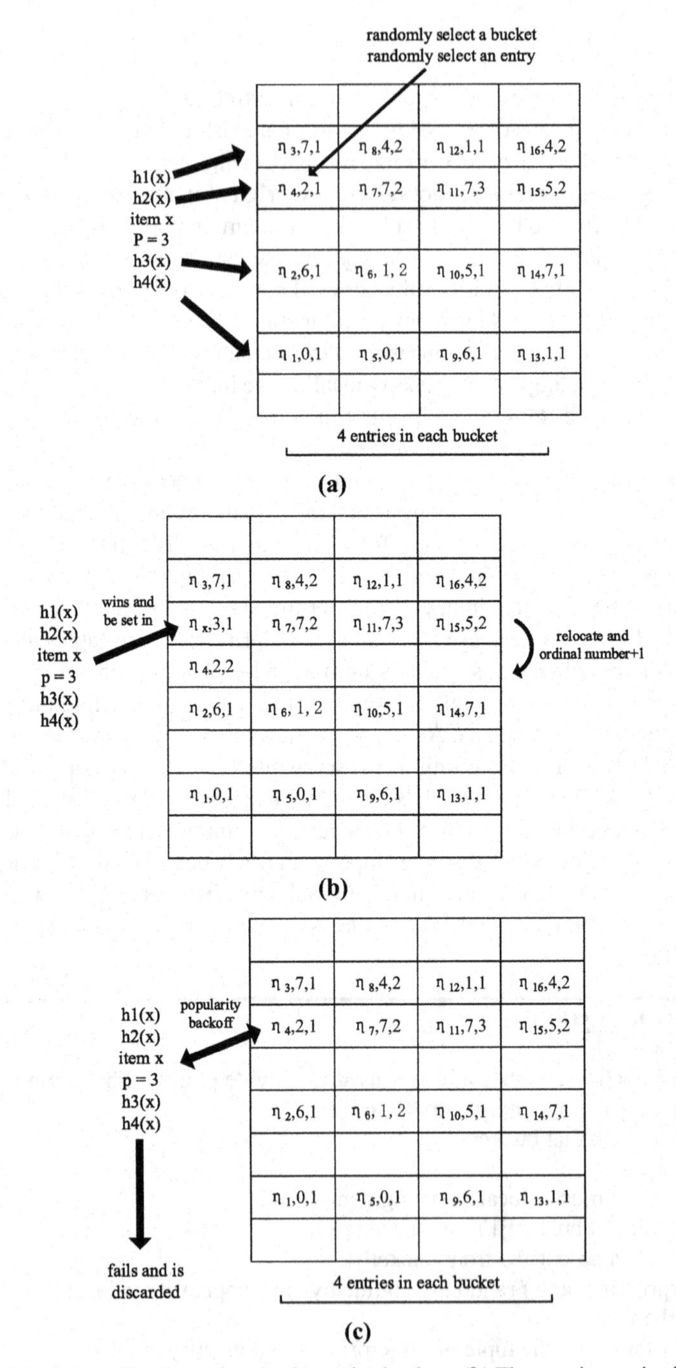

Fig. 3. (**a**) A full-filled PCF for an item's alternative buckets. (**b**) The new item wins in popularity backoff. (**c**) The new item fails in popularity backoff and is discarded

There are four significant features using more hash functions and popularity backoff:

- More collisions may arise, but space utilization will increase, because most items have more positions than before. What is showed in the experiment is that the load factor of a popularity cuckoo filter is bigger than the standard cuckoo filter with the same b entries in each bucket.
- The filters can accommodate more identical items, especially when they have high calorific popularity values, because available alternative buckets are positively correlated to numbers of hash functions. The standard cuckoo filter is suitable for applications that insert the same item for $2b$ times at most (b is the bucket size), the same indicator is $k * b$ times in popularity cuckoo filter (k is the number of hash functions), because there are two indexes of buckets for each item and both will become overloaded.
- Items with high popularity values are more likely to stay in the filter, which is because they have more alternative index of buckets and have a higher chance of be kept in popularity backoff algorithm in collisions.
- When an item is moved to the last bucket but needs to be reinserted while adding the next item, it can attempt to reload in the first bucket. This helps optimize storage locations and improve space utilization.

3.2 Lookup

The lookup process of a popularity cuckoo filter is a bit easier than a standard cuckoo filter, what is shown in Algorithm 2. For a given item x and popularity value, the algorithm first calculates the possible hash functions and fingerprint of x and then identifies all indexes of buckets. If there is a match in any of the fingerprints in the buckets, the popularity cuckoo filter will give a true result, otherwise it will give a false result. As long as bucket overflow doesn't occur, the false negative possibility will be absent. In an additional way, if no popularity value is applied, the algorithms will check for all the entries possible with all hash functions. The average time-complexity of each lookup is O(1). Algorithm 2 shows how to look up an item.

Algorithm 2: Lookup(x)

f = fingerprint(x);
i_1~i_x = hashlib(f);
if bucket[i_1~i_x] has f's tuple **then**
    return True;
return False;

3.3 Delete

Traditional Bloom filters lack the ability to delete, meaning that to eliminate a single item, the entire filter must be reconstructed. Standard Cuckoo filters support deletion by removing corresponding fingerprints from entries in buckets, which are mostly simpler compared to other filters with analogous deletion methods, like counting Bloom filters, quotient filters and so on [32]. Counting Bloom filters need added counters and quotient filters need to move the sequence of fingerprints to occupy the empty entry after deletion to preserve their bucket structure. Similar to standard cuckoo filters, the deletion process of popularity cuckoo filters checks all candidate buckets for a given item and remove one matched fingerprint and the tuple it belongs to from buckets if any fingerprint matches in any bucket. In the "false deletion" problem, if two items collide on fingerprint f, share one candidate bucket but have two different popularity values, the filter can still treat them as separate items. For those sharing the same fingerprint and the same popularity value like items x and y, they will have all the same numbers of hash functions, the same tuples and the same candidate buckets. The filter will just remove one matched tuple of them, like x as an example. After x is deleted, y can still match x's lookup action because they have the same fingerprint and popularity value. It's a tiny but not negligible false positive behavior, the probability is bounded by bit size of fingerprints and range of popularity values. Another noteworthy fact is that, an uninserted item cannot be tried to delete from buckets, either the structure cannot find it and will give an error, or in a worse case, a different item which happens to share the same fingerprint may be deleted as the target item. The average time-complexity of each delete is O(1). Algorithm 3 shows how to delete an item.

Algorithm 3: Delete(x)

```
f = fingerprint(x);
i_1~i_x = hashlib(f);
if bucket[i_1~i_x] has f then
    remove a copy of f's tuple from this bucket;
    return True;
return False;
```

3.4 Theoretical Analysis

We use following several factors to compare standard cuckoo filters and popularity cuckoo filters (Table 1).

Minimum Fingerprint Size: Considering a popularity Cuckoo filter having m buckets and b entries per bucket, it can accept $m * b$ items in total. For a tuple of each item, the distinguishable items are fingerprint and popularity value, the ordinal number is only used to confirm which hash function corresponds to the index of current buckets. The depriving process is similar to [13]. The popularity value is difficult to think about for each item, we use average value P as an alternative to approximately research. Firstly, we derive the probability that a given set of q items collide in the same buckets. The probability they have the same fingerprints is $(\frac{1}{2^f})^{q-1}$. And all the alternative buckets are all the same, which occurs with probability $\lceil log_2 P \rceil / m$. Therefore, the probability of such q items sharing the all the same alternative buckets is $(\frac{\lceil log_2 P \rceil}{m} * \frac{1}{2^f})^{q-1}$. Now suppose we have an empty table with m buckets and we want to insert n random items, keeping bucket size c a constant. If we try to map $\lceil log_2 P \rceil * b + 1$ items into the same $\lceil log_2 P \rceil$ buckets, insertion will fail. This gives us the minimum probability of failure. Since there are $\binom{n}{\lceil log_2 P \rceil * b + 1}$ possible sets of $\lceil log_2 P \rceil * b + 1$ items out of n items,

Table 1. Notation used for analysis in sections

ε	target false positive rate
α	the load factor of the filter
n	number of items in dataset
f	fingerprint length in bits
p	popularity value length in bits
d	ordinal number length in bits
b	number of entries per bucket
m	number of buckets
n	number of all items in total dataset
P	average popularity value per item

we can expect a certain number of groups of $\lceil log_2 P \rceil * b + 1$ items to collide during construction, denoted by:

$$\binom{n}{[log_2 P] * b + 1} \left(\frac{[log_2 P]}{m * 2^f} \right)^{[log_2 P]*b} = \Omega\left(\frac{n}{2^{[log_2 P]bf}} \right) \quad (3)$$

In a loose way, we can conclude that the $2^{[log_2 P]bf}$ must be $\Omega(n)$ to avoid a non-trivial probability of failure. Therefore, the fingerprint size must be $f = \Omega(logn/[log_2 P]b)$ bits.

Upper Bound of False Positive Possibility: Suppose that a certain item can be placed in average k different buckets, the probability that a query is matched against another stored fingerprint and its popularity value, returns a false-positive match is at most $\frac{1}{2^f} * \frac{1}{2^p}$. Each buckets have b entries, so the match will make $k * b$ times in total. The upper bound of the total probability of a false fingerprint hit is:

$$1 - \left(1 - \frac{1}{2^f} * \frac{1}{2^p} \right)^{kb} \approx \frac{kb}{2^{f+p}} \quad (4)$$

Compared with the upper bound $\frac{2b}{2^f}$ of standard Cuckoo filters, the possibility becomes smaller because k is usually smaller than 2^p.

Space Cost: Each item is stored in a popularity cuckoo filter as a tuple with a fingerprint, a number to indicate its popularity and can be used to calculate its number of hash functions, and an ordinal number. The size of popularity value p is equal to the ordinal number d. So, storing a tuple of an item averagely cost $f + p + d = f + 2p$. Then the amortized space cost for each item is $\frac{f+2p}{\alpha}$ bits.

Popularity Value Riddling: Items with high popularity values have more alternative buckets and are easier to win in probability backoff algorithm. Finally, after enough insertion and removal processes, the items that remain in the filters will have a higher average popularity value than the items that were in the original dataset or were removed.

Supposed an item has popularity value p_1 and another has p_2, a simple idea is that the probability for the first item to win is $p_1/(p_1+p_2)$, in which the probabilities of winning are proportional. In more complex situations, the algorithms can be freely edited to satisfy different probability distributions.

Maximum Capacity: Cuckoo filters have a limited (is the same for both standard and popularity cuckoo filters) capacity for items, while bloom filters can accommodate almost unlimited items with an increasing false positive rate at the expense. Once the maximum possible load factor has been reached, the probability of failure in insertions will be progressively higher. The load factor α will be test in Sect. 4.

Limited Duplicates: The maximum times that a given item can be inserted into a popularity cuckoo filter is similarly to standard Cuckoo filters, which is determined by the number of hash functions $k = \lceil log_2 P \rceil$ related to every item. The difference is that k in standard Cuckoo filters is 2 for all items but individually different for items in popularity Cuckoo filters. Inserting the same item $k * b + 1$ times will cause a failure, because b entries in each of k buckets have been all non-empty, which is similar to counter overflow of a counting Bloom filter [10].

3.5 Potential Improvement

- Savings on hash function: Hash functions often used is limited. An example is the hash functions used in a python module named hashlib: more algorithms may be available on personal platform, but the functions are guaranteed to exist in Windows, Linux and any other platforms is only 14. In a situation where functions are not reused, 14 hash functions only can be used to support filters with $2^{14} = 16384$ buckets at most. There must be a way to reuse limited hash functions for filters big enough, some related work [29, 30] has researched this target and achieved remarkable results, but how to combine them and popularity cuckoo filters together is still a worth-researching problem.
- Semi-sorting: Semi-sorting is very useful way to save bits used for entries stored in buckets [10, 13]. It averagely saves 1 bit for each 4-bits fingerprint in each bucket. Note that this permutation-based encoding is a type of lookup table method in essence, which requires extra encoding/decoding tables and indirections on each lookup [13]. So, in order to attain fast lookup speed, it is crucial to ensure the encoding/decoding table compact enough to be accommodated within the computer's cache capacity. Higher spatial efficiency is one of the most important parameters of popularity cuckoo filters because of its huge tuples, but this limitation makes the entries of buckets must be kept to be 4 or any other sizes small enough. How to get inspiration from semi-soring to improve sorting and storage is also a challenge.
- Frequent hash-calculations: Each time we want to know an item's hash value to calculate the index of buckets, the fingerprints stored in entries will be hashed at the biggest possible time rather than using the ordinal number, which is just used to "pick" a hash value from the hash list. A significant improvement is to combine ordinal numbers, hash functions and popularity values of items to minimize the number of calculations as much as possible, and strike a balance between time and space efficiency.

4 Experiment

In this section we compare the insertion/query/deletion performance of popularity Cuckoo filters with standard Cuckoo filters. We have implemented an almost 200 lines of python code for popularity cuckoo filter. In the following, we denote a standard cuckoo filter as "CF", a popularity cuckoo filter as "PCF". We describe our experimental settings and present the results of evaluations. The source code of this paper can be found on *github.com/AzureYuki/PopularityCF*.

For insertion, load factor α is evaluate with $\alpha = \frac{itemsstoredsuccessfully}{filter'scapacity}$. Other metrics include insertion failures time (IF), insertion time consumption, the query/deletion time consumption and the popularity riddling of items. The experiments were carried out using a host that has an Intel Core i5-8250 CPU with 8GB RAM. Test dataset is made with adjustable strings with randomly combined characters.

The default setting of Popularity Cuckoo filters in experiment is $f = 24$, $p = 12$, $d = 12$, $m = 2^{12}$=1024, $n = b * m$, the max time of relocation in CF is 50. All the hash functions come from a Python module named hashlib. To compare the performance of Popularity Cuckoo filters and Cuckoo filters in the same condition, we set the same f, b, m, and total capacity as Popularity Cuckoo filters for Cuckoo filters. To compare the performance of PCF and CF in the same condition, for CF, we set the same f, m, n and total capacity as PCF. Additionally, we are not considering the semi-sorting [10, 13, 14] Cuckoo filter.

- *Insertion failures:* When the number of entries in each bucket, i.e., b varies from 1 to 64, as illustrated in Fig. 4(a), the insertion failure of PCF has trend of increasing, because the demo code will mark the item as failed and discard them when the item fails in popularity backoff, which means they would not be relocated. On the contrary, increasing b results in a significant decrease first and then a light increase for the number of insertion failures of CF.
- *Insertion time consumption:* As the value of b increases from 1 to 64, shown in Fig. 4(b), the time it takes for PCF to insert items increases rapidly. This is because each time an item is loaded or reloaded, the hash algorithms must be recalled in order to calculate the indexes of buckets. The number of hash functions within the hash algorithm increases with larger popularity values, which results in a significant amount of time being consumed. On the other hand, inserting items in CF only requires the fingerprint to be invariably hashed once, which results in a much shorter insertion time consumption.
- *Load factor:* The load factor of PCF is initially higher but later relatively lower, and of CF is initially lower but later relatively higher. As b begins to increase, the load factor of PCF increases slowly as b gets larger. In Fig. 4(c), when the value of b is not less than 4, and the load factor of CF remains close to 100%. By setting the value of b to 64, the load factor of PCF becomes 99.59%. The main reason for this is that, to prevent long calculation times, the demo code will label items that fail in popularity backoff as failed and discard them. These items will not be relocated. This means the items distribution will be not sufficient enough and a lower load factor of the filter, but can eliminate lengthy backoff-reloading operation chains.

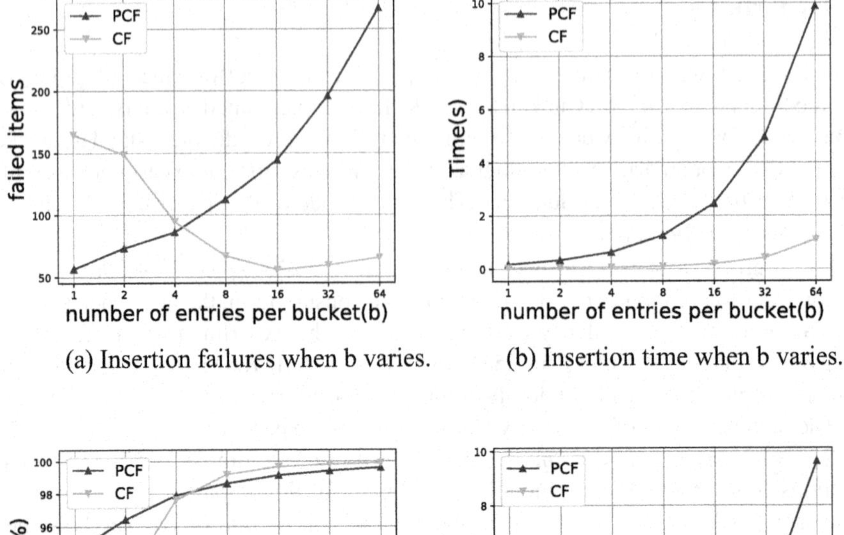

(a) Insertion failures when b varies.

(b) Insertion time when b varies.

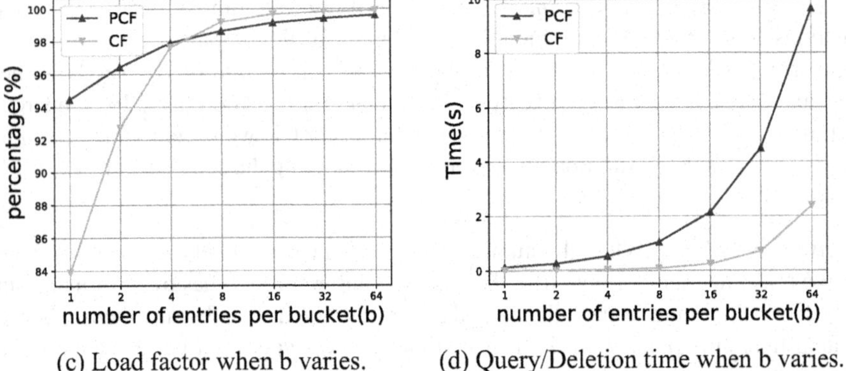

(c) Load factor when b varies.

(d) Query/Deletion time when b varies.

Fig. 4. Compare the performance between PCF and CF

- *Query/Deletion time consumption:* The algorithms for these two operations are almost identical, with the only difference being whether or not to remove the tuple of items from the entries. The broken lines shown in Fig. 4(d) represent the time taken to query or delete each item in the dataset once from the filters that have been structured. Based on the same reason in *Insertion time consumption*, lines show a similar pattern of change. The query/deletion time consumption in CF increases a bit faster in the later stage, i.e., from $b = 16$ to $b = 64$, compared to insertion consumption.

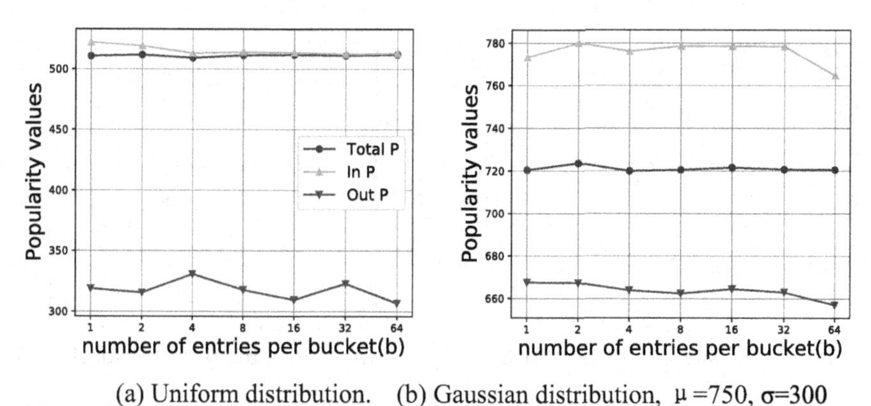

(a) Uniform distribution. (b) Gaussian distribution, μ =750, σ=300

Fig. 5. Popularity riddling with different dataset when b varies.

- *Popularity value riddling:* This metric comes from the average popularity of three kinds of items: 1). Items in total dataset 2). Items that were successfully loaded in filters 3) items that were discarded out of filters. In Fig. 5(a), the popularity values of items in test dataset are evenly distributed in range 0–1023 from the algorithm, so the "Total *P*" in Fig. 5 remains close to 512, but the difference is inconspicuous. In Fig. 5(b), by changing the dataset distribution from uniform to Gaussian with mean of 750 and standard deviation of 300, the results indicate that it effectively distributed and retained items with varying values of popularity.
- *Popularity value riddling for different parameters:* Furthermore, we test several potentially influential different indicators for this kind of riddling, including the distribution of data's popularity values, the number of entries per bucket (*b*) and the multiple of the dataset of relative capacity.

In Fig. 6, the vertical axis is the average popularity values of items and the horizontal axis is the multiple of the relative capacity (k) of dataset. By using bigger dataset to insert and let items "collide" sufficiently in the filter, we can see the average popularity values of items kept in filters increasing with k and the effects are obvious on the dataset which following uniform and Gaussian distribution, because the values are more uniform. When b is fixed, the best performance improvement appears when there are neither too many, like the Poisson and Chi-square distribution, nor too few, like $k = 1$, hot items, in which case the query frequencies or the popularity values are appropriately skewed (Fig. 7).

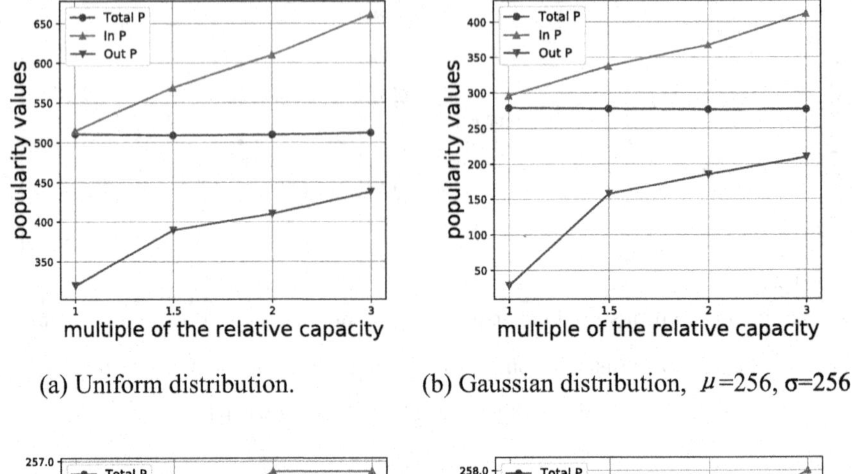

(a) Uniform distribution. (b) Gaussian distribution, μ=256, σ=256

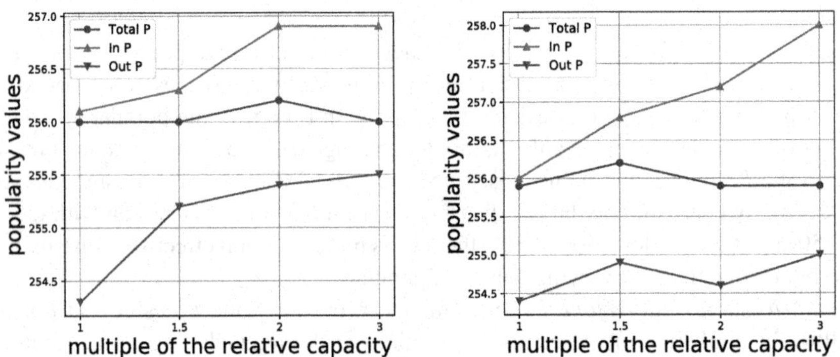

(c) Poisson distribution, μ=256. (d) Chi-square distribution, degree Freedom=256

Fig. 6. Popularity riddling with different distributions when b = 4.

If we vary the value of *b*, the effects remain almost unchanged. Varying *b* only changes the capacity of filters and other like false positive possibility, but does not contribute to the improvement of average popularity values. Note that the horizontal axis represents the multiple of relative capacity, if the capacity is fixed and then enlarge *b*, the performance will even be impaired.

- The difference among lines illustrates that the popularity backoff algorithm can separate items with varying levels of popularity and ensuring that popular items are consistently filtered. While dealing with dataset with more skewed query frequencies, the performance gain of popularity Cuckoo filter will improve.

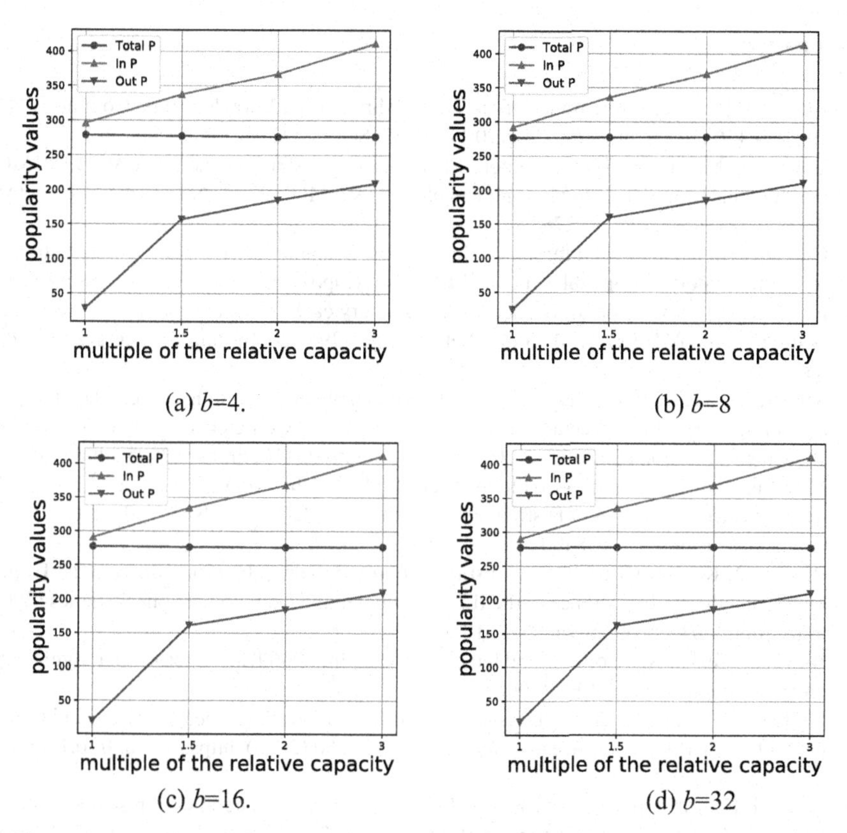

Fig. 7. Popularity riddling on Gaussian distribution ($\mu = 256$, $\sigma = 256$) with varying b.

5 Conclusions

Popularity cuckoo filters are a new data structure for items with varying frequencies. Popularity cuckoo filters improve upon standard cuckoo filters in three ways: (1) bigger load factor; (2) lower false positive possibility; (3) special algorithms for popularity of items like hash functions decided by popularities of items and popularity backoff. Our exploration suggests that the popularity cuckoo filters can perform well under reasonable query frequency and membership likelihood models. For the future work, it would be interesting to further investigate the combination of semi-sorting [10] and popularity.

Acknowledgement. This work is supported by National Natural Science Foundation of China under Grant No. 62002378, as well as partially funded by the Research Funding of NUDT under Grant ZK20-3.

References

1. Bloom, B.H.: Space/time trade-offs in hash coding with allowable errors. Commun. ACM **13**, 422–426 (1970). https://doi.org/10.1145/362686.362692
2. Cohen, S., Matias, Y.: Spectral bloom filters. In: Proceedings of the 2003 ACM SIGMOD International Conference on Management of Data (SIGMOD 2003). New York (2003). https://doi.org/10.1145/872757.872787
3. Chazelle, B., Kilian, J., Rubinfeld, R., Tal, A.: The Bloomier filter: an efficient data structure for static support lookup tables. In: ACM-SIAM Symposium on Discrete Algorithms (2004)
4. Fan, L., Cao, P., Almeida, J., Broder, A.Z.: Summary cache: a scalable wide-area Web cache sharing protocol. IEEE/ACM Trans. Netw. **8**, 281–293 (2000). https://doi.org/10.1109/90.851975
5. Kumar, A., Xu, J., Zegara, E.W.: Efficient and scalable query routing for unstructured peer-to-peer networks. In: Proceedings IEEE 24th Annual Joint Conference of the IEEE Computer and Communications Societies (2005). https://doi.org/10.1109/infcom.2005.1498343
6. Mitzenmacher, M.: Compressed bloom filters. In: Proceedings of the Twentieth Annual ACM Symposium on Principles of Distributed Computing (PODC 2001), New York (2001). https://doi.org/10.1145/383962.384004
7. Rhea, S.C., Kubiatowicz, J.: Probabilistic location and routing. In: Proceedings of the Twenty-First Annual Joint Conference of the IEEE Computer and Communications Societies (2002). https://doi.org/10.1109/infcom.2002.1019375
8. Bruck, J., Jie, G., Anxiao, J.: Weighted bloom filter. In: 2006 IEEE International Symposium on Information Theory. IEEE (2006)
9. Ye, F., Luo, H., Lu, S., Zhang, L.: Statistical en-route filtering of injected false data in sensor networks. IEEE J. Select. Areas Commun. **23**, 839–850 (2005). https://doi.org/10.1109/jsac.2005.843561
10. Bonomi, F., Mitzenmacher, M., Panigrahy, R., Singh, S., Varghese, G.: Presented at the An Improved Construction for Counting Bloom Filters (2006). https://doi.org/10.1007/11841036_61
11. Bender, M.A., et al.: Don't thrash. Proc. VLDB Endow **5**, 1627–1637 (2012). https://doi.org/10.14778/2350229.2350275
12. Kleyko, D., Rahimi, A., Gayler, R.W., Osipov, E.: Autoscaling Bloom filter: controlling trade-off between true and false positives. Neural Comput. Appl. **32**, 3675–3684 (2019). https://doi.org/10.1007/s00521-019-04397-1
13. Fan, B., Andersen, D.G., Kaminsky, M., Mitzenmacher, M.D.: Cuckoo filter. In: Proceedings of the 10th ACM International on Conference on emerging Networking Experiments and Technologies, New York (2014). https://doi.org/10.1145/2674005.2674994
14. Pagh, R., Rodler, F.F.: Cuckoo hashing. J. Algorithms **51**, 122–144 (2004). https://doi.org/10.1016/j.jalgor.2003.12.002
15. Breslau, L., Cao, P., Fan, L., Phillips, G., Shenker, S.: Web caching and Zipf-like distributions: evidence and implications. In: Proceedings of the Eighteenth Annual Joint Conference of the IEEE Computer and Communications Societies (IEEE INFOCOM 1999), Conference on Computer Communications. The Future is Now (Cat. No. 99CH36320) (1999). https://doi.org/10.1109/infcom.1999.749260
16. Liu, S., Kang, L., Chen, L., Ni, L.: Distributed incomplete pattern matching via a novel weighted bloom filter. In: 2012 IEEE 32nd International Conference on Distributed Computing Systems (2012). https://doi.org/10.1109/icdcs.2012.24
17. Luo, L., Guo, D., Ma, R.T.B., Rottenstreich, O., Luo, X.: Optimizing bloom filter: challenges, solutions, and comparisons. IEEE Commun. Surv. Tutorials. **21**, 1912–1949 (2019). https://doi.org/10.1109/comst.2018.2889329

18. Charikar, M., Chen, K., Farach-Colton, M.: Presented at the Finding Frequent Items in Data Streams (2002). https://doi.org/10.1007/3-540-45465-9_59
19. Hua, Y., Xiao, B., Veeravalli, B., Feng, D.: Locality-sensitive bloom filter for approximate membership query. IEEE Trans. Comput. **61**, 817–830 (2012). https://doi.org/10.1109/tc.2011.108
20. Alexander, H., Khalil, I., Cameron, C., Tari, Z., Zomaya, A.: Cooperative web caching using dynamic interest-tagged filtered bloom filters. IEEE Trans. Parallel Distrib. Syst. **26**, 2956–2969 (2015). https://doi.org/10.1109/tpds.2014.2363458
21. Sun, B., Luo, L., Li, S., Chen, Y., Guo, D.: The parallelized cuckoo filter for cold data representation. In: 2021 IEEE 23rd International Conference on High Performance Computing & Communications; 7th International Conference on Data Science & Systems; 19th International Conference on Smart City; 7th International Conference on Dependability in Sensor, Cloud & Big Data Systems & Application (HPCC/DSS/SmartCity/DependSys) (2021). https://doi.org/10.1109/hpcc-dss-smartcity-dependsys53884.2021.00055
22. Dayan, N., Twitto, M.: Chucky: a succinct cuckoo filter for LSM-tree. In: Proceedings of the 2021 International Conference on Management of Data, New York (2021). https://doi.org/10.1145/3448016.3457273
23. Breslow, A.D., Jayasena, N.S.: Morton filters: fast, compressed sparse cuckoo filters. VLDB J. **29**, 731–754 (2019). https://doi.org/10.1007/s00778-019-00561-0
24. Einziger, G., Friedman, R.: Counting with TinyTable. In: Proceedings of the 17th International Conference on Distributed Computing and Networking, New York (2016). https://doi.org/10.1145/2833312.2833449
25. Fountoulakis, N., Khosla, M., Panagiotou, K.: The multiple-orientability thresholds for random hypergraphs. Combinator. Probab. Comp. **25**, 870–908 (2015). https://doi.org/10.1017/s0963548315000334
26. Dietzfelbinger, M., Weidling, C.: Balanced allocation and dictionaries with tightly packed constant size bins. Theor. Comput. Sci. **380**, 47–68 (2007). https://doi.org/10.1016/j.tcs.2007.02.054
27. Fan, B., Andersen, D., Kaminsky, M.: MemC3: compact and concurrent MemCache with dumber caching and smarter hashing. In: Symposium on Networked Systems Design and Implementation (2013)
28. Fu, P., Luo, L., Guo, D., Zhao, X., Li, S., Wang, H.: Jump filter: a dynamic sketch for big data governance. J. Softw. **34**(3) (2022)
29. Fu, P., Luo, L., Li, S., Guo, D., Cheng, G., Zhou, Y.: The vertical cuckoo filters: a family of insertion-friendly sketches for online applications. In: 2021 IEEE 41st International Conference on Distributed Computing Systems (ICDCS). IEEE (2021)
30. Li, S., Luo, L., Guo, D., Zhao, Y.: Stable cuckoo filter for data streams. In: 2021 IEEE 27th International Conference on Parallel and Distributed Systems (ICPADS). IEEE (2021)
31. Luo, L., Fu, P., Li, S., Guo, D., Zhang, Q., Wang, H.: Ark Filter: A General and Space-Efficient Sketch for Network Flow Analysis IEEE/ACM Transactions on Networking
32. Fu, P., Luo, L., Guo, D., Li, S., Zhou, Y.: A Shifting Filter Framework for Dynamic Set Queries IEEE/ACM Transactions on Networking

Solving Client Dropout in Federated Learning via Client Similarity Discovery and Gradient Supplementation Mechanism

Maoxuan Yan[1,2](✉), Qingcai Luo[2], Bo Zhang[1](✉), and Shanbao Sun[2]

[1] School of Information Science and Engineering, University of Jinan, Jinan, China
ise_zhangb@ujn.edu.cn

[2] Shandong Inspur Science Research Institute Co., Ltd., Jinan, China
yanmaoxuan@stu.ujn.edu.cn, {luoqc,sunshb}@inspur.com

Abstract. In the realm of practical applications of federated learning, an issue arises wherein the performance suffers due to passive client disconnections during the federated training process, caused by factors such as resource limitations or network disruptions. This paper introduces a More Precise Similarity Discovery and Gradient Supplementation(MPSDGS) algorithm, which tackles the problem of passive client dropout in federated learning by employing precise clustering techniques to identify similar clients. It further leverages the gradients of clients whose data distribution closely aligns with the disconnected clients, effectively supplementing the disconnected client gradients. The algorithm's efficacy is verified through experimental evaluations conducted on real-world datasets, namely MNIST, CIFAR10, and CIFAR100. The experimental findings reveal that, under the same non-independent and identically distributed data partitioning approach for MNIST and CIFAR10 datasets, MPSDGS achieves notable accuracy enhancements. Specifically, at disconnection rates of 0.3, 0.5, and 0.7, the MPSDGS algorithm improves the accuracy of the MNIST dataset by 1.33%, 1.49%, and 1.35%, respectively. Similarly, for the CIFAR10 dataset, the algorithm enhances accuracy by 1.09%, 1.25%, and 1.6%, respectively, at the aforementioned disconnection rates. Remarkably, MPSDGS exhibits comparable excellence in performance on the CIFAR100 dataset.

Keywords: Federated learning · Client Dropout · Client Similarity Discovery · Gradient Supplementation · Clustering

1 Introduction

In recent years, the rapid development of mobile devices and the Internet of Things has brought about an enormous amount of distributed data and intelligent devices. In this era of data explosion, traditional centralized machine learning methods face challenges such as privacy leakage and heavy burdens on data centers [1]. Federated learning, as a distributed machine learning paradigm,

Z. Tari et al. (Eds.): ICA3PP 2023, LNCS 14491, pp. 446–457, 2024.
https://doi.org/10.1007/978-981-97-0808-6_26

addresses the drawbacks of traditional centralized approaches by coordinating client devices with decentralized data to train models while ensuring the security and privacy of local data throughout the training process [2]. However, in federated learning, participants are often mobile or edge devices such as smartphones, sensor nodes, or IoT devices [3]. Due to the instability of mobile networks, device power limitations, and other external factors, client dropout have become a major factor affecting the performance and stability of federated learning. When a client disconnects during the model update process, the incomplete model update cannot be submitted to the server, thereby affecting the accuracy and convergence of the global model. Furthermore, client dropout lead to the instability of federated learning systems by compromising the continuous participation of clients and the integrity and security of their data. Federated learning has found wide applications in areas such as healthcare, intellgent transportation, and the Internet of Things [4], but the presence of disconnection issues limits its effectiveness and reliability. Therefore, addressing client dropout to improve model accuracy, convergence, and the stability and data privacy of federated learning systems is a current research hotspot and challenge.

In the context of the aforementioned problem, recent research can be categorized into two approaches. The first involves integrating techniques from other domains into federated learning and adopting an active client selection strategy to improve convergence efficiency. However, this approach demands high computational resources and yields unsatisfactory performance in both federated training and testing, while also lacking the ability to execute different data selection criteria. The second approach combines federated learning with cryptographic techniques, ensuring data privacy but incurring significant communication overhead and leading to a complex architecture with limited model availability. To overcome these challenges, we revaluate the issue of client dropout, taking into account the balance between model availability, convergence, and data privacy. Additionally, we draw inspiration from the insights provided in reference [5], this paper proposes an algorithm to tackle the client dropout issue in the scenario where only a subset of clients actively participate in federated learning. It achieves this by leveraging clustering to accurately discover similar clients and using the gradients from the clients whose data distribution closely matches that of the disconnected clients to supplement the missing gradients. The algorithm first employs clustering [6] to partition non-independent and identically distributed (Non-IID) [7] data and allocate data to clients within each cluster. It then calculates the similarity between clients using the Pearson correlation coefficient [8] and iteratively updates the weights of similarity during each round. In the federated training process, the server selects gradients from the client with the most similar data distribution to complement the missing gradients of the disconnected client. Experimental results demonstrate that the proposed algorithm exhibits excellent accuracy performance, ensuring the stability and data privacy of the entire federated learning process, and achieves significant improvements in accuracy compared to previous experiments.

Our main contributions are as follows:

1. We propose the More Precise Similarity Discovery and Gradient Supplementation (MPSDGS) algorithm, which leverages the idea of precise gradient similarity discovery and gradient supplementation based on clustering to address the issue of client dropout in federated learning. This novel approach strikes a balance between model availability, convergence, and data privacy.
2. Successfully combining both outdated and up-to-date similarity measures with varying weights enables us to achieve accurate assessment of similarity between clients. Combining this update strategy with the Pearson correlation coefficient yields excellent results.
3. Through extensive research on related work and improving baseline algorithms, we validate the superior performance of the MPSDGS algorithm in handling client dropout issues in federated learning using real-world datasets.

2 Related Work

The existing research primarily focuses on addressing client disconnection issues in federated learning through active selection of participating clients and integration with cryptographic techniques such as secure sharing [9,10] and differential privacy [11]. In 2020, Huang et al. [12] dynamically selected clients using a multiarmed bandit approach in each round. Wang et al. [13] trained an additional deep reinforcement learning model to evaluate the value of clients in each round and selected the K clients with the highest value for federated learning training. However, this method requires additional training of deep models on top of the federated model, demanding higher computational resources. Ribero et al. [14] proposed a simple and efficient method for updating the global model under communication constraints by introducing Optimal Process Sampling into the federated learning system. They proposed a strategy where clients only transmit their computed model weights to the central server when the norm of their model update exceeds a predetermined communication threshold, indicating informative updates. Additionally, they introduced a non-trivial estimator for model updates that do not meet the communication threshold, providing an effective approach for balancing exploration accuracy and communication trade offs. In 2020, Wang et al. [15] introduced the experience-driven control framework FAVOR, combining reinforcement learning with federated learning. FAVOR described the data distribution on a device based on the uploaded model weights and the training data distribution observed on the device. They also proposed a DQN-based mechanism to learn the selection of device subsets in each communication round, maximizing rewards, encouraging improved validation accuracy, penalizing the use of more communication rounds. The authors demonstrated through experiments that using clustering algorithms for device selection, instead of random selection, can help balance data distribution and accelerate convergence. In 2021, researchers from the University of Michigan proposed the Oort federated selection framework [16] for identifying and selecting valuable participants to join the federated learning process. Oort evaluates the data from edge devices and selects devices that can help the model converge faster. Additionally, considering the significant differences in computational capabilities and network conditions among participants, Oort evaluates the training speed of edge devices and selects devices with faster training speeds for participation. In 2022, Wu et al. [17] proposed the FedPNS algorithm, a probabilistic node selection algorithm for non-independent and identically distributed data. They identified and excluded local updates that are detrimental to global convergence by examining the relationship between local gradients and global gradients, finding the optimal subset of participating nodes for local updates in each global round. This allows for dynamic changes in the selection prob ability of each node, promoting faster model convergence.

In recent years, with the emergence of privacy preserving computing concepts and the practical application of related technologies, techniques such as secure sharing [9,10] and differential privacy [11] have gained wider application. As an important branch of modern cryptography, secure sharing is not only a crucial means of data security protection but also a fundamental application technology for secure multi-party computation. Secure sharing divides secret information into multiple secret shares and distributes them among multiple custodians to achieve risk diversification and intrusion tolerance. Generally, a secure sharing scheme consists of a secret sharing algorithm and a secret reconstruction algorithm, involving three roles: the secret distributor, the secret share holders, and the receivers. The secret distributor holds the secret information and is responsible for executing the secret sharing algorithm and distributing secret shares to the secret share holders. The receiver is the party attempting to reconstruct the secret information. When the receiver wants to reconstruct the secret information, they collect secret shares from an authorized group of secret share holders and execute the secret reconstruction algorithm to compute the secret information. With sufficient secret shares, the secret information can be recovered. A participant can assume multiple roles simultaneously [18]. In 2022, Shao et al. [19]proposed the Disconnection-Resilient Secure Federated Learn ing (DReS-FL) framework based on Lagrange Coded Computing (LCC) [20] to ad dress non-iid data and client disconnection issues. The key idea is to use Lagrange coding to securely share private datasets among clients, enabling each client to receive an encoded version of the global dataset, and the computation of local gradi ents on this dataset is unbiased. To correctly decode the gradients on the server, the gradient function must be a polynomial in a finite field. They constructed a Polynomial Integer Neural Network (PINN) to enable the framework. Theoretical analysis demonstrates that the DReS-FL framework is resilient to client dropout and provides strong privacy guarantees. While this framework effectively addresses data heterogeneity and client disconnection issues in federated learning, the gradients in the Polynomial Integer Neural Network (PINN) experience exponential growth, hindering training of large models and resulting in significant communication over-head.

Differential Privacy (DP) is a privacy definition based on rigorous mathematical theory, aiming to ensure that attackers cannot infer sensitive information about individuals based on differences in query outputs by adding noise to mask the differences between neighboring data points. In any differential privacy algorithm, randomness is essential, and deterministic algorithms cannot satisfy the statistical in distinguishability required for differential privacy protection [18]. Differential privacy achieves privacy protection solely through the addition of noise, which incurs no additional computational overhead. However, it still impacts the availability of model data to some extent. Designing solutions that better balance privacy and utility is a key focus in the practical application of differential privacy [21]. In 2021, Lu et al. [21] proposed a user oriented privacy preserving method for federated learning that supports client dropout. This method supports a privacy preserving federated learning system with parameter protection and tolerance for client dropout. It is based on the sequential composability property of differential privacy and adds noise from the user's perspective. By handling dropout in federated learning differential privacy methods, this approach can tolerate client dropout. Experimental results demonstrate that the proposed method achieves a balance between model utility and privacy.

3 MPSDGS Algorithm

This paper proposes an MPSDGS algorithm that addresses the issue of client disconnection in federated learning by accurately identifying similar clients through clustering and using the model gradients from the most closely matching and similar clients to supplement the gradients of dropout clients. The 20 clients are initially divided into five clusters, with four clients in each cluster. The dataset is then divided into non-iid data, taking the CIFAR10 dataset as an example. The CIFAR10 dataset consists of 10 classes of data, and two classes of data are randomly assigned to each cluster. The server selects 500 data points from each class, totaling 1000 data points, and assigns them to each client in the cluster, ensuring that each client has unique data. This experimental setup satisfies the non-iid setting. Figure 1 delineates the intricate execution process of the algorithm MPSDGS.

Due to the randomness of the initial model and the stochastic nature of mini-batch gradient descent (MBGD) in local model computation, a single similarity score computed in a specific round does not provide accurate similarity information. Therefore, the server maintains and updates the average similarity score between client i and j based on all the similarity scores computed so far. In this module, a new update method is proposed to more accurately identify similar clients by incorporating the similarity score from the previous round and adjusting the weight allocation in the formula for the average similarity score. This update method aims to bring the average similarity score closer to a more accurate direction, thus refining the similarity scores between clients and assisting the server in discovering the clients that have the closest and most similar data distribution to each client. The specific computational formula is as follows:

$$P_t^{i,j} = \begin{cases} \frac{1}{2}P_{t-1}^{i,j} + \frac{L_{t-1}^{i,j}-1}{2L_{t-1}^{i,j}}p_{t-1}^{i,j} + \frac{1}{2L_{t-1}^{i,j}}p_t^{i,j} & i,j \in M_t \\ \frac{1}{2}P_t^{i,j} + \frac{1}{2}p_{t-1}^{i,j} & \text{otherwise} \end{cases} \tag{1}$$

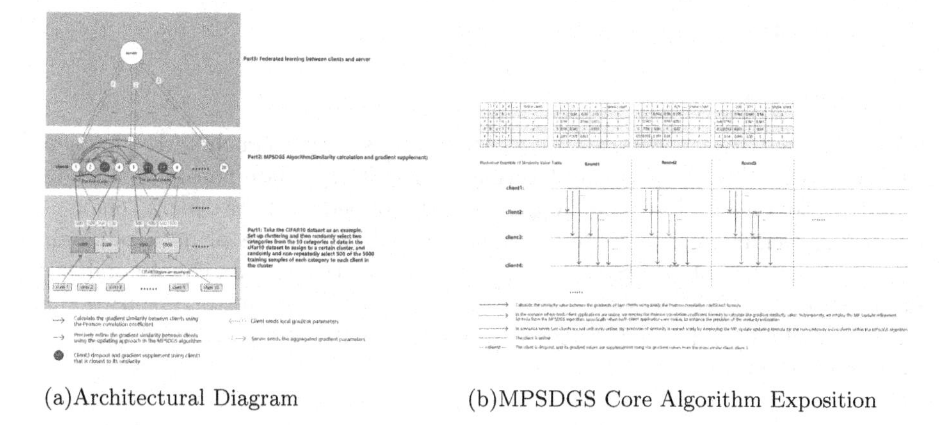

(a)Architectural Diagram (b)MPSDGS Core Algorithm Exposition

Fig. 1. The algorithm architecture is visually depicted through an illustrative diagram a, accompanied by a comprehensive explanation b of the detailed process.

In this formula, $L_{t-1}^{i,j}$represents the number of rounds before the round where clients i and j have not dropped out. $P_{t-1}^{i,j}$denotes the average similarity score from the previous round,$p_t^{i,j}$represents the similarity value between clients i and j in the current round, and $p_{t-1}^{i,j}$represents the similarity value between clients i and j in the previous round.

Furthermore, in the module for calculating the similarity between clients, we replace cosine similarity with Pearson correlation coefficient. For each pair of online clients, i and j, the server calculates their similarity value based on the local model gradients they upload.denoted as $p_t^{i,j}=P\left(\Delta_t^i, \Delta_t^j\right)$ Pearson correlation coefficient is a method used to measure the degree of linear correlation between two variables, the computation formula is as follows, with values ranging from -1 to 1. A value of -1 indicates a perfect negative correlation, 0 indicates no correlation, and 1 indicates a perfect positive correlation. Geometrically, the Pearson correlation coefficient is an improvement over cosine similarity [22].

$$P\left(\Delta_t^i, \Delta_t^j\right) = \frac{\sum_{i=1,j=1}^{n}\left(\Delta_t^i - \mu_i\right) - \left(\Delta_t^j - \mu_j\right)}{\sqrt{\sum_{i=1}^{n}\left(\Delta_t^i - \mu_i\right)^2}\sqrt{\sum_{j=1}^{n}\left(\Delta_t^j - \mu_j\right)^2}} \tag{2}$$

Where Δ_t^i and Δ_t^j are the individual values of the two variables being compared,μ_i and μ_j denote the means of the respective variables,n is the total number of similarity value.

Through extensive ablation experiments [23], we have discovered that the Pearson correlation coefficient can effectively complement our proposed method for updating the average similarity scores. This combination greatly assists the model in achieving faster and more accurate convergence.

4 Experiment

4.1 Datasets

We evaluated our proposed algorithm on several benchmark datasets: MNIST [24], CIFAR10 [25], and CIFAR100 [25]. To simulate non-i.i.d. data distribution, we created a total of 20 clients and grouped them into 5 clusters using clustering techniques. Each cluster was assigned two classes of data randomly selected from the dataset. Specifically, the server distributed an equal number of data samples from the two classes to each client within the cluster, ensuring that each client had unique data. This setup satisfied the requirement of non-i.i.d. data distribution.

4.2 Model Architecture

For the MNIST dataset, we employ a CNN model consisting of two 5×5 convolutional layers, one fully connected layer with 320 units and ReLU activation, one fully connected layer with 50 units, and a final output layer with softmax activation. The first convolutional layer has 10 channels, the second convolutional layer has 20 channels, and both convolutional layers are followed by 2×2 max pooling. For the CIFAR10 dataset, we utilize a CNN model comprising two 5×5 convolutional layers, three fully connected layers with ReLU activation, and a final output layer with softmax activation. For the CIFAR100 dataset, we employ a CNN model consisting of two 5×5 convolutional layers, three fully connected layers with ReLU activation, a Dropout layer, and a final output layer with softmax activation.

4.3 Baselines

In this experiment, we selected the FL-FDMS method from reference [5] as the baseline. We also considered the FL_Full method from reference [5] as the performance upper limit, representing an ideal scenario where clients remain fully engaged without any dropout throughout the process. We compared the performance of the two aforementioned methods with our proposed MPSDGS on the MNIST and CIFAR10 datasets. Additionally, to demonstrate the performance and applicability of the MPSDGS algorithm, we conducted separate experiments with MPSDGS on the CIFAR100 dataset.

4.4 Performance Comparison

We initially compared the accuracy performance of each algorithm under different dropout rates $\alpha \in \{0.3, 0.5, 0.7\}$. Figure 2 illustrates the accuracy curves on the MNIST and CIFAR10 datasets.

We observed that MPSDGS outperforms FL-FDMS in terms of test accuracy. Moreover, MPSDGS achieves comparable or even superior accuracy performance to FL_Full on the MNIST dataset and the CIFAR10 dataset with $\alpha = 0.5$.

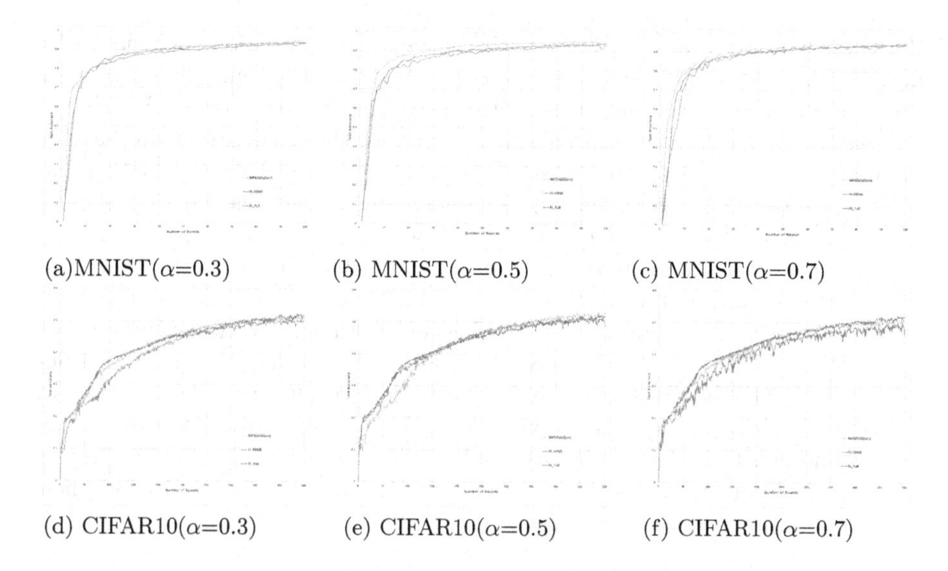

(a)MNIST(α=0.3) (b) MNIST(α=0.5) (c) MNIST(α=0.7)

(d) CIFAR10(α=0.3) (e) CIFAR10(α=0.5) (f) CIFAR10(α=0.7)

Fig. 2. The accuracy curves of the MPSDGS algorithm on the MNIST dataset and CIFAR10 dataset under different dropout rates $\alpha \in \{0.3, 0.5, 0.7\}$.

MPSDGS exhibits improvements of 1.33%, 1.49%, and 1.35% in terms of dropout rates $\alpha \in \{0.3, 0.5, 0.7\}$ on the MNIST dataset, and improvements of 1.09%, 1.25%, and 1.6% on the CIFAR10 dataset under the same dropout rates $\alpha \in \{0.3, 0.5, 0.7\}$, as shown in Table 1.

Table 1. Performance Comparison.

Dataset	MNIST			CIFAR10		
	α=0.3	α=0.5	α=0.7	α=0.3	α=0.5	α=0.7
MPSDGS	94.90%	95.03%	94.99%	51.54%	53.08%	51.69%
FL-FDMS	93.57%	93.54%	93.64%	50.45%	51.83%	50.09%
FL_Full	94.04%			52.45%		

4.5 Ablation Experiment

To further analyze the improvements brought by the combination of the MP_Update in our proposed algorithm MPSDGS and the Pearson correlation coefficient, we present quantitative results of MPSDGS on the MNIST and CIFAR10 datasets with a dropout rate of α=0.5, as shown in Table 2. Figure 3 illustrates the accuracy curves on the MNIST and CIFAR10 datasets, Here, we employ two methods as baselines. The first baseline method, Baseline 1, combines the CSM_Update from FL-FDMS with the Pearson correlation coefficient used in MPSDGS algorithm. The second baseline method, Baseline 2, combines the MP_Update of MPSDGS algorithm with the cosine similarity used in FL-FDMS. To conduct the ablation experiment, we compare the accuracy of the baseline methods with our proposed MPSDGS algorithm that utilizes the MP_Update and Pearson correlation coefficient. The accuracy of Baseline 1 is 93.33% and 52.05% on the MNIST and CIFAR10 datasets, respectively. The accuracy of Baseline 2 is 93.43% and 50.97% on the respective datasets. The MPSDGS algorithm achieves an accuracy of 95.03% and 53.08% on the MNIST and CIFAR10 datasets, respectively. Compared to the baseline methods, the accuracy improvements of our MPSDGS algorithm are 1.7% and 0.7% on the MNIST dataset and 1.03% and 2.11% on the CIFAR10 dataset. These results strongly demonstrate that our algorithm can more accurately identify the client with data distribution closest and most similar to the dropped client, thereby mitigating the impact of client dropout and achieving accuracy performance equivalent to full client participation in the training process.

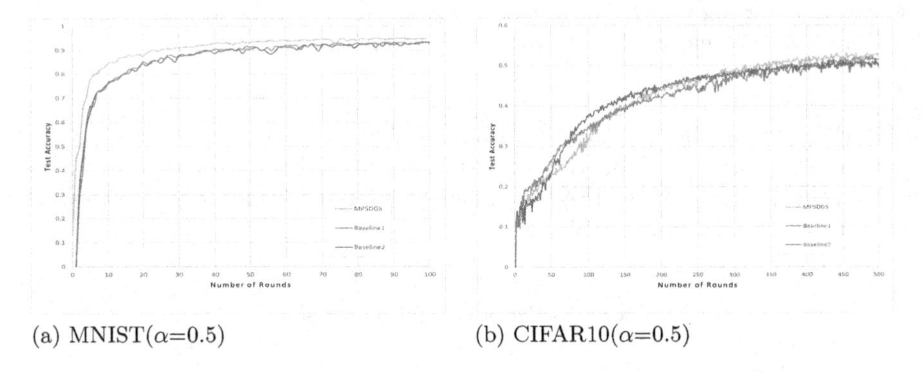

(a) MNIST(α=0.5) (b) CIFAR10(α=0.5)

Fig. 3. The accuracy curves of MPSDGS algorithm, Baselin1 and Baseline2 on the MNIST and CIFAR10 datasets with a dropout rate of α=0.5.

Table 2. Ablation Experiment.

Dataset	MNIST	CIFAR10
Baseline1	93.33%	52.05%
Baseline2	93.43%	50.97%
MPSDGS	95.03%	53.08%

In addition, to further validate the performance and wide applicability of our proposed MPSDGS algorithm, we conducted experimental verification on the CIFAR100 dataset, which consists of 100 classes. In this experiment, we combined the original experimental approach with the approach described in reference [26]. Accuracies were computed by taking the average local accuracies for all clients every 10 rounds, ensuring more accurate and clear experimental results. Figure 4 shows the accuracy curve, indicating that MPSDGS exhibits excellent performance on the CIFAR100 dataset. The highest accuracies achieved were 53.53%, 51.86%, and 51.09% for dropout rates $\alpha \in \{0.3, 0.5, 0.7\}$, respectively, as shown in Table 3. These results collectively demonstrate the superiority of our proposed MPSDGS algorithm and its effectiveness in handling client dropout issues.

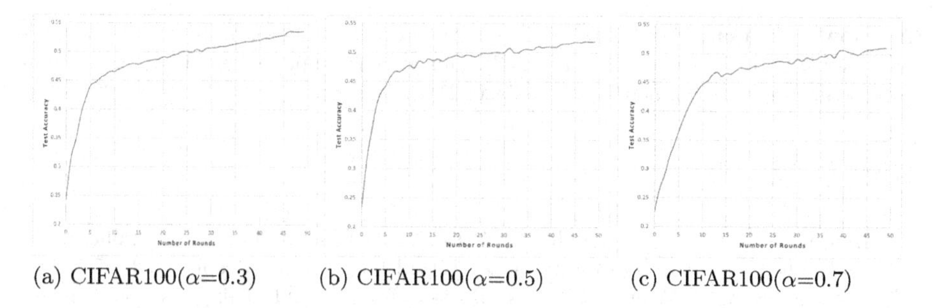

(a) CIFAR100(α=0.3) (b) CIFAR100(α=0.5) (c) CIFAR100(α=0.7)

Fig. 4. The accuracy curves of the MPSDGS algorithm on the CIFAR100 dataset under different dropout rates $\alpha \in \{0.3, 0.5, 0.7\}$.

Table 3. The accuracy of the MPSDGS algorithm on CIFAR100.

Dataset	CIFAR100		
	$\alpha = 0.3$	$\alpha = 0.5$	$\alpha = 0.7$
MPSDGS	53.53%	51.86%	51.09%

5 Conclusion

This paper proposes a novel algorithm, MPSDGS, which addresses the issue of client dropout in scenarios where only a passive fraction of clients participate in federated learning. Our approach utilizes clustering to accurately identify clients and supplements the gradients of dropped clients with the gradients of the most similar clients whose data distributions closely match those of the dropped clients. By providing privacy protection to local datasets, MPSDGS successfully resolves the client dropout problem in federated learning. Extensive experimental results demonstrate the excellent accuracy and convergence of our method. Therefore, MPSDGS is an effective approach for tackling the client dropout problem in federated learning, and it also provides a fresh perspective amidst the existing research focused largely on active client selection and cryptographic methods for handling dropout issues.

Future research directions could explore how the algorithm maintains its performance and superiority in settings with a larger number of clients and massive datasets. Additionally, investigating the applicability of the algorithm in devices with limited computational power or storage resources would be beneficial. Furthermore, integrating other similarity metrics to achieve more comprehensive and accurate similarity measurements opens up new avenues for advancing the potential of federated learning in practical applications.

Acknowledgement. We would like to express our gratitude for the insightful feedback provided by the reviewers of ICA3PP. This work is supported by Shandong Provincial Natural Science Foundation(NO.ZR2022MF264).

References

1. Bharati, S., Mondal, M., Podder, P., Prasath, V.: Federated learning: applications, challenges and future scopes. Inter. J. Hybrid Intell. Syst. (Preprint), 1–17 (2022)
2. Zhou, X., Sun, Y., Wang, D., Ge, H.: Survey of federated learning research Chinese journal of network and information. Security **7**(5), 77–92 (2021)
3. Lim, W.Y.B., et al.: Federated learning in mobile edge networks: a comprehensive survey. IEEE Commun. Surv. Tutorials **22**(3), 2031–2063 (2020)
4. Yang, Q., Tong, Y., Wang, Y., et al.: A survey of federated learning algorithms in swarm intelligence. J. Intel. Sci. Technol. **4**(1), 29–44 (2022)
5. Wang, H., Xu, J.: Combating client dropout in federated learning via friend model substitution (2023)
6. Hartigan, J.A., Wong, M.A.: Algorithm as 136: A k-means clustering algorithm. J. Royal Statist. Soc.. Series c (Appli. Statist.) **28**(1), 100–108 (1979)
7. Zhao, Y., Li, M., Lai, L., Suda, N., Civin, D., Chandra, V.: Federated learning with non-iid data. arXiv preprint arXiv:1806.00582 (2018)
8. Rodgers, J.L., Nicewander, W.A.: Thirteen ways to look at the correlation coefficient. American Statist., 59–66 (1988)
9. Shamir, A.: How to share a secret. Commun. ACM **22**(11), 612–613 (1979)
10. Blakley, G.R.: Safeguarding cryptographic keys. In: Managing Requirements Knowledge, International Workshop on, pp. 313–313. IEEE Computer Society (1979)
11. Dwork, C., McSherry, F., Nissim, K., Smith, A.: Calibrating noise to sensitivity in private data analysis. In: Halevi, S., Rabin, T. (eds.) TCC 2006. LNCS, vol. 3876, pp. 265–284. Springer, Heidelberg (2006). https://doi.org/10.1007/11681878_14
12. Huang, T., Lin, W., Wu, W., He, L., Li, K., Zomaya, A.Y.: An efficiency-boosting client selection scheme for federated learning with fairness guarantee. IEEE Trans. Parallel Distrib. Syst. **32**(7), 1552–1564 (2020)
13. Wang,H., Kaplan,Z., Niu,D., et al.: Optimizing federated learning on non-iid data with re-inforcement learning. In: IEEE INFOCOM 2020-IEEE Conference on Computer Communications IEEE, pp.1698-1707(2020)
14. Ribero, M., Vikalo, H.: Communication-efficient federated learning via optimal client sampling. arXiv preprint arXiv:2007.15197 (2020)

15. Wang, H., Kaplan, Z., Niu, D., Li, B.: Optimizing federated learning on non-iid data with reinforcement learning. In: IEEE INFOCOM 2020-IEEE Conference on Computer Communications, pp. 1698–1707. IEEE (2020)
16. Lai, F., Zhu, X., Madhyastha, H.V., Chowdhury, M.: Oort: efficient federated learning via guided participant selection. In: OSDI, pp. 19–35 (2021)
17. Wu, H., Wang, P.: Node selection toward faster convergence for federated learning on non-iid data. IEEE Trans. Netw. Sci. Eng. **9**(5), 3099–3111 (2022)
18. China Information and Communication Research Institute, Alibaba (China) Co. , Ltd. , Beijing Digital Bamboo Technology Co. , Ltd. Privacy Protection Computing Technology Research Report (2020)
19. Shao, J., Sun, Y., Li, S., Zhang, J.: Dres-fl: dropout-resilient secure federated learning for non-iid clients via secret data sharing. arXiv preprint arXiv:2210.02680 (2022)
20. Zhu, J., Li, S.: Generalized lagrange coded computing: a flexible computation-communication tradeoff. In: 2022 IEEE International Symposium on Information Theory (ISIT), pp. 832–837. IEEE (2022)
21. Lu, H., Wang, L.: User-oriented data privacy preserving method for federated learning that supports user disconnection. Netinfo Sec. **21**(3), 64–71 (2021)
22. Luo, C., Zhan, J., Xue, X., Wang, L., Ren, R., Yang, Q.: Cosine normalization: using cosine similarity instead of dot product in neural networks. In: Kůrková, V., Manolopoulos, Y., Hammer, B., Iliadis, L., Maglogiannis, I. (eds.) ICANN 2018. LNCS, vol. 11139, pp. 382–391. Springer, Cham (2018). https://doi.org/10.1007/978-3-030-01418-6_38
23. Meyes, R., Lu, M., de Puiseau, C.W., Meisen, T.: Ablation studies in artificial neural networks. arXiv preprint arXiv:1901.08644 (2019)
24. Deng, L.: The mnist database of handwritten digit images for machine learning research [best of the web]. IEEE Signal Process. Mag. **29**(6), 141–142 (2012)
25. Krizhevsky, A., Hinton, G., et al.: Learning multiple layers of features from tiny images (2009)
26. Collins, L., Hassani, H., Mokhtari, A., Shakkottai, S.: Exploiting shared representations for personalized federated learning. In: International Conference on Machine Learning, pp. 2089–2099. PMLR (2021)

A KPIs-Based Reliability Measuring Method for Service System

Shuwei Yan, Zhiying Cao(✉), Xiuguo Zhang(✉), Peipeng Wang, and Zhiwei Chen

School of Information Science and Technology, Dalian Maritime University, Dalian 116026, China
{yan_shw,czysophy,zhangxg,wpp7,czw1120211464}@dlmu.edu.cn

Abstract. Distributed systems may experience various abnormalities due to the influence of sub-node status and complex network environment. Failure to respond in advance will lead to system failure and heavy losses. Therefore, it is very important to obtain real-time reliability values of the service system. This paper proposes a KPIs-based service system reliability measurement method. First, a KPIs feature selection method based on improved GMM is proposed to filter the KPIs data collected by the server and extract some KPIs features that play a key role in reliability measurement. Then, a reliability measurement model based on LightGBM is constructed, and the Focal Loss function is also introduced to further solve the model performance problem caused by data imbalance. Finally, a hyperparameter tuning method based on the climbing evolutionary strategy algorithm and the improved Bootstrap verification method is proposed to train the constructed reliability measurement model. Using the trained Focal-LightGBM model can quickly measure the real-time reliability of the service system. Experiments on public data sets show that compared with other reliability measurement methods, the method in this paper has better results on accuracy, MSE, R2-score, measurement time and other indicators.

Keywords: reliability measurement · Gaussian Mixture Models (GMM) · climbing evolutionary strategy · Bootstrap verification · Focal loss-Light Gradient Boosting Machine (Focal - LightGBM)

1 Introduction

The reliability of the service system refers to the ability of the service system to perform the required functions within a unit of time. It is a measure of the stable operation of the service system and the correct output [1]. In the field of intelligent operation and maintenance, the reliability measurement of the service system is a difficult problem to solve. On the one hand, because the reliability values do not all have time series characteristics (such as periodicity, trend, etc.), in some scenarios (such as bank Trading system) using the traditional time series prediction method to measure the reliability of the service system is less effective. On the other hand, without considering the call conflicts of internal nodes in the

Z. Tari et al. (Eds.): ICA3PP 2023, LNCS 14491, pp. 458–477, 2024.
https://doi.org/10.1007/978-981-97-0808-6_27

service system, there will be thousands of feature dimensions collected by a large service system at the same time, some of which are strongly related to reliability metrics, but there are also It is difficult to distinguish the two from the characteristic indicators of part of the interference reliability measurement.

Most of the existing reliability measurement methods rely on time-series analysis of reliability values to measure, without considering the influence of multi-dimensional KPIs characteristic data inside the service system on the reliability of the service system, and the results obtained are not comprehensive and accurate. However, the reliability measurement method based on KPIs is difficult to deal with high-dimensional KPIs data, and it is prone to problems such as gradient explosion and error accumulation. The LightGBM model can capture the mapping relationship between multidimensional data and target values, and can solve the above problems by establishing a tree structure to measure the real-time reliability value of the service system.

This paper first proposes a KPIs feature selection method based on cross-window and GMM, and based on this method, the high-dimensional KPIs data of the service system is selected, then a reliability measurement model Focal - LightGBM is proposed, and finally a method based on The hyperparameter tuning method of the climbing evolutionary strategy and improved Bootstrap verification is used to tune the hyperparameters of the constructed reliability measurement model. The contributions of this paper are summarized as follows:

(1) This paper proposes a feature selection method for KPIs based on cross-window and GMM. By adding feature cross windows to reduce the randomness and contingency of the original GMM algorithm, feature data strongly related to reliability can be extracted effectively.
(2) This article combines the Focal loss function with the LightGBM model to propose a reliability measurement model Focal-LightGBM, The structural characteristics of the Focal loss function and LightGBM model allow Focal LightGBM to use histograms to accelerate model construction without considering the adverse effects of data balance on gradients [2].
(3) This paper proposes a model hyperparameter tuning method based on evolution strategy and improved Bootstrap validation. On the one hand, based on the gradient-free evolutionary strategy algorithm, the parameters are optimized heuristically, avoiding redundant searches, and improving the speed of parameter adjustment. On the other hand, the verification method is improved, which ensures the uniform distribution of sample data and improves the accuracy of parameter adjustment.

2 Related Work

2.1 Deep Learning Programme

At present, the reliability measurement methods of service systems are mostly based on statistical ideas or deep learning ideas. Statistics-based reliability measurement methods are simple, practical, and highly modifiable, so many scholars and experts favor statistical models. For example, Fulin Gao et al. [3] built a decision tree based on the Gini coefficient to evaluate project risk and judge its reliability on the basis of knowing the probability of occurrence of various situations, but this kind of method is difficult to predict the value of continuity And a lot of preprocessing work is required for time-ordered data. Shridhar Allagi et al. [4,5] constructed a reliability measurement model by describing a Markov process with implicit unknown parameters and determining the implicit parameters of the process from observable parameters. However, such methods cannot effectively utilize contextual information, and the effect is limited. Nguyen et al. [6] proposed a reliability prediction model based on a 3-parameter sigmoid function, which is an extended form of the 2-parameter sigmoid function obtained by adding a growth rate controller, and is essentially an improved non- homogeneous Poisson process. Some methods [7,8] generate prediction results based on the mining analysis of historical experience data by building item similarity models. Although this method is simple and effective, it cannot predict situations other than historical experience.

2.2 Statistical Programme

With traditional statistical models, deep learning methods do not require data processing such as feature selection and dimension compression, and can extract time series features through neural networks, providing a new solution for reliability measurement. In recent years, many reliability measurement methods based on deep learning have emerged. For example, Jiahui Li et al. [9] proposed a topological neural network model TAN (Topology-Aware Neural), which uses explicit path modeling layer and implicit cross modeling layer respectively. Capturing path features and endpoint intersection features can learn the influence of network topology on prediction results, but the disadvantage is that it cannot handle complex dynamic topological structures. Wei-Chang Yeh [10] and others used the memory function of LSTM (Long Short - Term Memory) neurons to model time series, but the LSTM model itself is difficult to handle high-dimensional data. In order to solve this problem, Wang [11] and others chose to combine LSTM and convolutional neural network model CNN (Convolutional Neural Networks) to use CNN's convolutional layer and pooling layer for feature extraction to alleviate the impact of high-dimensional data on LSTM. Training stress.

3 Method

This paper proposes a service system reliability measurement method based on KPIs, and proposes a KPIs feature selection method based on cross-window and GMM, which can effectively extract feature data strongly related to reliability. A reliability measurement model Focal - LightGBM is proposed to measure the reliability of the service system, so that the model can be better trained in unbalanced data sets. A hyperparameter tuning method based on climbing evolutionary strategy algorithm and improved Bootstrap verification is proposed to tune the hyperparameters of the reliability measurement model to further improve the reliability measurement effect of the model.

3.1 Method Framework

The overall method framework of this paper is shown in Fig. 1. Firstly, a KPIs feature selection method based on cross-window and GMM is proposed. This method is used to select the multi-dimensional KPIs data of the service system, and then Kalman filter is used to correct the KPIs data obtained by feature selection, and then according to the modified KPIs features Data and target values build a service system reliability measurement model based on Focal - LightGBM, and propose a hyperparameter tuning method based on climbing evolutionary strategy algorithm and improved Bootstrap verification to perform hyperparameter tuning on the constructed Focal-LightGBM model Excellent, using the tuned Focal-LightGBM model to achieve accurate measurement of service system reliability.

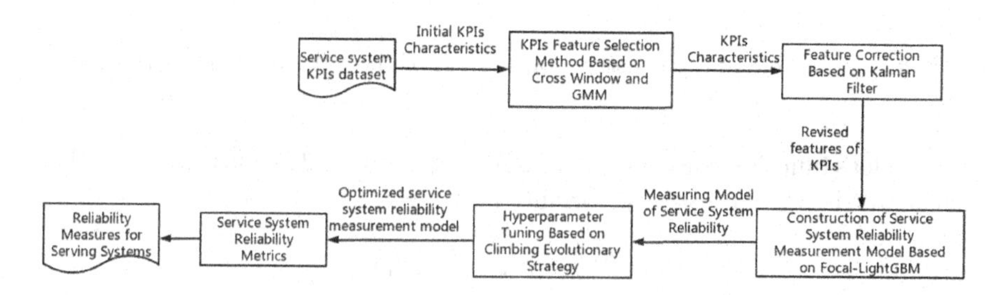

Fig. 1. Method framework

3.2 Reliability Measurement Model Focal - Light GBM

First, based on the exclusive feature bundling algorithm (Exclusive Feature Bundling, EFB), the input KPIs feature data is further compressed, and the mutually exclusive KPI features are bound into one KPI feature, and the unilateral gradient sampling algorithm (Gradient-based One -Side Sampling, GOSS) discard some samples that are not helpful for computing information gain. Then build a decision tree based on the depth-limited Leaf-wise algorithm, and use the histogram algorithm (Histogram) to find the best split point of the tree. Each time you add a tree, you actually learn a new function to fit the last predicted residual. When loss function is less than the set threshold or the tree grows to the maximum height, it stops splitting. Finally, the output corresponding to each tree is added up to obtain the service system reliability measurement value. The lightGBM tree structure construction process is shown in Fig. 2.

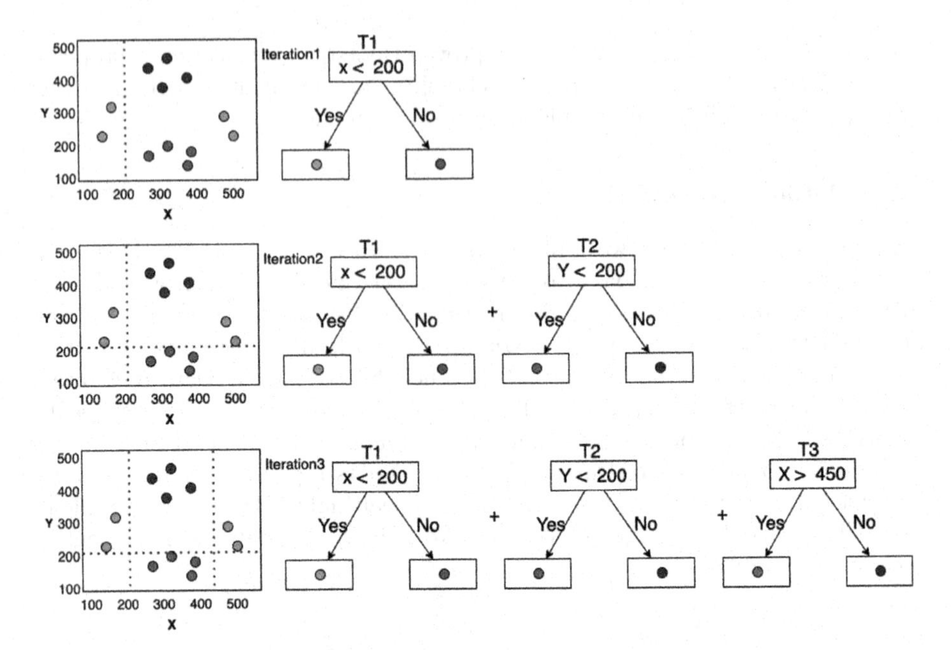

Fig. 2. Example of the construction process of the lightGBM tree structure

Focal loss function consists of a modulating factor and an entropy. The adjustment factor is M_i expressed as formula (1,2):

$$F_i = (y_i \times P_{(\hat{y}_i)} + (1 - y_i) \times (1 - P_{(\hat{y}_i)})) \tag{1}$$

$$M_i = (1 - F_i)^g \tag{2}$$

In the above formula y_i, it represents the real value, $\hat{y}_i$ represents the predicted value, and represents the probability, which is often represented by the sigmod function, which is a constant. For samples with accurate classification, there $F_i \to 1, M_i \to 0$ is, for samples with inaccurate classification $1-F_i) \to 1, M_i \to 1$. That is, compared with other loss functions, focal loss does not change the loss for samples with inaccurate classification, and the loss for samples with accurate classification will become smaller. Overall, it is equivalent to increasing the weight of inaccurate samples in the loss function.

entropy $H(y)$ is shown in formula (3):

$$H(y) = (y_i \times \log_P(\hat{y}_i) + (1 - y_i) \times \log_P((1 - \hat{y}_i))) \tag{3}$$

Focal The overall representation of the loss function is shown in formula (4):

$$obj = \sum_{i=0}^{n} -(\alpha \times y_i + (1 - \alpha) \times (1 - y_i)) \times M_i \times H(y) \tag{4}$$

3.3 Feature Selection of KPIs

KPIs Feature Selection Method Based on Cross Window and GMM. The GMM algorithm is a dimensionality reduction method based on Gaussian distribution. GMM assumes that a probability distribution of any shape can be approximated by a linear combination of multiple single Gaussian distributions. In order to reduce the randomness of GMM, this paper adds the concept of feature cross window on the basis of GMM algorithm, and proposes a KPIs feature selection method based on cross window and GMM. The basic steps of this method are as follows : Set all the features of the KPIs dataset as basic features, copy the basic features as noise features, and randomly shuffle the data order in the noise features, and finally merge the basic features and noise features horizontally to establish a feature cross window (see Fig 3). The role of the noise feature is to create random errors during iterations to avoid the same output for each iteration. Since the basic features are further simplified in each iteration, the size of the intersection window is reduced after each iteration.

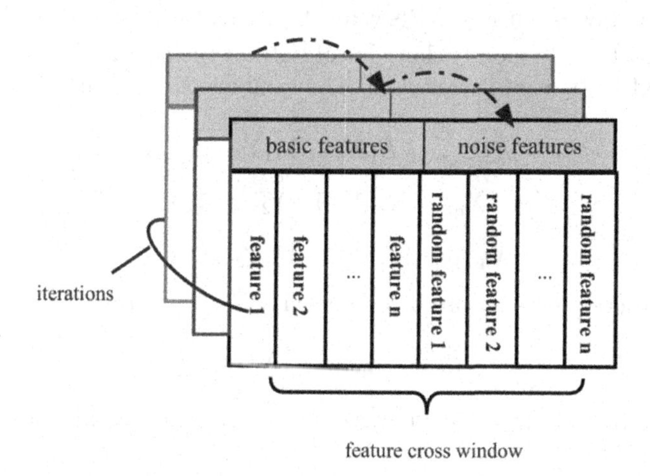

Fig. 3. Structure of Feature Cross Window

Calculate the importance of each feature of the KPIs data set I according to formula (5). In order to reduce the error, take the mean value of multiple calculation results. Since random noise has been created in the cross window, the results of each calculation are different and will not be invalid. operation.

$$I_i = \frac{\sum_{j=1}^{m} \sigma(S_j)[i]}{m} \tag{5}$$

In the above formula i is the feature index, σ is the sample log-likelihood probability of GMM, S is the feature cross window created in step 2, and m is the number of iterations.

According to the formula (6), the features O whose importance is greater than the I mean are selected as the result of feature selection. Since S half of the features in are noise features , and these features are artificially created and do not exist in the actual data set , the output should not include this part of the content.

$$O = \{i|I_i > \frac{\sum_{j=1}^{n} I_j}{n}\} \tag{6}$$

where n is the capacity of S. O select the result for the final feature.

Feature Correction Based on Kalman Filter. There are some discrete values and outliers in the KPIs data after feature selection, and Kalman filter is needed for data correction. The Kalman filter corrects the error by making a linear approximation to the nonlinear equation, which will make the approximation of the system higher and the effect better. The process can be divided into two steps: prediction and correction.

Kalman filtering uses the predicted value of KPIs at the previous moment $\hat{X}_{n-1|n}$ to modify the current KPIs value $\hat{X}_{n|n}$, and then uses the predicted value of the current KPIs to modify the value of KPIs at the next moment $\hat{X}_{n+1|n}$ until the entire KPIs data set is traversed. The symbols of each part are as follows:

$$\hat{X}_{n-1|n} = E(X_{n-1}|Y_1, Y_2, ..., Y_n) \tag{7}$$

$$\hat{X}_{n|n} = E(X_n|Y_1, Y_2, ..., Y_n) \tag{8}$$

$$\hat{X}_{n+1|n} = E(X_{n+1}|Y_1, Y_2, ..., Y_n) \tag{9}$$

The prediction process can be expressed as formula (10):

$$\hat{X}_{n+1|n} = F_n \hat{X}_{n|n} \tag{10}$$

where F_n is the state transition matrix. The correction process can be expressed as formula (11):

$$\hat{X}_{n+1|n+1} = \hat{X}_{n+1|n} + K_{n+1}(Y_{n+1} - H_{n+1}\hat{X}_{n+1|n}) \tag{11}$$

where K_n is the Kalman gain, Y_n is the measurement error, and H_n is the observed value.

3.4 Construction of Reliability Measurement Model Based on Focal - LightGBM

Data Preprocessing. In real scenarios , the service system is affected by various factors, and the collected data may contain a large number of missing values. In order to weaken the data offset, this paper uses linear interpolation to process the missing values of KPIs data. First, the slope is calculated according to the values before and after the missing value, and then the missing KPI value is filled according to the slope. When x_k missing, the slope b is expressed as formula (12):

$$b = \frac{x_{k+1} - x_{k-1}}{t_{k+1} - t_{k-1}} \tag{12}$$

At this point the missing can be imputed as :

$$x_k = x_{k-1} + b \times (t_k - t_{k-1}) \tag{13}$$

Target Value Calculation. According to the formula (14), the reliability value is calculated by using the response time rt, the success rate sr and the number of calls, where cnt and β are constants. α is used to adjust the contribution of the response time index to the reliability value calculation. It is in the range of 0-1, and the default is 0.1. The smaller α the response time index is, the greater the influence on the reliability value calculation is. β is the threshold of the success rate, which is customarily set to 0.5. At that time, $sr < \beta; \psi < 0$, means that the reliability of the service system is low at this time.

$$\psi = e^{-1*(e^{-1*rt}+\alpha)*rt} * (sr - \beta) * cnt \tag{14}$$

the reliability value calculated by the formula (14) has positive and negative points, in order to describe the reliability of the system more clearly, the reliability value is mapped to the range of 0-1, and it is necessary to use the formula (15) for data standardization. After data standardization, the order of magnitude of the reliability index is consistent, which is suitable for comprehensive evaluation.

$$\psi^* = ceil(\frac{\psi - \psi_{min}}{\psi_{max} - \psi_{min}}) \tag{15}$$

$\psi^*$is the service system reliability value.

Construction of Reliability Measurement Model. Input the processed feature data and target values into the Focal-LightGBM model for training. During the training process, the Focal - LightGBM model will continuously reduce the residual error, and the measurement result is getting closer and closer to the real value.

3.5 Model Hyperparameter Tuning Based on Evolution Strategy

Improved Bootstrap Validation. Bootstrap verification first samples the training set and test set from the data set in proportion, and then uses these data to verify the degree of adaptation between the parameter sample and the model sr (it is generally accepted that the accuracy of the parameter instantiation model is used to measure the degree of adaptation [12,13]), after the verification is completed, the output sr mean is the verification result. In order to solve the problem of uneven sampling of data set samples in the traditional bootstrap method, this method adds a self-increasing random seed in each round of verification, and ensures the uniformity of sampling when the process is controllable by establishing a pseudo-random distribution. The improved bootstrap check is shown in Algorithm 1.

Algorithm 1. bootstrap validation

Require: hyperparameter sample $param$, target model$model$, data set $data$
Ensure: How well the hyperparameters fit the model sr

1: $sr \Leftarrow 0, seed \Leftarrow 0$
2: **for** $j = 1 \rightarrow M$ **do**
3: md ⇐model($param$)
4: set_seed($seed$)
5: $test\_set$ ⇐sample(len($data$),len($data$) $*$ $scale$) $scale$ scale Indicates the split ratio of the train-test set
6: $train\_set \Leftarrow data - test\_set$
7: md.fit($train\_set$)
8: $sr \Leftarrow sr + md$.evaluate($test\_set$)
9: $seed \Leftarrow seed + 1$ random seed slide forward
10: **end for**
11: **return** sr/M

Climbing Evolutionary Strategy Algorithm. In the evolutionary strategy algorithm, a complete iteration is called a generation, and a candidate solution is called an individual. The new candidate solution generated by each iteration is called offspring, and the solution obtained by selection and used to generate offspring is called parent. The climbing evolutionary strategy is a greedy search algorithm for local optimization, which essentially involves iteratively adjusting a normal distribution for optimization. The first generation of parents is a hyperparameter individual randomly selected from a hyperparameter set according to a normal distribution.

The mutation operator σ is used to generate offspring through the parent generation, and the mutation operator itself can also be mutated, σ as shown in formula (16).

$$\sigma \in (0, s] \tag{16}$$

Among them S is the maximum capacity of the hyperparameter set. The larger the mutation operator, the larger the step size in the search process, the larger the scanning range in the iteration, and the smaller the possibility of falling into a local optimal solution. Therefore, the first generation of mutation The operator is set to S. The offspring is generated by the mutation of the parent, and the generation method of the offspring is shown in formula (17).

$$x = m_t + \sigma_t y, y \sim N(0, 1) \tag{17}$$

Among them m_t,σ_t respectively represent the parent generation and mutation operator in the t-th iteration, which y obey the standard normal distribution with 0 as the mean and 1 as the standard deviation. It can be found from the formula (14) that the y positive or negative of, determines the individual evolution direction. The σ_t greater the difference, the greater the difference between the parent and offspring, which σ_t determines the speed of individual evolution.

During the evolution process, compare the hyperparameters of the offspring with the hyperparameters of the parent, and select the one with the higher bootstrap verification index as the parent of the next iteration, otherwise discard it directly, and adjust the mutation operator according to the comparison result σ. According to the "1/5 successful rule" [14] criterion, if less than 1/5 of the offspring are stronger than the parent, it indicates that the search process is about to converge. At this time σ_t, the decrease will make the generated offspring closer to the global optimal solution; otherwise there is no convergence, increase it at this time σ_t to further speed up the convergence speed. The calculation method in the decision-making process is shown σ_t in formula (18).

$$\sigma = \sigma \times exp(\frac{1}{3} \times \frac{ps - p\_{target}}{1 - p\_{target}}) \tag{18}$$

Among them ps, it represents the ratio of the children who are stronger than the parent to the total number of children, p_target which is the convergence threshold, which is set to 1/5. Evolutionary Algorithm 2 is shown.

Algorithm 2. Hyperparameter individual selection

Require: parent hyperparameters *parent*,kid hyperparameters *kid*, mutation operator σ

Ensure: next parent hyperparamoters *parent*, next kid hyperparameters σ

1: $fp \Leftarrow$bootstrap($parent$) fp indicates the result of the parental test
2: $fk \Leftarrow$bootstrap(kid) fk indicates the result of the parental test
3: $p\_target \Leftarrow 1/5$ p_target is used to calculate σ
4: **if** $fp < fk$ **then** fp against with fk
5: $\quad$ $parent \Leftarrow kid$ the winner becomes the new parent
6: $\quad$ $ps \Leftarrow 1$ ps is used to calculate σ
7: **else**
8: $\quad$ $ps \Leftarrow 0$
9: **end if**
10: σ calculate by formula (15)
11: **return** $parent, \sigma$

Early studies on the convergence of evolutionary algorithms [15,16] show that if the number of iterations of the algorithm is large enough, and the individual evolves sufficiently, theoretically the offspring and the parent can converge to the optimal solution at the same time, which is equivalent to that from multiple The initial point is optimized for gradient in parallel. When the initial point is dense enough, the global optimal point can be found. Therefore, this method judges whether the optimization is completed by comprehensively judging whether the number of algorithm iterations exceeds the maximum threshold and whether the mutation operator is infinitely approaching 0 (that is, whether the individual loses the ability to mutate).

When the evolutionary algorithm converges, the offspring and the parent converge to the optimal solution at the same time, and the hyperparameters of the offspring are equal to the hyperparameters of the parent. At this time, the hyperparameters of the offspring or the hyperparameters of the parent are output as the tuning result.

4 Experiment

In this part, we have done a lot of experiments to verify the effectiveness of the method proposed in this paper, and selected the multi-dimensional KPIs data set provided by the 2020 AIOps competition "Microservice Application System Fault Discovery and Root Cause Location" to evaluate the method in this paper. A detailed description of the datasets used can be found in Table 1.

Table 1. AIOps -2020 Dataset Description.

describe	value
total data volume	5,617
training data volume	4,500
test data volume	1,117
data dimension	2,080
target value range	0-10

Firstly, the effectiveness of the feature selection method based on cross-window and GMM proposed in this paper is verified, then the effectiveness of the hyperparameter tuning method based on climbing evolutionary strategy proposed in this paper is verified, and finally the reliability measurement method based on KPIs proposed in this paper is verified. effectiveness.

4.1 Feature Selection Based on Cross Window and GMM

In order to verify the effectiveness of the feature selection method based on cross-window and GMM, the method proposed in this paper, the original GMM algorithm, and the feature selection method proposed in [17,18] were used for experiments.

This paper uses widely used indicators, namely accuracy rate, mean square error loss (MSE) and coefficient of determination (R2-score) to evaluate the effectiveness of the proposed method and the comparison methods in feature selection.

Accuracy is the ratio of the number of correct samples predicted by the model to the total number of samples. See the formula (19) for the calculation method, where TF is the number of samples that the model predicts correctly. TP is the number of samples that the model predicts incorrectly. The higher the accuracy rate, the better the prediction effect of the model.

$$precision = \frac{TP}{TP + FP} \tag{19}$$

MSE is the mean of the squared differences between actual and predicted values. See the formula (20) for the calculation method, where $N$is the number of samples predicted by the experiment, P_i is the actual reliability value in the sample, and R_i is the reliability value predicted by the experimental method. The smaller the mean square error loss, the better the prediction effect of the model.

$$MSE = \frac{1}{N}\sum_{i=1}^{N}(P_i - R_i)^2 \tag{20}$$

R2-score reflects the proportion of all variation of the dependent variable that can be explained by the independent variable through the regression relationship. See the formula (21) for the calculation method. Among them, N is the number of experimentally predicted samples, P_i is the reliability value predicted by the experimental method, and R_i is the actual reliability value in the sample.$\hat{R}_i$ Indicates the average of the actual reliability values. The closer the R2 score is to 1, the better the independent variable explains the dependent variable in the regression analysis, and the better the model learning effect is.

$$R2\_score = 1 - \frac{\sum_{i=1}^{N}(P_i - R_i)^2}{\sum_{i=1}^{N}(P_i - \hat{R}_i)^2} \tag{21}$$

According to the features extracted by different feature selection algorithms, reliability measurement models based on decision tree [3], GBDT algorithm [19] and AdaBoost algorithm [20] were respectively established for experiments. For the sake of fairness, the parameter settings of all feature selection algorithms are exactly the same, and all the experimental data below are the average of the results obtained from 5 consecutive experiments. The comparison of different models is shown in Table 2-4.

It can be seen from Table 2-4 that the feature selection results of the method proposed in this paper are better than the original GMM algorithm and other common feature selection methods in all indicators. However, the feature selection method based on principal component analysis proposed in the literature [18] is ineffective, which shows that the principal component is not the core

Table 2. Decision tree algorithm.

describe	This paper	GMM	Literature [17]	Literature [18]
Accuracy	**0.93230**	0.90715	0.91102	0.56866
MSE	**0.06769**	0.09284	0.08897	0.19402
R2 -score	**0.90405**	0.86841	0.87389	0.80270

Table 3. GBDT algorithm

describe	This paper	GMM	Literature [17]	Literature [18]
Accuracy	**0.82011**	0.80851	0.80464	0.67117
MSE	**0.17988**	0.19148	0.20116	0.41392
R2 -score	**0.74505**	0.72860	0.71489	0.41334

Table 4. AdaBoost algorithm

describe	This paper	GMM	Literature [17]	Literature [18]
Accuracy	**0.91682**	0.89748	0.90715	0.79497
MSE	**0.08317**	0.10251	0.09284	0.50676
R2 -score	**0.88211**	0.85470	0.86841	0.28175

element of multidimensional KPIs data. Density-based dimensionality reduction methods (such as literature [17]) and probability distribution-based dimensionality reduction methods (such as the method proposed in this paper) can well select KPIs indicators that are strongly related to reliability from KPIs data, and this paper proposes The method also introduces the concept of feature cross-window, and the feature selection is more accurate by repeatedly eliminating errors in iterations.

4.2 Hyperparameter Tuning Based on Climbing Evolutionary Strategy

In order to verify the effectiveness of the hyperparameter tuning method of the reliability measurement model of service system based on climbing evolutionary strategy proposed in this paper, the methods of this paper, literature [21], literature [22] and literature [23] were used to perform hyperparameter tuning of several common reliability metric models. Decision tree algorithm [3], KNN algorithm [24] and XGBoost algorithm [25] are used as reference models for experimentation (Tables 5, 6 and 7).

Hyperparameter value range:

Table 5. Decision tree algorithm hyperparameters and value range

parameter symbol	parameter explanation Ranges	parameter	symbol
criterion	feature selection method	gini, entropy	criterion
splitter	Feature division point selection method	best,random	splitter
max_depth	the maximum depth of the tree	[10,100]	max_depth
min_samples_split	The minimum number of node subdivision samples	[2,20]	min_samples_split
min_samples_leaf	The minimum number samples for a leaf node	[1,20]	min_samples_leaf

Table 6. KNN algorithm hyperparameters and value range

parameter symbol	parameter explanation	Ranges	parameter symbol
n_neighbors	K value	[1, 10]	n_neighbors
algorithm	radius nearest neighbor	auto,ball_tree, kd_tree,brute	algorithm
leaf_size	Threshold for the number of leaf nodes	[10,100]	leaf_size

Table 7. KNN algorithm hyperparameters and value range

parameter symbol	parameter explanation	Ranges	parameter symbol
min_child_weight	The smallest sample weight sum in the child node	[0,20]	min_child_weight
subsample	Weak learner training sample proportion	[0,1]	subsample
max_depth	the maximum depth of the tree	[10,100]	max_depth
colsample_bytree	Ratio to random sample features when building weak learner	[0,1]	colsample_bytree
scale_pos_weight	Regulator	[0,1]	scale_pos_weight

Evaluation Indicators: The Tuning Performance Indicator (TPI) of hyperparameter tuning methods is compared by comparing the measured accuracy of models instantiated by hyperparameter sets [26]. See the formula (22) for the calculation method, where TP is the number of samples that the model measures correctly, and FP is the number of samples that the model measures incorrectly.

$$TPI = precision = \frac{TP}{TP + FP} \tag{22}$$

In order to avoid the impact of random errors on the experiment, all the following experimental data are the average value of the results obtained from five consecutive experiments. The comparison of different tuning methods is shown in Table 8 and 9.

Table 8. TPI comparison of each method under the AIOps -2020 dataset

	This paper	Literature [21]	Literature [22]	Literature [23]
decision tree	**0.89361**	0.87427	0.83365	0.87427
KNN	**0.67504**	0.67117	0.67117	0.67117
XGBoost	**0.94777**	0.94390	0.94390	0.92843

Table 9. Comparison of parameter adjustment time of each method under the AIOps -2020 dataset

	This paper	Literature [21]	Literature [22]	Literature [23]
decision tree	**193.84**	4441.69	7140.72	01162.665
KNN	**316.28**	52196.62	14731.85	2086.105
XGBoost	**1611.925**	48767.25	1231748.21	14285.07

From Tables 8 and 9 that in terms of TPI, experiments on all models show that the hyperparameter tuning method based on the climbing evolutionary strategy proposed in this paper is superior to other hyperparameter tuning methods. This is because the improved Bootstrap method in this paper can be verified while maintaining the uniform distribution of sample data, so the tuning effect is better. However, the TPI obtained by each parameter tuning method on different models is quite different (as shown in Table 8), which indicates that the model itself determines the upper limit of TPI. When the model has not been effectively trained on the training data set, by adjusting the hyperparameters It has been impossible to significantly increase TPI.

In terms of parameter tuning time, the hyperparameter tuning method based on the climbing evolutionary strategy proposed in this paper is significantly better than other hyperparameter tuning methods. This shows that compared with all other tuning methods, the tuning method proposed in this paper has the least iteration rounds, the fastest convergence speed, and the highest tuning efficiency. In addition, as the complexity of the model increases, the time efficiency of all parameter tuning methods will decrease (as shown in Table 9). Improve tuning efficiency.

To sum up, the hyperparameter tuning method based on the climbing evolutionary strategy proposed in this paper has obvious advantages compared with other hyperparameter tuning methods in terms of TPI and time efficiency.

4.3 Reliability Measurement Performance Evaluation

The characteristics of KPIs selected through the above experiments are shown in Table 10, and the hyperparameter values of the optimized Focal - LightGBM model are shown in Table 11.

Table 10. Comparison of parameter adjustment time of each method under the AIOps -2020 dataset

KPIs name	Category of KPIs	Number of KPIs
db	Oracle database	142
redis	message middleware	110
os	operating system	133
docker	run container	13

Table 11. Optimized model hyperparameter values

parameter symbol	parameter explanation	value
min_child_weight	The smallest sample weight sum in the child node	0.3
learning_rate	learning rate	0.1
max_depth	the maximum depth of the tree	30
boosting	weak learner type	"gbdt"
colsample_bytree	Ratio to random sample features when building a weak learner	0.8

In order to verify the performance of the reliability measurement method proposed in this paper, Place the method in this paper in Fig. 4(a) and the original LightGBM algorithm in Fig. 4(b), literature [27] in Fig. 4(c), literature [28] in Fig. 4(d), literature [29] in Fig. 4(e), literature [30] in Fig. 4(f) were used for reliability measurement experiments under the same conditions. The experiments in this section still use the evaluation indicators precision, MSE, and R2-score in Chap. 4.1 for comparison. Literature [27] is a probabilistic label tree algorithm for multi-label classification, and literature [28] is a gradient boosting algorithm based on natural gradients. Literature [29] is an ensemble learning framework for large-scale highly imbalanced classification, and literature [30] is a probabilistic graphical model based on Bayesian theory. The red scattered points in the figure indicate the situation of prediction errors.

Table 12. Comparison of the method in this paper with other reliability measurement methods

	Accuracy	MSE	R2 - score	training time/s
Method in this paper	0.94197	0.05802	0.91775	7.98055
Original LightGBM algorithm	0.92263	0.077369	0.89034	6.64894
Literature [27]	0.72920	738.419	−1045.563	1.28014
Literature [28]	0.82011	0.18568	0.73682	1582.27201
Literature [29]	0.88201	0.12379	0.82455	22.62347
Literature [30]	0.85686	703.50096	−996.07338	116.60224

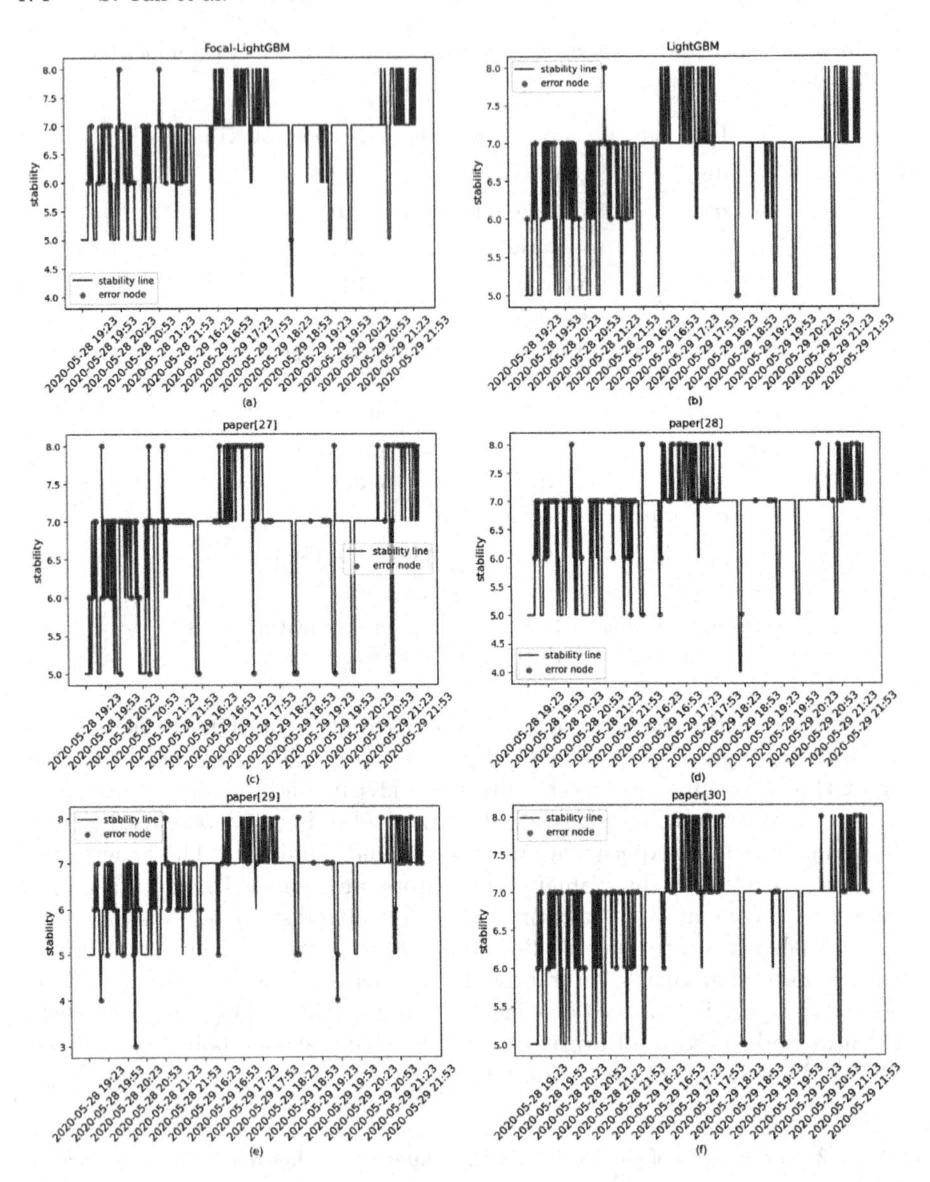

Fig. 4. Comparison of prediction results between this method and other reliability measurement methods

From Fig. 4 that the reliability measurement method proposed in this paper has 30 measurement error points, while the original LightGBM algorithm has 40, the literature [27] is 140, the literature [28] is 93, and the literature [29] is 61, and literature [30] is 74. This shows that after data selection, correction and model tuning, Focal - LightGBM has the least number of measurement errors compared with other reliability measurement methods, and the measurement results are closer to the real value, especially compared to the original LightGBM model, in terms of accuracy, MSE, R2 - score indicators have all improved. The MSE and R2 -score indicators of literature [27,30] are abnormal, which shows that these models have poor measurement effect and low recall rate, which are difficult to apply to actual scenarios . The accuracy rate of literature [29] is relatively high, but in extreme cases, the measurement value deviates significantly from the real value (for example, the actual value at 20:23 on May 28, 2020 is 7, and the measurement value is 3). Maintenance personnel make wrong judgments about the state of the service system.

From Table 12, under the premise that the accuracy rate, MSE, and R2-score indicators are all dominant, the time consumption of the method in this paper is lower than most reliability measurement methods. The literature [27] has the highest time efficiency (1.28014 s), but its low time consumption is at the expense of low accuracy (0.72920) . On the whole, the method in this paper achieves a balance between accuracy and time efficiency, and is more suitable for practical application scenarios.

Based on the above analysis, it can be shown that the method in this paper can have both high accuracy and high time efficiency of service system reliability measurement. This is mainly because the method in this paper is based on the improved GMM feature selection method and the Kalman filter at the data processing level, which can effectively extract special indicators that are strongly related to the reliability of the service system. The fluctuation is obvious, the false alarm rate is low, and it is suitable for the reliability measurement of the service system. In terms of reliability measurement model construction, considering the impact of uneven sample distribution on reliability measurement in the real environment of the service system, The focal loss function desensitizes the model measurement and sample distribution and improves the measurement accuracy of the model. In terms of hyperparameter tuning of the reliability measurement model, the hyperparameter tuning method based on climbing evolutionary strategy is used to optimize the hyperparameters of the model to further improve the measurement accuracy. In summary, the method in this paper has obvious advantages over other methods in the reliability measurement of service systems.

5 Conclusion

This paper proposes a reliability measurement method based on KPIs, which can measure the reliability of the service system in real time according to the status of various KPIs indicators in the service system. On the one hand, feature selection is performed on KPIs data by adding feature cross windows to the GMM algorithm, and the KPIs indicators that are strongly related to the reliability of the service system are screened out, and then the Kalman filter is used to correct the selected KPIs indicators On the other hand, considering the characteristics of the service system KPIs data itself, the Focal loss function is used to construct the reliability measurement model of the service system. Finally, in order to further improve the performance of the model, a hyperparameter tuning method is designed by taking advantage of the characteristics of the evolutionary strategy algorithm. Experimental results on public datasets show that the improvements in the above three aspects proposed in this paper are superior to most of the same types of methods in their respective fields. In general, the reliability measurement method proposed in this paper improves the accuracy and time efficiency of reliability measurement, and helps the operation and maintenance personnel to grasp the real-time operation status of the service system.

Acknowledgements. This work is supported by Liaoning Province Applied Basic Research Program Project (Grant No. 2023JH2/101300195).

References

1. Zhao, Y., Zhang, X., Shang, Z., Cao, Z.: A novel hybrid method for kpi anomaly detection based on VAE and SVDD. Symmetry **13**(11), 2104 (2021)
2. Lin, T., Goyal, P., Girshick, B., He, K.: Piotr Dollár: focal loss for dense object detection. IEEE Trans. Pattern Anal. Mach. Intell. **42**(2), 318–327 (2020)
3. Gao, F., Tan, S., Shi, H., Tao, Y., Song, B.: Improved ensemble feature selection based on DT for KPI prediction. IEEE Access **9**, 136861–136871 (2021)
4. S. Allagi, P. Surasura . Predicting Reliability of Web Services Using Hidden Markov Model. FICTA (1) pp. 169-179 (2020)
5. Wang, H., Fei, H., Yu, Q., Zhao, W., Yan, J., Hong, T.: A motifs-based Maximum Entropy Markov Model for realtime reliability prediction in System of Systems. J. Syst. Softw. **151**, 180–193 (2019)
6. Cuong, N., Huynh, Q.: New non-homogeneous Poisson process software reliability model based on a 3-parameter S-shaped function. IET Softw. **16**(2), 214–232 (2022)
7. Zheng, Z., Li, X., Tang, M., Xie, F., Lyu, M.: Web Service QoS Prediction via Collaborative Filtering: A Survey. IEEE Trans. Serv. Comput. **15**(4), 2455–2472 (2022)
8. Song, Y.: Web service reliability prediction based on machine learning. Comput. Stand. Interfaces **73**, 103466 (2021)
9. Li, J., Wu, H., Chen, J., He, Q., Hsu, C.: Topology-aware neural model for highly accurate QoS prediction. IEEE Trans. Parallel Distributed Syst. **33**(7), 1538–1552 (2022)
10. Yeh, W., Du, C., Tan, S., Forghani-elahabad, M.: Application of LSTM based on the BAT-MCS for binary-state network approximated time-dependent reliability problems. Reliab. Eng. Syst. Saf. **235**, 108954 (2023)
11. Wang, H., Yang, Z., Yu, Q., Hong, T., Lin, X.: Online reliability time series predation via convolutional neural network and long short term MEM over for service-oriented systems. Knowl. Based Syst. **159**, 132–147 (2018)
12. Acharki, N., Bertoncello, A., Garnier, J.: Robust prediction interval estimation for Gaussian processes by cross-validation method. Comput. Stat. Data Anal. **178**, 107597 (2023)
13. Soper, D.: Hyperparameter optimization using successful halving with greedy cross validation. Algorithms **16**(1), 17 (2023)
14. I.Rechenberg 1973. Evolutions strategy – Optimierung technischer Systeme nach Prinzipien der biologischen Evolution, Frommann-Holzboog

15. G. Tollo, F. Lardeux, J. Maturana, F. Saubion: An Experimental Study of Adaptive Control for Evolutionary Algorithms. CoRR abs/1409.1715 (2014)
16. A. Torn, A. Zilinskas: Global Optimization. Lecture Notes in Computer Science 350, Springer 1989, ISBN 3-540-50871-6
17. H. Du, S. Chen, H. Niu, Y. Li: Application of DBSCAN clustering algorithm in evaluating students' learning status. CIS 2021: 372-376
18. Xia, Z., Chen, Y., Xu, C.: Multiview PCA: a methodology of feature extraction and dimension reduction for high-order data. IEEE Trans. Cybern. **52**(10), 11068–11080 (2022)
19. F. JH. Greedy Function Approximation: A Gradient Boosting Machine[J]. Annals Stat. **29**(5):1189-1232 (2001)
20. Freund, Y., Schapire, R.E.: A decision-theoretic generalization of on-line learning and an application to boosting. J. Comput. Syst. Sci. **55**(1), 119–139 (1997)
21. Ahmad, G., Fatima, H., Ullah, S., Saidi, A., Imdadullah, A.: Efficient medical diagnosis of human heart diseases using machine learning techniques with and without GridSearchCV. IEEE Access **10** 80151–8017 3 (2022)
22. akkala, H.R., Khanduri, V., Singh, A., Somepalli, S.N., Maddineni, R., Patra, S: Kyphosis Disease Prediction with help of RandomizedSearchCV and AdaBoosting. ICCCNT, pp. 1–5 (2022)
23. Kassem, M., Hadidy, M.: Optimal multiplicative Bayesian search for a lost target. Appl. Math. Comput. **247**, 795–802 (2014)
24. Loredo-Pong, V., Morales-Rodriguez, M.L., Patricia Diaz-Zavala, N., Rangel-Valdez, N., Sosa-Sevilla, J.E.: A Design and analysis of classification models for the gelification of alkoxybenzoates using the kNN algorithm. Int . J. Comb. Optim. Probl. Inform. **13**(2), 58–64 (2022)
25. Chen, T., Guestrin, C.: XGBoost: a scalable tree boosting system. In: KDD pp. 785–794 (2016)
26. Duan, K., Keerthi, S., Poo, A.: Evaluation of simple performance measures for tuning SVM hyperparameters. Neurocomputing **51**, 41–59 (2003)
27. Wydmuch, M., Jasinska-Kobus, K., Thiruvenkatachari, D., Dembczynski, K.: Online probabilistic label trees. In: AISTATS, pp. 1801–1809 (2021)
28. Duan, T., et al.: NGBoost: natural gradient boosting for probabilistic prediction. In: ICML, pp. 2690–2700 (2020)
29. Kokel, H., Odom, P., Yang, S., Natarajan, S.: A unified framework for knowledge intensive gradient boosting: leveraging human experts for noisy sparse domains. In: AAAI, pp. 4460–4468 (2020)
30. Prokhorenkova, L., Gusev, G., Vorobev, A., Dorogush, A., Gulin, A.: CatBoost: unbiased boosting with categorical features. In: NeurIPS, pp. 6639–6649 (2018)

A Seasonal Decomposition-Based Hybrid-BHPSF Model for Electricity Consumption Forecasting

Xiaoyong Tang(✉), Juan Zhang, Ronghui Cao, Wenzheng Liu, and Li Yang

School of Computer and Communication Engineering,
Changsha University of Science and Technology, Changsha 410114, China
tangxy@csust.edu.cn

Abstract. A reasonable balance between energy consumption and production may be achieved with accurate electricity consumption forecasts, which assist in laying down operational expenses and resource waste. However, electricity consumption data exhibits nonlinearity, high volatility, and susceptibility to various factors. Existing forecasting schemes inadequately account for these traits, resulting in weak performance. This paper proposes a novel hybrid model (Hybrid-BHPSF) based on seasonal decomposition to address this issue. The proposed model incorporates a new BHPSF algorithm that effectively captures data patterns with noticeable variations. Initially, the electricity consumption data is segmented into multiple subsequences with distinct characteristics, and the BHPSF algorithm predicts the subsequences exhibiting clear trends. Subsequently, due to the ability of LightGBM to handle flat and nonlinear data, it is embedded into the model to process the remaining sequences that fluctuate irregularly within a certain range. We have evaluated our proposed model using four distinct datasets, and the results indicate that it outperforms existing models across different prediction horizons.

Keywords: Electricity demand prediction · Artificial Intelligence · Seasonal and trend decomposition · Big data

1 Introduction

The electric power industry profoundly influences national production, the livelihood of residents, and the overall economy [1]. Its primary role is ensuring a consistent supply of sustainable, reliable, high-quality electrical energy. With the evolving industrial society, there are increasing demands for more reliable and higher-quality power supply [2]. However, electricity is special among commodities due to its inability to be easily stored. This means that accurate power demand forecasting is essential [3]. Therefore, it is critical to accurately forecast electricity demand based on historical demand patterns for energy management, policy development, and investment decisions [4].

The demand for electricity in various industries is rapidly increasing with the advent of the third information wave. As a result, power-related data is becoming

Z. Tari et al. (Eds.): ICA3PP 2023, LNCS 14491, pp. 478–490, 2024.
https://doi.org/10.1007/978-981-97-0808-6_28

increasingly large due to the popularity of information technology. Furthermore, the development of various artificial intelligence technologies provides an effective tool for analyzing big data [5–7]. Extracting necessary information from large-scale electric power data using appropriate technology for accurate power consumption prediction is a popular research direction [8].

Due to the importance of electricity demand, various methods have been proposed [9,10]. Electricity consumption data is often non-stationary, nonlinear, and influenced by factors like seasonality and holidays. Hence, many studies treat it as time series data and build prediction models based on historical data and relevant factors. Zheng et al. [11] proposed a multi-scale prediction algorithm based on the Time-Frequency Variational Autoencoder. Alvarez et al. [12] proposed a pattern sequence similarity-based prediction algorithm (PSF) that searches for certain patterns across the dataset using clustering methods to predict time series. Perez et al. [13] proposed a scalable bigPSF that uses temporal proximity weighting to obtain the final prediction results.

Furthermore, combined predictive models have advanced significantly in recent years [14,15], enhancing final performance through the amalgamation of multiple model strengths. Guo et al. [16] proposed the hybrid ARIMA-SVR model for building electricity consumption prediction. Lu et al. [17] developed a hybrid model using ICEEMDAN, MOGWO, and SVM for predicting daily electricity demand during the pandemic. Xia et al. [18] developed a hybrid framework integrating innovative data preprocessing, multi-objective optimization, and deep neural networks. Zulfiqar et al. [19] proposed a hybrid model merging multivariate empirical modal decomposition, adaptive differential evolution, and SVM.

Accurate short-term electricity forecasting helps balance power systems, preventing supply-demand imbalances and electricity interruptions. Medium-term electricity forecasting helps plan power supply in advance, reducing supply shortages. Long-term electricity forecasting supports strategic power system planning, promoting the transition to clean energy sources, and reducing greenhouse gas emissions. Hence, short-term, medium-term, and long-term electricity forecasting has its unique significance and roles. Currently, most electricity forecasting research focuses on short-term predictions [14,15,20]. For longer-range forecasting, two common strategies include direct multi-step [21,22] and iterative multi-step [23] methods. Direct multi-step predicts multiple future time units in a single step but is unsuitable for long-term predictions. In contrast, iterative multi-step predicts the next step by adding the predicted value of the previous step to the training set, but this approach may lead to error accumulation.

Considering the limitations revealed in the aforementioned analysis of various models, we propose an innovative hybrid model to enhance prediction accuracy. Initially, we perform data decomposition. In contrast to previous research, our aim is for each decomposed sequence to exhibit unique characteristics, rather than being split into multiple stable sequences. Subsequently, we model based on these distinctive characteristics to maximize the advantages of each model and achieve an effective model combination. In general, this paper proposes a novel hybrid model (Hybrid-BHPSF) for forecasting electricity consumption across different ranges. The main contribution can be summarized as follows:

(1) We design a power time series prediction method called the BHPSF algorithm. It improved the original PSF with Bayesian optimization and hierarchical clustering, and considers the relationship between parameters.
(2) We propose a novel hybrid model (Hybrid-BHPSF). This model first extracts the ingredients related to the trend in the data and uses the BHPSF algorithm for accurate prediction. Then it removes the seasonal components, and finally embeds the LightGBM to handle the remaining ingredients in the data.
(3) We conduct experiments on the four real datasets. The experimental results show that Hybrid-BHPSF shows a better effect than other related algorithms in each forecast range.

The rest of the paper is organized as follows: Sect. 2 analyzes the dataset characteristics. Section 3 describes the proposed prediction model. Section 4 presents experimental results. Finally, Sect. 5 summarizes the paper.

2 Dataset

Electricity data is closely related to our daily life and the development of society, so it often shows some obvious characteristics with changes in time. This article analyzes the power data sets in multiple regions, showing that most of the data sets show a certain rule every day. As shown in Fig. 1, the daily electricity curve shows a clear peak low valley characteristics. Therefore, we use the daily electricity consumption as a sample, and the electricity consumption of each hour as the specific feature of the sample, and each sample contains 24 features. Over time, although the changes in the amount of electricity in different regions are different, they generally contain irregular periodicity and trend. This is because the electricity consumption is closely related to the local climate environment and social development, and it is also affected by some special events.

Considering the complex nature of electricity consumption data, proper preprocessing is vital to mitigate its complexity. In this paper, we employ the classical time series decomposition method to decompose the original data into three sub-series: the trend component, which captures long-term development trends; the seasonal component, which reflects cyclic fluctuations in fixed periods; and the residual component, which represents the irregularities after trend and seasonal components removal.

3 Proposed Model

Our model primarily decomposes the input data into various components and then builds models specifically targeting each component. Subsequently, the outputs from these individual models are combined to obtain the final prediction.

Next, we first introduce the BHPSF algorithm and then systematically describe the proposed Hybrid-BHPSF model.

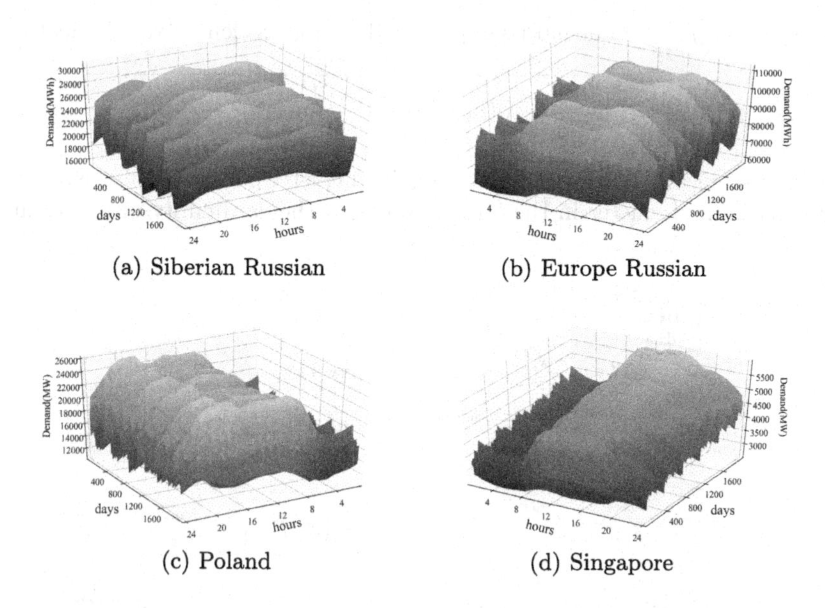

(a) Siberian Russian

(b) Europe Russian

(c) Poland

(d) Singapore

Fig. 1. The characteristics of electricity consumption datasets.

3.1 The BHPSF Algorithm

PSF is an effective general-purpose multi-output approach, which is specially used to handle the time sequence of any length predictive range. We develop the BHPSF algorithm based on this foundation to enhance the accuracy of power big data prediction. BHPSF mainly includes the following steps.

(1) The data used for clustering is standardized. Because the dimensions of datasets may be different, to minimize the influence of dimensions and improve the accuracy and speed of prediction, normalization is performed using Eq. 1.

$$X' = \frac{X - X_{min}}{X_{max} - X_{min}} \tag{1}$$

(2) Hierarchical clustering is used to group and label data. Hierarchical clustering builds a binary tree by measuring category similarity, is robust to initial value selection, and accommodates clusters of any shape. Here, we use divisive hierarchical clustering, initially grouping all samples into one cluster and iteratively dividing them based on predefined rules until each forms a single cluster or a termination condition is met. We calculate the distance between samples using Eq. 2 and apply Eq. 3 to split new clusters.

$$d(x, y) = \sqrt{\sum_{i=1}^{n} (x_i - y_i)^2} \tag{2}$$

where, x and y represent data samples, and x_i and y_i denote sample features.

$$d(u,v) = \sqrt{\frac{|v|+|s|}{T}d(v,s)^2 + \frac{|v|+|t|}{T}d(v,t)^2 - \frac{|v|}{T}d(s,t)^2} \tag{3}$$

where, $T = |v|+|s|+|t|$, u is the newly joined cluster consisting of clusters s and t, v is an unused cluster in the forest, and $|*|$ is the cardinality of its argument.

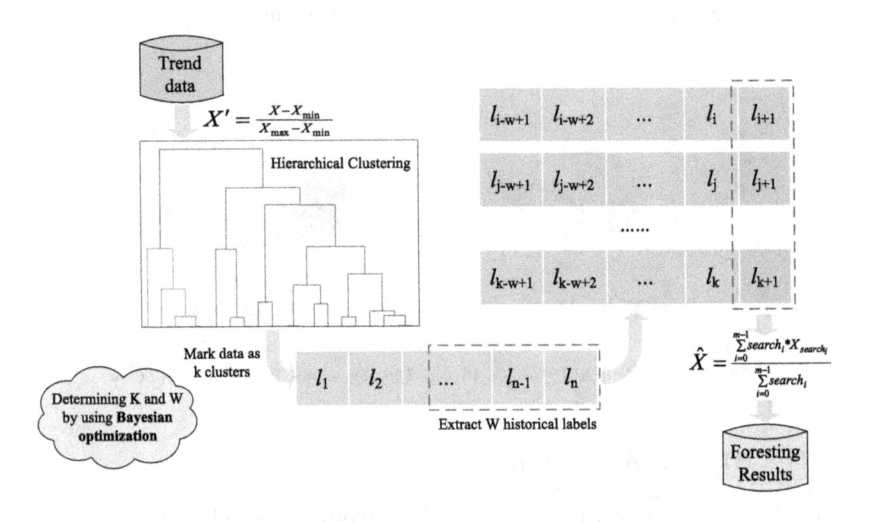

Fig. 2. Framework of the BHPSF algorithm.

Following the application of hierarchical clustering, a sequence S is produced, with each element representing the cluster number to which the corresponding sample belongs. The length of S is equal to the number of samples in the original dataset, and the index of each element in S corresponds to the index of the corresponding sample in the original dataset.

(3) Bayesian optimization is utilized to select the optimal hyperparameters K and W. In the PSF algorithm, the choice of W is closely related to K, and the optimal window size W corresponding to various clustering numbers K is also different, so K and W should be considered together. However, traditional grid search is computationally intensive and slow to converge, while random search may overlook crucial values. To address this, we employ Bayesian optimization in this paper to find the best K and W combination.

(4) The pattern strings are extracted, searched, and then weighted to average the searched values to get the predicted results. The last W values of S form the pattern sequence sq, which is searched within S to find a matching sequence sq'. Once the matching is successful, the index of the value immediately following sq' is recorded in a list called *search*. If *search* is empty, W minus one, and the process is reiterated. Otherwise, the corresponding actual values

in the historical data X are retrieved using the index numbers in *search*, and a weighted average can be calculated using Eq. 4 to obtain the final prediction.

$$\hat{X} = \frac{\sum_{i=0}^{m-1} search_i * X_{search_i}}{\sum_{i=0}^{m-1} search_i} \tag{4}$$

where, m is the count of successful matches of sq' in S, and $\hat{X}$ is the final prediction result. $search_i$ denotes the index of the target value in S, while X_{search_i} is the actual historical data sample value corresponding to index $search_i$.

3.2 The Hybrid-BHPSF Model

The Hybrid-BHPSF model is depicted in Fig. 3. Next, we provide a systematic overview of the model in three steps (Fig. 2).

The first step involves decomposing the original data. In this stage, we decompose data to separate distinct characteristics, simplifying the dataset. We employ classical decomposition to split the original data into three components: trend, seasonal, and residual, as shown in Eq. 5, representing various data traits.

$$y_t = S_t + T_t + R_t \tag{5}$$

where y_t is the original data, S_t is the seasonal component, T_t is the trend component, and R_t is the residual component.

The second step is to design distinct processing approaches for the characteristics based on the features of each component.

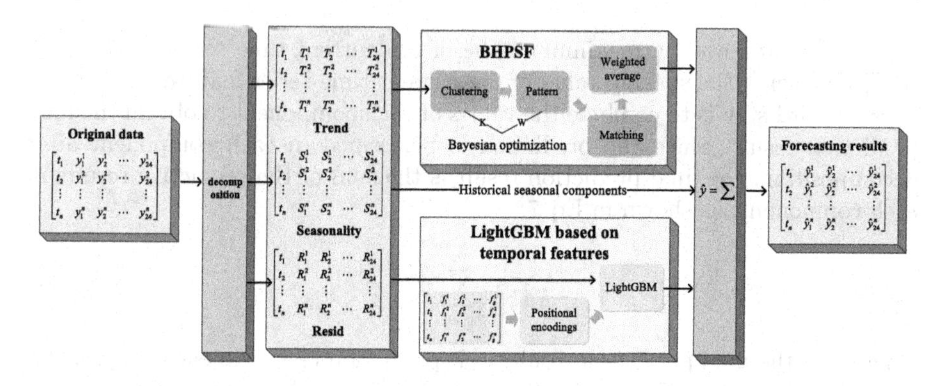

Fig. 3. Framework of the Hybrid-BHPSF model.

For the trend component, we employ the BHPSF algorithm for prediction. On the one hand, the trend component obtained through decomposition often exhibits distinct pattern information with evident regularity. On the other hand, the BHPSF algorithm is based on the original PSF algorithm and is good at processing data with fixed modes. Additionally, it chooses a cluster algorithm that is more suitable for the current environment and considers the interrelationships among multiple parameters to optimize the parameter selection process. Therefore, using the BHPSF algorithm to forecast the trend component readily achieves high-precision predictions.

The seasonal component is isolated from the data after removing the trend components, which shows the cyclic data fluctuation patterns. We consider the seasonal component as a repeating pattern, meaning it remains consistent across all cycles. Therefore, we extract the seasonal component before making predictions and reintegrate it into the results after the prediction process.

For the residual component, we employ LightGBM [24] for prediction. The residual component represents what remains after removing the trend and seasonal patterns, often exhibiting fluctuations within a specific range. LightGBM excels at predicting relatively stable data without clear trends. Therefore, we use it to predict the residual component.

We start by constructing a set of time series features, decomposing date-time stamps into different time units such as year, month, day, and week. This assists the model in recognizing patterns across various time dimensions. To preserve the adjacency of temporally adjacent periodic features, we encode them using sine and cosine functions (Eq. 6). We also label holidays and create binary features to identify whether it is holiday. Once the temporal features are established, we input them into LightGBM along with the residual component for prediction.

$$\begin{aligned} feature_i^{sin} &= sin\frac{2\pi i}{T} \\ feature_i^{cos} &= cos\frac{2\pi i}{T} \end{aligned} \tag{6}$$

where, T represents the maximum value of the current time series feature. $i \in [1, T]$ represents the specific value of the current time series feature.

The third step is to combine the values of each component to obtain the final prediction result. Since the prediction is performed in each component after decomposition, the final prediction result is the sum of the predictive results in each component, as shown in Eq. 7.

$$\hat{Y}_t = \sum_{c=1}^{n} \hat{y}_c \tag{7}$$

where $\hat{Y}$ is the final prediction result, $\hat{y}_c$ is prediction results in each component, and c is the number of components decomposed from the original data.

4 Experiments

4.1 Data Description

This article employs four datasets related to electricity demand, originating from Siberia and European Russia, as well as Poland and Singapore.

The two datasets from Russia[1] cover the period from September 1, 2006, to November 22, 2011, with measurements taken every hour. The dataset from Poland[2] spans from January 1, 2012, to December 31, 2022, with hourly measurements. Similarly, the dataset from Singapore[3] includes hourly measurements from January 1, 2003, to December 31, 2022.

4.2 Quality Measures

To evaluate the performance of the Hybrid-BHPSF model, root mean square error (RMSE), mean absolute error (MAE) and mean absolute percentage error (MAPE) are considered. The lower values of these statistics indicate that the difference between actual and predicted values is relatively small. Their formulas are shown below.

$$\begin{aligned} MAPE(\%) &= 100 \cdot \frac{1}{n} \sum_{h=1}^{n} \frac{|y_h - \hat{y}_h|}{y_h} \\ MAE &= \frac{1}{n} \sum_{h=1}^{n} |y_h - \hat{y}_h| \\ RMSE &= \sqrt{\frac{1}{n} \sum_{h=1}^{n} |y_h - \hat{y}_h|^2} \end{aligned} \tag{8}$$

4.3 Analysis of BHPSF Results

This section aims to verify the effect of BHPSF, compared with the original PSF [12] and the latest bigPSF [13]. Figure 4 illustrates their MAPE. It can be

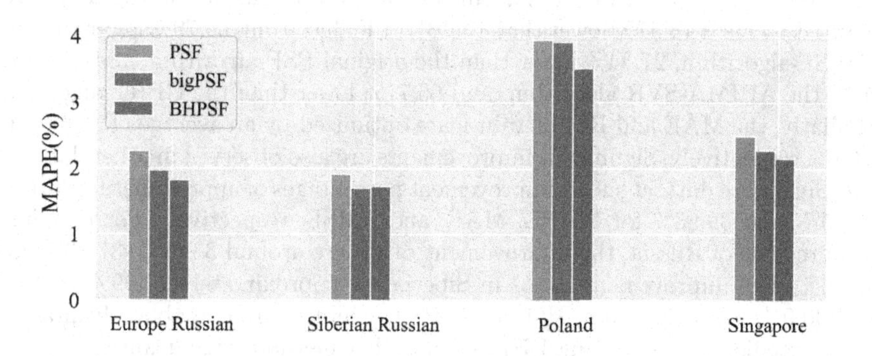

Fig. 4. Forecasting effects of BHPSF, PSF and bigPSF on different datasets.

[1] https://www.kaggle.com/irinachuchueva.
[2] https://www.entsoe.eu/.
[3] https://www.emcsg.com/.

found that in the four real-world power data, BHPSF is better than the original PSF algorithm, and only slightly inferior to bigPSF in the Siberian dataset. Among them, BigPSF achieves a MAPE of 1.70%, while BHPSF achieves 1.72%. It is worth noting that BHPSF outperforms bigPSF significantly in runtime. We compare the time taken by BHPSF and bigPSF to make predictions on various datasets in Table 1, and it is obvious that BHPSF takes the least time to run and has higher efficiency. In summary, we choose BHPSF as an important part of the proposed model for further research.

4.4 Analysis of Hybrid-BHPSF Results

In this part, we comprehensively evaluate Hybrid-BHPSF on four real datasets (Table 2), and compared with BHPSF, Hybrid-BHPSF shows the best performance in all datasets. In addition, it is also compared with PSF [12], bigPSF [13], GBRT [13], ARIMA-SVR [16]. It can be seen that the Hybrid-BHPSF clearly outperformed all of them, with the lowest MAPE, MAE and RMSE.

Table 1. The average runtime of each prediction for BHPSF and bigPSF (s).

Dataset	bigPSF	BHPSF
Singapore	2.79	2.16
Poland	2.40	0.89
European Russia	2.02	0.24
Siberia Russia	1.92	0.22

Notably, the Poland dataset demonstrates the superiority of Hybrid-BHPSF, with its MAPE being only 1.84%, which is the lowest value for all the considered methods. This is 18.47% lower than the BHPSF algorithm, 26.79% lower than the bigPSF algorithm, 27.34% lower than the original PSF algorithm, 48.84% lower than the ARIMA-SVR algorithm, and 66.71% lower than the GBRT algorithm. Similarly, the MAE and RMSE values are optimized by an average of 30.4% and 33.4%, respectively. Significant improvements are also observed in other datasets. The Singapore dataset shows improvement percentages of approximately 35.17%, 36.45%, and 39.93% for MAPE, MAE, and RMSE, respectively. For the European regions of Russia, the improvement ratios are around 38.69%, 37.73%, and 43.14%. The improvement ratios in Siberia are approximately 27.99%, 27.21%, and 30.91%. In conclusion, Hybrid-BHPSF exhibits a remarkable advantage in future prediction, surpassing BHPSF and other methods significantly.

Table 2. Forecasting results of the proposed and related models on distinct datasets.

Dataset	Measure	GBRT	ARIMA-SVR	PSF	bigPSF	BHPSF	**Hybrid-BHPSF**
Singapore	MAPE	3.99%	4.81%	2.49%	2.28%	2.16%	**1.84%**
	MAE	281.13	282.44	151.63	138.65	130.98	**111.33**
	RMSE	245.52	369.18	238.45	158.59	150.87	**125.62**
Poland	MAPE	8.62%	5.61%	3.95%	3.92%	3.52%	**2.87%**
	MAE	1656.74	1073.25	773.36	769.03	684.99	**563.10**
	RMSE	1806.11	1312.81	1309.46	900.36	799.62	**643.64**
Siberian Russian	MAPE	2.38%	2.95%	1.91%	1.70%	1.73%	**1.47%**
	MAE	528.22	643.01	421.22	378.87	383.17	**328.73**
	RMSE	601.33	766.00	536.32	436.74	440.84	**367.54**
Europe Russian	MAPE	3.11%	2.40%	2.26%	1.97%	1.82%	**1.37%**
	MAE	2662.53	2006.26	1943.13	1699.12	1563.10	**1189.69**
	RMSE	3070.67	2421.75	2749.49	1998.16	1833.26	**1323.45**

4.5 Multi-step Ahead Prediction

To comprehensively assess the predictive capability of Hybrid-BHPSF, we conduct experiments across 10 distinct prediction ranges. Figure 5 depicts the changes in MAPE, MAE and RMSE for each dataset across different prediction ranges. It is evident that our algorithm, Hybrid-BHPSF, consistently maintains optimal performance under all circumstances. This demonstrates that the proposed model not only exhibits high accuracy in single-step forecasting but also holds significant potential for multi-step forecasting applications. In practical use, a combination of single-step and multi-step forecasting results can be utilized to assess future electricity consumption. If the forecasted results exceed the planned electricity consumption, policies to restrict electricity usage can be implemented.

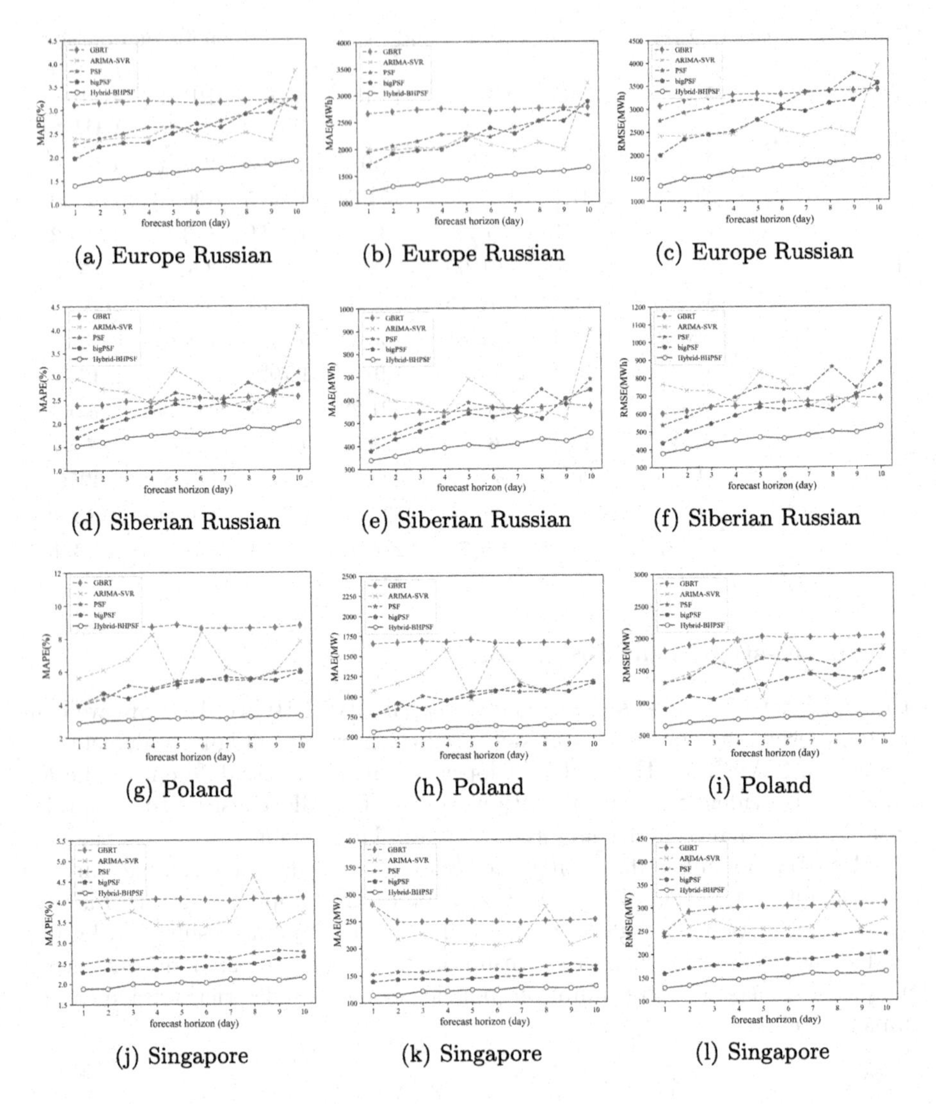

(a) Europe Russian (b) Europe Russian (c) Europe Russian

(d) Siberian Russian (e) Siberian Russian (f) Siberian Russian

(g) Poland (h) Poland (i) Poland

(j) Singapore (k) Singapore (l) Singapore

Fig. 5. Results of different forecast horizons.

5 Conclusion

The paper proposes a Hybrid-BHPSF model that combines seasonal decomposition to improve electricity consumption forecasting accuracy. Initially, Hybrid-BHPSF extracts trend change and seasonal fluctuation characteristics from the original data and considers the remaining part as the residual. Subsequently, a BHPSF algorithm is designed to fit the trending part of the data accurately, and LightGBM is embedded to predict the remaining components. In addition, we adopted a direct multi-step strategy for predicting electricity consumption one day in advance and a combination of direct multi-step and iterative multi-step strategies for longer-range predictions, which significantly reduced prediction error. We have conducted comprehensive experiments on multiple real datasets and various prediction horizons. The experimental results consistently demonstrate that Hybrid-BHPSF outperforms other comparative methods across all scenarios, exhibiting superior predictive performance.

Acknowledgements. This work was supported in part by the National Natural Science Foundation of China under Grant 61972146 and 62372064, as well as by the Hunan Provincial Natural Science Foundation of China under Grant 2021JJ40612.

References

1. Zugno, M., Morales, J.M., Pinson, P., Madsen, H.: A bilevel model for electricity retailers' participation in a demand response market environment. Energy Econ. **36**, 182–197 (2013)
2. Zhang, K., Ni, J., Yang, K., Liang, X., Ren, J., Shen, X.S.: Security and privacy in smart city applications: challenges and solutions. IEEE Commun. Mag. **55**(1), 122–129 (2017)
3. Hwang, J., Suh, D., Otto, M.O.: Forecasting electricity consumption in commercial buildings using a machine learning approach. Energies **13**(22), 5885 (2020)
4. Bouktif, S., Fiaz, A., Ouni, A., Serhani, M.A.: Single and multi-sequence deep learning models for short and medium term electric load forecasting. Energies **12**(1), 149 (2019)
5. Chen, C., Li, K., Zhongyao, C., Piccialli, F., Hoi, S.C., Zeng, Z.: A hybrid deep learning based framework for component defect detection of moving trains. IEEE Trans. Intell. Transp. Syst. **23**(4), 3268–3280 (2020)
6. Li, Y., Li, K., Chen, C., Zhou, X., Zeng, Z., Li, K.: Modeling temporal patterns with dilated convolutions for time-series forecasting. ACM Trans. Knowl. Disc. Data (TKDD) **16**(1), 1–22 (2021)
7. Zou, X., Zhou, L., Li, K., Ouyang, A., Chen, C.: Multi-task cascade deep convolutional neural networks for large-scale commodity recognition. Neural Comput. Appl. **32**(10), 5633–5647 (2020)
8. Wang, S., Song, A., Qian, Y.: Predicting smart cities' electricity demands using k-means clustering algorithm in smart grid. Comput. Sci. Inf. Syst. **20**, 657–678 (2023)
9. Imani, M.H., Bompard, E., Colella, P., Huang, T.: Forecasting electricity price in different time horizons: an application to the Italian electricity market. IEEE Trans. Ind. Appl. **57**(6), 5726–5736 (2021)
10. Tang, Z., Yin, H., Yang, C., Yu, J., Guo, H.: Predicting the electricity consumption of urban rail transit based on binary nonlinear fitting regression and support vector regression. Sustain. Urban Areas **66**, 102690 (2021)
11. Zheng, K., et al.: A multi-scale electricity consumption prediction algorithm based on time-frequency variational autoencoder. IEEE Access **9**, 90937–90946 (2021)
12. Alvarez, F.M., Troncoso, A., Riquelme, J.C., Ruiz, J.S.A.: Energy time series forecasting based on pattern sequence similarity. IEEE Trans. Knowl. Data Eng. **23**(8), 1230–1243 (2010)

13. Pérez-Chacón, R., Asencio-Cortés, G., Martínez-Álvarez, F., Troncoso, A.: Big data time series forecasting based on pattern sequence similarity and its application to the electricity demand. Inf. Sci. **540**, 160–174 (2020)
14. Zhang, T., Tang, Z., Wu, J., Du, X., Chen, K.: Short term electricity price forecasting using a new hybrid model based on two-layer decomposition technique and ensemble learning. Electr. Power Syst. Res. **205**, 107762 (2022)
15. Zhu, G., Peng, S., Lao, Y., Su, Q., Sun, Q.: Short-term electricity consumption forecasting based on the EMD-Fbprophet-LSTM method. Math. Probl. Eng. **2021**, 1–9 (2021)
16. Guo, N., Chen, W., Wang, M., Tian, Z., Jin, H.: Appling an improved method based on Arima model to predict the short-term electricity consumption transmitted by the internet of things (IoT). Wirel. Commun. Mob. Comput. **2021**, 1–11 (2021)
17. Lu, H., Ma, X., Ma, M.: A hybrid multi-objective optimizer-based model for daily electricity demand prediction considering COVID-19. Energy **219**, 119568 (2021)
18. Xia, Y., Wang, J., Wei, D., Zhang, Z.: Combined framework based on data preprocessing and multi-objective optimizer for electricity load forecasting. Eng. Appl. Artif. Intell. **119**, 105776 (2023)
19. Zulfiqar, M., Kamran, M., Rasheed, M., Alquthami, T., Milyani, A.: Hyperparameter optimization of support vector machine using adaptive differential evolution for electricity load forecasting. Energy Rep. **8**, 13333–13352 (2022)
20. Talavera-Llames, R., Pérez-Chacón, R., Troncoso, A., Martínez-Álvarez, F.: MV-kWNN: a novel multivariate and multi-output weighted nearest neighbours algorithm for big data time series forecasting. Neurocomputing **353**, 56–73 (2019)
21. Galicia, A., Talavera-Llames, R., Troncoso, A., Koprinska, I., Martínez-Álvarez, F.: Multi-step forecasting for big data time series based on ensemble learning. Knowl. Based Syst. **163**, 830–841 (2019)
22. Galicia, A., Torres, J.F., Martínez-Álvarez, F., Troncoso, A.: A novel spark-based multi-step forecasting algorithm for big data time series. Inf. Sci. **467**, 800–818 (2018)
23. Ribeiro, M.H.D.M., da Silva, R.G., Ribeiro, G.T., Mariani, V.C., dos Santos Coelho, L.: Cooperative ensemble learning model improves electric short-term load forecasting. Chaos Solitons Fractals **166**, 112982 (2023)
24. Ke, G., et al.: LightGBM: a highly efficient gradient boosting decision tree. In: Advances in Neural Information Processing Systems, vol. 30 (2017)

An Improved Model of PBFT with Anonymity and Proxy Based on Linkable Ring Signature

Zhuobiao Wang, Gengran Hu(✉), and Lin You

School of Cyberspace, Hangzhou Dianzi University, Hangzhou 310018, China
wangzhuobiao@hdu.edu.cn, grhu@hdu.edu.cn

Abstract. At present, PBFT has become an important consensus mechanism in the blockchain due to its high efficiency. In PBFT, a primary node must be chosen. If the primary node is attacked, PBFT needs to select a new node as the primary node. The continuous occurence of such attacks makes PBFT become much less efficient. Considering the importance of the primary node, this paper proposes an improved model of PBFT with anonymity and proxy. The linkable ring signature technology is utilized to hide the primary node in the ordinary nodes, which can significantly reduce the risk of the primary node being attacked. This countermeasure improves the security of PBFT to some extent. Moreover, in this scheme, each node has its corresponding proxy node. When a node is not online for participating in the PBFT, its proxy node can be authorized to complete the whole steps in the consensus, which improves the reliability of PBFT.

Keywords: PBFT · linkable ring signature · proxy node · blockchain

1 Introduction

Since Satoshi Nakamoto proposed the concept of Bitcoin [10], the blockchain has been keeping developing. In recent years, the blockchain has evolved from the original digital currency to various application systems. The consensus mechanism [11] solves the problem of decentralization in the blockchain. At present, the commonly used consensus algorithms are PoW, PoS [13] and PBFT [1]. The alliance chains [7,8] are partially decentralized, and the nodes in the alliance chains can be supervised. Therefore, PBFT in the alliance chains needs to be studied more deeply.

PBFT uses the Byzantine protocol, which can tolerate a certain number of malicious nodes. PBFT requires communication between nodes to convey messages and reach consensus. In PBFT, there exists a primary node and other

This work was supported by the National Natural Science Foundation of China (No. 61602143, No. 61772166), the Key Program of the Natural Science Foundation of Zhejiang Province of China under Grant LZ17F020002, and the Postdoctoral Research Project of Zhejiang Province of China under Grant ZJ2022115.

Z. Tari et al. (Eds.): ICA3PP 2023, LNCS 14491, pp. 491–502, 2024.
https://doi.org/10.1007/978-981-97-0808-6_29

nodes. The primary node starts the consensus, and the other nodes will conduct two rounds of voting to complete the consensus. PBFT can reach a correct consensus as long as the number of votes received is greater than 2/3 of the total number of nodes. At present, one main problem of PBFT is the communication complexity, which is $O(n^3)$. When the number of nodes increases, the communication complexity will be quite large. Therefore, PBFT is suitable for limited number of nodes to reach consensus. Nowadays, PBFT is also combined with PoW/PoS to form mixed consensus mechanisms such as PeerCensus [3] and Algorand [5]. In such mechanisms, PoW/PoS is used to filter nodes and PBFT is utilized to reach the agreement. Therefore, it is important to improve PBFT for the development of the alliance chains.

Rivest et al. [12] proposed the first ring signature scheme in 2001. In a ring signature scheme, the verifier can do the verification without knowing the signer's real identity. In 2002, the first identity-based ring signature scheme was proposed by Zhang et al. [14]. Linkable ring signatures add a link label to ring signatures, which can used to decide whether two signatures are from the same user. Liu et al. proposed the first practical linkable ring signature scheme [9] in 2004.

Researchers have made several improvements to address the shortcomings of PBFT. In the literature [15], the reputation value was introduced and the random forest algorithm was utilized to divide the nodes into high, medium and low clusters. Then by selecting the primary node from high reputation nodes and excluding the malicious nodes from low reputation nodes, the security of PBFT can be improved. In the literature [2], a voting algorithm was used, and the node with the higher number of votes will become the primary node in the next round, which reduced the chance of malicious nodes chosen to be the primary node. The works above use voting and other methods to reduce the probability of malicious nodes becoming the primary node. However, these works do not consider the possibility that the nodes will be attacked by malicious nodes. In the literature [4], ring signature was introduced to PBFT to improve the security of transactions. We think the security of nodes should be also considered, where the linkable ring signature is suitable in this case. Thus in this paper, the security and reliability perspectives of PBFT are mainly considered. By using linkable ring signature to hide the primary node, the probability of the primary node being attacked can be reduced. In addition, by considering the authorized proxy node affiliated to each original node, the proxy node can participate in the consensus on behalf of its corresponding original node.

In this paper, an improvement of PBFT is proposed, which uses proxy nodes and linkable ring signature to improve the security and reliability of PBFT. In Sect. 2, the basic procedures of PBFT, linkable ring signature and ECDSA signature are described. And the improvement of PBFT with anonymity and proxy is introduced in Sect. 3. Then the security of this scheme is analyzed from three aspects in Sect. 4. This improvement of PBFT is analyzed on the efficiency and compared with other related schemes for attributes in Sect. 5. The conclusion is drawn in Sect. 6. The main contributions are as follows:

1. In this model, each node can select a proxy node. When the original node is not online for participating in the consensus due to some network issue, its proxy node can be authorized to help to complete the consensus, which ensures that the PBFT process can be finished successfully. This improves the reliability of PBFT.
2. The classical linkable ring signature is utilized to substitute the original signature in PBFT, which can hide the identity of the primary node, therefore reducing the probability of the primary node being attacked by malicious nodes. This improves the security of PBFT to some extent. To make the anonymity and proxy consistent, we improve the classical linkable ring signature to ensure that the node and its proxy node share the same link label.

2 Preliminaries

2.1 PBFT

PBFT [1] is one of the current efficient consensus mechanisms in the blockchain. In PBFT, there exists a primary node and other nodes. The primary node is used to initiate the consensus, and the other nodes reach the consensus by sending and checking messages. PBFT mainly consists of five steps as follows:

- **Request**: The message m is sent from the client to the primary the node.
- **Pre-Prepare**: On receiving the message m, the primary node assigns it a serial number, sign the prepared message by its private key and sends it to the other nodes.
- **Prepare**: After receiving the prepared message, the other nodes verify the validity of the signature and the legitimacy of the message. When a node finishes the verification successfully, it starts to prepare the message, and broadcasts the message to the whole network. Then each node collects the broadcast preparation messages, and if the number of legitimate messages received is greater than 2/3 of the whole number of nodes, the preparation voucher will be formed.
- **Commit**: After each node generates the preparation credentials, it calculates the commitment message and broadcast it. Similarly, the nodes need to collect the promise messages, and the ratio of legitimate messages should be greater than 2/3 to fulfill the promise to the message m.
- **Answer**: After all nodes complete the commitment, the results are returned to the client, and the client confirms that the message m finally reaches the consensus.

2.2 Linkable Ring Signature

Compared to previous signatures, ring signatures do not need a manager, and all members form a ring, where the private key of the signer and the public keys of the other ring members are utilized to generate the signature. This signature can be publicly verified for correctness, but other members in the ring can not decide the signer's identity. By using linkable ring signatures, it is possible to determine whether two signatures are signed by the same user. This characteristic is called linkability, which is achieved by adding a label generated by each user. The specific steps of the classic linkable ring signature in [9] (or LSAG ring signature) are as follows:

- **Key generation**:
  1. Let $\mathbb{G} = \langle g \rangle$ be a cyclic group, with the order of this group set to be q.
  2. Let $H_1 : \{0,1\}^* \rightarrow \mathbb{Z}_q$, $H_2 : \{0,1\}^* \rightarrow \mathbb{G}$ be two cryptographically secure hash functions.
  3. Let n be th number of users. Each user select its private key randomly $x_i \in \mathbb{Z}_q$, and compute the public key $y_i = g^{x_i}$.
- **Signature generation**:
  1. Given the message m, the set of public keys of the ring members is $L = \{y_1, y_2, ..., y_n\}$, and the signer's private key is x_π.
  2. The signer calculates $h = H_2(L) \in \mathbb{G}$; $\tilde{y} = h^{x_\pi}$.
  3. The signer selects $u \in_R \mathbb{Z}_q$, and calculate $C_{\pi+1} = H_1(L, \tilde{y}, m, g^u, h^u)$.
  4. For $i = \pi + 1, ..., n, 1, ..., \pi - 1$, the signer selects $t_i \in_R \mathbb{Z}_q$, and calculates $C_{i+1} = H_1(L, \tilde{y}, m, g^{t_i} y_i^{C_i}, h^{t_i} \tilde{y}^{C_i})$.
  5. The signer calculates $t_\pi = u - x_\pi C_\pi$.
  6. Then the signature is $\sigma = (C_1, t_1, ..., t_n, \tilde{y})$.
- **Signature verification**:
  1. The verifier obtains the signature $\sigma = (C_1, t_1, ..., t_n, \tilde{y})$, and the public key list $L = \{y_1, y_2, ..., y_n\}$.
  2. The verifier calculate $h = H_2(L)$.
  3. For $i = 1, ..., n - 1$, the verifier calculates $C_{i+1} = H_1(L, \tilde{y}, m, g^{t_i} y_i^{C_i}, h^{t_i} \tilde{y}^{C_i})$.
  4. The verifer judges whether the following holds:

$$C_1 = H_1(L, \tilde{y}, H(m), g^{t_n} y_n^{C_n}, h^{t_n} \tilde{y}^{C_n}).$$

     If this equation holds, the verification succeeds, otherwise it fails.

2.3 ECDSA Signature

ECDSA [6] is a digital signature proposed in 1999, whose secutity is derived from the hardness of the discrete logarithm problem over elliptic curves. The detailed steps of ECDSA signature are as follows:

- **Key generation**:
  1. For large prime p, let q be the order of base point G in elliptic curve E over F_p.
  2. Select a positive integer d s.t. $1 < d < q - 1$, calculate $P = dG$.
  3. Let d and P be the private key and public key, respectively.

- **Signature generation**: For message m, the signer randomly selects an integer k s.t. $0 \leq k < q$ and calculates $kG = (x, y)$

$$r = x \bmod q, \quad s = (H(m) + dr)k^{-1} \bmod q.$$

Then (r, s) is the signature of message m, where H is a hash function.
- **Signature verification**: The verifier calculates

$$t_1 = H(m)s^{-1} \bmod q, \quad t_2 = rs^{-1} \bmod q, t_1G + t_2P = (x', y')$$

Then the signature verification passes if and only if $r = x' \bmod q$ holds.

It can be seen that the recovery of the private key d from the public key P is actually equivalent to solving the following discrete logarithm problem over elliptic curves.

DLP (Discrete Logarithm Problem): Let $\mathbb{G} = \langle g \rangle$ be a cyclic group of order q with generator g, given $\beta \in \mathbb{G}$, the discrete logarithm problem is to find the unique integer $\alpha \in [1, q)$ satisfying $g^\alpha = \beta$.

3 An Improved Model of PBFT with Anonymity and Proxy

Considering the existing problems of PBFT, we propose an improved model of PBFT with anonymity and proxy. First, we assign each node a proxy node, which can help to complete the consensus when the original node is not online. Then by adopting the classical linkable ring signature, the identity of the primary node

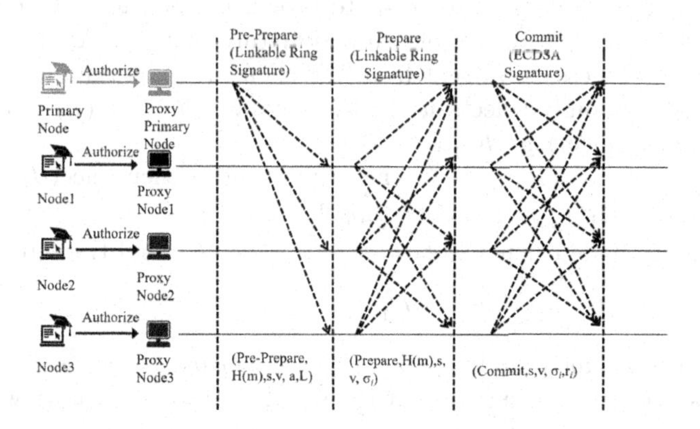

Fig. 1. Proxy and Anonymity-Based PBFT Process.

can be hidden, therefore reducing the probability of malicious nodes attacking the primary node. More specifically, a linkable ring signature was used in both of Pre-Prepare and Prepare, and an ECDSA signature is used in Commit.

Figure 1 shows the workflow of the improved model of PBFT.

3.1 Proxy Node Selection

For each node, its proxy node can be chosen. In this process, the node and the proxy node need to authenticate each other.

- **Parameter generation**: For cyclic group $\mathbb{G} = \langle g \rangle$ with order q, and two hash functions $H_1 : \{0,1\}^* \to \mathbb{Z}_q$, $H_2 : \{0,1\}^* \to \mathbb{G}$, each node generates the private key $x_i \in \mathbb{Z}_q$ and the public key $y_i = g^{x_i}$. Then each proxy node also selects the private key $x_i^{'} \in \mathbb{Z}_q$ and the public key $y_i^{'} = g^{x_i^{'}}$.
- **Authorization information sharing**: The node sends the authorization information to the proxy node and use ECDSA signature to sign the authorization information m_s. Then the node sends the authorization information m_s and its signature to its proxy node.
- **Verification**: If the authentication is successful, the public key of the proxy node is considered to be correct. Then the primary node collects and verifies the legitimacy of the authorization certificates of all the nodes and proxy nodes, and adds the public keys of the nodes and proxy nodes to the public key list after the verification succeeds. The final public key set forming the ring is $L = \{y_1, y_1^{'}, y_2, y_2^{'}, ..., y_n, y_n^{'}\}$.

3.2 Pre-Prepare

In this phase, the primary node needs to broadcast messages. To hide the primary node's identity, the linkable ring signature is used. After the primary node receives the message m sent from the client, it signs the message m. The primary node randomly selects $u \in_R \mathbb{Z}_q$, compute $h = H_2(L)$, and calculate $C_{\pi+1} = H_1(L, \tilde{y}, H_1(m), g^u, h^u)$.

For $i = \pi + 1, ..., n, 1, ..., \pi - 1$, the primary node then select $t_i \in_R \mathbb{Z}_q$, and calculate $C_{i+1} = H_1(L, \tilde{y}, H_1(m), g^{t_i}(y_i y_i^{'})^{C_i}, h^{t_i}\tilde{y}^{C_i})$.

The primary node calculates $t_\pi = u - (x_\pi + x_\pi^{'})C_\pi$ and the signature is

$$\sigma = (C_1, t_1, ..., t_n, \tilde{y}).$$

Finally, the primary node sends $(Pre-Prepare, H_1(m), s, v, \sigma, L)$ to the other nodes, where s and v represent the serial number and the view number of the current status.

3.3 Prepare

In the preparation phase, all the nodes except the primary node need to broadcast messages. Thus to make the primary node anonymous, the linkable ring signature is be used, so the nodes and proxy nodes need to negotiate unique labels.

Thus, when the node collects the broadcast messages, it can judge whether the messages are from the same node identification, assuming that an original node and its proxy node share the same identification according to the label. The label negotiation process is as follows:

Each node calculates $h = H_2(L)$, generates the label h^{x_π} and sends it to the proxy node. The proxy node receives h^{x_π} and compute $\tilde{y} = h^{x_\pi + x_\pi'}$ by multiplying $h^{x_\pi'}$. Then the proxy node sends the result $\tilde{y} = h^{x_\pi + x_\pi'}$ back to the node. Both of them form a unique tag $\tilde{y} = h^{x_\pi + x_\pi'}$.

On receiving the Pre-Prepare messages, the other nodes do the verification.

Given the signature σ, for $i = 1, ..., n-1$, the verifier calculates $C_{i+1} = H_1(L, \tilde{y}, H_1(m), g^{t_i}(y_i y_i')^{C_i}, h^{t_i}\tilde{y}^{C_i})$.

Finally, the verifier checks $C_1 = H_1(L, \tilde{y}, H(m), g^{t_n}(y_n y_n')^{C_n}, h^{t_n}\tilde{y}^{C_n})$. If the equation holds, the verification succeeds, and the received message is believed to be untampered.

After the verification succeeds, each node generates the prepare message $(Prepare, H(m), s, v, \sigma_i)$, and the preparation message is broadcast on the whole network. In the prepare message,

$$\sigma_i = (C_1, t_1, ..., t_n, \tilde{y_i}), \quad t_i = u - (x_i + x_i')C_i.$$

Each node collects the prepared messages and compares the labels $\tilde{y}_i$. For two Prepare messages, if $\tilde{y}_{i_1} \neq \tilde{y}_{i_2}$, then the two prepared messages are considered to be sent from different nodes. If a node receives $\geq 2f+1$ legitimate messages (f is the number of malicious nodes), then the node generates the prepare credential.

3.4 Commit

After the node forms the preparation credential, it calculate the commitment message $(Commit, s, v, \sigma_i, r_i)$. Since this procedures needs each node to send massages to each other node, then it is not necessary to reach the anonymity of the primary node. Thus the ECDSA signature is used for signing the commitment. The promise message will be broadcast. The node also writes the message m to the local log.

The nodes collect the commitment messages and then verify. If the number of successfully verified promised messages is not smaller than $2f + 1$, the commitment to the message is fulfilled. Then each node returns the confirmed message, which completes the consensus.

When the node can not participate in the consensus, the proxy node can use the authorization from the original node to join in the consensus as a representative, and when the primary node can not participate, the proxy primary node can also initiate the consensus, which greatly reduces the number of view changes and improves the reliability of the consensus.

3.5 Functionality Comparison

We compare the functionality of our scheme with other improved schemes of PBFT in Table 1. Our scheme can not only hide the primary node, but also arrow the participation of the proxy node, which improves the security and reliability of PBFT to some extent.

Table 1. The comparison of the function of the our scheme and other PBFT (PN represents Primary Node)

	Anonymity of PN	Proxy of Each Node	Randomness in Generating PN	Feekback in PN Chosen
Our scheme	√	√	×	×
Ziyang et al. [15]	×	×	×	√
Chen et al. [2]	×	×	√	×
Fang et al. [4]	√	×	×	×

4 Security Analysis

For the improved model of PBFT with anonymity and proxy, we mainly give the security analysis for the used linkable ring signature, since the security of the other parts can be derived from classical PBFT. The detailed process of the security analysis mostly refers to [9], which consists of unforgeability, anonymity and linkability.

4.1 Unforgeability

Theorem 1. *Under the random oracle model, if DLP is hard, the linkable ring signature used in Sect. 3 is unforgeable.*

Proof. Assume $\mathcal{A}$ is an PPT adversary, who can do q_H queries to H_1, and H_2 and q_S queries to $\mathcal{SO}$, can efficiently forge the a valid signature, i.e.,

$$Pr[\mathcal{A}(L) \rightarrow (m, \sigma) : \mathcal{V}(L, m, \sigma) = 1] > \frac{1}{poly(k)},$$

where $poly(k)$, q_H and q_S are polynomials on the security parameter k, and $\mathcal{SO}$ is a signature oracle. According to the query of $\mathcal{A}$, it returns a valid signature.

We build a polynomial time simulator $\mathcal{M}$, which uses the ability of adversary $\mathcal{A}$ to forge signatures, then to solve DLP of at least one of the public key product $y_i y_i^{'}$ in L with non-negligible probability. Notice that $L = \{y_1, y_2, ..., y_n, y_1^{'}, y_2^{'}, ..., y_n^{'}\}$, the forged signature regarding L is expressed as $\sigma = (C_1, t_1, ..., t_n, \tilde{y})$. Then the following two conditions for signature verification must be satisfied at the same time:

$$C_{i+1} = H_1(L, \tilde{y}, H(m), g^{t_i}(y_i y_i^{'})^{C_i}, h^{t_i}\tilde{y}^{C_i}) \text{ for } 1 \leq i \leq n-1,$$

$$C_1 = H_1(L, \tilde{y}, H(m), g^{t_n}(y_n y_n^{'})^{C_n}, h^{t_n}\tilde{y}^{C_n}).$$

Let $\mathcal{T}$ be the Turing script that the simulator $\mathcal{M}$ invokes $\mathcal{A}$ to obtain. According to the proof in [9], the probability of guessing a random predictor is negligible. Thus for each valid forged signature built with the script $\mathcal{T}$, the n equations are checked to verify the signature. Let $X_{i_1},...,X_{i_n}$ be all the queries verifying the ith forgery. and π be the index corresponding to the last validation query, then we have $X_{i_n} = H_1(L, \tilde{y}, H(m), g^{t_{\pi-1}}(y_{\pi-1}y'_{\pi-1})^{C_{\pi-1}}, h^{t_{\pi-1}}\tilde{y}^{C_{\pi-1}})$.

If $i_1 = l$ and π is mentioned above, then call that the adversary $\mathcal{A}$ outputs a (l, π)-forgery. Let $\mathcal{T}'$ be the rewound transcript $\mathcal{T}$ before the l-th query. It is not hard to see that the probability of forging signatures σ and σ' respectively using $\mathcal{T}$ and $\mathcal{T}'$ is not negligible.

For the two (l, π)-forgeries according to $\mathcal{T}$ and $\mathcal{T}'$, we obtain the following relationships:

$$g^u = g^{t_\pi}(y_\pi y'_\pi)^{C_\pi} = g^{t_\pi+(x_\pi+x'_\pi)C_\pi},\ h^v = h^{t_\pi}\tilde{y}^{C_\pi} = h^{t_\pi+(r_\pi+r'_\pi)C_\pi},$$

$$g^u = g^{t'_\pi}(y_\pi y'_\pi)^{C'_\pi} = g^{t'_\pi+(x_\pi+x'_\pi)C'_\pi},\ h^v = h^{t'_\pi}\tilde{y}^{C'_\pi} = h^{t'_\pi+(r_\pi+r'_\pi)C'_\pi}.$$

By the above formulas, it can be obtained that

$$x_\pi + x'_\pi = \frac{t'_\pi - t_\pi}{C_\pi - C'_\pi} \bmod q,\quad r_\pi + r'_\pi = \frac{t'_\pi - t_\pi}{C_\pi - C'_\pi} \bmod q.$$

From the above equations, we know that $x_\pi + x'_\pi$ can be calculated, which means the DLP of the public key product $y_\pi y'_\pi$ is solved. However, this contradict with DLP difficulty hypothesis, which completes the proof of unforgeability.

4.2 Anonymity

For signature $\sigma = (C_1, t_1, ..., t_n, \tilde{y})$, it is easy to find that the selected $t_1, ..., t_n$ are uniformly random. In addition, since C_1 is the output of the cryptographically secure hash function H_1, and H_1 is regarded as a random oracle, then C_1 is also uniformly random. For the label $\tilde{y}$, it needs to solve the DLP to find the solution $x_\pi + x'_\pi$, which leaks the signer's identification. Since the DLP difficulty hypothesis holds, the label $\tilde{y}$ shows no information about the signer's identity. Thus the probability of matching the real signer among n users by tha attacker is not greater than $1/n$, which means that this scheme is anonymous.

4.3 Linkability

Since the original node and its proxy node share the same identification, if two signatures $\sigma = (C_1, t_1, ..., t_n, \tilde{y})$ and $\sigma' = (C'_1, t'_1, ..., t'_n, \tilde{y}')$ are generated by the same node identification, but $\tilde{y} \neq \tilde{y}'$. The labels of the two signatures are $\tilde{y} = h^{x_\pi+x'_\pi}$, $\tilde{y}' = h^{x_{\pi'}+x'_{\pi'}}$, then we have $h^{x_\pi+x'_\pi} \neq h^{x_{\pi'}+x'_{\pi'}} \bmod p$. Notice that $h = H_2(L)$ is also in the group $\mathbb{G}$, then it must be $x_\pi + x'_\pi \neq x_{\pi'} + x'_{\pi'} \bmod q$.

Since these two signatures are generated by the same node identification, then we know $y_\pi = y_{\pi'}$ and $y'_\pi = y'_{\pi'}$, or $y_\pi = y'_{\pi'}$ and $y'_\pi = y_{\pi'}$. Both of the two cases result in

$$g^{x_\pi + x'_\pi} = g^{x'_\pi + x'_{\pi'}} \bmod p,$$

which implies

$$x_\pi + x'_\pi = x_{\pi'} + x'_{\pi'} \bmod q.$$

Thus we have a contradictory, which means it must be $y = \tilde{y}$, completing the analysis of linkability.

5 Efficiency Analysis

For the improved model of PBFT with anonymity and proxy, the computational efficiency was analyzed and compared with the relative PBFTs. It is not hard to see that the communication overhead of this scheme is also $O(n^2)$, which is similar to that of the original PBFT.

Table 2 shows the signature time and verification time of the relative PBFTs, where E represents the time cost of the exponentiation operation of the group, T represents that of the multiplication operation of the group, h_1 is the cost of hash function H_1, and S represents that of the modular inversion operation of the multiplication group. By Table 2, the time cost of our scheme is greater than the previous schemes. However, our scheme can improve the security and reliability.

Table 2. Time overhead comparison

Scheme	Signature time	Verification time
Our scheme	$(2n-1)T + (4n-1)E + nh_1$	$2nT + 4nE + nh_1$
PBFT	$E + h_1$	E
Fang et al. [4]	$E + 2T + h_1 + S$	$3E + T + h_1$

Subsequently, we conduct experiments on the linkable ring signature part of this scheme. The signature generation and signature verification steps are run with 5 to 100 nodes. In our experiment, p is a 256-bit prime number and the group $\mathbb{G}$ is set to be the cyclic group generated by the base point in the elliptic curve SECP256k1.

Python is used to run our experiments. The computer processor is Intel (R) Core (TM) i5-6500 CPU @3.20 GHz with 8 GB RAM and the operating system is Windows 10. In our experiment, a 256bit string was used as the input for the signature, making it more similar to the real case of PBFT.

In Fig. 2, the time of signature generation and verification in our scheme is compared. When the number of nodes increases, both of the signature time and verification time increase linearly. At the same time, there is not much difference between node signature time and verification time, while the verification is slightly faster than signature when the number of nodes is large.

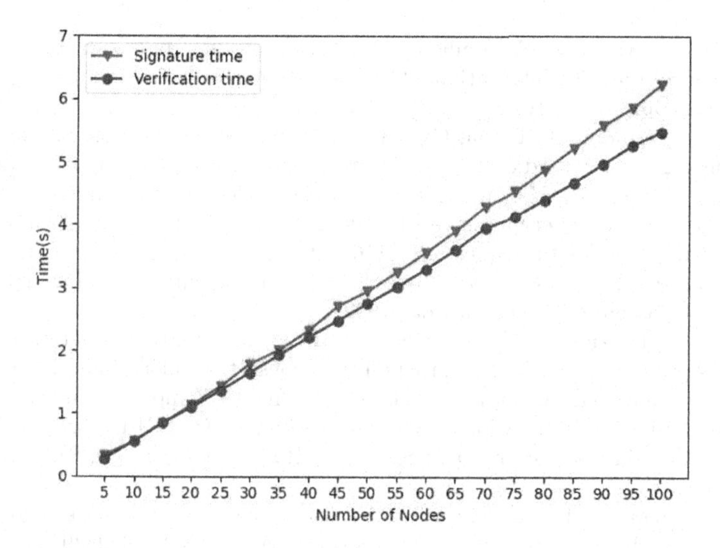

Fig. 2. Signature time and verification time.

6 Conclusion

In this paper, an improved model of PBFT with anonymity and proxy is proposed. An proxy node can be assigned to each original node, to participate in the consensus When the original is not online. At the same time, the linkbale ring signature is used to hide the primary node's identity, which protects the primary node from being attacked by the malicious nodes. This improves the security and reliability of PBFT to some extent. Moreover, we improve the LSAG ring signature to ensure that the node and its proxy node have unique labels. And we prove that the improved linkable ring signature also has the characteristics of unforgeability, anonymity and linkability. Finally, compared with the original PBFT, the improved model requires more computing overhead but is more secure and reliable. However, there are still space for the improvement of this paper, such as how to allocate rewards for nodes and proxy nodes, and whether there is a way to verify signatures more efficiently. We leave these issues as open questions to be addressed in the future.

References

1. Castro, M., Liskov, B., et al.: Practical byzantine fault tolerance. In: OsDI, vol. 99, pp. 73–186 (1999)
2. Chen, Z., Tan, M., Lei, P.: Improved scheme of practical byzantine fault tolerance algorithm based on voting mechanism. In: 2022 IEEE International Conference on Advances in Electrical Engineering and Computer Applications (AEECA), pp. 386–390. IEEE (2022)

3. Decker, C., Seidel, J., Wattenhofer, R.: Bitcoin meets strong consistency. In: Proceedings of the 17th International Conference on Distributed Computing and Networking, pp. 1–10 (2016)
4. Fang, Y., Jianqiu, D., Linhu, C., Chongyi, L.: An improved pbft blockchain consensus algorithm based on ring signature. Comput. Eng. **45**(11), 32–36 (2019)
5. Gilad, Y., Hemo, R., Micali, S., Vlachos, G., Zeldovich, N.: Algorand: scaling byzantine agreements for cryptocurrencies. In: Proceedings of the 26th Symposium on Operating Systems Principles, pp. 51–68 (2017)
6. Johnson, D., Menezes, A., Vanstone, S.: The elliptic curve digital signature algorithm. Certicom research, Canada (1999)
7. Li, K., Li, H., Hou, H., Li, K., Chen, Y.: Proof of vote: a high-performance consensus protocol based on vote mechanism & consortium blockchain. In: 2017 IEEE 19th International Conference on High Performance Computing and Communications; IEEE 15th International Conference on Smart City; IEEE 3rd International Conference on Data Science and Systems (HPCC/SmartCity/DSS), pp. 466–473. IEEE (2017)
8. Li, Y., Qiao, L., Lv, Z.: An optimized byzantine fault tolerance algorithm for consortium blockchain. Peer-to-Peer Network. Appl. **14**, 2826–2839 (2021)
9. Liu, J.K., Wei, V.K., Wong, D.S.: Linkable spontaneous anonymous group signature for ad hoc groups. In: Wang, H., Pieprzyk, J., Varadharajan, V. (eds.) Information Security and Privacy: 9th Australasian Conference, ACISP 2004, Sydney, Australia, 13–15 July 2004. Proceedings, vol. 9, pp. 325–335. Springer, Heidelberg (2004). https://doi.org/10.1007/978-3-540-27800-9_28
10. Nakamoto, S.: Bitcoin: a peer-to-peer electronic cash system. Decentralized business review (2008)
11. Nguyen, G.T., Kim, K.: A survey about consensus algorithms used in blockchain. J. Inf. Process. Syst. **14**(1), 101–128 (2018)
12. Rivest, R.L., Shamir, A., Tauman, Y.: How to leak a secret. In: Boyd, C. (ed.) Advances in Cryptology-ASIACRYPT 2001: 7th International Conference on the Theory and Application of Cryptology and Information Security Gold Coast, Australia, 9–13 December 2001 Proceedings, vol. 7, pp. 552–565. Springer, Heidelberg (2001). https://doi.org/10.1007/3-540-45682-1_32
13. Saleh, F.: Blockchain without waste: proof-of-stake. Rev. Finan. Stud. **34**(3), 1156–1190 (2021)
14. Zhang, F., Kim, K.: Id-based blind signature and ring signature from pairings. In: Zheng, Y. (ed.) Advances in Cryptology-ASIACRYPT 2002: 8th International Conference on the Theory and Application of Cryptology and Information Security Queenstown, New Zealand, 1–5 December 2002, Proceedings, vol. 8, pp. 533–547. Springer, Heidelberg (2002). https://doi.org/10.1007/3-540-36178-2_33
15. Ziyang, W., Juan, W., Yaning, L., Wei, W.: Improvement of PBFT consensus mechanism based on credibility. In: 2021 18th International Computer Conference on Wavelet Active Media Technology and Information Processing (ICCWAMTIP), pp. 93–96. IEEE (2021)

Author Index

Z. Tari et al. (Eds.): ICA3PP 2023, LNCS 14491, pp. 503–504, 2024.
https://doi.org/10.1007/978-981-97-0808-6

S

T

W

X

Y

Z

Printed in the United States
by Baker & Taylor Publisher Services